Statistical Tests for Mixed Linear Models

Statistical Tests for Mixed Linear Models

ANDRÉ I. KHURI
University of Florida, Gainesville, Florida

THOMAS MATHEW
BIMAL K. SINHA
University of Maryland, Baltimore, Maryland

A Wiley-Interscience Publication
JOHN WILEY & SONS, INC.
New York • Chichester • Weinheim • Brisbane • Singapore • Toronto

This book is printed on acid-free paper. ♾

Published simultaneously in Canada.

Library of Congress Cataloging-in-Publication Data:
Khuri, André I., 1940–
Statistical tests for mixed linear models / André I. Khuri, Thomas Mathew, Bimal K. Sinha.
p. cm. -- (Wiley series in probability and statistics. Applied probablility and statistics section)
"A Wiley–Interscience publication."
Includes bibliographical references and indexes.
ISBN 0-471-15653-1 (cloth : alk. paper)
1. Linear models (Statistics) 2. Statistical hypothesis testing. I. Mathew, Thomas, 1955– . II. Sinha, Bimal K., 1946– . III. Title. IV. Series: Wiley series in probability and statistics. Applied probability and statistics.
QA279.K484 1998
519.5'6--dc21 97-22363
CIP

Printed in the United States of America.
10 9 8 7 6 5 4 3 2 1

In memory of Paul Khuri, a Caring Brother — André I. Khuri

To my parents, K. T. Mathew and Aleyamma Mathew — Thomas Mathew

To the loving memory of my mother, Mrs. Jogmaya Sinha — Bimal K. Sinha

Contents

Preface

There is a vast literature on the analysis of linear models with mixed and random effects (i.e., variance component models). Even though most books on linear models and experimental designs discuss mixed and random models to some extent, books devoted entirely to this topic have appeared only recently. The books by Rao and Kleffe (1988) and Searle, Casella, and McCulloch (1992) deal exclusively with point estimation and prediction problems in variance component models, and the book by Burdick and Graybill (1992) deals with interval estimation of variance components. The object of the present book is to address an important aspect not covered in the above books, namely, *hypothesis testing.*

In practical applications involving a variance component model, testing the significance of the fixed effects and variance components is an important part of data analysis. The result of such tests can be of practical interest in itself. Furthermore, if an effect is not significant, it may be possible to use a reduced model, involving fewer parameters, for a subsequent analysis. Data that are typically analyzed using a mixed model can be balanced or unbalanced. Data with equal numbers of observations in the subclasses are described as *balanced* data, whereas those with unequal numbers of observations in the subclasses, including perhaps some that contain no observations at all (empty subclasses or empty cells), are called *unbalanced* data. Frequently, when an experiment is run under several treatment combinations, the resulting data are in general unbalanced. This includes the especially troublesome situation in which no data are available on some treatment combinations, which is commonly referred to as the *missing-cell* problem.

The analysis of balanced data is fairly straightforward since the analysis of variance (ANOVA) decomposition of the total sum of squares is unique and the different sums of squares in such a decomposition provide tests for various hypotheses of interest. In contrast to balanced data, the analysis of unbalanced data is much more complicated. The main difficulty stems from the fact that in an unbalanced data situation, the partitioning of the total sum of squares can be done in a variety of ways; hence, there is no unique way to write the ANOVA table, which is in sharp contrast with the balanced data situation. Furthermore, the sums of squares in an ANOVA table are not in general independent or distributed as multiples of chi-squared variates. Consequently, little is known about exact or optimum tests for a general unbalanced random or mixed model, except in a few special unbalanced situations. For the most part, and until fairly recently, drawing inference concerning the parameters of an unbalanced model was based on approximate procedures. These procedures are not always reliable and it is not known how well they perform with regard to accuracy of nominal significance levels of tests, coverage probabilities of confidence intervals, or powers of tests.

Wald (1947) was the first to introduce exact tests for random one-way and two-way crossed classification without interaction models. Exact or optimum tests for many other variance component models, however, remained largely unknown. Fortunately, a surge of activity in the area of exact and optimum inference for such models has materialized in just the last few years. The main purpose of this book is to compile the available results in this area into a single volume so that they will be more easily accessible to research workers, students, and practitioners. Such an undertaking is a manifestation of the importance and significance of this area in statistical research.

The book contains ten chapters. Chapter 1 gives an introduction to the development of Wald's variance component test in a general framework, and a brief discussion of the nature of optimum tests. Throughout the book, optimality of a test mostly refers to its being either locally or uniformly best among the class of relevant *invariant* tests. Chapter 2 deals with balanced data analysis and the optimality of the standard F-tests. It is shown that for balanced data, the F-tests constructed from the ANOVA table are optimum. Chapter 3 describes a procedure for measuring the degree of imbalance of an unbalanced data set. The effect of imbalance on data analysis is clearly brought out using a variety of examples and problems. A general method for determining the effect of imbalance is also included. This method is based on a certain technique for generating designs having a specified degree of imbalance. Chapter 4 covers exact and optimum tests for the unbalanced one-way model and the unbalanced random two-way crossed classification and two-fold nested models. This chapter also provides a foundation of the methodology for the derivation of exact tests to be discussed in later chapters. Chapter 5 presents an exact test for a general unbalanced random model with imbalance occurring only in the last stage of the design. The analysis of the unbalanced two-way mixed model, covering both exact and optimum tests, is the subject of Chapter 6. Chapter 7 deals with hypothesis tests that are designed to recover inter-block information in block designs. Several such tests are derived and compared in this chapter. The analysis of balanced and unbalanced split-plot designs under mixed and random models is the topic of Chapter 8. Exact and optimum tests are presented for testing the different hypotheses that can arise in this context. Chapter 9 deals with some applications of the recent concept of a generalized P-value for testing hypotheses in some mixed models with seemingly nonexistent exact tests of any kind.

The topics covered in Chapters 7, 8, and 9 are based on very recent developments. In Chapter 10, a summary of the available testing procedures for multivariate mixed and random models is presented. This chapter also includes a procedure for testing the adequacy of multivariate Satterthwaite's approximation. Except for Chapter 1, the book contains exercises given at the end of each chapter. Detailed solutions to selected exercises in Chapters 2–10 are provided in the Appendix. In addition, a general bibliography is included for the convenience of the reader.

The book is intended to provide the state of the art with regard to its subject area. Researchers in analysis of variance, experimental design, variance components analysis, linear models, and other related areas can benefit from its comprehensive coverage of the statistical literature. The book can therefore be very useful to graduate students who plan to do their research in the area of mixed models. Familiarity with standard ANOVA procedures is a required prerequisite for using this book. Such a prerequisite will be adequate for understanding the various exact tests derived in the book. However, for comprehending the various optimum tests and their derivations, the reader must be familiar with the various concepts of optimality and the standard approaches for deriving optimum tests, as described, for example, in Lehmann (1986). Readers not familiar with the optimality concepts can omit the derivations and discussions concerning optimum tests.

Given the increased interest in, and applicability of, mixed and random models, the book is designed to provide a comprehensive account of all the available results on hypothesis testing concerning these models. It is therefore hoped that it will contribute to the development of this area, and enhance its exposure and usefulness.

The authors acknowledge the facilities and support received from the Department of Statistics, University of Florida, Gainesville, and the Department of Mathematics and Statistics, University of Maryland Baltimore County. Bimal Sinha is grateful to his elder brothers for their parental care during his school days. He is also thankful to Professors S. B. Chaudhuri, J. K. Ghosh, and P. K. Sen for their support and encouragement throughout his academic career.

A. I. KHURI
T. MATHEW
B. K. SINHA

CHAPTER 1

Nature of Exact and Optimum Tests in Mixed Linear Models

1.1. INTRODUCTION

The central theme of this book, *tests for unbalanced mixed linear models*, can be accomplished in essentially two ways: (i) **exact** tests without any regard to optimality, based mostly on the central F-distribution, and (ii) **optimum** tests without any regard to the distributional simplicity of the underlying test statistics.

For many balanced as well as unbalanced models, exact tests for certain fixed effects and variance components can be derived using the analysis of variance (ANOVA) decomposition. For balanced models, fixed, mixed, or random, this can be readily established in a unique way from the standard ANOVA table showing different sources of variation along with their respective sums of squares; see Montgomery (1991, Chapter 8), Searle (1971, Chapter 9), or Chapter 2 in this book. For unbalanced fixed effects models also, one can use the ANOVA table to derive exact tests of hypotheses concerning the various fixed effects on the basis of standard linear model theory. Furthermore, for certain unbalanced random or mixed effects models, it is possible to use an ANOVA decomposition to derive exact tests of fixed effects as well as variance components. This is precisely what Wald (1947) originally developed, for a related but different problem, and was further advanced in Seely and El-Bassiouni (1983). In the next section, we provide the details of this approach with some examples. Applications of Wald's (1947) procedure and its modifications due to Seely and El-Bassiouni (1983) appear at different places in this book (see Chapters 4, 6, and 8).

The derivation of optimum tests for *balanced* models depends on the usual assumption of independence and normality of the random effects and the error components. In this case, the distribution of the data vector $\mathbf{y}$ is a member of the multiparameter exponential family (see Mathew and Sinha, 1988a; formula (1.3.13) below). One can then use a standard theory (see Lehmann, 1986) to derive optimum tests. It should be mentioned that the nature of optimality (discussed in Section 1.3 below), namely, *uniformly most powerful (UMP)*, *uniformly most powerful unbiased (UMPU)*, *uniformly most powerful invariant (UMPI)*, or *uniformly most powerful invariant unbiased (UMPIU)*, essentially depends on the dimension of the parameters being tested. Moreover, it turns out that in balanced models optimum tests do *always* coincide with the ANOVA-based exact tests, whenever such tests exist. Many examples of balanced models along with various optimum tests are given in Chapter 2.

In the case of unbalanced models, while the multiparameter exponential nature of the underlying distribution of $\mathbf{y}$ still holds (under the assumption of normality of $\mathbf{y}$), existence of a uniformly optimum test such as a UMP, UMPU, UMPI, or UMPIU test is rather rare even when a single parameter (like one variance component) is to be tested. Thus, for example, in the case of a one-way random effects model with *unequal* treatment replications, there is no uniformly optimum test for $H_0 : \sigma_\tau^2 = 0$, where σ_τ^2 is the unknown treatment variance (see Das and Sinha, 1987; see also Chapter 4). If this happens, our target is to derive what are known as locally optimum tests such as locally best unbiased (LBU), locally best invariant (LBI), and locally best invariant unbiased (LBIU) tests. Examples of such tests appear throughout this book (see Chapters 2, 4, 6, and 8).

In Section 1.3 we discuss briefly the procedures to derive optimum tests. For more details, we refer to Ferguson (1967), Eaton (1983), Lehmann (1986), and Kariya and Sinha (1989).

1.2. EXACT F-TESTS

Consider the following mixed linear model for an $N \times 1$ vector of observations $\mathbf{y}$:

$$\mathbf{y} = \boldsymbol{X}_1\boldsymbol{\alpha}_1 + \boldsymbol{X}_2\boldsymbol{\alpha}_2 + \boldsymbol{X}_3\boldsymbol{\alpha}_3 + \mathbf{e}, \tag{1.2.1}$$

where $\boldsymbol{\alpha}_1$ is a vector of effects of primary interest, fixed or random, $\boldsymbol{\alpha}_2$ is a vector of fixed effects, $\boldsymbol{\alpha}_3$ is a vector of random effects, and $\mathbf{e}$ is the vector of experimental error terms. Here $\boldsymbol{X}_1$, $\boldsymbol{X}_2$, and $\boldsymbol{X}_3$ are known design matrices. We assume that $\boldsymbol{\alpha}_3$ and $\mathbf{e}$ are independently distributed as $\boldsymbol{\alpha}_3 \sim N(\mathbf{0}, \boldsymbol{\Gamma})$ and $\mathbf{e} \sim N(\mathbf{0}, \sigma_e^2\boldsymbol{I}_N)$, where $\boldsymbol{\Gamma}$ is an unknown positive definite matrix. When $\boldsymbol{\alpha}_1$ is a vector of random effects, we shall also assume that $\boldsymbol{\alpha}_1 \sim N(\mathbf{0}, \sigma_1^2\boldsymbol{I})$ and is distributed independently of $\boldsymbol{\alpha}_3$ and $\mathbf{e}$. We are interested in testing a hypothesis concerning $\boldsymbol{\alpha}_1$, when $\boldsymbol{\alpha}_1$ is a vector of fixed effects or random effects. Note that when $\boldsymbol{\alpha}_1$ is random, our interest will be to test the significance of the variance component σ_1^2.

We note that model (1.2.1) is quite general and the various fixed, mixed and random effects models used in the analysis of variance are special cases of it. If our main interest is in the effects represented by $\boldsymbol{\alpha}_1$, $\boldsymbol{\alpha}_2$ will be the vector of all other effects that are fixed and $\boldsymbol{\alpha}_3$ will be the vector of all other effects that are random. Of course, the covariance matrix $\boldsymbol{\Gamma}$ associated with

$\boldsymbol{\alpha}_3$ will have a special structure depending on the assumptions on the various random effects included in $\boldsymbol{\alpha}_3$.

We shall explore the possibility of constructing an exact F-test for testing a hypothesis concerning $\boldsymbol{\alpha}_1$. The approach that we shall describe is based on an idea originally due to Wald (1947), and extended by Seely and El-Bassiouni (1983). Wald's original procedure dealt with model (1.2.1) with $\boldsymbol{\alpha}_1$ random and $\boldsymbol{\alpha}_3$ absent, and Wald derived a confidence interval for the ratio σ_1^2/σ_e^2, which can immediately be used to construct a test for testing H_0: $\sigma_1^2 = 0$. The test coincides with the F-test for testing the significance of $\boldsymbol{\alpha}_1$, when $\boldsymbol{\alpha}_1$ is a vector of fixed effects, and the test is commonly referred to as Wald's variance component test. For Wald's idea to work, a rank condition is necessary concerning the matrices $\boldsymbol{X}_1$, $\boldsymbol{X}_2$ and $\boldsymbol{X}_3$ in (1.2.1). This will be made explicit in our derivation that follows.

We shall first give an algebraic derivation of the F-test for testing a hypothesis concerning $\boldsymbol{\alpha}_1$ and then identify the components in the test statistic with appropriate sums of squares in an ANOVA decomposition based on model (1.2.1). Let $r = \text{rank}(\boldsymbol{X}_2 : \boldsymbol{X}_3)$ and let $\boldsymbol{Z}$ be an $N \times (N - r)$ matrix satisfying $\boldsymbol{Z}'[\boldsymbol{X}_2 : \boldsymbol{X}_3] = \mathbf{0}$ and $\boldsymbol{Z}'\boldsymbol{Z} = \mathbf{I}_{N-r}$. Thus the columns of $\boldsymbol{Z}$ form an orthonormal basis for the orthogonal complement of the vector space generated by the columns of $[\boldsymbol{X}_2 : \boldsymbol{X}_3]$. Let $\boldsymbol{u} = \boldsymbol{Z}'\boldsymbol{y}$. Then the model for $\boldsymbol{u}$ is

$$\boldsymbol{u} = \boldsymbol{Z}'\boldsymbol{X}_1\boldsymbol{\alpha}_1 + \boldsymbol{Z}'\boldsymbol{e}. \tag{1.2.2}$$

It follows from standard linear model theory (see Searle, 1987, Chapter 8) that the sum of squares due to $\boldsymbol{\alpha}_1$, computed from model (1.2.2), is the same as the sum of squares due to $\boldsymbol{\alpha}_1$ adjusted for $\boldsymbol{\alpha}_2$ and $\boldsymbol{\alpha}_3$, computed from the original model (1.2.1). We shall denote this sum of squares by $SS(\alpha_1|\alpha_2, \alpha_3)$. Note in particular that the computation of $SS(\alpha_1|\alpha_2, \alpha_3)$ can be accomplished by an ANOVA decomposition based on model (1.2.1) and hence does not require an explicit computation of the matrix $\boldsymbol{Z}$. Also, the error sum of squares computed from models (1.2.1) and (1.2.2) coincide. We shall denote this error sum of squares by $SS(e)$. Note that $SS(\alpha_1|\alpha_2, \alpha_3)$ and $SS(e)$ are independently distributed. Since the only effect involved in model (1.2.2) is $\boldsymbol{\alpha}_1$, tests of hypothesis concerning $\boldsymbol{\alpha}_1$ can be accomplished using model (1.2.2). Clearly, this argument will work only if $\boldsymbol{Z}'\boldsymbol{X}_1 \neq \mathbf{0}$. Furthermore, if $\boldsymbol{\alpha}_1$ is a vector of fixed effects, in order to test a linear hypothesis concerning $\boldsymbol{\alpha}_1$ using only model (1.2.2), an appropriate estimability condition must be met, in which case the test will coincide with the standard F-test from the linear model theory for a fixed effects model.

We shall now derive an F-test for testing H_0: $\sigma_1^2 = 0$ when $\boldsymbol{\alpha}_1$ is random. Toward this, we shall first give expressions for $SS(\alpha_1|\alpha_2, \alpha_3)$ and $SS(e)$ using $\mathbf{u}$ following model (1.2.2). These expressions are given by

$$\begin{aligned} SS(\alpha_1|\alpha_2, \alpha_3) &= \mathbf{u}'\boldsymbol{Z}'\boldsymbol{X}_1(\boldsymbol{X}_1'\boldsymbol{Z}\boldsymbol{Z}'\boldsymbol{X}_1)^{-}\boldsymbol{X}_1'\boldsymbol{Z}\mathbf{u} \\ SS(e) &= \mathbf{u}'\mathbf{u} - SS(\alpha_1|\alpha_2, \alpha_3), \end{aligned} \tag{1.2.3}$$

where the superscript "$-$" denotes a generalized inverse. Let $P_{(\boldsymbol{X}_2:\boldsymbol{X}_3)}$ denote the orthogonal projection matrix onto the vector space generated by the columns of $[\boldsymbol{X}_2 : \boldsymbol{X}_3]$. From the definition of the matrix $\boldsymbol{Z}$ it follows that $\boldsymbol{Z}\boldsymbol{Z}' = \mathbf{I}_N - P_{(\boldsymbol{X}_2:\boldsymbol{X}_3)}$. Let m_1 and m_e respectively denote the degrees of freedom associated with $SS(\alpha_1|\alpha_2, \alpha_3)$ and $SS(e)$. Then,

$$\begin{aligned} m_1 &= \text{rank}(\boldsymbol{Z}'\boldsymbol{X}_1) = \text{rank}[\boldsymbol{X}_1'(\mathbf{I}_N - P_{(\boldsymbol{X}_2:\boldsymbol{X}_3)})\boldsymbol{X}_1] \\ &= \text{rank}(\boldsymbol{X}_1 : \boldsymbol{X}_2 : \boldsymbol{X}_3) - \text{rank}(\boldsymbol{X}_2 : \boldsymbol{X}_3) \\ m_e &= N - \text{rank}(\boldsymbol{X}_1 : \boldsymbol{X}_2 : \boldsymbol{X}_3). \end{aligned} \tag{1.2.4}$$

As was already pointed out, in order to compute $SS(\alpha_1|\alpha_2, \alpha_3)$ and $SS(e)$, computation of the matrix $\boldsymbol{Z}$ is not necessary; these sums of squares can be obtained from an ANOVA decomposition based on model (1.2.1). Noting that $SS(\alpha_1|\alpha_2, \alpha_3)$ and $SS(e)$ are quadratic forms in $\mathbf{u}$, their expected values are given by the following expressions: (Recall our assumptions that $\boldsymbol{\alpha}_1 \sim N(\mathbf{0}, \sigma_1^2\boldsymbol{I})$ and $\mathbf{e} \sim N(\mathbf{0}, \sigma_e^2\boldsymbol{I}_N)$).

$$\begin{aligned} \text{E}(SS(\alpha_1|\alpha_2, \alpha_3)) &= m_1\sigma_e^2 + \text{tr}[\boldsymbol{X}_1'(\mathbf{I}_N - P_{(\boldsymbol{X}_2:\boldsymbol{X}_3)})\boldsymbol{X}_1]\sigma_1^2 \\ \text{E}(SS(e)) &= m_e\sigma_e^2. \end{aligned} \tag{1.2.5}$$

Thus, to test H_0: $\sigma_1^2 = 0$, we can use the statistic F_1 given by

$$F_1 = \frac{SS(\alpha_1|\alpha_2, \alpha_3)/m_1}{SS(e)/m_e}. \tag{1.2.6}$$

In other words, as long as $\boldsymbol{Z}'\boldsymbol{X}_1 \neq \mathbf{0}$ (i.e., $m_1 \neq 0$), H_0: $\sigma_1^2 = 0$ can be tested using an exact F-test. Furthermore, when $\boldsymbol{\alpha}_1$ is a vector of fixed effects, the above F-test is the same test that will be used to test the significance of the fixed effects. Also implicit in the above development is the assumption that $m_e \neq 0$. These observations are actually given in Seely and El-Bassiouni (1983). These authors refer to the above F-test based on F_1 as Wald's variance component test.

For special cases of model (1.2.1), Wald's test has been derived by several authors (Thompson, 1955a, b; Spjϕtvoll, 1967; Portnoy, 1973). Throughout this book, whenever possible, we have outlined Wald-type exact F-tests for various models and hypotheses (see Chapters 4 and 6).

We now present an example which is investigated in detail in Chapter 4.

Example 1.2.1. Consider the random two-way model with interaction given by

$$y_{ijk} = \mu + \tau_i + \beta_j + (\tau\beta)_{ij} + e_{ijk}, \tag{1.2.7}$$

$i = 1, \ldots, v$, $j = 1, \ldots, b$, $k = 1, \ldots, n_{ij}$, where μ is a fixed effect, and the rest of the effects are random with the usual normality and independence assumptions. Let $n_{..} = \sum_{i=1}^{v} \sum_{j=1}^{b} n_{ij}$ and let $\mathbf{1}_{n_{..}}$ denote the $n_{..} \times 1$ vector of ones. We can write the model in matrix form as

$$\mathbf{y} = \mathbf{1}_{n_{..}} \mu + \boldsymbol{X}_1 \boldsymbol{\tau} + \boldsymbol{X}_2 \boldsymbol{\beta} + \boldsymbol{X}_3 (\boldsymbol{\tau\beta}) + \mathbf{e}, \tag{1.2.8}$$

where $\mathbf{y}$ is the vector of observations, $\boldsymbol{X}_1$, $\boldsymbol{X}_2$, and $\boldsymbol{X}_3$ are known design matrices, and $\boldsymbol{\tau}$, $\boldsymbol{\beta}$, and $(\boldsymbol{\tau\beta})$ are, respectively, the vectors consisting of τ_i's, β_j's and $(\tau\beta)_{ij}$'s. Moreover, $\boldsymbol{\tau}$, $\boldsymbol{\beta}$, $(\boldsymbol{\tau\beta})$, and $\mathbf{e}$ are assumed to be independently distributed as $\boldsymbol{\tau} \sim N(\mathbf{0}, \sigma_\tau^2 \mathbf{I}_v)$, $\boldsymbol{\beta} \sim N(\mathbf{0}, \sigma_\beta^2 \mathbf{I}_b)$, $(\boldsymbol{\tau\beta}) \sim N(\mathbf{0}, \sigma_{\tau\beta}^2 \mathbf{I}_{bv})$, and $\mathbf{e} \sim N(\mathbf{0}, \sigma_e^2 \mathbf{I}_{n_{..}})$. Let n_0 denote the number of nonzero n_{ij}'s. Then, the following results can be verified (see Lemma 4.3.1 in Chapter 4).

1. $\text{Rank}(\boldsymbol{X}_3) = n_0$.
2. $\text{Rank}(\boldsymbol{X}_1 : \boldsymbol{X}_3) = \text{rank}(\boldsymbol{X}_2 : \boldsymbol{X}_3) = \text{rank}(\boldsymbol{X}_1 : \boldsymbol{X}_2 : \boldsymbol{X}_3) = \text{rank}(\boldsymbol{X}_3)$.
3. The vector space generated by the columns of $\boldsymbol{X}_i$ $(i = 1, 2)$ is a subspace of the vector space generated by the columns of $\boldsymbol{X}_3$.

Note that the error sum of squares $SS(\boldsymbol{e})$ has degrees of freedom $m_e = n_{..} - n_0$. We assume that $m_e > 0$. The following observations can now be made.

1. The degrees of freedom associated with $SS(\tau\beta|\tau, \beta)$ is $n_0 - b - v + 1$. Hence, if $n_0 - b - v + 1 > 0$, a Wald-type F-test can be constructed for testing $H_{\tau\beta} : \sigma_{\tau\beta}^2 = 0$, based on $SS(\tau\beta|\tau, \beta)$ and $SS(e)$. If $n_{ij} \geq 1$ for all i and j, then $n_0 = bv$ and the degrees of freedom associated with $SS(\tau\beta|\tau, \beta)$ is $(v - 1)(b - 1)$.
2. For testing $H_\tau : \sigma_\tau^2 = 0$, or H_0: $\sigma_\beta^2 = 0$, a Wald-type F-test does not exist. This is due to the fact that $SS(\tau|\tau\beta, \beta)$ and $SS(\beta|\tau\beta, \tau)$ are zeros. This follows from property 3 above concerning the vector spaces generated by the columns of $\boldsymbol{X}_1$ and $\boldsymbol{X}_2$.

The above example is analyzed in detail in Chapter 4. It should be noted that even though Wald-type F-tests do not exist for testing H_0: $\sigma_\tau^2 = 0$ and H_0: $\sigma_\beta^2 = 0$, other exact F-tests do exist for testing these hypotheses. This is also discussed in Chapter 4.

1.3. OPTIMALITY OF TESTS

We first discuss briefly the basic framework of the Neyman-Pearson theory of testing of statistical hypotheses. Let

$$\mathcal{F}(\boldsymbol{\theta}) = \{f(x|\boldsymbol{\theta})|\boldsymbol{\theta} \in \Theta\}, \quad x \in \mathcal{X} \tag{1.3.1}$$

be a class of probability densities of a random variable X on a subset $\mathcal{X}$ of a Euclidean space. Throughout this book, $f(x|\boldsymbol{\theta})$ stands for an appropriate normal density of an observation vector or a matrix $\mathbf{X}$, and $\boldsymbol{\theta}$ denotes the associated unknown parameters such as the mean and dispersion of $\mathbf{X}$. Then, quite generally, a testing problem is described by two hypotheses H_0 and H_1 as

$$H_0 : \boldsymbol{\theta} \in \Theta_0 \ \ versus \ \ H_1 : \boldsymbol{\theta} \in \Theta_1, \tag{1.3.2}$$

where Θ_0 and Θ_1 are two disjoint subsets of Θ. A test or a test function is a measurable function $\phi(x)$ from $\mathcal{X}$ into $[0, 1]$, which denotes the probability of rejecting H_0 when x is observed. In particular, if $\phi(x) = 1$ for $x \in w$, and 0 otherwise, H_0 is rejected whenever $x \in w$, and w is called the rejection region of H_0. The average probability of a test function $\phi(x)$ for rejection under $\boldsymbol{\theta}$ is defined by

$$P(\phi, \boldsymbol{\theta}) = E_{\boldsymbol{\theta}}[\phi(x)] = \int_{\mathcal{X}} \phi(x) f(x|\boldsymbol{\theta})\, dx, \tag{1.3.3}$$

and this is known as the power function of the test $\phi(x)$ when $\boldsymbol{\theta} \in \Theta_1$. Our goal, in the Neyman-Pearson theory, is to find a test function $\phi(x)$ which maximizes $P(\phi, \boldsymbol{\theta})$ for $\boldsymbol{\theta} \in \Theta_1$ in some sense in the class $\mathcal{C}_\alpha$ of level α tests defined by

$$\mathcal{C}_\alpha = \{\phi | \phi \ \ is \ \ a \ \ test, \ \ sup_{\boldsymbol{\theta} \in \Theta_0} P(\phi, \boldsymbol{\theta}) \leq \alpha\}. \tag{1.3.4}$$

In the above, α is called a level of significance, and it corresponds to the maximum average probability of an incorrect decision under the test ϕ when H_0 is true. The quantity $\{sup_{\boldsymbol{\theta} \in \Theta_0} P(\phi, \boldsymbol{\theta})\}$ is defined to be the size of the test. When a test maximizing power uniformly in $\boldsymbol{\theta} \in \Theta_1$ exists, it is called a *uniformly most powerful* (UMP) test. The following generalized Neyman-Pearson Lemma is often used to maximize power.

Lemma 1.3.1. *(Lehmann, 1986, p. 96)*. Let $f_1, \ldots, f_{m+1}$ be real-valued integrable functions defined on $\mathcal{X}$, and suppose that for given constants $c_1, \ldots, c_m$, there exists a test function ϕ satisfying

$$\int \phi f_i \, dx = c_i, \ \ i = 1, \cdots, m. \tag{1.3.5}$$

Let $\mathcal{D}$ be the class of tests for which (1.3.5) holds. Then there exists a test ϕ_0 in $\mathcal{D}$ that maximizes

$$\int \phi f_{m+1} \, dx. \tag{1.3.6}$$

If a test ϕ^* in $\mathcal{D}$ is of the form

$$\begin{aligned} \phi^*(x) &= 1 \ \ if \ \ f_{m+1}(x) > \sum_{i=1}^{m} k_i f_i(x), \\ &= 0 \ \ if \ \ f_{m+1}(x) \leq \sum_{i=1}^{m} k_i f_i(x) \end{aligned} \tag{1.3.7}$$

for some constants $k_1, \ldots, k_m$, then ϕ^* maximizes $\int \phi f_{m+1}\, d\mu$ in $\mathcal{D}$. If a test ϕ_0^* in $\mathcal{D}$ satisfies (1.3.7) with $k_i \geq 0$ $(i = 1, \ldots, m)$, then it maximizes $\int \phi f_{m+1}\, d\mu$ in the class of tests satisfying

$$\int \phi_0 f_i \, dx \leq c_i, \quad i = 1, \cdots, m. \tag{1.3.8}$$

□

For $m = 1$ with $f_1(x) = f(x|\boldsymbol{\theta}_0)$ and $f_2(x) = f(x|\boldsymbol{\theta}_1)$, where $\boldsymbol{\theta}_0 \in \Theta_0$ and $\boldsymbol{\theta}_1 \in \Theta_1$, the above lemma reduces to the celebrated Neyman-Pearson (N-P) Lemma.

There are plenty of applications of the N-P Lemma in the entire area of tests of hypotheses. However, it turns out that, except in some simple problems, there exists no test ϕ_0 that maximizes the power $P(\phi, \boldsymbol{\theta})$ uniformly in $\boldsymbol{\theta} \in \Theta_1$. In other words, a UMP test exists very rarely. A class of interesting problems for which a UMP test readily exists deals with testing a simple null hypothesis $H_0 : \theta = \theta_0$ against a composite alternative $H_1 : \theta > \theta_0$ in the probability density function (*pdf*) $f(x|\theta)$, which is a member of a *one-parameter exponential family*, given by

$$f(x|\theta) = Q(\theta) h(x) e^{T(x)\theta}, \quad x \ \ real, \quad \theta \in (a, b), \tag{1.3.9}$$

where $h(x)$ is the *pdf* of X when $\theta = 0$, $Q(\theta) = [\int h(x) e^{T(x)\theta}\, dx]^{-1}$, and the interval (a, b) defines the natural parameter space for which the above integral exists. Here, without any loss of generality, 0 and θ_0 are assumed to be interior points of (a, b). Examples of such $f(x|\theta)$ include *normal with unknown mean and known variance*, *normal with known mean and unknown variance*, *exponential*, and *gamma with known shape and unknown scale* distributions. In each case the UMP test is based on $T(x)$, and is one-sided. In fact, under the model (1.3.9), this result holds even when H_0 is enlarged to H_0^*: $\theta \leq \theta_0$. Of course, if $(x_1, x_2, \ldots, x_n)$ is a random sample from (1.3.9), $T(x)$ stands for $\sum_{i=1}^n T(x_i)$.

As mentioned before, a UMP test exists rather rarely, and in most applications we may have to be satisfied with tests other than UMP tests. Thus, for example, in the context of the one-parameter exponential family defined above, a UMP test does not exist if the null hypothesis is H_0 and the one-sided alternative $H_1 : \theta > \theta_0$ is changed to a two-sided alternative $H_2 : \theta \neq \theta_0$. Moreover, even for one-sided alternatives, a UMP test does not exist in problems involving more than one unknown parameter (for example, a test for a normal mean with an unknown variance or for a normal variance with an unknown mean). When this happens, a standard compromise is to restrict the class $\mathcal{C}_\alpha$ to a suitable subclass by a certain criterion, and try to find a test that maximizes the power within this subclass. The following two criteria to restrict $\mathcal{C}_\alpha$ are well known.

1. *Unbiasedness, Similarity*. The test functions of size α in this subclass are required to be *similar* in the sense that $P(\phi, \boldsymbol{\theta}) = \alpha$ for all $\boldsymbol{\theta} \in \Theta_0$,

and also *unbiased* in the sense that $P(\phi, \boldsymbol{\theta}) \geq \alpha$ for all $\boldsymbol{\theta} \in \Theta_1$. This is briefly discussed below. For details, see Lehmann (1986), Ferguson (1967).

2. *Invariance.* The test functions in this subclass are required to satisfy certain special forms depending on the group structure of $f(x|\boldsymbol{\theta})$ and the nature of the hypotheses H_0 and H_1. This is briefly discussed below. For details, see Lehmann (1986), Ferguson (1967).

In addition, we should also mention the following two criteria as methods of test construction without any regard to optimality.

1. *Likelihood Ratio Test (LRT).* This is a very general procedure which works in almost any situation, and depends on the maximum likelihood estimators of the relevant unknown parameters under both the null and the alternative hypotheses.
2. *Studentization.* This is an ad hoc method of test construction which works well when the parameter of interest is scalar, and is based on a suitable studentization of an estimator of the parameter.

1.3.1. Uniformly Most Powerful Similar and Uniformly Most Powerful Unbiased Tests

Consider the testing problem given in (1.3.2). Let

$$C_\alpha^U = \{\phi | P(\phi, \boldsymbol{\theta}) \leq \alpha, \ \ for \ \ \boldsymbol{\theta} \in \Theta_0; \ \ P(\phi, \boldsymbol{\theta}) \geq \alpha, \ \ for \ \ \boldsymbol{\theta} \in \Theta_1\} \tag{1.3.10}$$

be the class of level α unbiased tests. In most cases, the power function $P(\phi, \boldsymbol{\theta})$ of a test procedure ϕ is a continuous function of $\boldsymbol{\theta}$. In such a situation, whenever a test function ϕ in $\mathcal{C}_\alpha$, defined in (1.3.4), is also unbiased, it must satisfy

$$P(\phi, \boldsymbol{\theta}) = \alpha \ \ for \ \ all \ \ \boldsymbol{\theta} \in \mathcal{B}(\Theta_0, \Theta_1), \tag{1.3.11}$$

where $\mathcal{B}(\Theta_0, \Theta_1)$ is the boundary of Θ_0 and Θ_1. Such a test ϕ is known as *similar* on the boundary. Defining

$$\mathcal{C}_\alpha^S = \{\phi | P(\phi, \boldsymbol{\theta}) = \alpha \ \ for \ \ all \ \ \boldsymbol{\theta} \in \mathcal{B}(\Theta_0, \Theta_1)\}, \tag{1.3.12}$$

which is the class of level α similar (on the boundary) tests, and noting that $\mathcal{C}_\alpha^U$, the class of unbiased tests, is a subclass of $\mathcal{C}_\alpha^S$, a standard practice is to derive a UMP test in this bigger class of tests (i.e., a UMPS test) which is then trivially UMP in the class $\mathcal{C}_\alpha^U$. In other words, a UMPU test may be obtained by first deriving a UMPS test. This procedure is especially useful in problems involving nuisance (i.e., unspecified) parameters. A characterization of tests which are similar on the boundary of Θ_0 and Θ_1 is often achieved by what is known as *Neyman Structure*. Roughly speaking, we first find a sufficient statistic $U(x)$ for unspecified or nuisance parameters under $\boldsymbol{\theta} \in \mathcal{B}(\Theta_0, \Theta_1)$, and then require that a test function has its conditional size equal to α, given the sufficient statistic $U(x)$, so that the requirement that $\phi \in \mathcal{C}_\alpha^S$ is trivially satisfied. In most cases $U(x)$ is complete under $\boldsymbol{\theta} \in \mathcal{B}(\Theta_0, \Theta_1)$ so that $\phi \in \mathcal{C}_\alpha^S$ holds only if ϕ has its conditional size equal to α. A UMPS test is then derived by maximizing the conditional power of a test function ϕ, given the sufficient statistic $U(x)$, by the usual application of the Neyman-Pearson Lemma. The standard one-sided Student's t-test for the normal mean when the variance is unknown and the standard one-sided χ^2-test for the normal variance when the mean is unknown are simple examples of UMPS tests (see Lehmann, 1986).

Quite generally, suppose $f(x|\boldsymbol{\theta})$ is a member of a *multiparameter exponential family*, and is given by

$$f(x|\boldsymbol{\theta}) = Q(\boldsymbol{\theta})h(x)e^{\eta_1(\boldsymbol{\theta})T_1(x)+\ldots+\eta_m(\boldsymbol{\theta})T_m(x)}, \tag{1.3.13}$$

where $\boldsymbol{\theta} = (\theta_1, \ldots, \theta_k)$ denotes a k-dimensional vector parameter, $h(x)$ is the *pdf* of X when $\boldsymbol{\eta} = \mathbf{0}$, and $Q(\boldsymbol{\theta}) = [\int h(x)e^{\eta_1(\boldsymbol{\theta})T_1(x)+\ldots+\eta_m(\boldsymbol{\theta})T_m(x)}\,dx]^{-1}$ is the normalizing constant. In this context, $\boldsymbol{\eta}(\boldsymbol{\theta}) = (\eta_1(\boldsymbol{\theta}), \ldots, \eta_m(\boldsymbol{\theta}))'$ stands for a reparameterization of the original parameter $\boldsymbol{\theta}$, and is called the natural parameter of the exponential family (1.3.13). In most problems, the dimension m of $\boldsymbol{\eta}$ is equal to the dimension k of $\boldsymbol{\theta}$, and the transformation $\boldsymbol{\theta} \to \boldsymbol{\eta}(\boldsymbol{\theta})$ is one-to-one. When this happens, the natural parameter space Θ for which $\int h(x)e^{\eta_1(\boldsymbol{\theta})T_1(x)+\ldots+\eta_k(\boldsymbol{\theta})T_k(x)}\,dx$ is finite is a k-dimensional open rectangle (which is assumed to contain $\mathbf{0}$ without any loss of generality). In this setup, the tests of various hypotheses are stated in terms of η's. Thus, for testing $\eta_1 = \eta_1^0$ against $H_1 : \eta_1 > \eta_1^0$ when the rest of the η's are unspecified, a *UMPS* test exists very naturally, which is then also a UMPU test. The one-sided t-test and the one-sided χ^2-test mentioned above are examples of this situation.

There are some problems where m is greater than k in which case the natural parameters $(\eta_1, \ldots, \eta_m)$, which are generated by the k-dimensional vector parameter $\boldsymbol{\theta}$, become functionally dependent, and the resultant exponential family is commonly known as a *curved* exponential family. Under this situation, the existence of a UMPS or UMPU test is not guaranteed, and most often such tests do not exist. This is primarily because of a lack of completeness of $U(x)$ under $\boldsymbol{\theta} \in \mathcal{B}(\Theta_0, \Theta_1)$. One such problem is discussed at the end of this section (*unbalanced one-way random effects model*).

When the derivation of a UMPS test is not possible (i.e., a UMPS test does not exist), a UMPU test can still exist. Thus, in the context of the one-parameter exponential family given by (1.3.9), if the problem is to test $H_0 : \theta = \theta_0$ against $H_2 : \theta \neq \theta_0$, a UMPS test does not exist while a UMPU test always exists. Similarly, in the context of the multiparameter exponential family (1.3.13) with $m = k$ and under the assumption that $\boldsymbol{\theta} \to \boldsymbol{\eta}$ is one-to-one, if the problem is to test $H_0 : \eta_1 = \eta_1^0$ against $H_2 : \eta_1 \neq \eta_1^0$, a UMPS test does not exist, although a UMPU test can be found as follows (the approach is similar for the model (1.3.9)). If a test function ϕ is unbiased, it essentially means that the power function $P(\phi, \boldsymbol{\eta})$ of ϕ has a minimum at $\eta_1 = \eta_1^0$, irrespective of the values of the other parameters. For the multiparameter exponential family given in (1.3.13), it holds quite generally that $P(\phi, \boldsymbol{\eta})$ is differentiable with respect to $\boldsymbol{\eta}$, and that the process of differentiation can be carried out under the appropriate integral sign. The above condition that ϕ has a minimum power at $\eta_1 = \eta_1^0$ then implies

$$\int_{\mathcal{X}} \left\{ \frac{\partial}{\partial \eta_1} [f(x|\eta_1, \cdots, \eta_k)] | \eta_1 = \eta_1^0 \right\} dx = 0, \quad for \;\; all \;\; \eta_2, \cdots, \eta_k. \tag{1.3.14}$$

Let $\mathcal{C}_\alpha^U *$ denote the class of tests satisfying (1.3.10) (with $\mathcal{B}(\Theta_0, \Theta_1) = \{\eta_1^0\}$) and (1.3.14), which is obviously bigger than $\mathcal{C}_\alpha^U$ but smaller than $\mathcal{C}_\alpha^S$. We then use the generalized Neyman-Pearson Lemma (Lemma 1.3.1) with

$$\begin{aligned} f_1 &= f(x|\eta_1^0, \eta_2, \cdots, \eta_k), \\ f_2 &= \frac{\partial}{\partial \eta_1} [f(x|\eta_1, \cdots, \eta_k)] | \eta_1 = \eta_1^0, \\ f_3 &= f(x|\eta_1, \eta_2, \cdots, \eta_k), \end{aligned} \tag{1.3.15}$$

and derive a UMP test in $\mathcal{C}_\alpha^U *$, using the "conditional approach" by conditioning on a sufficient statistic $\boldsymbol{U}(\boldsymbol{x}) = (T_2(x), \ldots, T_k(x))$ for $(\eta_2, \ldots, \eta_k)$ under H_0. The resultant test is a two-sided test based on $T_1(x)$, conditioned on $\boldsymbol{U}(\boldsymbol{x})$, and in most cases can be further simplified to yield an unconditional test by invoking what is known as Basu's Theorem (see Theorem 2, Lehmann (1986), page 191). Two-sided Student's t-test for a normal mean with an unknown variance and two-sided χ^2-test for a normal variance with an unknown mean are common examples of UMPU tests (see Lehmann, 1986). Throughout the above, x stands for $(x_1, x_2, \ldots, x_n)$, a random sample from (1.3.13), and $T_i(x) = \sum_{j=1}^n T_i(x_j)$, for all i.

When UMP or $UMPU$ tests do not exist, we use a local expansion of the power function of a test around the null hypothesis and the generalized Neyman-Pearson Lemma to derive a LB or LBU test. Thus, if $f(x|\theta)$ involves only one unknown parameter θ and the problem is to test $H_0 : \theta = \theta_0$ against $H_1 : \theta > \theta_0$, a LB test ϕ is derived by maximizing (see Rao, 1973) the leading term of its *local* power, given by

$$\int \phi(x) [\frac{\partial f(x|\theta)}{\partial \theta} | \theta = \theta_0] \, dx \tag{1.3.16}$$

subject to the usual size condition, namely,

$$\int \phi(x) f(x|\theta_0) \, dx = \alpha. \tag{1.3.17}$$

Similarly, under the same setup, when the alternative is both-sided (i.e., H_2 : $\theta \neq \theta_0$), an LBU test is derived by again maximizing the leading term of its *local* power, namely,

$$\int \phi(x) \left[\frac{\partial^2 f(x|\theta)}{\partial \theta^2} | \theta = \theta_0\right] dx \tag{1.3.18}$$

subject to (1.3.17) and the unbiasedness implication

$$\int \phi(x) \left[\frac{\partial f(x|\theta)}{\partial \theta} | \theta = \theta_0\right] dx = 0. \tag{1.3.19}$$

Some examples of LB and LBU tests appear in Rao (1973) and Lehmann (1986).

1.3.2. Uniformly Most Powerful Invariant and Locally Most Powerful or Locally Best Invariant Tests

Many problems of hypothesis testing in the context of mixed linear models are either multiparameter in nature or else involve a single parameter (such as a fixed effect or a variance component) in the presence of many unknown parameters (unspecified fixed effects and variance components), which are known as nuisance parameters. While UMPU tests may exist in some problems of the latter type, they rarely exist in problems of the former type. It is well known that invariance is perhaps the only useful and successful tool that allows us to consider the optimality of tests in a variety of multiparameter testing problems.

For a general theory on invariance, we refer to Lehmann (1986), Ferguson (1967), Eaton (1983), Muirhead (1982), and Kariya and Sinha (1989). To describe the invariance principle in the context of tests of hypotheses, consider the setup in (1.3.1) in the multiparameter case, which can be written as

$$\mathcal{P}_{\boldsymbol{\theta}} = \left\{P_{\boldsymbol{\theta}} | \frac{dP_{\boldsymbol{\theta}}}{dx} = f(x|\boldsymbol{\theta}),\ \ f(x|\boldsymbol{\theta}) \in \mathcal{F}(\boldsymbol{\theta})\right\} \tag{1.3.20}$$

and the two hypotheses (null and alternative) $H_0 : \boldsymbol{\theta} \in \Theta_0$ and $H_1 : \boldsymbol{\theta} \in \Theta_1$. Let $\mathcal{G}$ be a group of transformations from $\mathcal{X}$ onto $\mathcal{X}$ such that

$$P_{\boldsymbol{\theta}} \in \mathcal{P}_{\boldsymbol{\theta}} \rightarrow gP_{\boldsymbol{\theta}} \in \mathcal{P}_{\boldsymbol{\theta}} \ \ for\ \ any\ \ \boldsymbol{\theta}, \tag{1.3.21}$$

where $gP_{\boldsymbol{\theta}}(A) = P_{\boldsymbol{\theta}}(g^{-1}(A))$ for a Borel set A of $\mathcal{X}$ (for a definition of a group, see page 15 of Kariya and Sinha, 1989). When (1.3.21) holds, we say that the group $\mathcal{G}$ keeps the family $\mathcal{P}_{\boldsymbol{\theta}}$ invariant. The relation (1.3.21) along with the *identifiability* of $\mathcal{P}_{\boldsymbol{\theta}}$ means that there exists a parameter $\boldsymbol{\theta}'$ in Θ such that $gP_{\boldsymbol{\theta}} = P_{\boldsymbol{\theta}'}$, and we can write the correspondence between $\boldsymbol{\theta}$ and $\boldsymbol{\theta}'$ by $\boldsymbol{\theta}' = \bar{g}\boldsymbol{\theta}$. If it holds that

$$\bar{g}(\Theta_0) = \Theta_0, \quad \bar{g}(\Theta_1) = \Theta_1, \quad for \ any \ \bar{g} \in \bar{\mathcal{G}}, \tag{1.3.22}$$

we say that the testing problem is left invariant under the group $\mathcal{G}$. Here $\bar{\mathcal{G}}$ is the induced group of transformations with elements $\bar{g}$. When a testing problem remains invariant under a group $\mathcal{G}$, it is natural to require that a test function ϕ also remains invariant in the sense that

$$\phi(gx) = \phi(x) \quad for \ all \ g \in \mathcal{G} \ and \ x \in \mathcal{X}, \tag{1.3.23}$$

and we restrict our attention to the class of invariant level α tests $\mathcal{C}_\alpha^I$. An invariant function $T(x)$ is said to be *maximal* invariant if it satisfies

$$\begin{aligned} T(gx) &= T(x) \quad for \ all \ g \in \mathcal{G} \ and \ x \in \mathcal{X} \\ T(x_1) &= T(x_2) \quad implies \ x_1 = gx_2 \ for \ some \ g \in \mathcal{G}. \end{aligned} \tag{1.3.24}$$

It can be shown that if a test function ϕ is invariant under $\mathcal{G}$, then ϕ must depend on x through $T(x)$; in other words, ϕ can be expressed as

$$\phi(x) = \psi(T(x)), \tag{1.3.25}$$

where ψ is a test function. It turns out that in such a situation, the power of an invariant test function $\phi(x)$ depends on $\boldsymbol{\theta}$ only through a special function $\tau(\boldsymbol{\theta})$, which is known as a maximal invariant parameter.

The problem of testing H_0 versus H_1 under the model (1.3.20) is now reduced by invariance as follows. Since all invariant tests can be expressed as functions of $T(x)$ whose distribution depends on $\tau(\boldsymbol{\theta})$, the null and the alternative hypotheses are expressed as

$$H_0' : \tau \in \tau(\Theta_0) \quad versus \quad H_1' : \tau \in \tau(\Theta_1), \tag{1.3.26}$$

and our goal in this reduced problem is to derive a test $\psi_0(T(x))$ in $\mathcal{C}_\alpha^I$, the class of level α invariant tests, which would maximize in some sense the power function

$$E_{\boldsymbol{\theta}}[\psi(T(x))] \tag{1.3.27}$$

with respect to $\psi(T(x))$. We can use the (generalized) Neyman-Pearson Lemma to solve this problem. A test $\phi_0 = \psi_0(T(x))$ in $\mathcal{C}_\alpha^I$ is called a UMPI test if for any $\phi = \psi(T(x))$ in $\mathcal{C}_\alpha^I$

$$E_{\boldsymbol{\theta}}[\psi_0(T(x))] \geq E_{\boldsymbol{\theta}}[\psi(T(x))] \quad for \ all \ \boldsymbol{\theta} \in \Theta_1, \tag{1.3.28}$$

and a test $\phi_0 = \psi_0(T(x))$ is called LBI if there exists an open neighborhood $\tilde{\Theta}_0$ of Θ_0 such that for any $\phi \in \mathcal{C}_\alpha^I$,

$$E_{\boldsymbol{\theta}}[\psi_0(T(x))] \geq E_{\boldsymbol{\theta}}[\psi(T(x))] \quad for \ all \ \boldsymbol{\theta} \in \{\tilde{\Theta}_0 - \Theta_0\}. \tag{1.3.29}$$

To summarize, the following steps are usually taken for the analysis of an invariant testing problem.

1. Given the underlying testing problem, first find a group $\mathcal{G}$ which keeps the testing problem invariant.
2. Choose a convenient maximal invariant $T(x)$ under $\mathcal{G}$.
3. Derive the null and nonnull distributions of $T(x)$.
4. Use (generalized) Neyman-Pearson Lemma to derive UMPI or LBI tests as appropriate, using the criteria of unbiasedness and similarity, if necessary. This is important because the basic testing problem, even after reduction by invariance, may still retain the characteristics of an ordinary testing problem involving two-sided alternatives or the presence of nuisance parameters. The resulting tests are then called UMPIU (UMPIS) or LBIU (LBIS) tests.

Step 3 above often creates difficulties because in many cases a maximal invariant $T(x)$ is too complicated to treat analytically. This happens in many multiparameter problems, including tests for fixed effects and variance components in complicated mixed effects models. It should be noted that it is not necessary to explicitly identify a maximal invariant $T(x)$ in many problems, neither it is necessary to know the null and the nonnull distributions of $T(x)$ separately. After all, to derive a UMPI or an LBI test based on $T(x)$, what we really need is the ratio of the distributions of a maximal invariant under the null and the alternative hypotheses. In fact, for the LBI test, we simply need a local expansion of this ratio around the null hypothesis. In such cases, what is now widely known as Wijsman's (Wijsman, 1967; Kariya and Sinha, 1989) representation of the probability ratio $R(T(x))$ of a maximal invariant $T(x)$ (more popularly as Wijsman's Representation Theorem) is extremely useful. There are several versions of this representation depending upon conditions on the group, which leaves the underlying testing problem invariant, but all of them have the same general form, which is given in (A.1.1) in Appendix 1.1, at the end of this chapter.

We now describe some special features of the testing problems which arise in the context of mixed linear models, and clarify some points with regard to the optimality of tests.

Example 1.3.1. One-Way Random Effects Model. This model, which is very basic and appears in Chapters 2, 4, and 6, deals with v random treatment effects and a set of $n. = n_1 + \ldots + n_v$ observations following the structure

$$y_{ij} = \mu + \tau_i + e_{ij}, \quad i = 1, \cdots, v; \quad j = 1, \cdots, n_i, \tag{1.3.30}$$

where μ is fixed, $\tau_1, \ldots, \tau_v$ are independent normal with mean 0 and variance σ_τ^2, and e_{ij}'s are independent normal with mean 0 and variance σ_e^2. The problem is to test

$$H_\tau : \sigma_\tau^2 = 0 \textit{ versus } H_1 : \sigma_\tau^2 > 0. \tag{1.3.31}$$

In the balanced case (i.e., when $n_1 = \ldots = n_v$), the distribution of the data vector $\mathbf{y}$ satisfies (1.3.13) with $m = k = 3$, and consequently, there is a UMPS (UMPU) test which is also a UMPI test for H_τ (this is the usual ANOVA F-test; see Chapter 2). In the unbalanced case, however, while the pattern in (1.3.13) still holds, the dimension m becomes much larger than 3 (its actual value depends on the nature of treatment replications $n_1, \ldots, n_v$), and it turns out that there is no UMPI test. However, an LBI test, which is different from the valid traditional F-test (Wald-type exact test) does exist (see formula (4.2.8) in Chapter 4 for an expression of the LBI test and Section 6.2.2 in Chapter 6 for its derivation). It may be noted that the underlying group $\mathcal{G}$ here consists of elements $g = (a, b)$ with $a > 0$ and b real, and the transformation is given by $y_{ij} \to a(y_{ij} + b)$.

Example 1.3.2. Two-Way Random Effects Model Without Interaction. Such a model is given by

$$y_{ijk} = \mu + \tau_i + \beta_j + e_{ijk}, i = 1, \cdots, v; \;\; j = 1, \cdots, b; \;\; k = 1, \cdots, n_{ij}, \tag{1.3.32}$$

where μ is fixed, τ_i's are independent normal with mean 0 and variance σ_τ^2, β_j's are independent normal with mean 0 and variance σ_β^2, and e_{ijk}'s are independent normal with mean 0 and variance σ_e^2. The problem is to test $H_\tau : \sigma_\tau^2 = 0$ against $H_1 : \sigma_\tau^2 > 0$. In the balanced case (i.e., when the n_{ij}'s are all equal), the distribution of the data vector $\mathbf{y}$ satisfies the model (1.3.13) with $k = 4$, and the usual F-test, based on the ratio of the treatment sum of squares to the error sum of squares, turns out to be a UMPS (UMPU) test as well as a UMPI test. In the unbalanced case, while the pattern in (1.3.13) still holds, m is much larger than 4, and there does not exist a UMPU or UMPI test. Moreover, an LBIU test exists only under the assumption of an equireplicate and equiblock design; that is, $\sum_i n_{ij}$'s are all equal (equiblock) and $\sum_j n_{ij}$'s are all equal (equireplicate). This test is described in Section 4.3 of Chapter 4. The group $\mathcal{G}$ here is the same as described in Example 1.3.1.

Example 1.3.3. Two-Way Mixed Effects Model Without Interaction. The model here is the same as given above in (1.3.32) except that there are two distinct situations: The treatment effects $\tau_1, \ldots, \tau_v$ are fixed while the β_j's are random, or the treatment effects $\tau_1, \ldots, \tau_v$ are random while the β_j's are fixed. Accordingly, there are two testing problems in this context: (i) $H_\tau : \tau_1 = \ldots = \tau_v = 0$ against $H_1 : \tau_i$'s are unequal in the first situation, and (ii) $H_\tau : \sigma_\tau^2 = 0$ against $H_1 : \sigma_\tau^2 > 0$ in the second situation. Again, in the balanced case, the standard F-test based on the ratio of the treatment sum of squares to the error sum of squares easily turns out to be an optimum test in both cases (see Chapter 2). However, in the unbalanced case, although the general nature of the multiparameter exponential family (1.3.13) holds for the distribution of the data vector $\mathbf{y}$ in both the situations, an optimum test for the first kind of null hypothesis exists only under some restrictive conditions. Thus, for H_τ in the former case where invariance is essential to derive an optimum test, there is no UMPIU test. An LBIU test was derived by Mathew and Sinha (1988b) under the assumption that the underlying design is a *balanced incomplete block design (BIBD)*. This BIBD assumption allows consideration of a group of transformations well beyond the usual translation and shift (as mentioned in the above two cases). Section 6.2 in Chapter 6 provides details of this result and a description of the LBIU test. For H_τ in the latter case, an optimum invariant test always exists, and the nature of optimality (i.e., UMPIU or LBIU) depends on the eigenvalues of the associated $\boldsymbol{C}$ matrix (see Section 6.2 of Chapter 6 for details of the description of the group $\mathcal{G}$ and the tests).

It should be mentioned that there are no optimum tests for hypotheses concerning fixed effects and variance components for unbalanced two-way mixed or random models *with* interaction. Finally, we remark that optimum tests may exist for hypotheses about fixed effects as well as variance components in unbalanced split-plot models in which the whole plot treatments can be replicated an unequal number of times while the split-plot treatments appear exactly once within each whole plot. Such tests are described in Chapter 8.

We conclude this section with a general observation that under the usual mixed linear model

$$\mathbf{y} \sim N(\boldsymbol{X\beta}, \Sigma), \tag{1.3.33}$$

if the testing problem is concerned with the variance components (i.e., Σ), the group $\mathcal{G}$ of transformations keeping the testing problem invariant always includes the subgroup $\mathcal{H}$ given by

$$\mathbf{y} \rightarrow c(\mathbf{y} + \boldsymbol{X\alpha}), \quad c > 0, \quad \boldsymbol{\alpha} \ \textit{arbitrary}. \tag{1.3.34}$$

Thus, if $\boldsymbol{P}$ is the orthogonal projection matrix onto the column space of $\boldsymbol{X}$, and $\boldsymbol{Z}$ satisfies $\boldsymbol{Z}'\boldsymbol{Z} = \boldsymbol{I}$ and $\boldsymbol{I} - \boldsymbol{P} = \boldsymbol{ZZ}'$, invariance under $\mathcal{H}$ means that we can restrict our attention to tests depending only on $\boldsymbol{Z}'\mathbf{y}$, the so-called error functions. Exact nature of $\mathcal{G}$, in addition to $\mathcal{H}$, then depends on the particular testing problem concerning Σ (see Chapters 4 and 6).

Remark 1.3.1. Throughout this book, our emphasis is on the derivation and presentation of tests for fixed effects and variance components, which are optimum (locally or uniformly) or nearly so in the class of appropriate invariant tests. It will be seen that quite frequently these tests are simple, explicit, and easy to implement. Whenever exact tests are available in the literature, whether optimum or not, we have mentioned them. Because of the difficult nature of the theory underlying the derivation of optimum invariant tests, some readers may find these derivations and discussions of their optimum properties rather difficult. However, for a practitioner, there should not be any problem to identify and use these tests in applications. We should also point out that, except in some simple cases, the standard likelihood ratio tests (LRTs) are not readily applicable in the context of mixed linear models, and often such LRTs are neither exact nor easily available. For example, even in the simplest case of a one-way balanced random effects model, the exact distribution of the LRT statistic under the null hypothesis of no treatment effects is not known and can only be simulated. For a one-way unbalanced random effects model, the derivation of the LRT itself is extremely complicated, not to mention its null distribution and its implementation in practice.

REMARK 1.3.2. Regarding tests for variance components in the context of a mixed model, for the most part, the book deals with the null hypothesis H_0: a variance component $= 0$. This is by no means an implication that a test for a nonzero value of a variance component is not important. The fact is such a problem cannot be tackled by a routine procedure, and must be dealt with carefully. We have devoted an entire chapter (Chapter 9) to discuss some special techniques using the concept of a generalized P-value to solve these problems.

APPENDIX 1.1

Distribution of a Maximal Invariant $T(x)$: Wijsman's Representation Theorem

The probability ratio $R[T(x)]$ of a maximal invariant $T(x)$ is given by

$$R[T(x)] = \frac{dP^T_{\boldsymbol{\theta}_1}}{dP^T_{\boldsymbol{\theta}_0}}(T(x)) = \frac{\int_{\mathcal{G}} f(gx|\boldsymbol{\theta}_1)\chi(g)\nu(dg)}{\int_{\mathcal{G}} f(gx|\boldsymbol{\theta}_0)\chi(g)\nu(dg)}. \quad \text{(A.1.1)}$$

In the above, $T(x)$ is a maximal invariant, $P^T_{\boldsymbol{\theta}}$ is the distribution of $T(x)$ under $\boldsymbol{\theta}$, $f(x|\boldsymbol{\theta})$ is a density with respect to a relatively invariant measure μ, $\mathcal{G}$ is a locally compact group, ν is a left-invariant measure on $\mathcal{G}$, and $\chi(g)$ is a multiplier function of μ.

In applications to mixed linear models with the usual assumption of normality of all relevant random effects and the error components, μ stands for the Lebesgue measure dx, $f(x|\boldsymbol{\theta})$ for the (joint) normal density of x, and $\chi(g)$ for the inverse of the Jacobian of transformation from $x \to gx$. In most cases, $\mathcal{G}$ would be a simple locally compact group of translation and scale shift like $x \to a(x+b)$ for b real and $a > 0$ (in the univariate case) or $\mathbf{x} \to \boldsymbol{A}(\mathbf{x}+\mathbf{b})$ for $\boldsymbol{A}$ a nonsingular matrix and $\mathbf{b}$ a vector of real constants (in the multivariate case). A left-invariant measure ν in the former case is given by $dbda/a$, and in the latter case by $db|\mathbf{A}|^{-(N+1)/2}\,d\mathbf{A}$ if $\mathbf{x}$ is an $N \times 1$ vector. When $\mathcal{G}$ stands for a (finite) permutation group of p elements, ν is a discrete uniform measure (a probability distribution) assigning equal weights $1/p!$ to each of the $p!$ members in $\mathcal{G}$. If $\mathcal{G}$ is an orthogonal group, construction of ν is also well known (see Eaton, 1983). Examples of UMPI (UMPIS, UMPIU) and LBI (LBIU, LBIS) tests abound in the statistics literature, especially in the context of testing multiparameter hypotheses (see Lehmann, 1986; Ferguson, 1967; Eaton, 1983; Muirhead, 1982; Kariya and Sinha, 1989).

BIBLIOGRAPHY

Das, R. and Sinha, B. K. (1987). "Robust optimum invariant unbiased tests for variance components." In: *Proc. of the Second International Tampere Conference in Statistics* (T. Pukkila, S. Puntanen, Eds.), University of Tampere, Finland, 317–342.

Eaton, M. L. (1983). *Multivariate Statistics*. Wiley, New York.

Ferguson, T. S. (1967). *Mathematical Statistics*. Academic Press, New York.

Kariya, T. and Sinha, B. K. (1989). *Robustness of Statistical Tests*. Academic Press, Boston.

Lehmann, E. L. (1986). *Testing Statistical Hypotheses*, Second Edition. Wiley, New York.

Mathew, T. and Sinha, B. K. (1988a). "Optimum tests for fixed effects and variance components in balanced models." *Journal of the American Statistical Association*, 83, 133–135.

Mathew, T. and Sinha, B. K. (1988b). "Optimum tests in unbalanced two-way models without interaction". *The Annals of Statistics*, 16, 1727–1740.

Montgomery, D. C. (1991). *Design and Analysis of Experiments*, Third Edition. Wiley, New York.

Muirhead, R. J. (1982). *Aspects of Multivariate Statistical Theory*. Wiley, New York.

Portnoy, S. (1973). "On recovery of intra-block information." *Journal of the American Statistical Association*, 68, 384–391.

Rao, C. R. (1973). *Linear Statistical Inference and Its Applications*. Wiley, New York.

Searle, S. R. (1971). *Linear Models*. Wiley, New York.

Searle, S. R. (1987). *Linear Models for Unbalanced Data*. Wiley, New York.

Seely, J. F. and El-Bassiouni, Y. (1983). "Applying Wald's variance component test." *The Annals of Statistics*, 11, 197–201.

Spjøtvoll, E. (1967). "Optimum invariant tests in unbalanced variance components models." *The Annals of Mathematical Statistics*, 38, 422–428.

Thompson, W. A. (1955a). "The ratio of variances in a variance components model." *The Annals of Mathematical Statistics*, 26, 325–329.

Thompson, W. A. (1955b). "On the ratio of variances in the mixed incomplete block model." *The Annals of Mathematical Statistics*, 26, 721–733.

Wald, A. (1947). "A note on regression analysis." *The Annals of Mathematical Statistics*, 18, 586–589.

Wijsman, R. A. (1967). "Cross-section of orbits and their applications to densities of maximal invariants." In: *Fifth Berkeley Symposium on Mathematical Statistics and Probability, I*, University of California, Berkeley, 389–400.

CHAPTER 2

Balanced Random and Mixed Models

2.1. INTRODUCTION

The purpose of this chapter is to provide a comprehensive account of the main results on statistical tests of relevant hypotheses for *balanced* random and mixed models. Examples of such models abound in the statistical literature and also in applications. The familiar one-way classification, two-way classification, both crossed and nested, and in general m-way classification with and without interactions of various levels (i.e., two-factor, three-factor, etc.) represent such cases. A variety of examples of mixed and random models involving *treatments/varieties, mice/men, cows/bulls, mice/diets, treatments/crosses,* and so forth, can be found in Searle (1971). For most of the practical hypotheses in a balanced mixed model, optimum tests like uniformly most powerful unbiased (UMPU), uniformly most powerful invariant (UMPI), and uniformly most powerful invariant unbiased (UMPIU) do exist and coincide with the standard F-tests from the associated ANOVA-tables (Herbach, 1959; Spjøtvoll, 1967; Seifert, 1978, 1979; Arnold, 1981; Humak, 1984; Das and Sinha, 1987; Mathew and Sinha, 1988a). There are, however, some important hypotheses even in simple balanced models for which the above kind of optimum tests do not exist, though fortunately exact unbiased tests of Bartlett–Scheffé type can be easily constructed (Bartlett, 1936; Scheffé, 1943; Seifert, 1979, 1981).

The organization of this chapter is as follows. In Section 2.2 we provide notations and definitions of a balanced model along with a few standard examples. Some important properties of balanced models are stated in Section 2.3. In Section 2.4 we give the relevant distribution theory for balanced mixed models. Derivation of optimum tests for fixed effects and variance components is given in Section 2.5. Section 2.6 contains some results on approximate and exact tests. Here we provide details of Satterthwaite's approximation (Satterthwaite, 1941, 1946) and a discussion of tests of Bartlett–Scheffé type to cover situations where exact optimum tests in the traditional sense cannot be found.

2.2. BALANCED MODELS — NOTATIONS AND DEFINITIONS

A general balanced model can be expressed as

$$y_\theta = \sum_{i=0}^{\nu+1} g^{(i)}_{\theta_i(\bar{\theta}_i)}, \tag{2.2.1}$$

where $\theta = \{k_1, k_2, \ldots, k_s\}$ is a complete set of subscripts that identify a typical response y, where $k_j = 1, 2, \ldots, a_j$ $(j = 1, 2, \ldots, s)$. The total number of observations is $N = \prod_{i=1}^{s} a_i$. The term $g^{(i)}_{\theta_i(\bar{\theta}_i)}$ denotes the ith effect in the model, where $\bar{\theta}_i$ and θ_i are the corresponding sets of rightmost and nonrightmost bracket subscripts, respectively. The last term in model (2.2.1) is identified as a random experimental error. By the set of rightmost bracket subscripts for the ith effect we mean those subscripts that do not nest any other subscripts of that effect. The grouping of these subscripts is indicated by using parentheses. If ψ_i denotes the complete set of subscripts for the ith effect, then the set of nonrightmost bracket subscripts, θ_i, is the complement of $\bar{\theta}_i$ with respect to ψ_i. Note that ψ_i is the empty set for $i = 0$ since the corresponding effect is the grand mean. In this case, both θ_i and $\bar{\theta}_i$ are empty. For $i = \nu + 1$, ψ_i is equal to θ since the corresponding effect is the experimental error (see Examples 2.2.1–2.2.3 below).

As will be seen later in this chapter, the rightmost brackets for a given balanced model play an important role in determining the sums of squares and degrees of freedom associated with the effects in the model. Note that superscript i for the ith effect in model (2.2.1) is needed for identification purposes: If two particular values of ψ_i and ψ_j are equal for $i \neq j$, then to differentiate between their corresponding effects we use $g^{(i)}_{\theta_i(\bar{\theta}_i)}$ and $g^{(j)}_{\theta_j(\bar{\theta}_j)}$, respectively.

Definition 2.2.1 (partial mean). A partial mean of the response y_θ is the average of y_θ over the entire range of values of a particular subset of θ. Partial means are denoted by the same symbol as the one for the response except that the subscripts that have been averaged out are omitted. A superscript is added for identification purposes.

Definition 2.2.2 (admissible mean). A partial mean is admissible if, whenever a nested subscript appears, all the subscripts that nest it appear also.

Definition 2.2.3 (component). A component associated with an admissible mean is a linear combination of admissible means obtained by selecting all those admissible means which are yielded by the mean in question when some, all, or none of its rightmost bracket subscripts are omitted in all possible ways. Whenever an odd number of subscripts is omitted, the mean is given a negative sign, and whenever an even number of subscripts is omitted, the mean is given a positive sign (the number zero is considered even).

The total number of components for a given balanced model is the same as the number of admissible means. We denote the component corresponding to the ith admissible mean $y^{(i)}_{\theta_i(\bar{\theta}_i)}$ by $C^{(i)}_{\theta_i(\bar{\theta}_i)}$, $i = 0, 1, 2, \ldots, \nu + 1$. A general expression for $C^{(i)}_{\theta_i(\bar{\theta}_i)}$ is given by

$$C^{(i)}_{\theta_i(\bar{\theta}_i)} = \sum_{j=0}^{\nu+1} \lambda_{ij} y^{(j)}_{\theta_j(\bar{\theta}_j)}, \quad i = 0, 1, \cdots, \nu + 1, \tag{2.2.2}$$

where $\lambda_{ij} = -1, 0$, or 1. The values -1 and 1 are attained whenever an odd number or an even number of subscripts is omitted from $\bar{\theta}_i$, respectively. It is easy to show that if ν_i is the number of subscripts in the rightmost bracket $\bar{\theta}_i$, then the number of admissible means that make up the component $C^{(i)}_{\theta_i(\bar{\theta}_i)}$ in formula (2.2.2) is equal to 2^{ν_i} $(i = 0, 1, \ldots, \nu + 1)$.

Model (2.2.1) can be written in vector form as

$$\mathbf{y} = \sum_{i=0}^{\nu+1} \mathbf{H}_i \boldsymbol{\beta}_i, \tag{2.2.3}$$

where the vector $\boldsymbol{\beta}_i$ consists of the elements of $g^{(i)}_{\theta_i(\bar{\theta}_i)}$, $i = 0, 1, \ldots, \nu + 1$, and the matrix $\mathbf{H}_i$ is expressible as a Kronecker (direct) product of the form

$$\mathbf{H}_i = \otimes_{j=1}^{s} \mathbf{L}_{ij}, \quad i = 0, 1, \cdots, \nu + 1, \tag{2.2.4}$$

where $\mathbf{L}_{ij}$ is either a vector of ones or an identity matrix defined as

$$\mathbf{L}_{ij} = \begin{matrix} \mathbf{I}_{a_j}, & k_j \in \psi_i \\ \mathbf{1}_{a_j}, & k_j \notin \psi_i. \end{matrix} \quad i = 0, 1, \cdots, \nu + 1; \quad j = 1, 2, \cdots, s. \tag{2.2.5}$$

Here, ψ_i is the set of subscripts associated with the ith effect $(i = 0, 1, \ldots, \nu + 1)$, and $\mathbf{I}_{a_j}$ and $\mathbf{1}_{a_j}$ are, respectively, the $a_j \times a_j$ identity matrix and the $a_j \times 1$ vector of ones $(j = 1, 2, \ldots, s)$. Note that the Kronecker product of two matrices, $\mathbf{A} : m \times n = (a_{ij})$ and $\mathbf{B} : r \times s$, is a matrix of order $mr \times ns$ whose $(i, j)^{th}$ entry is the matrix $a_{ij}\mathbf{B}$.

In order to illustrate the application of the aforementioned notation and definitions, let us consider the following well-known examples of balanced models.

Example 2.2.1. One-Way Classification Model. A common application of this model involves a comparison of a few treatment effects $\tau_1, \tau_2, \ldots, \tau_v$ based on observations y_{ij} following the model

$$y_{ij} = \mu + \tau_i + e_{i(j)}, \tag{2.2.6}$$

where $i = 1, 2, \ldots, v$; $j = 1, 2, \ldots, n$, and the $e_{i(j)}$'s represent error terms. The error term $e_{i(j)}$ in the above model is usually written as e_{ij} once the *nesting* is understood. This model is balanced because the replication of each treatment is the same, namely n. Using the general model format as in equation (2.2.1), we have

$$y_{ij} = g^0 + g^{(1)}_{(i)} + g^{(2)}_{i(j)}, \tag{2.2.7}$$

where $g^{(1)}_{(i)} = \tau_i$ and $g^{(2)}_{i(j)} = e_{i(j)}$. The admissible means are $y^{(0)} = \bar{y}_{..} = \sum_{i,j} y_{ij}/nv$, $y^{(1)}_{(i)} = \bar{y}_{i.} = \sum_j y_{ij}/n$, and $y^{(2)}_{i(j)} = y_{ij}$, and the corresponding components are

$$\begin{aligned} C^{(0)} &= y^{(0)} \\ C^{(1)}_{(i)} &= y^{(1)}_{(i)} - y^{(0)} \\ C^{(2)}_{i(j)} &= y^{(2)}_{i(j)} - y^{(1)}_{(i)}. \end{aligned} \tag{2.2.8}$$

Furthcrmore, the vector form of model (2.2.6) is

$$\mathbf{y} = \mathbf{H}_0\beta_0 + \mathbf{H}_1\boldsymbol{\beta}_1 + \mathbf{H}_2\boldsymbol{\beta}_2, \tag{2.2.9}$$

where

$$\begin{aligned} \mathbf{H}_0 &= \mathbf{1}_v \otimes \mathbf{1}_n, & \beta_0 &= \mu \\ \mathbf{H}_1 &= \mathbf{I}_v \otimes \mathbf{1}_n, & \boldsymbol{\beta}_1 &= (\tau_1, \tau_2, \cdots, \tau_v)' \\ \mathbf{H}_2 &= \mathbf{I}_v \otimes \mathbf{I}_n, & \boldsymbol{\beta}_2 &= (e_{1(1)}, e_{1(2)}, \cdots, e_{v(n)})'. \end{aligned} \tag{2.2.10}$$

The hypotheses of interest in this model are: $H_\mu : \mu = 0$ and $H_\tau : \sigma^2_\tau$ (treatment variability) =0, under the assumption that the τ_i's are random.

Example 2.2.2. Two-Way Crossed Classification Without Interaction Model. There are plenty of applications of this model, which is typically used to compare two unrelated sets of quantities known as "treatments" and "blocks." If y_{ij} denotes the observation from the ith treatment in the jth block, then we have the model

$$y_{ij} = \mu + \tau_i + \beta_j + e_{ij}, \tag{2.2.11}$$

where $i = 1, 2, \ldots, v$; $j = 1, 2, \ldots, b$, and the e_{ij}'s represent the error terms. This model can be written as

$$y_{ij} = g^{(0)} + g^{(1)}_{(i)} + g^{(2)}_{(j)} + g^{(3)}_{(ij)}, \tag{2.2.12}$$

where $g^{(0)} = \mu$, $g^{(1)}_{(i)} = \tau_i$, $g^{(2)}_{(j)} = \beta_j$, and $g^{(3)}_{(ij)} = e_{ij}$. As before, the admissible means are $y^{(0)} = \bar{y}_{..} = \sum_{ij} y_{ij}/vb$, $y^{(1)}_{(i)} = \bar{y}_{i.} = \sum_j y_{ij}/b$, $y^{(2)}_{(j)} = \bar{y}_{.j} = \sum_i y_{ij}/v$, and $y^{(3)}_{(ij)} = y_{ij}$, and the corresponding components are

$$\begin{aligned} C^{(0)} &= y^{(0)} \\ C^{(1)}_{(i)} &= y^{(1)}_{(i)} - y^{(0)} \\ C^{(2)}_{(j)} &= y^{(2)}_{(j)} - y^{(0)} \\ C^{(3)}_{(ij)} &= y^{(3)}_{(ij)} - y^{(1)}_{(i)} - y^{(2)}_{(j)} + y^{(0)}. \end{aligned} \tag{2.2.13}$$

In vector form, model (2.2.11) is written as

$$\mathbf{y} = \mathbf{H}_0\beta_0 + \mathbf{H}_1\boldsymbol{\beta}_1 + \mathbf{H}_2\boldsymbol{\beta}_2 + \mathbf{H}_3\boldsymbol{\beta}_3, \tag{2.2.14}$$

where

$$\begin{aligned} \mathbf{H}_0 &= \mathbf{1}_v \otimes \mathbf{1}_b, & \beta_0 &= \mu \\ \mathbf{H}_1 &= \mathbf{I}_v \otimes \mathbf{1}_b, & \boldsymbol{\beta}_1 &= (\tau_1, \tau_2, \cdots, \tau_v)' \\ \mathbf{H}_2 &= \mathbf{1}_v \otimes \mathbf{I}_b, & \boldsymbol{\beta}_2 &= (\beta_1, \beta_2, \cdots, \beta_b)' \\ \mathbf{H}_3 &= \mathbf{I}_v \otimes \mathbf{I}_b, & \boldsymbol{\beta}_3 &= (e_{11}, e_{12}, \cdots, e_{vb})'. \end{aligned} \tag{2.2.15}$$

There are two variations of this model: mixed and random. In the former case, the block effects $\beta_1, \ldots, \beta_b$ are taken as fixed and the treatment effects $\tau_1, \ldots, \tau_v$ are taken as random with mean 0 and variance σ^2_τ, independently of the random error components, while in the latter case both β's and τ's are taken as independent random variables with mean 0 and variances σ^2_β and σ^2_τ, respectively. The two hypotheses of interest in the former case are: $H_\beta : \beta_1 = \ldots = \beta_b$ and $H_\tau : \sigma^2_\tau = 0$, while in the latter case the hypotheses are: $H_\beta : \sigma^2_\beta = 0$ and $H_\tau : \sigma^2_\tau = 0$.

Example 2.2.3. Two-Way Crossed Classification With Interaction Model. This is similar to Example 2.2.2 with the addition of an interaction term $(\tau\beta)_{ij}$ and the same number of observations in each cell (that is, ij combination), which results in the model

$$y_{ijk} = \mu + \tau_i + \beta_j + (\tau\beta)_{ij} + e_{ij(k)}. \tag{2.2.16}$$

Here, $i = 1, 2, \ldots, v$; $j = 1, 2, \ldots, b$; $k = 1, 2, \ldots, n$. The error term $e_{ij(k)}$ in the above model is usually written in the statistical literature as e_{ijk}. Equivalently, we can write

$$y_{ijk} = g^{(0)} + g^{(1)}_{(i)} + g^{(2)}_{(j)} + g^{(3)}_{(ij)} + g^{(4)}_{ij(k)}, \tag{2.2.17}$$

where $g^{(0)} = \mu$, $g^{(1)}_{(i)} = \tau_i$, $g^{(2)}_{(j)} = \beta_j$, $g^{(3)}_{(ij)} = (\tau\beta)_{ij}$, and $g^{(4)}_{ij(k)} = e_{ij(k)}$. The admissible means $y^{(0)}, y^{(1)}_{(i)}, y^{(2)}_{(j)}, y^{(3)}_{(ij)} = \bar{y}_{ij.}$ are defined appropriately as before, and $y^{(4)}_{ij(k)} = y_{ijk}$. The corresponding components are

$$
\begin{aligned}
C^{(0)} &= y^{(0)} \\
C^{(1)}_{(i)} &= y^{(1)}_{(i)} - y^{(0)} \\
C^{(2)}_{(j)} &= y^{(2)}_{(j)} - y^{(0)} \\
C^{(3)}_{(ij)} &= y^{(3)}_{(ij)} - y^{(1)}_{(i)} - y^{(2)}_{(j)} + y^{(0)} \\
C^{(4)}_{ij(k)} &= y^{(4)}_{ij(k)} - y^{(3)}_{(ij)}.
\end{aligned}
\tag{2.2.18}
$$

The vector form of model (2.2.16) is

$$
\mathbf{y} = \mathbf{H}_0\beta_0 + \mathbf{H}_1\boldsymbol{\beta}_1 + \mathbf{H}_2\boldsymbol{\beta}_2 + \mathbf{H}_3\boldsymbol{\beta}_3 + \mathbf{H}_4\boldsymbol{\beta}_4, \tag{2.2.19}
$$

where

$$
\begin{aligned}
\mathbf{H}_0 &= \mathbf{1}_v \otimes \mathbf{1}_b \otimes \mathbf{1}_n, & \beta_0 &= \mu \\
\mathbf{H}_1 &= \mathbf{I}_v \otimes \mathbf{1}_b \otimes \mathbf{1}_n, & \boldsymbol{\beta}_1 &= (\tau_1, \tau_2, \cdots, \tau_v)' \\
\mathbf{H}_2 &= \mathbf{1}_v \otimes \mathbf{I}_b \otimes \mathbf{1}_n, & \boldsymbol{\beta}_2 &= (\beta_1, \beta_2, \cdots \beta_b)', \\
\mathbf{H}_3 &= \mathbf{I}_v \otimes \mathbf{I}_b \otimes \mathbf{1}_n, & \boldsymbol{\beta}_3 &= ((\tau\beta)_{11}, (\tau\beta)_{12}, \cdots, (\tau\beta)_{vb})' \\
\mathbf{H}_4 &= \mathbf{I}_v \otimes \mathbf{I}_b \otimes \mathbf{I}_n, & \boldsymbol{\beta}_4 &= (e_{11(1)}, e_{11(2)}, \cdots, e_{vb(n)})'.
\end{aligned}
\tag{2.2.20}
$$

Note that model (2.2.16) becomes unbalanced in case there are unequal numbers of observations in the different cells.

Again, there are two variations of this model: mixed and random. In the former case, the block effects $\beta_1, \ldots, \beta_b$ are taken as fixed while the treatment effects $\tau_1, \ldots, \tau_v$ and the interaction effects $(\tau\beta)_{11}, \ldots, (\tau\beta)_{vb}$ are taken as independent random effects with mean 0 and variances σ^2_τ and $\sigma^2_{\tau\beta}$ respectively. Here, the main hypotheses of interest are: $H_{\tau\beta} : \sigma^2_{\tau\beta} = 0$ (absence of interaction between blocks and treatments) and $H_\tau : \sigma^2_\tau = 0$ (absence of significant differences among treatment main effects). In the latter case all the effects (except μ) are taken as independent random variables with mean 0 and variances σ^2_β, σ^2_τ, and $\sigma^2_{\tau\beta}$. The hypotheses of interest here concern σ^2_β (block variance), σ^2_τ (treatment variance), and $\sigma^2_{\tau\beta}$ (interaction variance).

Example 2.2.4. Two-Fold Nested Model. Let us consider the model

$$
y_{ijk} = \mu + \tau_i + \beta_{i(j)} + e_{ij(k)}, \tag{2.2.21}
$$

where τ_i is the nesting effect ($i = 1, 2, \ldots, v$), $\beta_{i(j)}$ is the nested effect ($j = 1, 2, \ldots, b$); and $e_{ij(k)}$ is a random error ($k = 1, \ldots, n$). The error term $e_{ij(k)}$

in the above model is usually written as e_{ijk} when the effect of nesting is understood. This model is balanced since the ranges v, b, and n are the same for all values of subscripts i, j, k, respectively. Using the general format for a balanced model, we obtain the model

$$y_{ijk} = g^{(0)} + g^{(1)}_{(i)} + g^{(2)}_{i(j)} + g^{(3)}_{ij(k)}, \tag{2.2.22}$$

where $g^0 = \mu$, $g^{(1)}_{(i)} = \tau_i$, $g^{(2)}_{i(j)} = \beta_{i(j)}$, and $g^{(3)}_{ij(k)} = e_{ij(k)}$. The admissible means $y^{(0)}, y^{(1)}_{(i)}, y^{(2)}_{i(j)}$ are defined analogously, and $y^{(3)}_{ij(k)} = y_{ijk}$. The corresponding components are

$$\begin{aligned} C^{(0)} &= y^{(0)} \\ C^{(1)}_{(i)} &= y^{(1)}_{(i)} - y^{(0)} \\ C^{(2)}_{i(j)} &= y^{(2)}_{i(j)} - y^{(1)}_{(i)} \\ C^{(3)}_{ij(k)} &= y^{(3)}_{ij(k)} - y^{(2)}_{i(j)}. \end{aligned} \tag{2.2.23}$$

The vector form of model (2.2.21) is

$$\mathbf{y} = \mathbf{H}_0\beta_0 + \mathbf{H}_1\boldsymbol{\beta}_1 + \mathbf{H}_2\boldsymbol{\beta}_2 + \mathbf{H}_3\boldsymbol{\beta}_3, \tag{2.2.24}$$

where

$$\begin{aligned} \mathbf{H}_0 &= \mathbf{1}_v \otimes \mathbf{1}_b \otimes \mathbf{1}_n, & \beta_0 &= \mu \\ \mathbf{H}_1 &= \mathbf{I}_v \otimes \mathbf{1}_b \otimes \mathbf{1}_n, & \boldsymbol{\beta}_1 &= (\tau_1, \tau_2, \cdots, \tau_v)' \\ \mathbf{H}_2 &= \mathbf{I}_v \otimes \mathbf{I}_b \otimes \mathbf{1}_n, & \boldsymbol{\beta}_2 &= (\beta_{1(1)}, \beta_{1(2)}, \cdots \beta_{v(b)})' \\ \mathbf{H}_3 &= \mathbf{I}_v \otimes \mathbf{I}_b \otimes \mathbf{I}_n, & \boldsymbol{\beta}_3 &= (e_{11(1)}, e_{11(2)}, \cdots, e_{vb(n)})'. \end{aligned} \tag{2.2.25}$$

Under the assumption of a random effects model, the only linear hypothesis about the fixed effect is $H_\mu : \mu = 0$. The other two hypotheses of interest are: $H_\tau : \sigma^2_\tau = 0$ and $H_\beta : \sigma^2_{\beta(\alpha)} = 0$, where $\sigma^2_{\beta(\alpha)}$ is the variance component associated with $\beta_{i(j)}$.

Example 2.2.5. In this example, we have a combination of crossed and nested effects. Consider, therefore, the following model:

$$y_{ijkl} = \mu + \tau_i + \beta_j + (\tau\beta)_{ij} + \delta_{j(k)} + (\tau\delta)_{j(ik)} + e_{ijk(l)}, \tag{2.2.26}$$

where $i = 1, 2, \ldots, v$; $j = 1, 2, \ldots, b$; $k = 1, 2, \ldots, n$; $l = 1, 2, \ldots, p$. Here, subscripts i and j are crossed, but subscript k is nested within j, and subscript l is nested with i, j, and k. Therefore $\delta_{j(k)}$ is a nested effect and $(\tau\delta)_{j(ik)}$ is an interaction effect involving τ_i and $\delta_{j(k)}$. Note that the rightmost bracket for $(\tau\delta)_{j(ik)}$ consists of subscripts i and k since they do not nest each other.

It may be noted that the error term $e_{ijk(l)}$ is usually written in the statistical literature as e_{ijkl}. In this case, using the general format for a balanced model, we have

$$y_{ijkl} = g^{(0)} + g^{(1)}_{(i)} + g^{(2)}_{(j)} + g^{(3)}_{(ij)} + g^{(4)}_{j(k)} + g^{(5)}_{j(ik)} + g^{(6)}_{ijk(l)}, \tag{2.2.27}$$

where $g^{(0)} = \mu, g^{(1)}_{(i)} = \tau_i, g^{(2)}_{(j)} = \beta_j, g^{(3)}_{(ij)} = (\tau\beta)_{ij}, g^{(4)}_{j(k)} = \delta_{j(k)}, g^{(5)}_{j(ik)} = (\tau\delta)_{j(ik)}$, and $g^{(6)}_{ijk(l)} = e_{ijk(l)}$. The admissible means $y^{(0)}, y^{(1)}_{(i)}, y^{(2)}_{(j)}, y^{(3)}_{(ij)}, y^{(4)}_{j(k)}, y^{(5)}_{j(ik)}, y^{(6)}_{ijk(l)}$ are defined analogously, and the corresponding components are

$$\begin{aligned}
C^{(0)} &= y^{(0)} \\
C^{(1)}_{(i)} &= y^{(1)}_{(i)} - y^{(0)} \\
C^{(2)}_{(j)} &= y^{(2)}_{(j)} - y^{(0)} \\
C^{(3)}_{(ij)} &= y^{(3)}_{(ij)} - y^{(1)}_{(i)} - y^{(2)}_{(j)} + y^{(0)} \\
C^{(4)}_{j(k)} &= y^{(4)}_{j(k)} - y^{(2)}_{(j)} \\
C^{(5)}_{j(ik)} &= y^{(5)}_{j(ik)} - y^{(3)}_{(ij)} - y^{(4)}_{j(k)} + y^{(2)}_{(j)} \\
C^{(6)}_{ijk(l)} &= y^{(6)}_{ijk(l)} - y^{(5)}_{j(ik)}.
\end{aligned} \tag{2.2.28}$$

The vector form of model (2.2.27) is

$$\mathbf{y} = \mathbf{H}_0\beta_0 + \mathbf{H}_1\boldsymbol{\beta}_1 + \mathbf{H}_2\boldsymbol{\beta}_2 + \mathbf{H}_3\boldsymbol{\beta}_3 + \mathbf{H}_4\boldsymbol{\beta}_4 + \mathbf{H}_5\boldsymbol{\beta}_5 + \mathbf{H}_6\boldsymbol{\beta}_6, \tag{2.2.29}$$

where

$$\begin{aligned}
\mathbf{H}_0 &= \mathbf{1}_v \otimes \mathbf{1}_b \otimes \mathbf{1}_n \otimes \mathbf{1}_p, & \beta_0 &= \mu \\
\mathbf{H}_1 &= \mathbf{I}_v \otimes \mathbf{1}_b \otimes \mathbf{1}_n \otimes \mathbf{1}_p, & \boldsymbol{\beta}_1 &= (\tau_1, \tau_2, \cdots, \tau_v)' \\
\mathbf{H}_2 &= \mathbf{1}_v \otimes \mathbf{I}_b \otimes \mathbf{1}_n \otimes \mathbf{1}_p, & \boldsymbol{\beta}_2 &= (\beta_1, \beta_2, \cdots \beta_b)' \\
\mathbf{H}_3 &= \mathbf{I}_v \otimes \mathbf{I}_b \otimes \mathbf{1}_n \otimes \mathbf{1}_p, & \boldsymbol{\beta}_3 &= ((\tau\beta)_{11}, (\tau\beta)_{12}, \cdots, (\tau\beta)_{vb})' \\
\mathbf{H}_4 &= \mathbf{1}_v \otimes \mathbf{I}_b \otimes \mathbf{I}_n \otimes \mathbf{1}_p, & \boldsymbol{\beta}_4 &= (\delta_{1(1)}, \delta_{1(2)}, \cdots, \delta_{b(n)})' \\
\mathbf{H}_5 &= \mathbf{I}_v \otimes \mathbf{I}_b \otimes \mathbf{I}_n \otimes \mathbf{1}_p, & \boldsymbol{\beta}_5 &= ((\tau\delta)_{1(11)}, (\tau\delta)_{1(12)}, \cdots, (\tau\delta)_{b(vn)})' \\
\mathbf{H}_6 &= \mathbf{I}_v \otimes \mathbf{I}_b \otimes \mathbf{I}_n \otimes \mathbf{I}_p, & \boldsymbol{\beta}_6 &= (e_{111(1)}, e_{111(2)}, \cdots, e_{vbn(p)})'.
\end{aligned} \tag{2.2.30}$$

Under the assumption of a random effects model, the hypotheses of interest here concern all the variance components.

We note from these examples that the writing of a model requires knowledge of the nesting and nonnesting (crossed) relationships that exist among the subscripts. For this purpose it would be convenient to identify the so-called population structure for a given experimental situation. This population structure provides a description of all nesting and nonnesting relationships that exist among the factors (or subscripts). To describe such relationships we adopt the following convention: If i and j, for example, are crossed subscripts, then this fact is denoted by separating i and j using parentheses, such as $(i)(j)$. If, however, j is nested within i, then we separate i and j by using a colon, such as $i : j$, where the subscript appearing to the right of the colon is the nested subscript and the one to its left is the nesting subscript. Brackets can be used to provide separation if there are several subscripts with both nesting and nonnesting relationships. For example, the population structures associated with Examples 2.2.1–2.2.5 are shown below.

Example	Population Structure
2.2.1	$i : j$
2.2.2	$(i)(j)$
2.2.3	$[(i)(j)] : k$
2.2.4	$i : j : k$
2.2.5	$[(i)(j : k)] : l$

Note that in Example 2.2.3, k is nested within both i and j; in Example 2.2.4, k is nested within j, which is nested within i; and in Example 2.2.5, k is nested within j, which is crossed with i, and l is nested within i, j, k.

The advantage of describing the population structure for a given experimental situation is to enable us to easily identify the admissible means, and hence the corresponding components, and consequently the complete model for the experiment. This follows from the fact that the components correspond in a one-to-one manner to the effects in the complete model.

2.3. BALANCED MODEL PROPERTIES

In this section we list several results that provide additional properties associated with balanced models. The proofs of these results can be found in Zyskind (1962). See also Smith and Hocking (1978), Khuri (1982), and Seifert (1979).

The sum of squares associated with the ith effect for model (2.2.1), expressed as a quadratic form $\mathbf{y}'\boldsymbol{P}_i\mathbf{y}$ $(i = 0, 1, \ldots, \nu + 1)$, is defined as

$$\mathbf{y}'\boldsymbol{P}_i\mathbf{y} = \sum_{\theta} \left[C^{(i)}_{\theta_i(\bar{\theta}_i)} \right]^2, \quad i = 0, 1, \cdots, \nu + 1. \tag{2.3.1}$$

This sum of squares can be also written as (Zyskind, 1962, p. 120)

$$\mathbf{y}'\boldsymbol{P}_i\mathbf{y} = \sum_{j=0}^{\nu+1} \lambda_{ij} \sum_{\theta} \left[y^{(j)}_{\theta_j(\bar{\theta}_j)} \right]^2, \quad i = 0, 1, \cdots, \nu + 1, \tag{2.3.2}$$

where the λ_{ij}'s are the same as in formula (2.2.2).

The matrix $\boldsymbol{P}_i$ in formula (2.3.1) is idempotent for $i = 0, 1, \ldots, \nu$, and is such that $\boldsymbol{P}_i\boldsymbol{P}_j = 0$, $i \neq j$, $\sum_{i=0}^{\nu+1} \boldsymbol{P}_i = \boldsymbol{I}_N$ (see Smith and Hocking, 1978).

Furthermore, we have the following properties, expressed as Lemmas 2.3.1–2.3.4 (see Khuri, 1982, pp. 2910–11):

Lemma 2.3.1. The matrix $\boldsymbol{P}_i$ is expressible as

$$\mathbf{P}_i = \sum_{j=0}^{\nu+1} \frac{\lambda_{ij}}{b_j} \mathbf{A}_j, \quad i = 0, 1, \cdots, \nu+1, \tag{2.3.3}$$

where

$$b_j = \begin{cases} \prod_{k_l \notin \psi_j} a_l, & \text{if } \psi_j \neq \theta \\ 1, & \text{if } \psi_j = \theta, \end{cases} \qquad j = 0, 1, \cdots, \nu+1 \tag{2.3.4}$$

$$\mathbf{A}_j = \mathbf{H}_j \mathbf{H}'_j, \quad j = 0, 1, \cdots, \nu+1, \tag{2.3.5}$$

ψ_j is the set of subscripts for the jth effect in model (2.2.1) with θ being the complete set of subscripts, and a_l is the range of subscript k_l $(l = 1, \ldots, s)$.

We note that formula (2.3.3) is of the same form as formula (2.2.2) with $\boldsymbol{P}_i$ used instead of the ith component $C^{(i)}_{\theta_i(\bar{\theta}_i)}$ and $\mathbf{A}_j/b_j$ instead of the jth admissible mean $y^{(j)}_{\theta_j(\bar{\theta}_j)}$. □

Lemma 2.3.2. Let m_i be the rank of $\boldsymbol{P}_i$. Then m_i is the same as the number of degrees of freedom for the ith source of variation in model (2.2.1), and is equal to

$$m_i = \left[\prod_{k_j \in \theta_i} a_j\right] \left[\prod_{k_j \in \bar{\theta}_i} (a_j - 1)\right], \quad i = 0, 1, \cdots, \nu+1, \tag{2.3.6}$$

where θ_i and $\bar{\theta}_i$ are, respectively, the sets of nonrightmost and rightmost bracket subscripts for the ith effect, and a_j is the range of subscript k_j $(j = 1, 2, \ldots, s)$. If θ_i is the empty set, then $\prod_{k_j \in \theta_i} a_j = 1$. If both θ_i and $\bar{\theta}_i$ are empty, then $m_0 = 1$. □

Lemma 2.3.3.

$$\frac{\mathbf{A}_j}{b_j} = \sum_{\psi_i \subset \psi_j} \mathbf{P}_i, \quad j = 0, 1, \cdots, \nu+1, \tag{2.3.7}$$

where the summation extends over subscript i such that $\psi_i \subset \psi_j$. □

Lemma 2.3.4.

$$\mathbf{A}_j \mathbf{P}_i = \kappa_{ij} \mathbf{P}_i, \tag{2.3.8}$$

$i, j = 0, 1, \ldots, \nu + 1$, where

$$\kappa_{ij} = \begin{array}{l} 0, \; if \;\; \psi_i \not\subset \psi_j \\ b_j, \; if \;\; \psi_i \subset \psi_j. \end{array} \qquad \square \tag{2.3.9}$$

We now illustrate the applications of the above results with the help of an example.

Example 2.3.1. Consider the model

$$y_{ijk} = g^{(0)} + g^{(1)}_{(i)} + g^{(2)}_{i(j)} + g^{(3)}_{(k)} + g^{(4)}_{(ik)} + g^{(5)}_{i(jk)}, \tag{2.3.10}$$

where subscript j is nested within subscript i and subscripts i and k are crossed ($i = 1, 2, \ldots, v$, $j = 1, 2, \ldots, b$; $k = 1, 2, \ldots, n$). See (2.2.17), (2.2.22), and (2.2.27) for descriptions of similar models. In this case,

$$\begin{aligned} \mathbf{H}_0 &= \mathbf{1}_v \otimes \mathbf{1}_b \otimes \mathbf{1}_n \\ \mathbf{H}_1 &= \mathbf{I}_v \otimes \mathbf{1}_b \otimes \mathbf{1}_n \\ \mathbf{H}_2 &= \mathbf{I}_v \otimes \mathbf{I}_b \otimes \mathbf{1}_n \\ \mathbf{H}_3 &= \mathbf{1}_v \otimes \mathbf{1}_b \otimes \mathbf{I}_n \\ \mathbf{H}_4 &= \mathbf{I}_v \otimes \mathbf{1}_b \otimes \mathbf{I}_n \\ \mathbf{H}_5 &= \mathbf{I}_v \otimes \mathbf{I}_b \otimes \mathbf{I}_n. \end{aligned} \tag{2.3.11}$$

Furthermore, by Lemma 2.3.1, we have

$$\begin{aligned} \mathbf{P}_0 &= \frac{\mathbf{A}_0}{b_0} = \frac{\mathbf{H}_0\mathbf{H}_0'}{vbn} \\ \mathbf{P}_1 &= \frac{\mathbf{A}_1}{b_1} - \frac{\mathbf{A}_0}{b_0} = \frac{\mathbf{H}_1\mathbf{H}_1'}{bn} - \frac{\mathbf{H}_0\mathbf{H}_0'}{vbn} \\ \mathbf{P}_2 &= \frac{\mathbf{A}_2}{b_2} - \frac{\mathbf{A}_1}{b_1} = \frac{\mathbf{H}_2\mathbf{H}_2'}{n} - \frac{\mathbf{H}_1\mathbf{H}_1'}{bn} \\ \mathbf{P}_3 &= \frac{\mathbf{A}_3}{b_3} - \frac{\mathbf{A}_0}{b_0} = \frac{\mathbf{H}_3\mathbf{H}_3'}{vb} - \frac{\mathbf{H}_0\mathbf{H}_0'}{vbn} \\ \mathbf{P}_4 &= \frac{\mathbf{A}_4}{b_4} - \frac{\mathbf{A}_1}{b_1} - \frac{\mathbf{A}_3}{b_3} + \frac{\mathbf{A}_0}{b_0} \\ &= \frac{\mathbf{H}_4\mathbf{H}_4'}{b} - \frac{\mathbf{H}_1\mathbf{H}_1'}{bn} - \frac{\mathbf{H}_3\mathbf{H}_3'}{vb} + \frac{\mathbf{H}_0\mathbf{H}_0'}{vbn} \\ \mathbf{P}_5 &= \frac{\mathbf{A}_5}{b_5} - \frac{\mathbf{A}_2}{b_2} - \frac{\mathbf{A}_4}{b_4} + \frac{\mathbf{A}_1}{b_1} \\ &= \mathbf{H}_5\mathbf{H}_5' - \frac{\mathbf{H}_2\mathbf{H}_2'}{n} - \frac{\mathbf{H}_4\mathbf{H}_4'}{b} + \frac{\mathbf{H}_1\mathbf{H}_1'}{bn}. \end{aligned} \tag{2.3.12}$$

We note that

$$\begin{aligned}
\frac{\mathbf{A}_0}{b_0} &= \mathbf{P}_0 \\
\frac{\mathbf{A}_1}{b_1} &= \mathbf{P}_0 + \mathbf{P}_1 \\
\frac{\mathbf{A}_2}{b_2} &= \mathbf{P}_0 + \mathbf{P}_1 + \mathbf{P}_2 \\
\frac{\mathbf{A}_3}{b_3} &= \mathbf{P}_0 + \mathbf{P}_3 \\
\frac{\mathbf{A}_4}{b_4} &= \mathbf{P}_0 + \mathbf{P}_1 + \mathbf{P}_3 + \mathbf{P}_4 \\
\frac{\mathbf{A}_5}{b_5} &= \mathbf{P}_0 + \mathbf{P}_1 + \mathbf{P}_2 + \mathbf{P}_3 + \mathbf{P}_4 + \mathbf{P}_5.
\end{aligned} \tag{2.3.13}$$

This provides a verification to Lemma 2.3.3. It can also be verified, using Lemma 2.3.4, that

$$\begin{aligned}
\mathbf{A}_1\mathbf{P}_1 &= bn\mathbf{P}_1, \ \ since \ \ b_1 = bn, \psi_1 = \{i\} \\
\mathbf{A}_1\mathbf{P}_2 &= \mathbf{0}, \ \ since \ \ \psi_2 = \{i, j\} \not\subset \psi_1 \\
\mathbf{A}_1\mathbf{P}_3 &= \mathbf{0}, \ \ since \ \ \psi_3 = \{k\} \not\subset \psi_1 \\
\mathbf{A}_2\mathbf{P}_1 &= n\mathbf{P}_1, \ \ since \ \ b_2 = n, \ \psi_1 \subset \psi_2 \\
\mathbf{A}_2\mathbf{P}_3 &= \mathbf{0}, \ \ since \ \ \psi_3 \not\subset \psi_2 \\
\mathbf{A}_3\mathbf{P}_2 &= \mathbf{0}, \ \ since \ \ \psi_2 \not\subset \psi_3 \\
\mathbf{A}_3\mathbf{P}_3 &= vb\mathbf{P}_3, \ \ since \ \ b_3 = vb \\
\mathbf{A}_3\mathbf{P}_4 &= \mathbf{0}, \ \ since \ \ \psi_4 = \{i, k\} \not\subset \psi_3 \\
\mathbf{A}_5\mathbf{P}_3 &= \mathbf{P}_3, \ \ since \ \ b_5 = 1, \ \ \psi_3 \subset \psi_5, \ \ where \ \ \psi_5 = \{i, j, k\}.
\end{aligned} \tag{2.3.14}$$

2.4. BALANCED MIXED MODELS: DISTRIBUTION THEORY

Suppose that model (2.2.1) has some fixed effects and at least one random effect besides the experimental error. In this case, the model is called a balanced mixed model. Without loss of generality, we consider that the effects associated with $i = 0, 1, \ldots, \nu - p$ are fixed while those corresponding to $i = \nu - p + 1, \ \nu - p + 2, \ldots, \nu + 1$ are random, where p is a nonnegative integer not exceeding ν. Model (2.2.1) can then be expressed as

$$\mathbf{y} = \mathbf{X}\mathbf{g} + \mathbf{Z}\mathbf{h}, \tag{2.4.1}$$

where

$$\mathbf{X}\mathbf{g} = \sum_{i=0}^{\nu-p} \mathbf{H}_i \boldsymbol{\beta}_i \tag{2.4.2}$$

is the fixed portion of the model and

$$\mathbf{Zh} = \sum_{i=\nu-p+1}^{\nu+1} \mathbf{H}_i \boldsymbol{\beta}_i \tag{2.4.3}$$

is the random portion. Note that when $p = 0$, the model (2.4.1) is fixed, and when $p = \nu$, the model is completely random. Moreover, $\boldsymbol{\beta}_{\nu+1}$ always stands for the vector of experimental errors and consequently $\mathbf{H}_{\nu+1}$ is an identity matrix of order $N \times N$.

The following assumptions concerning the effects in model (2.4.1) will be considered in the remainder of this chapter.

Mixed Model Assumptions

(a) $\boldsymbol{\beta}_0, \boldsymbol{\beta}_1, \ldots, \boldsymbol{\beta}_{\nu-p}$ are fixed unknown parameter vectors.

(b) $\boldsymbol{\beta}_{\nu-p+1}, \boldsymbol{\beta}_{\nu-p+2}, \ldots, \boldsymbol{\beta}_{\nu+1}$ are independent normally distributed random vectors with zero means and variance–covariance matrices given by $Var(\boldsymbol{\beta}_i) = \sigma_i^2 \mathbf{I}_{c_i}$, where c_i is given by

$$c_i = \begin{cases} \prod_{k_j \in \psi_i} a_j, & i = \nu-\text{p}+1, \nu-\text{p}+2, \ldots, \nu \\ N, & i = \nu+1. \end{cases} \tag{2.4.4}$$

Note that c_i is the number of columns of $\mathbf{H}_i$, $i = 0, 1, \ldots, \nu + 1$. By formula (2.3.4), it is also equal to $\frac{N}{b_i}$, where N is the total number of observations. In the context of a mixed linear model, our primary interest is to test hypotheses concerning the fixed effects $\mathbf{Xg}$ and the variance components $\sigma^2_{\nu-p+1}, \ldots, \sigma^2_\nu$.

On the basis of the Mixed Model Assumptions, we have the following theorem (see Khuri, 1982, p. 2916).

Theorem 2.4.1. Let $\mathbf{y}'\mathbf{P}_i\mathbf{y}$ be the sum of squares associated with the ith source of variation ($i = 0, 1, \ldots, \nu + 1$) in model (2.4.1). If the Mixed Model Assumptions are valid, then we have the following results:

(a) $\mathbf{y}'\mathbf{P}_0\mathbf{y}, \mathbf{y}'\mathbf{P}_1\mathbf{y}, \ldots, \mathbf{y}'\mathbf{P}_{\nu+1}\mathbf{y}$ are mutually independent and $\mathbf{y}'\mathbf{P}_i\mathbf{y}/\delta_i$ is distributed as a noncentral chi-squared variate with m_i degrees of freedom, which is equal to the rank of $\mathbf{P}_i$, and a noncentrality parameter given by $\zeta_i = \mathbf{g}'\mathbf{X}'\mathbf{P}_i\mathbf{Xg}/\delta_i$, where

$$\delta_i = \sum_{j \in w_i} b_j \sigma_j^2. \tag{2.4.5}$$

Here, b_j is the number defined in formula (2.3.4) and w_i is the set

$$w_i = \{j : \nu - p + 1 \leq j \leq \nu + 1 | \psi_i \subset \psi_j\}, \tag{2.4.6}$$

where, if we recall, ψ_i is the set of subscripts associated with the ith effect ($i = 0, 1, \ldots, \nu + 1$) in the model.

(b) $\zeta_i = 0$ for $i = \nu - p + 1, \nu - p + 2, \ldots, \nu + 1$, that is, if the ith effect is random, then $\mathbf{y}'\mathbf{P}_i\mathbf{y}/\delta_i$ has the central chi-squared distribution.

(c) $E(\mathbf{y}'\mathbf{P}_i\mathbf{y}) = \mathbf{g}'\mathbf{X}'\mathbf{P}_i\mathbf{X}\mathbf{g} + m_i\delta_i$, $i = 0, 1, \ldots, \nu + 1$. □

Example 2.4.1. Consider again the balanced two-way crossed classification with interaction model mentioned in Example 2.2.3. Suppose that $g_{(i)}^{(1)}$ is a fixed effect and $g_{(j)}^{(2)}$ is a random effect. In this case,

$$\begin{aligned} \mathbf{Xg} &= \mathbf{H}_0\boldsymbol{\beta}_0 + \mathbf{H}_1\boldsymbol{\beta}_1 \\ \mathbf{Zh} &= \mathbf{H}_2\boldsymbol{\beta}_2 + \mathbf{H}_3\boldsymbol{\beta}_3 + \mathbf{H}_4\boldsymbol{\beta}_4. \end{aligned} \tag{2.4.7}$$

The model's variance components are σ_2^2, σ_3^2, and σ_4^2, which are the same as σ_β^2, $\sigma_{\tau\beta}^2$, and σ_e^2. Under the Mixed Model Assumptions, $\mathbf{y}'\mathbf{P}_0\mathbf{y}/\delta_0$ and $\mathbf{y}'\mathbf{P}_1\mathbf{y}/\delta_1$ are distributed as noncentral chi-squared variates with $m_0 = 1$ and $m_1 = v - 1$ degrees of freedom, respectively. The corresponding noncentrality parameters are

$$\zeta_0 = \frac{\mathbf{g}'\mathbf{X}'\mathbf{P}_0\mathbf{X}\mathbf{g}}{\delta_0} \tag{2.4.8}$$

$$\zeta_1 = \frac{\mathbf{g}'\mathbf{X}'\mathbf{P}_1\mathbf{X}\mathbf{g}}{\delta_1}, \tag{2.4.9}$$

where

$$\delta_0 = vn\sigma_2^2 + n\sigma_3^2 + \sigma_4^2 \tag{2.4.10}$$

$$\delta_1 = n\sigma_3^2 + \sigma_4^2. \tag{2.4.11}$$

Furthermore, $\mathbf{y}'\mathbf{P}_2\mathbf{y}/\delta_2$, $\mathbf{y}'\mathbf{P}_3\mathbf{y}/\delta_3$, and $\mathbf{y}'\mathbf{P}_4\mathbf{y}/\delta_4$ are distributed as central chi-squared variates with $m_2 = b - 1$, $m_3 = (v - 1)(b - 1)$, and $m_4 = vb(n - 1)$ degrees of freedom, respectively, where

$$\begin{aligned} \delta_2 &= vn\sigma_2^2 + n\sigma_3^2 + \sigma_4^2 \\ \delta_3 &= n\sigma_3^2 + \sigma_4^2 \\ \delta_4 &= \sigma_4^2. \end{aligned} \tag{2.4.12}$$

It should be noted that the numerators of ζ_0 and ζ_1 in (2.4.8) and (2.4.9) can be easily obtained by substituting the mean $E(\mathbf{y}) = \mathbf{Xg}$ for $\mathbf{y}$ in the formulas for $\mathbf{y}'\mathbf{P}_0\mathbf{y}$ and $\mathbf{y}'\mathbf{P}_1\mathbf{y}$, respectively. By formula (2.3.1) these sums of squares are

$$\begin{aligned} \mathbf{y}'\mathbf{P}_0\mathbf{y} &= \sum_{i,j,k} \left[y^{(0)}\right]^2 = vbn\left[y^{(0)}\right]^2 \\ \mathbf{y}'\mathbf{P}_1\mathbf{y} &= \sum_{i,j,k} \left[y_{(i)}^{(1)} - y^{(0)}\right]^2 \\ &= bn\sum_{i=1}^{v} \left[y_{(i)}^{(1)} - y^{(0)}\right]^2. \end{aligned} \tag{2.4.13}$$

By replacing $y^{(0)}$ and $y_{(i)}^{(1)} - y^{(0)}$ by their corresponding means, we obtain

$$\begin{aligned}\mathbf{g}'\mathbf{X}'\mathbf{P}_0\mathbf{X}\mathbf{g} &= vbn\left[g^{(0)} + \bar{g}.^{(1)}\right]^2 \\ \mathbf{g}'\mathbf{X}'\mathbf{P}_1\mathbf{X}\mathbf{g} &= bn\sum_{i=1}^{v}\left[g_{(i)}^{(1)} - \bar{g}.^{(1)}\right]^2,\end{aligned} \tag{2.4.14}$$

where

$$\bar{g}.^{(1)} = \frac{1}{v}\sum_{i=1}^{v} g_{(i)}^{(1)}. \tag{2.4.15}$$

Example 2.4.2. Consider the model used in Example 2.3.1. Suppose that all the effects are random. The model's variance components are $\sigma_1^2, \sigma_2^2, \sigma_3^2, \sigma_4^2$, and σ_5^2. In this case, $\mathbf{y}'\mathbf{P}_i\mathbf{y}/\delta_i$ has the central chi-squared distribution with m_i degrees of freedom ($i = 1, 2, \ldots, 5$), where

$$\begin{aligned}\delta_1 &= bn\sigma_1^2 + n\sigma_2^2 + b\sigma_4^2 + \sigma_5^2, \qquad m_1 = v - 1 \\ \delta_2 &= n\sigma_2^2 + \sigma_5^2, \qquad m_2 = v(b-1) \\ \delta_3 &= vb\sigma_3^2 + b\sigma_4^2 + \sigma_5^2, \qquad m_3 = n - 1 \\ \delta_4 &= b\sigma_4^2 + \sigma_5^2, \qquad m_4 = (v-1)(n-1) \\ \delta_5 &= \sigma_5^2, \qquad m_5 = v(b-1)(n-1).\end{aligned} \tag{2.4.16}$$

2.5. DERIVATION OF OPTIMUM TESTS

In this section, we explain the derivation of appropriate optimum tests for fixed effects $\mathbf{Xg}$ and the variance components σ_i^2's, $i = \nu - p + 1, \ldots, \nu$ in the context of the mixed model (2.4.1). Under the Mixed Model Assumptions and from Theorem 2.4.1, it follows readily that

$$\mathbf{y} \sim N\left[\mathbf{Xg}, \sum_{j=\nu-p+1}^{\nu+1} \sigma_j^2 \boldsymbol{H}_j\boldsymbol{H}_j'\right]. \tag{2.5.1}$$

However, $Var(\mathbf{y}) = \sum_{j=\nu-p+1}^{\nu+1} \sigma_j^2 \boldsymbol{H}_j\boldsymbol{H}_j'$ can be expressed as $\boldsymbol{V}$, where

$$\begin{aligned}\boldsymbol{V} &= \sum_{j=\nu-p+1}^{\nu+1} \sigma_j^2 \boldsymbol{H}_j\boldsymbol{H}_j' \\ &= \sum_{j=\nu-p+1}^{\nu+1} \sigma_j^2 \boldsymbol{A}_j\end{aligned}$$

$$= \sum_{j=\nu-p+1}^{\nu+1} (\sigma_j^2 b_j)(\mathbf{A}_j/b_j)$$

$$= \sum_{j=\nu-p+1}^{\nu+1} (\sigma_j^2 b_j)\left[\sum_{\psi_i \subset \psi_j} \mathbf{P}_i\right] \quad (by\ Lemma\ 2.3.3)$$

$$= \sum_{i=0}^{\nu+1} \mathbf{P}_i \left[\sum_{j \in w_i} b_j \sigma_j^2\right]$$

$$= \sum_{i=0}^{\nu+1} \delta_i \mathbf{P}_i \tag{2.5.2}$$

by the definition of δ_i in formula (2.4.5), where we have used the convention that $\sigma_i^2 = 0$ for $i = 0, \ldots, \nu - p$. Since the matrices $\mathbf{P}_i$'s are idempotent, that is, satisfy $\boldsymbol{P}_i\boldsymbol{P}_j = \mathbf{0}$, $i \neq j$, and $\sum_{i=0}^{\nu+1} \boldsymbol{P}_i = \boldsymbol{I}_N$, we readily obtain $[Var(\mathbf{y})]^{-1} = \sum_{i=0}^{\nu+1} \mathbf{P}_i/\delta_i$. Hence, under the assumption of independence and normality of all the random effects and the error components in model (2.4.1), the *exponent* in the joint probability density function (*pdf*) of $\mathbf{y}$ can be expressed as

$$(\mathbf{y} - \mathbf{Xg})'V^{-1}(\mathbf{y} - \mathbf{Xg}) = (\mathbf{y} - \mathbf{Xg})'\left(\sum_{i=0}^{\nu+1} \frac{\mathbf{P}_i}{\delta_i}\right)(\mathbf{y} - \mathbf{Xg})$$

$$= (\mathbf{y} - \mathbf{Xg})'\left(\sum_{i=0}^{\nu-p} \frac{\mathbf{P}_i}{\delta_i}\right)(\mathbf{y} - \mathbf{Xg}) + \mathbf{y}'\left(\sum_{i=\nu-p+1}^{\nu+1} \frac{\mathbf{P}_i}{\delta_i}\right)\mathbf{y}, \tag{2.5.3}$$

where we have used the fact that $\mathbf{g}'\mathbf{X}'\mathbf{P}_i\mathbf{Xg} = 0$ for $i = \nu - p + 1, \ldots, \nu + 1$ (see Theorem 2.4.1, part (b)). It then follows from (2.5.3) that the *linear* forms $\mathbf{P}_i\mathbf{y}$ for $i = 0, \ldots, \nu - p$ and the *quadratic* forms $\mathbf{y}'\mathbf{P}_i\mathbf{y}$ for $i = \nu - p + 1, \ldots, \nu + 1$ are jointly sufficient for the unknown parameters $\mathbf{Xg}$ (the fixed effects) and σ_i^2's, $i = \nu - p + 1, \ldots, \nu + 1$ (the variance components). However, these are not minimal sufficient. See the discussion below for a description of minimal sufficient statistics in this context. Moreover, using (2.5.3), the joint density of $\mathbf{y}$ can be written as

$$f(\mathbf{y}) = (2\pi)^{-N/2}\left(\prod_{j=0}^{\nu+1} \delta_j^{-m_j/2}\right)$$

$$exp\left(-\frac{1}{2}\left[\sum_{j=\nu-p+1}^{\nu+1} \frac{\mathbf{y}'\boldsymbol{P}_j\mathbf{y}}{\delta_j} + \sum_{i=0}^{\nu-p} \frac{(\mathbf{P}_i\mathbf{y} - \mathbf{P}_i\mathbf{Xg})'(\mathbf{P}_i\mathbf{y} - \mathbf{P}_i\mathbf{Xg})}{\delta_i}\right]\right)$$

$$= (2\pi)^{-N/2} \left(\prod_{j=0}^{\nu+1} \delta_j^{-m_j/2} \right)$$

$$exp \left(-\frac{1}{2} \left[\sum_{j=\nu-p+1}^{\nu+1} \frac{\mathbf{y}'\boldsymbol{P}_j\mathbf{y}}{\delta_j} + \sum_{i=0}^{\nu-p} \frac{(\boldsymbol{C}_i\mathbf{y} - \boldsymbol{C}_i\mathbf{X}\mathbf{g})'(\boldsymbol{C}_i\mathbf{y} - \boldsymbol{C}_i\mathbf{X}\mathbf{g})}{\delta_i} \right] \right), \tag{2.5.4}$$

where $\mathbf{P}_i = \boldsymbol{C}_i'\boldsymbol{C}_i$ and $\boldsymbol{C}_i$ is of order $m_i \times N$ and rank m_i, for $i = 0, \ldots, \nu - p$. The rows of $\boldsymbol{C}_i$ are orthonormal eigenvectors of $\boldsymbol{P}_i$ with eigenvalue equal to one. Hence, $\boldsymbol{C}_i\boldsymbol{C}_i' = \boldsymbol{I}_{m_i}$. In (2.5.4) above, we have used the easily verifiable fact (see (2.5.2)) that $|\boldsymbol{V}| = \prod_{j=0}^{\nu+1} \delta_j^{m_j}$. Incidentally, this representation of the joint density of $\mathbf{y}$ makes clear the distributional results mentioned in Theorem 2.4.1, part (a).

The functional form of $f(\mathbf{y})$ given above, which is always valid for balanced data, is fundamental and at the root of our subsequent discussion of optimal tests. Writing $\mathbf{C}_i\mathbf{X}\mathbf{g} = \boldsymbol{\gamma}_i$, for $i = 0, \ldots, \nu - p$, we can regard (2.5.4) as the canonical form of the joint density of $\mathbf{y}$, and the parameters δ_j's and $\boldsymbol{\gamma}_i$'s, which are range independent, as the canonical parameters. Since, obviously, the joint density (2.5.4) belongs to a multiparameter exponential family with the natural convex parameter space $\omega = \{\delta_j, \boldsymbol{\gamma}_i : \delta_j > 0,$ *for all* j; $\boldsymbol{\gamma}_i$ *in* R^{m_i}, *for all* $i\}$, it follows that the minimal sufficient statistics $v_j = \mathbf{y}'P_j\mathbf{y},\ j = \nu - p + 1, \ldots, \nu + 1$ and $\mathbf{w}_i = \boldsymbol{C}_i\mathbf{y},\ i = 0, \ldots, \nu - p$ are complete for the unknown canonical parameters. From the standard theory of tests for linear exponential families (Lehmann, 1986; Chapter 1), we can then conclude that optimal tests (i.e., UMP, UMPU, UMPI, UMPIU as the case may be) do exist and can be easily constructed as long as the null hypothesis of interest is either linear in η_j's for $j = \nu - p + 1, \ldots, \nu + 1$ where $\eta_j = (\delta_j)^{-1}$, or involves a $\boldsymbol{\gamma}_i$ for some $i = 0, \ldots, \nu - p$ such that the corresponding δ_i is equal to another δ_j for some $j = \nu - p + 1, \ldots, \nu + 1$. Below we elaborate on this point.

Typically, for many balanced models, the null hypothesis of interest on the estimable fixed effects $\mathbf{X}\mathbf{g}$ is reduced to testing $H_\gamma : \boldsymbol{\gamma}_i = 0$ for any given $i = 0, \ldots, \nu - p$ and that on a variance component σ_i^2 to testing $H_{\sigma^2} : \delta_j = \delta_{j'}$ against the one-sided alternative $\delta_j > \delta_{j'}$ for some $j \neq j' = \nu - p + 1, \ldots, \nu + 1$. In the former case, an exact optimum test can be easily derived if the associated δ_i from (2.5.4) coincides with one of the δ_j's for some $j = \nu - p + 1, \ldots, \nu + 1$, which happens quite often though not always. Assume $\delta_i = \delta_{\nu-p+1}$ so that the F-test based on $\mathbf{w}_i'\mathbf{w}_i/v_{\nu-p+1}$ is exact. If $\boldsymbol{\gamma}_i$ is a scalar, then from standard multiparameter exponential family results (Lehmann, 1986), it follows directly that this F-test is UMPU. However, if the dimension of $\boldsymbol{\gamma}_i$ is more than one, then it is well known that a UMPU does not exist. Nevertheless, it can be easily shown that the above F-test is UMPIU. In case of H_{σ^2}, it is an easy consequence of the nature of the multiparameter exponential family (2.5.4) and the null hypothesis H_{σ^2} that the exact F-test based on $v_j/v_{j'}$ is both UMPU and UMPIU (Mathew and Sinha, 1988).

We now illustrate the above points with some of the examples already given. Throughout the remainder of this chapter, we denote the error variance by σ_e^2.

Example 2.5.1. Consider the one-way classification model described in Example 2.2.1. A canonical form of this model involves three familiar quantities $v_1 = SS(\tau)$ (treatment sum of squares) with associated $\delta_1 = \sigma_e^2 + n\sigma_\tau^2$ and $m_1 = v - 1$, $v_2 = SS(e)$ (error sum of squares) with associated $\delta_2 = \sigma_e^2$ and $m_2 = v(n-1)$, and $w_0 = \boldsymbol{C}_0\mathbf{y} = \sqrt{N}\bar{y}_{..}$ with $\gamma_0 = \sqrt{N}\mu$ and $\delta_0 = n\sigma_\tau^2 + \sigma_e^2$. Here, as mentioned above, σ_e^2 represents the error variance, and $\eta_1 = 1/\delta_1$, $\eta_2 = 1/\delta_2$. Also, $N = vn$.

The main hypothesis of interest $H_\tau : \sigma_\tau^2 = 0$ corresponds to $\eta_1 = \eta_2$, and it follows trivially that the F-test based on $SS(\tau)/SS(e)$ is exact and optimum (UMPU, UMPIU). Because the dimension of γ is one and δ_0 coincides with δ_1, it follows directly that the F-test based on $\bar{y}_{..}^2/SS(\tau)$ is both UMPU and UMPIU for testing $H_\mu : \mu = 0$.

Example 2.5.2. Consider the mixed model version of Example 2.2.2 where the block effects are fixed and the treatments effects are random. A canonical form of this model involves $v_2 = SS(\tau)$ with associated $\delta_2 = \sigma_e^2 + b\sigma_\tau^2$ and $m_2 = v - 1$, $v_3 = SS(e)$ with associated $\delta_3 = \sigma_e^2$ and $m_3 = (v-1)(b-1)$, $w_0 = \boldsymbol{C}_0\mathbf{y} = \sqrt{N}\bar{y}_{..}$ with $\gamma_0 = \sqrt{N}\mu$, $\delta_0 = b\sigma_\tau^2 + \sigma_e^2$, and $\boldsymbol{w}_1 = \boldsymbol{C}_1\mathbf{y}$ where $\boldsymbol{C}_1'\boldsymbol{C}_1 = \mathbf{I}_v \otimes (\mathbf{I}_b - \mathbf{1}_b\mathbf{1}_b'/b)$ with $\boldsymbol{\gamma}_1 = (\mathbf{I}_b - \mathbf{1}_b\mathbf{1}_b'/b)\boldsymbol{\beta}$ and $\delta_1 = \sigma_e^2$. Here $\boldsymbol{\beta}$ stands for the vector of block effects, and, as before, $\eta_2 = 1/\delta_2$ and $\eta_3 = 1/\delta_3$.

The main hypotheses of interest here are $H_\tau : \sigma_\tau^2 = 0$ which corresponds to $\eta_2 = \eta_3$, and $H_\beta : \beta_1 = \ldots = \beta_b$ which corresponds to $H_\gamma : \boldsymbol{\gamma}_1 = \mathbf{0}$. Obviously, the F-test based on $SS(\tau)/SS(e)$ is exact and optimum (UMPU, UMPIU) for the first hypothesis and the F-test based on $SS(\beta)/SS(e)$ is exact and optimum (UMPIU) for the second hypothesis, where $SS(\beta) = \mathbf{y}'\mathbf{P}_1\mathbf{y}$ denotes the block sum of squares. The second assertion follows because δ_1 and δ_3 coincide under H_β and the dimension of $\boldsymbol{\gamma}_1$ is more than one. It may also be noted that, as in Example 2.5.1, the F-test based on $\bar{y}_{..}^2/SS(\tau)$ is UMPU and UMPIU for $H_\mu : \mu = 0$ because δ_0 given above coincides with δ_2 under H_μ.

Example 2.5.3. This is the random model version of Example 2.2.2. A canonical form of this model involves $v_1 = SS(\tau)$ with associated $\delta_1 = \sigma_e^2 + b\sigma_\tau^2$ and $m_1 = v - 1$, $v_2 = SS(\beta)$ with associated $\delta_2 = \sigma_e^2 + v\sigma_\beta^2$ and $m_2 = b - 1$, $v_3 = SS(e)$ with associated $\delta_3 = \sigma_e^2$ and $m_3 = (v-1)(b-1)$, and $w_0 = \boldsymbol{C}_0\mathbf{y} = \sqrt{N}\bar{y}_{..}$ with $\gamma_0 = \sqrt{N}\mu$ and $\delta_0 = b\sigma_\tau^2 + v\sigma_\beta^2 + \sigma_e^2$. The η's are the reciprocals of the δ's.

The main hypotheses of interest are $H_\tau : \sigma_\tau^2 = 0$ which corresponds to $\eta_1 = \eta_3$, and $H_\beta : \sigma_\beta^2 = 0$ which corresponds to $\eta_2 = \eta_3$. Clearly, the F-test based on $SS(\tau)/SS(e)$ is exact and optimum (UMPU, UMPIU) for H_τ while the F-test based on $SS(\beta)/SS(e)$ is exact and optimum (UMPU, UMPIU) for H_β.

Unlike in the previous cases, here no exact optimum test for $H_\mu : \mu = 0$ exists because δ_0 given above does not correspond to any δ_j under H_μ. We will discuss this problem in Section 2.6.

Example 2.5.4. This is the mixed model version of Example 2.2.3 described in Section 2.2 in which the block effects are fixed while the treatment and intercation effects are random. A canonical form of this model involves $v_2 = SS(\tau)$ with associated $\delta_2 = \sigma_e^2 + bn\sigma_\tau^2 + n\sigma_{\tau\beta}^2$ and $m_2 = (v-1)$, $v_3 = SS(\tau\beta)$ (interaction sum of squares) with associated $\delta_3 = \sigma_e^2 + n\sigma_{\tau\beta}^2$ and $m_3 = (v-1)(b-1)$, $v_4 = SS(e)$ with associated $\delta_4 = \sigma_e^2$ and $m_4 = vb(n-1)$, $w_0 = \boldsymbol{C}_0\mathbf{y} = \sqrt{N}\bar{y}_{...}$ with $\gamma_0 = \sqrt{N}\mu$ and $\delta_0 = bn\sigma_\tau^2 + n\sigma_{\tau\beta}^2 + \sigma_e^2$, and $\boldsymbol{w}_1 = \boldsymbol{C}_1\mathbf{y}$ where $\boldsymbol{C}_1'\boldsymbol{C}_1 = \mathbf{P}_1 = \mathbf{I}_v \otimes (\mathbf{I}_b - \frac{1}{b}\mathbf{1}_b\mathbf{1}_b') \otimes \frac{1}{n}\mathbf{1}_n\mathbf{1}_n'$ with $\boldsymbol{\gamma}_1 = (\mathbf{I}_b - \frac{1}{b}\mathbf{1}_b\mathbf{1}_b')\boldsymbol{\beta}$ and $\delta_1 = n\sigma_{\tau\beta}^2 + \sigma_e^2$. Here, $\boldsymbol{\beta}$ is the vector of block effects, and as before, the η's are the inverses of the δ's.

The main hypotheses of interest are (i) $H_{\tau\beta} : \sigma_{\tau\beta}^2 = 0$, which corresponds to $\eta_3 = \eta_4$, and (ii) $H_\tau : \sigma_\tau^2 = 0$, which corresponds to $\eta_2 = \eta_3$. Therefore, the F-test based on $SS(\tau\beta)/SS(e)$ is exact and optimum (UMPU, UMPIU) for $H_{\tau\beta}$ while the F-test based on $SS(\tau)/SS(\tau\beta)$ is exact and optimum (UMPU, UMPIU) for H_τ. The hypothesis of interest on fixed effects in this case is $H_\beta : \beta_1 = \ldots = \beta_b$, and it follows from the above discussion that the F-test based on $SS(\beta)/SS(\tau\beta)$ is exact and optimum (UMPIU) because δ_1 and δ_3 are exactly the same under H_β. Here $SS(\beta) = \boldsymbol{w}_1'\boldsymbol{w}_1$ stands for the usual block sum of squares.

Example 2.5.5. This is the random model version of Example 2.2.3. A canonical form of this model involves $v_1 = SS(\tau)$ with associated $\delta_1 = \sigma_e^2 + bn\sigma_\tau^2 + n\sigma_{\tau\beta}^2$ and $m_1 = (v-1)$, $v_2 = SS(\beta)$ with associated $\delta_2 = \sigma_e^2 + vn\sigma_\beta^2 + n\sigma_{\tau\beta}^2$ and $m_2 = b-1$, $v_3 = SS(\tau\beta)$ with associated $\delta_3 = \sigma_e^2 + n\sigma_{\tau\beta}^2$ and $m_3 = (v-1)(b-1)$, $v_4 = SS(e)$ with associated $\delta_4 = \sigma_e^2$ and $m_4 = vb(n-1)$. Moreover, for the only fixed effect component μ, we get $w_0 = \boldsymbol{C}_0\mathbf{y} = \sqrt{N}\bar{y}_{...}$ with $\gamma_0 = \sqrt{N}\mu$ and $\delta_0 = bn\sigma_\tau^2 + vn\sigma_\beta^2 + n\sigma_{\tau\beta}^2 + \sigma_e^2$.

The main hypotheses of interest are (i) $H_{\tau\beta} : \sigma_{\tau\beta}^2 = 0$, which corresponds to $\eta_3 = \eta_4$, and (ii) $H_\tau : \sigma_\tau^2 = 0$, which corresponds to $\eta_1 = \eta_3$. Clearly, the F-test based on $SS(\tau\beta)/SS(e)$ is exact and optimum (UMPU, UMPIU) for $H_{\tau\beta}$ while the F-test based on $SS(\tau)/SS(\tau\beta)$ is exact and optimum (UMPU, UMPIU) for H_τ. Another hypothesis of interest in this case is $H_\beta : \sigma_\beta^2 = 0$, and it follows from the above discussion that the F-test based on $SS(\beta)/SS(\tau\beta)$ is exact and optimum (UMPU, UMPIU).

It should, however, be noted that, as in Example 2.5.3, no exact optimum test of $H_\mu : \mu = 0$ exists in this case because obviously δ_0 is not equal to any δ_j. We will return to this example in Section 2.6.

Example 2.5.6. Consider the twofold nested random model described in Example 2.2.4. A canonical form of this model involves $v_1 = SS(\tau)$ with associated $\delta_1 = \sigma_e^2 + n\sigma_{\beta(\tau)}^2 + bn\sigma_\tau^2$ and $m_1 = (v-1)$, $v_2 = SS(\beta(\tau))$ with associated $\delta_2 = \sigma_e^2 + n\sigma_{\beta(\tau)}^2$ and $m_2 = v(b-1)$, and $v_3 = SS(e)$ with associated $\delta_3 = \sigma_e^2$ and $m_3 = vb(n-1)$. Here $SS(\beta(\tau))$ denotes the sum of squares for the nested effect. Moreover, for the only fixed effect μ, we get $w_0 = \boldsymbol{C}_0\mathbf{y} = \sqrt{N}\bar{y}_{...}$ with $\gamma_0 = \sqrt{N}\mu$ and $\delta_0 = bn\sigma_\tau^2 + n\sigma_{\beta(\tau)}^2 + \sigma_e^2$.

The main hypotheses of interest are (i) $H_{\beta(\tau)} : \sigma_{\beta(\tau)}^2 = 0$, which corresponds to $\delta_2 = \delta_3$, and (ii) $H_\tau : \sigma_\tau^2 = 0$, which is equivalent to $\delta_1 = \delta_2$. Clearly, the F-test based on $SS(\beta(\tau))/SS(e)$ is exact and optimum (UMPU, UMPIU) for $H_{\beta(\tau)}$ while the F-test based on $SS(\tau)/SS(\beta(\tau))$ is exact and optimum (UMPU, UMPIU) for H_τ. Another hypothesis of interest in this case is $H_\mu : \mu = 0$, and it follows from the above canonical form that the F-test based on $\bar{y}_{...}^2/SS(\tau)$ is exact and optimum (UMPU, UMPIU) because δ_0 is equal to δ_1.

Example 2.5.7. Consider the random model version of Example 2.2.5 given in Section 2.2. A canonical form of this model involves $v_1 = SS(\tau)$ with $\delta_1 = bnp\sigma_\tau^2 + np\sigma_{\tau\beta}^2 + p\sigma_{\tau\delta(\beta)}^2 + \sigma_e^2$ and $m_1 = v-1$, $v_2 = SS(\beta)$ with $\delta_2 = vnp\sigma_\beta^2 + np\sigma_{\tau\beta}^2 + vp\sigma_{\delta(\beta)}^2 + p\sigma_{\tau\delta(\beta)}^2 + \sigma_e^2$ and $m_2 = b-1$, $v_3 = SS(\tau\beta)$ (interaction sum of squares between τ and β) with associated $\delta_3 = np\sigma_{\tau\beta}^2 + p\sigma_{\tau\delta(\beta)}^2 + \sigma_e^2$ and $m_3 = (v-1)(b-1)$, $v_4 = SS(\delta(\beta))$ (nested factor sum of squares, δ within β) with $\delta_4 = vp\sigma_{\delta(\beta)}^2 + p\sigma_{\tau\delta(\beta)}^2 + \sigma_e^2$ and $m_4 = b(n-1)$, $v_5 = SS(\tau\delta(\beta))$ (nested interaction sum of squares, $(\tau\delta)$ nested within β) with $\delta_5 = p\sigma_{\tau\delta(\beta)}^2 + \sigma_e^2$ and $m_5 = b(v-1)(n-1)$, and finally $v_6 = SS(e)$ with associated $\delta_6 = \sigma_e^2$ and $q_6 = vbn(p-1)$. Moreover, for the only fixed effect component μ, we get $w_0 = \boldsymbol{C}_0\mathbf{y} = \sqrt{N}\bar{y}_{....}$ with $\gamma_0 = \sqrt{N}\mu$, $\delta_0 = bnp\sigma_\tau^2 + vnp\sigma_\beta^2 + np\sigma_{\tau\beta}^2 + vp\sigma_{\delta(\beta)}^2 + \sigma_e^2$ and $m_0 = 1$. As always, the η's are the reciprocals of the δ's.

The main hypotheses of interest are (i) $H_{\tau\beta} : \sigma_{\tau\beta}^2 = 0$, which corresponds to $\eta_3 = \eta_5$, (ii) $H_{\delta(\beta)} : \sigma_{\delta(\beta)}^2 = 0$, which is equivalent to $\eta_4 = \eta_5$, and (iii) $H_{\tau\delta(\beta)} : \sigma_{\tau\delta(\beta)}^2 = 0$, which is equivalent to $\eta_5 = \eta_6$. Therefore, the F-test based on $SS(\tau\beta)/SS(\tau\delta(\beta))$ is exact and optimum (UMPU, UMPIU) for $H_{\tau\beta}$, the F-test based on $SS(\delta(\beta))/SS(\tau\delta(\beta))$ is exact and optimum (UMPU, UMPIU) for $H_{\delta(\beta)}$, while the F-test based on $SS(\tau\delta(\beta))/SS(e)$ is exact and optimum for $H_{\tau\delta(\beta)}$.

Two other hypotheses of interest concern the variance components σ_τ^2 and σ_β^2. It is easy to verify that testing $H_\tau : \sigma_\tau^2 = 0$ is equivalent to $\eta_1 = \eta_3$ so that the F-test based on $SS(\tau)/SS(\tau\beta)$ is exact and optimum. However, it turns out that the hypothesis $H_\beta : \sigma_\beta^2 = 0$ is *not* equivalent to the equality of any two η_j's (or δ_j's). Therefore, we have an example of a situation involving a test of a variance component for which no exact or optimum test (in the usual sense) exists. We will return to this example in Section 2.6.

The above examples demonstrate that the general theory of multiparameter exponential families can be successfully applied to derive optimum tests for many interesting and useful hypotheses in a wide variety of balanced mixed and random models. We should recall, however, that whether we can achieve this goal or not depends very much on the nature of the null hypothesis H_0 in that it should be possible to express H_0 for a variance component σ_i^2 as the equality of two η_j's for $j = \nu - p + 1, \ldots, \nu + 1$. On the other hand, for a fixed effect hypothesis involving $\boldsymbol{\gamma}_i$ for some $i = 0, \ldots, \nu - p$, it should hold that the associated δ_i is equal to some δ_j for $j = \nu - p + 1, \ldots, \nu + 1$. That this is not always the case has been pointed out for a fixed effect hypothesis on μ in Examples 2.5.3 and 2.5.5; and for a variance component σ_β^2 in Example 2.5.7.

It is therefore clear that there are practical situations where exact optimum tests do *not* exist and there is a genuine need to derive other valid tests. In the next section, we discuss two procedures to tackle this problem. The first procedure is popularly known as Satterthwaite's approximation (Satterthwaite, 1941, 1946) and the second can be described to yield tests of Bartlett–Scheffé type (Bartlett, 1936; Scheffé, 1943; Mazuy and Connor, 1965; Linnik, 1966; Dobrohotov, 1968; Seifert, 1979, 1981).

2.5.1. A Numerical Example

We shall now apply the results of the preceding examples to a numerical example taken from Montgomery (1991, p. 172). The example deals with an engineer who is conducting an experiment on focus time in order to study the effect of the distance of an object from the eye on the focus time. Four different distances are of interest, implying that there are four *treatments* with *fixed* effects. There are five subjects available for the experiment, suggesting that there are five *blocks* with *random* effects. The data appears in a randomized block design as given in Table 2.1. Obviously this is an application of Example 2.2.2 and Example 2.5.2, and we have a balanced *mixed* model.

The *ANOVA* table, readily obtained using SAS (1989), takes the form in Table 2.2.

The upper 5% table F-values corresponding to 3 and 12 d.f.'s and 4 and 12 d.f.'s are, respectively, 3.49 and 3.26. We therefore reject both the null hypotheses $H_\tau : \tau_1 = \tau_2 = \tau_3 = \tau_4$ and $H_\beta : \sigma_\beta^2 = 0$ at 5% level of significance. Incidentally, it follows from the discussion in Example 2.5.2 that the F-test

Table 2.1. Distance by Subject Data on Focus Time

	Subject				
Distance	1	2	3	4	5
4	10	6	6	6	6
6	7	6	6	1	6
8	5	3	3	2	5
10	6	4	4	2	3

Source: D.C. Montgomery (1991). Reproduced with permission of John Wiley & Sons, Inc.

Table 2.2. ANOVA Table of Two-way Balanced Mixed Model

Source	d.f.	SS	Mean Square	F-Value
Distance	3	32.95	10.983	$F_\tau = 8.6143$
Subject	4	36.30	9.075	$F_\beta = 7.1176$
Distance*Subject (error)	12	15.30	1.275	

based on F_τ is $UMPIU$ and the F-test based on F_β is $UMPU$ and also $UMPIU$.

2.6. APPROXIMATE AND EXACT TESTS

In this section we first describe a procedure, due to Satterthwaite (1941, 1946), which leads to approximate tests.

2.6.1. Satterthwaite's Approximation

In situations where no exact optimum test exists for testing the significance of a variance component, Satterthwaite (1941, 1946) suggested using an approximate F-test by suitably combining more than two sums of squares from the ANOVA table. The advantage of this suggestion is its relative simplicity. However, the main disadvantages are that (i) the resulting size of the test cannot be controlled at the desired level of significance, (ii) the resultant test may be biased, and (iii) it is usually quite difficult to compute the actual size and power of this test. It may be mentioned that a measure as well as a test to evaluate the adequacy of this approximation was introduced by Khuri (1995 a, b).

To describe Satterthwaite's approximation quite generally, suppose we are interested in testing $H_0 : \sigma_i^2 = 0$ for some $i = \nu - p + 1, \ldots, \nu$ in the mixed model (2.4.1) and suppose in the notations of the canonical parameters given in the joint probability density function (pdf) (2.5.4) of $\mathbf{y}$ that H_0 is *not* equivalent to the equality of any two η_j's for $j = \nu - p + 1, \ldots, \nu + 1$. On the other hand, in terms of the mean squares $M_j = \mathbf{y}'\mathbf{P}_j\mathbf{y}/m_j$'s, suppose it holds that

$$E(M_r + \cdots + M_s) = k\sigma_i^2 + E(M_t + \cdots + M_u), \tag{2.6.1}$$

where k is a positive constant and $r, \ldots, s, \ldots, t, \ldots, u$ are appropriate indices from $\{\nu - p + 1, \ldots, \nu + 1\}$ (see Section 2.4). In such a situation, for testing $H_0 : \sigma_i^2 = 0$, Satterthwaite (1941, 1946) suggested using

$$F = \frac{M_r + \ldots + M_s}{M_t + \ldots + M_u} \sim F(f_1, f_2), \tag{2.6.2}$$

where

$$\begin{aligned} f_1 &= \frac{(M_r + \ldots + M_s)^2}{M_r^2/m_r + \ldots + M_s^2/m_s}, \\ f_2 &= \frac{(M_t + \ldots + M_u)^2}{M_t^2/m_t + \ldots + M_u^2/m_u}, \end{aligned} \tag{2.6.3}$$

and m_j is the number of degrees of freedom of M_j. Clearly, the above F-distribution is only approximate. The basis of Satterthwaite's argument is that both the numerator and the denominator of F as defined above in (2.6.2), when suitably normalized, can be individually approximated as central χ^2 variables. In a much later paper Gaylor and Hopper (1969) suggested a similar approximation of a difference of two mean squares by a central χ^2.

Analogously, for testing a fixed effect hypothesis $H_0 : \boldsymbol{\gamma}_i = \mathbf{0}$, for some $i = 0, \ldots, \nu - p$, suppose it holds that

$$E(M_i + M_r + \cdots + M_s) = k\boldsymbol{\gamma}_i'\boldsymbol{\gamma}_i + E(M_t + \cdots + M_u), \tag{2.6.4}$$

where k is a positive constant, and $M_r, \ldots, M_u$ denote the mean squares M_j's all arising from the η_j's for $j = \nu - p + 1, \ldots, \nu + 1$. Then one can proceed exactly in the same fashion as described above and construct an approximate F-test for $H_0 : \boldsymbol{\gamma}_i = \mathbf{0}$ based on

$$F = \frac{M_i + M_r + \ldots + M_s}{M_t + \ldots + M_u} \sim F(f_1, f_2), \tag{2.6.5}$$

where

$$\begin{aligned} f_1 &= \frac{(M_i + M_r + \ldots + M_s)^2}{M_i^2/m_i + M_r^2/m_r + \ldots + M_s^2/m_s}, \\ f_2 &= \frac{(M_t + \ldots + M_u)^2}{M_t^2/m_t + \ldots + M_u^2/m_u}. \end{aligned} \tag{2.6.6}$$

We now briefly explain the above procedure by means of a few examples.

Example 2.6.1. This is a continuation of Example 2.5.3 for a two-way crossed-classification random model without interaction. As noted before, in this example there does not exist an exact optimum test for $H_\mu : \mu = 0$. However, it is easy to verify that

$$E[N\bar{y}_{..}^2 + MS(e)] = vb\mu^2 + E[MS(\tau) + MS(\beta)], \tag{2.6.7}$$

where $MS(.)$, and so on, denote the respective mean squares. Comparing (2.6.7) with (2.6.4), it follows that we can use the approximate F-statistic for testing $H_\mu : \mu = 0$, where

$$F = \frac{N\bar{y}_{..}^2 + MS(e)}{MS(\tau) + MS(\beta)} \tag{2.6.8}$$

and the d.f.'s, f_1 and f_2, of the above F-test are given by

$$\begin{aligned} f_1 &= \frac{[N\bar{y}_{..}^2 + MS(e)]^2}{N^2\bar{y}^4 + (MS(e))^2/(v-1)(b-1)}, \\ f_2 &= \frac{[MS(\tau) + MS(\beta)]^2}{(MS(\tau))^2/(v-1) + (MS(\beta))^2/(b-1)}. \end{aligned} \tag{2.6.9}$$

Example 2.6.2. This is a continuation of Example 2.5.5. In this example also there does not exist an exact optimum test for $H_\mu : \mu = 0$. However, it is easy to verify that

$$E[N\bar{y}_{...}^2 + MS(\tau\beta)] = vbn\mu^2 + E[MS(\tau) + MS(\beta)]. \tag{2.6.10}$$

Comparing (2.6.10) with (2.6.4), it follows that we can use the approximate F-statistic for testing $H_\mu : \mu = 0$, where

$$F = \frac{N\bar{y}_{...}^2 + MS(\tau\beta)}{MS(\tau) + MS(\beta)} \tag{2.6.11}$$

and the d.f.'s, f_1 and f_2, of the above F-test are given by

$$\begin{aligned} f_1 &= \frac{[N\bar{y}_{...}^2 + MS(\tau\beta)]^2}{N^2\bar{y}_{...}^4 + (MS(\tau\beta))^2/(v-1)(b-1)}, \\ f_2 &= \frac{[MS(\tau) + MS(\beta)]^2}{(MS(\tau))^2/(v-1) + (MS(\beta))^2/(b-1)}. \end{aligned} \tag{2.6.12}$$

Example 2.6.3. This is a continuation of Example 2.5.7. Recall that in this unique example (see the discussion in Example 2.5.7) there does not exist an exact optimum test for a hypothesis about a variance component, namely, for $H_\beta : \sigma_\beta^2 = 0$. However, a close examination of different sums of squares and their expectations immediately reveals that

$$E[MS(\beta) + MS(\tau\delta(\beta))] = vnp\sigma_\beta^2 + E[MS(\tau\beta) + MS(\delta(\beta))]. \qquad (2.6.13)$$

But this is readily of the form (2.6.1). We can therefore use the test statistic

$$F = \frac{MS(\beta) + MS(\tau\delta(\beta))}{MS(\tau\beta) + MS(\delta(\beta))}, \qquad (2.6.14)$$

which is distributed as an approximate F with the d.f.'s, f_1 and f_2, given by

$$\begin{aligned} f_1 &= \frac{[MS(\beta) + MS(\tau\delta(\beta))]^2}{(MS(\beta))^2/(b-1) + (MS(\tau\delta(\beta)))^2/b(v-1)(n-1)}, \\ f_2 &= \frac{[MS(\tau\beta) + MS(\delta(\beta))]^2}{(MS(\tau\beta))^2/(v-1)(b-1) + (MS(\delta(\beta)))^2/b(n-1)}. \end{aligned} \qquad (2.6.15)$$

It is clear from the above examples that Satterthwaite's approximation is easy to apply and quite versatile in nature in that it can be used in a variety of situations. However, as noted previously, the procedure leads at most to an approximate F-test, and the approximation may not be justified in small samples.

We now proceed to describe another important procedure, developed mainly by Seifert (1979, 1981), which often yields exact unbiased tests in situations when exact optimum tests do not exist. These tests are commonly known as tests of *Bartlett–Scheffé* type because of their similarity to those derived by Bartlett (1936) and Scheffé (1943) for the famous *Behrens–Fisher* problem. Moreover, as Seifert (1979) pointed out, in addition to being unbiased, often such tests are also optimum in the restricted class of tests of Bartlett–Scheffé type. We refer the reader to Seifert (1979, 1981) for details.

2.6.2. Exact Unbiased Tests of Bartlett–Scheffé Type

Referring to the basic functional form of the joint density $f(\mathbf{y})$ given in (2.5.4), the starting points behind such tests are the linear forms

$$\begin{aligned} \boldsymbol{C}_i\mathbf{y} &\sim N_{m_i}(\boldsymbol{\gamma}_i, \delta_i\boldsymbol{I}_{m_i}), \quad i = 0, \cdots, \nu - p, \\ \boldsymbol{C}_j\mathbf{y} &\sim N_{m_j}(\mathbf{0}, \delta_j\boldsymbol{I}_{m_j}), \quad j = \nu - p + 1, \cdots, \nu + 1, \end{aligned} \qquad (2.6.16)$$

where $\boldsymbol{\gamma}_i$'s, δ_i's, and m_i's are all defined previously. Recall that $\boldsymbol{P}_j = \boldsymbol{C}_j'\boldsymbol{C}_j$ for $j = \nu - p + 1, \ldots, \nu + 1$. Note that, in the terminology of linear models, $\boldsymbol{C}_j\mathbf{y}$'s represent the so-called *error* functions for $j = \nu - p + 1, \ldots, \nu + 1$, and $\boldsymbol{C}_i\mathbf{y}$'s for $i = 0, \ldots, \nu - p$ correspond to estimable linear functions for $\mathbf{Xg}$ for model (2.4.1). Moreover, recall that these linear forms are the basis of our construction of ANOVA tests. Obviously, all linear functions $\boldsymbol{C}_i\mathbf{y}$'s and $\boldsymbol{C}_j\mathbf{y}$'s are independent.

Consider first the problem of testing $H_\gamma : \boldsymbol{\gamma}_i = \mathbf{0}$ for some $i = 1$, say, and suppose the usual condition of the existence of a $\delta_j = \delta_1$ for some $j = \nu - p + 1, \ldots, \nu + 1$ is not satisfied (so that an exact optimum test for H_γ does not exist). Suppose we can write the associated δ_1 as

$$\delta_1 = \sum_{j=\nu-p+1}^{\nu+1} g_j\delta_j = \sum\nolimits_1 g_j\delta_j + \sum\nolimits_2 g_j\delta_j. \tag{2.6.17}$$

where $\sum_1$ is the sum over all indices j such that $g_j > 0$ and $\sum_2$ is the sum over all indices j such that $g_j < 0$. Such a linear combination of the expectations δ_j's of the error mean squares $\mathbf{y}'\mathbf{P}_j\mathbf{y}/m_j$'s always exists, and is unique. We then try to choose two matrices $\boldsymbol{A}$ and $\boldsymbol{B}$ such that the linear forms $\boldsymbol{A}\mathbf{y}$ and $\boldsymbol{B}\mathbf{y}$ satisfy

$$\boldsymbol{A}\mathbf{y} \sim N_r\left(\boldsymbol{D}\boldsymbol{\gamma}_1,\ \left[\delta_1 + \sum\nolimits_2(-g_j)\delta_j\right]\boldsymbol{I}_r\right) \tag{2.6.18}$$

and

$$\boldsymbol{B}\mathbf{y} \sim N_f\left(\mathbf{0},\ \left[\sum\nolimits_1 g_j\delta_j\right]\boldsymbol{I}_f\right), \tag{2.6.19}$$

where the matrix $\boldsymbol{D} : r \times m_1$ satisfies : $\boldsymbol{D}\boldsymbol{D}' = \boldsymbol{I}_r$ for some $r \le m_1$, and $\boldsymbol{A}\mathbf{y}$ and $\boldsymbol{B}\mathbf{y}$ are independent. Such choices of $\boldsymbol{A}\mathbf{y}$ and $\boldsymbol{B}\mathbf{y}$ directly lead to tests of Bartlett–Scheffé type. Clearly, we can use the test statistic

$$F = \frac{\mathbf{y}'\boldsymbol{A}'\boldsymbol{A}\mathbf{y}/r}{\mathbf{y}'\boldsymbol{B}'\boldsymbol{B}\mathbf{y}/f} \tag{2.6.20}$$

for testing $H_\gamma : \gamma_1 = 0$, which is distributed as an exact central $F_{r,f}$ under H_γ. Moreover, since under the alternative $\boldsymbol{\gamma}_1 \neq \mathbf{0}$, $\mathbf{y}'\boldsymbol{A}'\boldsymbol{A}\mathbf{y}/[\delta_1 + \sum_2(-g_j)\delta_j]$ is distributed as a noncentral F with the noncentrality parameter $\boldsymbol{\gamma}_1'\boldsymbol{D}'\boldsymbol{D}\boldsymbol{\gamma}_1/[\delta_1 + \sum_2(-g_j)\delta_j]$, it follows that the above F-test is strictly unbiased (a test is defined to be strictly ubiased if its power is strictly more than its size under any deviation from the null hypothesis). Finding $\boldsymbol{A}$ and $\boldsymbol{B}$ such that the conditions (2.6.18) and (2.6.19) are satisfied essentially depends on a careful study of the underlying linear model.

To give a simple example, suppose

$$\begin{aligned}
\boldsymbol{C}_1\mathbf{y} &\sim N_{m_1}(\boldsymbol{\gamma}_1,\ [g_2\delta_2 - g_3\delta_3]\boldsymbol{I}_{m_1}),\\
\boldsymbol{C}_2\mathbf{y} &\sim N_{m_2}(\mathbf{0},\ \delta_2\boldsymbol{I}_{m_2}),\\
\boldsymbol{C}_3\mathbf{y} &\sim N_{m_3}(\mathbf{0},\ \delta_3\boldsymbol{I}_{m_3}),
\end{aligned} \tag{2.6.21}$$

where $g_2,\ g_3 > 0$, and $m_1 \le m_3$, and $\boldsymbol{C}_1\mathbf{y}$, $\boldsymbol{C}_2\mathbf{y}$, and $\boldsymbol{C}_3\mathbf{y}$ are independent. Then, comparing (2.6.21) with (2.6.17), we can take

$$\boldsymbol{A}\mathbf{y} = \boldsymbol{C}_1\mathbf{y} + (\sqrt{g_3})\boldsymbol{C}\boldsymbol{C}_3\mathbf{y} \sim N_{m_1}(\boldsymbol{\gamma}_1,\ g_2\delta_2\boldsymbol{I}_{m_1}) \tag{2.6.22}$$

and

$$\boldsymbol{B}\mathbf{y} = (\sqrt{g_2})\boldsymbol{C}_2\mathbf{y} \sim N_{m_2}(\mathbf{0},\ g_2\delta_2\boldsymbol{I}_{m_2}), \tag{2.6.23}$$

where the matrix $\boldsymbol{C} : m_1 \times m_3$ is chosen so as to satisfy $\boldsymbol{C}\boldsymbol{C}' = \boldsymbol{I}_{m_1}$, and use the exact F-test based on

$$F = \frac{[\boldsymbol{C}_1\mathbf{y} + (\sqrt{g_3})\boldsymbol{C}\boldsymbol{C}_3\mathbf{y}]'[\boldsymbol{C}_1\mathbf{y} + (\sqrt{g_3})\boldsymbol{C}\boldsymbol{C}_3\mathbf{y}]'/m_1}{[(\sqrt{g_2})\boldsymbol{C}_2\mathbf{y}]'[(\sqrt{g_2})\boldsymbol{C}_2\mathbf{y}]/m_2}. \tag{2.6.24}$$

The above idea can be used quite generally to construct an exact unbiased test for a variance component, say $\sigma^2_{\nu-p+1}$, for which an optimum test does not exist. Thus, if $\delta_{\nu-p+1} = h_{\nu-p+1}\sigma^2_{\nu-p+1} + \sum_{j=\nu-p+2}^{\nu+1} h_j\delta_j$ for some $h_{\nu-p+1} > 0$, and $h_{\nu-p+2}, \ldots, h_{\nu+1} \geq 0$, linear forms

$$\boldsymbol{A}\mathbf{y} \sim N_{f_1}(\mathbf{0},\ \left(\delta_1 + \sum\nolimits_2 (-h_j)\delta_j)\boldsymbol{I}_{f_1}\right) \tag{2.6.25}$$

and

$$\boldsymbol{B}\mathbf{y} \sim N_{f_2}\left(\mathbf{0},\ (\sum\nolimits_1 h_j\delta_j)\boldsymbol{I}_{f_2}\right), \tag{2.6.26}$$

where, as before, $\sum_1$ stands for the sum over indices j for which $h_j > 0$ and $\sum_2$ stands for those for which $h_j < 0$, and $\boldsymbol{A}\mathbf{y}$ and $\boldsymbol{B}\mathbf{y}$ are independent, lead to tests of Bartlett–Scheffé type. Obviously, $\boldsymbol{A}\mathbf{y}$ and $\boldsymbol{B}\mathbf{y}$ are constructed solely out of the error functions $\boldsymbol{C}_j\mathbf{y}$'s for $j = \nu - p + 1, \ldots, \nu + 1$.

We now illustrate the applications of the above procedure in Examples 2.5.3 and 2.5.5 for which no exact optimum test exists for H_μ, and show how we can construct exact unbiased tests of Bartlett–Scheffé type in these two problems. For some other useful applications (especially in the context of Example 2.5.7), we refer to Seifert (1979, 1981).

Example 2.6.4. This is a continuation of Example 2.5.3. Here we derive a Bartlett–Scheffé type test for $H_\mu : \mu = 0$. Without any loss of generality, assume $v \leq b$. Let

$$\boldsymbol{A}_1 = \frac{1}{2}(1, -1, 0, \cdots, 0)_v \otimes (1, -1, 0, \cdots, 0)_b. \tag{2.6.27}$$

In the above and also in Example 2.6.5, $(., \ldots, .)_k$ stands for a vector with k components. Then we can take $\boldsymbol{A}\mathbf{y}$ as

$$\boldsymbol{A}\mathbf{y} = \left(\frac{1}{\sqrt{(vb)}}1'_{vb} + \boldsymbol{A}_1\right)\mathbf{y} \sim N(\mu\sqrt{(vb)}, v\sigma^2_\beta + b\sigma^2_\tau + 2\sigma^2_e). \tag{2.6.28}$$

To define $\boldsymbol{B}\mathbf{y}$ appropriately, let $\boldsymbol{C}_b = [\frac{1}{\sqrt{b}}1_b : \boldsymbol{D}'_b]'$ and $\boldsymbol{C}_v = [\frac{1}{\sqrt{v}}1_v : \boldsymbol{D}'_v]'$ be orthogonal matrices of orders $b \times b$ and $v \times v$, respectively. Then we can take

$\boldsymbol{By}$ as

$$\begin{aligned}\boldsymbol{By} &= \boldsymbol{D}_v[\frac{1}{\sqrt{b}}(y_{1.}, \cdots, y_{v.})' + \frac{1}{\sqrt{v}}(y_{.1}, \cdots, y_{.v})'] \\ &\sim N_{v-1}(0, (v\sigma_\beta^2 + b\sigma_\tau^2 + 2\sigma_e^2)\boldsymbol{I}_{v-1}),\end{aligned} \tag{2.6.29}$$

independently of $\boldsymbol{Ay}$, where $y_{.j} = \sum_{i=1}^{v} y_{ij}$ and $y_{i.} = \sum_{j=1}^{b} y_{ij}$. These linear forms fulfill (2.6.25) and (2.6.26), and the F-test based on these linear forms, namely, $F = \frac{\boldsymbol{y'A'Ay}}{\boldsymbol{y'B'By}/(v-1)}$, provides an exact test of Bartlett–Scheffé type.

Example 2.6.5. This is a continuation of Example 2.5.5. Here, a Bartlett–Scheffé type test for $H_\mu : \mu = 0$ can be derived as follows. As before, assume without any loss of generality that $v \leq b$. Let

$$\boldsymbol{A}_1 = \frac{1}{2\sqrt{n}}(1, -1, 0, \cdots, 0)_v \otimes (1, -1, 0, \cdots, 0)_b \otimes 1_n'. \tag{2.6.30}$$

Now take $\boldsymbol{Ay}$ as

$$\boldsymbol{Ay} = (\frac{1}{\sqrt{(vbn)}} 1_{vbn}' + \boldsymbol{A}_1)\mathbf{y} \sim N(\mu\sqrt{(vbn)}, bn\sigma_\tau^2 + vn\sigma_\beta^2 + 2\sigma_e^2). \tag{2.6.31}$$

Analogously, as in the previous example, we take $\boldsymbol{By}$ as

$$\begin{aligned}\boldsymbol{By} &= \boldsymbol{D}_v[\frac{1}{\sqrt{(bn)}}(y_{1..}, \cdots, y_{v..})' + \frac{1}{\sqrt{(vn)}}(y_{.1.}, \cdots, y_{.v.})'] \\ &\sim N_{v-1}(0, (bn\sigma_\tau^2 + vn\sigma_\beta^2 + 2\sigma_e^2)\boldsymbol{I}),\end{aligned} \tag{2.6.32}$$

independently of $\boldsymbol{Ay}$, where $y_{i..} = \sum_{j,k} y_{ijk}$ and $y_{.j.} = \sum_{i,k} y_{ijk}$. As before, these linear forms fulfill (2.6.25) and (2.6.26), and the F-test based on $F = \frac{\boldsymbol{y'A'Ay}}{\boldsymbol{y'B'By}/(v-1)}$ provides an exact test of Bartlett–Scheffé type.

The reader should note that the use of exact unbiased tests in some problems may involve quite extensive and complicated computations (see Seifert, 1981, for an application involving Example 2.5.7) so that Satterthwaite's approximation is indeed a viable and much simpler alternative though it is only approximate.

EXERCISES

2.1. Consider the two-way crossed classification model

$$y_{ij} = \mu + \tau_i + \beta_j + e_{ij}, \ i = 1, 2, 3, 4; \ j = 1, 2, \cdots, 6.$$

It is assumed that τ_i, β_j, and e_{ij} are independently distributed as normal variates with zero means and variances σ_τ^2, σ_β^2, and σ_e^2, respectively. Let $\gamma_1 = \sigma_\tau^2/\sigma_e^2$, $\gamma_2 = \sigma_\beta^2/\sigma_e^2$. Find the UMVUE of $\lambda = (1+6\gamma_1)/(1+4\gamma_2)$. [Hint: $\bar{y}_{..}$, $SS(\tau)$, $SS(\beta)$, and $SS(e)$ are complete and sufficient statistics for μ, σ_τ^2, σ_β^2, and σ_e^2.]

2.2. An animal scientist recorded the weights at birth of male lambs. Each lamb was the progeny (offspring) of one of several rams (male sheep) that came from five distinct population lines, and each lamb had a different dam (a female sheep). The age of the dam was recorded as belonging to one of 3 categories, numbered 1 (1–2 years), 2 (2–3 years), and 3 (over 3 years). The animal scientist used the following model:

$$\begin{aligned} y_{ijkl} &= \mu + \alpha_i + \beta_j + \delta_{j(k)} + e_{ijk(l)}, \\ & \quad i = 1,2,3;\; j = 1,2,3,4,5;\; k = 1,2,3,4;\; l = 1,2,3, \end{aligned}$$

where: y_{ijkl} denotes the weight at birth of the lth lamb that is the offspring of the kth sire in the jth population line and of a dam belonging to the ith age category, α_i denotes the ith age effect, β_j denotes the jth line effect, $\delta_{j(k)}$ denotes the kth sire (ram) effect within the jth line, $e_{ijk(l)}$ is random error.

The age and line effects are fixed, and the sire effect is random. It is assumed that $\delta_{j(k)} \sim N(0, \sigma^2_{\delta(\beta)})$ and $e_{ijk(l)} \sim N(0, \sigma_e^2)$, and all the effects are independent. Note that the interaction effects $(\alpha\beta)_{ij}$ and $(\alpha\delta)_{j(ik)}$ are assumed to be negligible.

(a) Give the corresponding population structure.

(b) Give a test statistic for testing the significance of the age effect, then write an expression for its power function corresponding to an α-level of significance. Specify all the terms in this power function.

(c) Obtain an exact $100(1-\alpha)\%$ confidence interval for $\beta_j - \beta_{j'}$, $j \neq j'$.

(d) Obtain a $100(1-\alpha)\%$ confidence interval for $\lambda = \sigma^2_{\delta(\beta)} + \sigma_e^2$. Indicate if this interval is exact or approximate.

2.3. In an investigation of the variability of the strength of tire cord, preliminary to the establishment of control of cord-testing laboratories, the data in the table below were gathered from two plants using different manufacturing processes to make nominally the same kind of cord. Six bobbins of cord were selected at random from each plant. Adjacent pairs of breaks (to give duplicate measurements as nearly as possible) were made at 500-yard intervals over the length of each bobbin (the choice of 500 yard intervals was arbitrary). The coded raw data are the measured strengths recorded in 0.1 lb deviations from 21.5 lb.

	Distance	0 yd		500 yd		1000 yd		1500 yd		2000 yd	
	Adjacent breaks	1	2	1	2	1	2	1	2	1	2
	Bobbin 1	−1	−5	−2	−8	−2	3	−3	−4	0	−1
	2	1	10	1	2	2	2	10	−4	−4	3
Plant I	3	2	−3	5	−5	1	−1	−6	1	2	5
	4	6	10	1	5	0	5	−2	−2	1	1
	5	1	−8	5	−10	1	−5	1	−4	5	−5
	6	−1	−10	−8	−8	−2	2	0	−3	−8	−1
	Bobbin 7	10	8	−5	6	2	13	7	15	17	14
	8	9	12	6	15	15	12	18	16	13	10
Plant II	9	0	8	12	6	2	0	5	4	18	8
	10	5	9	2	16	15	5	21	18	15	11
	11	−1	−1	11	19	12	10	1	20	13	9
	12	7	16	15	11	12	12	8	12	22	11

(a) Give the population structure.

(b) Write down the complete model.

(c) Give the complete ANOVA table including the expressions for the expected mean squares.

(d) Test the significance of (i) plant effect, (ii) bobbin effect, and (iii) distance effect.

(e) Obtain a 95% confidence interval for $\sigma_b^2 + \sigma_d^2 + \sigma_e^2$, where these variance components are, respectively, for bobbin, distance, and error.

(f) Obtain a 95% confidence interval on σ_{bd}/σ_e, where, σ_{bd} is the square root of the variance component for the bobbin-distance interaction.

2.4. A manufacturer wants to investigate the variation of the quality of a product with regard to type A preproduction processes and type B preproduction processes. Factor A has 4 levels and each level has 5 sublevels. Factor B has 4 levels and each level has 6 sublevels. Each sublevel of each level of the A-factor is combined with each sublevel of each level of the B-factor. The same number of replications (3) is available for each sublevel combination.

(a) Write the complete model for this experiment.

(b) Give expressions for the expected mean squares assuming that the effects of A and B are fixed, while the remaining effects are random (the sublevels are chosen at random).

2.5. In an investigation of the can-making properties of tin plate, two methods of annealing were studied. Three coils were selected at random out of a supposedly infinite population of coils made by each of these two methods. From each coil, samples were taken from two particular locations, namely the head and tail of each coil. From each sample, two sets of cans were made up independently and from each set an estimate of the can life was obtained. The data are given below:

		Annealing Method 1			Annealing Method 2		
	Coils	1	2	3	4	5	6
Location 1	*Replication 1*	288	355	329	310	303	299
	Replication 2	295	369	343	282	321	328
Location 2	*Replication 1*	278	336	320	288	302	289
	Replication 2	272	342	315	287	297	284

(a) Write down the complete model.
(b) Obtain the expected mean square values and the degrees of freedom for all the effects in the model.
(c) Describe the distribution of each sum of squares in the ANOVA table.
(d) Compute all sums of squares in the ANOVA table and provide appropriate test statistics.

2.6. Consider the model

$$\begin{aligned} y_{ijkl} &= \mu + \alpha_i + \beta_{i(j)} + \gamma_{i(k)} + \delta_{ik(l)} + (\beta\gamma)_{i(jk)} + e_{ik(jl)}, \\ & i = 1,2,3;\ j = 1,2,3;\ k = 1,2,3,4;\ l = 1,2,3, \end{aligned}$$

where α_i is fixed, $\beta_{i(j)}$, $\gamma_{i(k)}$, $\delta_{ik(l)}$, and $(\beta\gamma)_{i(jk)}$ are random. It is assumed that $\beta_{i(j)} \sim N(0, \sigma_\beta^2)$, $\gamma_{i(k)} \sim N(0, \sigma_\gamma^2)$, $\delta_{ik(l)} \sim N(0, \sigma_\delta^2)$, $(\beta\gamma)_{i(jk)} \sim N(0, \sigma_{\beta\gamma}^2)$, and $e_{ijk(l)} \sim N(0, \sigma_e^2)$. All the effects are independent.

(a) Give the population structure for this model.
(b) Write down the complete ANOVA table including the expected mean squares.
(c) Give test statistics for testing the following hypotheses:
  (i) H_0: $\alpha_i = 0$ for all i versus H_1: $\alpha_i \neq 0$ for some i.
  (ii) H_0: $\sigma_\delta^2 = 2\sigma_e^2$ versus H_1: $\sigma_\delta^2 \neq 2\sigma_e^2$.
(d) Obtain a $100(1-\alpha)\%$ confidence region for the parameter vector $\boldsymbol{\theta} = (\theta_1, \theta_2)'$, where θ_1 and θ_2 are the expected values of the mean squares for $\beta_{i(j)}$ and $(\beta\gamma)_{i(jk)}$, respectively.
(e) Let $\hat{\sigma}_\beta^2$ be the ANOVA estimator of σ_β^2. Give an expression that can be used to compute the probability $P(\hat{\sigma}_\beta^2 < 0)$ in terms of $\Delta = \sigma_\beta^2/(3\sigma_{\beta\gamma}^2 + \sigma_e^2)$.

2.7. Give a listing of all the admissible means for the population structure $\{i : [(j)(k)]\} : l$.

2.8. Consider the three-fold nested model

$$\begin{aligned} y_{ijkl} &= \mu + \alpha_i + \beta_{i(j)} + \gamma_{ij(k)} + e_{ijk(l)}, \\ & \qquad i = 1, 2, \cdots, a;\ j = 1, 2, \cdots, b;\ k = 1, 2, \cdots, c;\ l = 1, 2, \cdots, n. \end{aligned}$$

The effect α_i is fixed, whereas the other effects are random. All random effects are assumed to be statistically independent with $\beta_{i(j)}$, $\gamma_{ij(k)}$, and $e_{ijk(l)}$ being distributed as $N(0, \sigma_\beta^2)$, $N(0, \sigma_\gamma^2)$, and $N(0, \sigma_e^2)$, respectively.

(a) Set up an ANOVA table complete with degrees of freedom, sums of squares, mean squares, expected mean sqaure values, and F-values.

(b) Let $\hat{\sigma}_\beta^2$ be the ANOVA estimator of σ_β^2. Give an expression that can be used to compute the probability $\mathrm{P}(\hat{\sigma}_\beta^2 < 0)$ for given values of σ_β^2, σ_γ^2 and σ_e^2.

(c) Give an estimate of $Var(\hat{\sigma}_\beta^2)$.

(d) Obtain an exact $100(1-\alpha)\%$ confidence interval for $\lambda = \alpha_1 - \alpha_2$.

2.9. Prove Lemma 2.3.3.

2.10. Prove Lemma 2.3.4.

2.11. Let $\boldsymbol{V} = \sum_{i=0}^{\nu+1} \delta_i \boldsymbol{P}_i$ where the matrices $\boldsymbol{P}_i$'s are idempotent with $\mathrm{rank}(\boldsymbol{P}_i) = m_i$ (see (2.5.2)). Prove that $|\boldsymbol{V}| = \prod_{j=0}^{\nu+1} \delta_j^{m_j}$.

2.12. Consider Example 2.2.1 or Example 2.5.1, which deals with the one-way random effects model. Write down the likelihood function from (2.2.6) and prove directly that the familiar treatment sum of squares and error sum of squares are independent scaled chi-squared variables with scale parameters δ_i's and degrees of freedom m_i's as given in Example 2.5.1. Hence conclude that the usual F-test for $H_0 : \sigma_\tau^2 = 0$ is UMPU as well as UMPIU.

2.13. Refer to Example 2.6.4. Prove that $\boldsymbol{Ay}$ and $\boldsymbol{By}$ are independent. Also, verify (2.6.29).

2.14. Refer to Example 2.6.5. Prove that $\boldsymbol{Ay}$ and $\boldsymbol{By}$ are independent. Also, verify (2.6.32).

BIBLIOGRAPHY

Arnold, S. F. (1981). *The Theory of Linear Models and Multivariate Analysis*. Wiley, New York.

Bartlett, M. S. (1936). "The information available in small samples." *Proceedings of the Cambridge Philosophical Society*, 32, 560–566.

Das, R. and Sinha, B. K. (1987). "Robust optimum invariant unbiased tests for variance components." In: *Proceedings of the Second International Tampere Conference in Statistics* (T. Pukkila, S. Puntanen, Eds.), University of Tampere, Tampere, Finland, 317–342.

Dobrohotov, I. S. (1968). "On the theory of a general linear hypothesis with unknown weights." *Trudy Mathematical Institute Steklov*, 104, 1–19.

Gaylor, D. W. and Hopper, F. N. (1969). "Estimating the degrees of freedom for linear combinations of mean squares by Satterthwaite's formula." *Technometrics*, 11, 691–706.

Herbach, L. H. (1959). "Properties of model-II type analysis of variance tests, A: Optimum nature of the F-test for model II in the balanced case." *The Annals of Mathematical Statistics*, 30, 939–959.

Humak, K. M. S. (1984). *Statistische Methoden dwe Modellbildung III: Statistische Inferenz fur Kovarianzparameter*. Akademie-Verlag, Berlin.

Khuri, A. I. (1982). "Direct products: A powerful tool for the analysis of balanced data." *Communications in Statistics, Theory and Methods*, 11, 2903–2920.

Khuri, A. I. (1995a). "A measure to evaluate the closeness of Satterthwaite's approximation." *Biometrical Journal*, 37, 547–563.

Khuri, A. I. (1995b). "A test to detect inadequacy of Satterthwaite's approximation in balanced mixed models." *Statistics*, 27, 45–54.

Lehmann, E. L. (1986). *Testing Statistical Hypotheses*, Second Edition. Wiley, New York.

Linnik, Ju. V. (1966). "Characterization of tests of the Bartlett–Scheffé type." *Trudy Mathematical Institute Steklov*, 79, 32–40.

Mathew, T. and Sinha, B. K. (1988). "Optimum tests for fixed effects and variance components in balanced models". *Journal of the American Statistical Association*, 83, 133–135.

Mazuy, K. K. and Connor, W. S. (1965). "Student's t in a two-way classification with unequal variances." *The Annals of Mathematical Statistics*, 36, 1248–1255.

Montgomery, D. C. (1991). *Design and Analysis of Experiments*, Third Edition. Wiley, New York.

SAS User's Guide: Statistics, 1989 Edition, SAS Institute, Inc., Cary, North Carolina.

Satterthwaite, F. E. (1941). "Synthesis of variance." *Psychometrika*, 6, 309–316.

Satterthwaite, F. E. (1946). "An approximate distribution of estimates of variance components." *Biometrics Bulletin*, 2, 110–114.

Scheffé, H. (1943). "On the solutions of the Behrens–Fisher problem, based on the t distribution." *The Annals of Mathematical Statistics*, 14, 35–44.

Searle, S. R. (1971). *Linear Models*. Wiley, New York.

Seifert, B. (1978). "A note on the UMPU character of a test for the mean in balanced randomized nested classification." *Statistics*, 9, 185–189.

Seifert, B. (1979). "Optimal testing for fixed effects in general balanced mixed classification models." *Mathematics Operations: Series Statistics*, 10, 237–255.

Seifert, B. (1981). "Explicit formulae of exact tests in mixed balanced ANOVA models." *Biometrical Journal*, 23, 535–550.

Smith, D. W. and Hocking, R. R. (1978). "Maximum likelihood analysis of the mixed model: The balanced case." *Communications in Statistics, Theory and Methods*, 7, 1253–1266.

Spjøtvoll, E. (1967). "Optimum invariant tests in unbalanced variance components models." *The Annals of Mathematical Statistics*, 38, 422–428.

Zyskind, G. (1962). "On structure, relation, Σ and expectation of mean squares." *Sankhyā, Series A*, 24, 115–148.

CHAPTER 3

Measures of Data Imbalance

3.1. INTRODUCTION

We recall from Chapter 2 that statistical tests for a balanced model have interesting properties. Under the assumptions of normality, homogeneity of variances, and independence of the model's random effects, the sums of squares in the associated ANOVA table are distributed independently as scaled chi-squared variates. For the most part, these sums of squares can be used to construct exact and optimum tests.

As will be shown in later chapters, the analysis of unbalanced models can be quite involved and complicated. In the first place, the distributional properties mentioned earlier for balanced models are no longer valid. Furthermore, the partitioning of the total sum of squares can be done in a variety of ways; hence there is no unique ANOVA table as is the case with balanced models. Imbalance can therefore have a major effect on data analysis, the extent of which depends on the degree of imbalance of the data set. The main purpose of this chapter is to develop appropriate measures to quantify this degree of imbalance.

Before we embark on the development of the measures of imbalance, we shall first provide an exposition of the effects of imbalance on data analysis. This will be demonstrated with reference to the unbalanced random one-way model.

3.2. THE EFFECTS OF IMBALANCE

The pattern of imbalance of a data set can have a profound effect on the efficiency of estimators and tests of significance concerning the parameters of a given model. Let us consider, for example, the unbalanced random one-way model

$$y_{ij} = \mu + \tau_i + e_{ij}, \tag{3.2.1}$$

Table 3.1. ANOVA for Model (3.2.1)

Source	Sum of Squares	d.f.	Expected Mean Squares
A	$\sum_{i=1}^{v} \frac{y_{i.}^2}{n_i} - \frac{y_{..}^2}{n.}$	$v-1$	$\frac{1}{v-1}\left(n. - \frac{1}{n.}\sum_{i=1}^{v} n_i^2\right)\sigma_\tau^2 + \sigma_e^2$
Error	$\sum_{i=1}^{v}\sum_{j=1}^{n_i} y_{ij}^2 - \sum_{i=1}^{v} \frac{y_{i.}^2}{n_i}$	$n. - v$	σ_e^2

where μ is a fixed unknown parameter, τ_i is the effect of level i of a factor, denoted by A, and e_{ij} is a random error. We assume that τ_i and e_{ij} are independently distributed such that $\tau_i \sim N(0, \sigma_\tau^2)$ and $e_{ij} \sim N(0, \sigma_e^2)$ ($i = 1, 2, \ldots, v$; $j = 1, 2, \ldots, n_i$). The corresponding ANOVA table is shown in Table 3.1 where $y_{i.} = \sum_{j=1}^{n_i} y_{ij}$, $i = 1, 2, \ldots v$; $y_{..}$ is the grand total, and $n. = \sum_{i=1}^{v} n_i$.

It can be easily verified that $SS(\tau)$ and $SS(e)$, the sums of squares for the main effect of A and the error term, respectively, are independent and that $SS(e)/\sigma_e^2$ has a central chi-squared distribution $\chi^2_{n.-v}$ with $n. - v$ degrees of freedom. It is not true, however, that $SS(\tau)$ is distributed as a scaled chi-squared variate, as in the balanced case. The reason for this is the following: $SS(\tau)$ can be expressed as a quadratic form, namely

$$SS(\tau) = \mathbf{y}'\mathbf{Q}\mathbf{y},$$

where $\mathbf{y}$ is the vector of observations, and

$$\mathbf{Q} = \bigoplus_{i=1}^{v} \left(\frac{\mathbf{J}_{n_i}}{n_i}\right) - \frac{\mathbf{J}_{n.}}{n.}. \tag{3.2.2}$$

Here, $\mathbf{J}$ denotes a square matrix of ones, and $\bigoplus$ is the symbol of direct sum of matrices, that is, $\oplus_{i=1}^{v} (\mathbf{J}_{n_i}/n_i) = \mathbf{diag}(\mathbf{J}_{n_1}/n_1, \mathbf{J}_{n_2}/n_2, \ldots, \mathbf{J}_{n_v}/n_v)$. The variance–covariance matrix of $\mathbf{y}$ is given by

$$\mathbf{\Sigma} = \sigma_\tau^2 \bigoplus_{i=1}^{v} \mathbf{J}_{n_i} + \sigma_e^2 \mathbf{I}_{n.}.$$

Let us now consider the matrix $\mathbf{Q\Sigma}$, which can be written as

$$\mathbf{Q\Sigma} = \left[\oplus_{i=1}^{v} \mathbf{J}_{n_i} - \frac{1}{n.}\mathbf{1}_{n.}\mathbf{a}'\right]\sigma_\tau^2 + \left[\oplus_{i=1}^{v}\left(\frac{\mathbf{J}_{n_i}}{n_i}\right) - \frac{\mathbf{J}_{n.}}{n.}\right]\sigma_e^2, \tag{3.2.3}$$

where $\mathbf{a}' = [n_1\mathbf{1}'_{n_1} : n_2\mathbf{1}'_{n_2} : \ldots : n_v\mathbf{1}'_{n_v}]$ and $\mathbf{1}$ denotes a vector of ones. It can be verified that $\mathbf{Q\Sigma}$ is not, in general, a multiple of an idempotent matrix.

Hence, $\mathbf{y}'\mathbf{Q}\mathbf{y}$ is not distributed as a scalar multiple of a chi-squared variate (see, for example, Searle, 1971, Chapter 2, Theorem 2, p. 57). However, $\mathbf{y}'\mathbf{Q}\mathbf{y}$ can be expressed as a linear combination of independent central chi-squared variates of the form

$$\mathbf{y}'\mathbf{Q}\mathbf{y} = \sum_{i=1}^{s} \lambda_i \chi^2_{m_i}, \tag{3.2.4}$$

where $\lambda_1, \lambda_2, \ldots, \lambda_s$ are the distinct nonzero eigenvalues of $\mathbf{Q}\boldsymbol{\Sigma}$ with multiplicities $m_1, m_2, \ldots, m_s$, respectively (see Johnson and Kotz, 1970, formula 23, p. 153). Note that $\lambda_i > 0$ for $i = 1, 2, \ldots, s$ since $\mathbf{Q}$ is positive semidefinite.

From formula (3.2.3) we note that if $\sigma^2_\tau = 0$, then $\mathbf{Q}\boldsymbol{\Sigma}/\sigma^2_e$ is an idempotent matrix of rank $v - 1$. In this case, $SS(\tau)/\sigma^2_e$ has a central chi-squared distribution with $v - 1$ degrees of freedom. It follows that the ratio $F = MS(\tau)/MS(e)$ provides a test statistic for testing the hypothesis $H_\tau : \sigma^2_\tau = 0$, where $MS(\tau)$ and $MS(e)$ are the mean squares for the main effect of A and the error term, respectively. Under H_τ, this ratio has an F-distribution with $v - 1$ and $n. - v$ degrees of freedom. Large values of F are significant.

In particular, if the data set is balanced, that is, $n_i = n$ for $i = 1, 2, \ldots, v$, then $\mathbf{Q}\boldsymbol{\Sigma}$ becomes

$$\begin{aligned}\mathbf{Q}\boldsymbol{\Sigma} &= \left[\mathbf{I}_v \otimes \mathbf{J}_n - \frac{1}{v}\mathbf{J}_v \otimes \mathbf{J}_n\right]\sigma^2_\tau + \left[\frac{1}{n}(\mathbf{I}_v \otimes \mathbf{J}_n) - \frac{1}{vn}(\mathbf{J}_v \otimes \mathbf{J}_n)\right]\sigma^2_e \\ &= \left[\frac{1}{n}(\mathbf{I}_v \otimes \mathbf{J}_n) - \frac{1}{vn}(\mathbf{J}_v \otimes \mathbf{J}_n)\right](n\sigma^2_\tau + \sigma^2_e),\end{aligned}$$

where $\otimes$ is the symbol of direct product of matrices. In this case, $\mathbf{Q}\boldsymbol{\Sigma}/(n\sigma^2_\tau + \sigma^2_e)$ is idempotent of rank $v - 1$, which implies that $\mathbf{y}'\mathbf{Q}\mathbf{y}/(n\sigma^2_\tau + \sigma^2_e)$ has a central chi-squared distribution with $v - 1$ degrees of freedom.

For a general unbalanced random one-way model, the ANOVA estimators of σ^2_τ and σ^2_e are given by

$$\begin{aligned}\hat{\sigma}^2_\tau &= \frac{1}{d}[MS(\tau) - MS(e)] \\ \hat{\sigma}^2_e &= MS(e),\end{aligned} \tag{3.2.5}$$

where

$$d = \frac{1}{v-1}\left(n. - \frac{1}{n.}\sum_{i=1}^{v} n_i^2\right).$$

We shall now investigate the effect of imbalance on the variance of $\hat{\sigma}^2_\tau$, the probability of a negative $\hat{\sigma}^2_\tau$, and the power of the test concerning the hypothesis $H_\tau : \sigma^2_\tau = 0$.

3.2.1. The Variance of $\hat{\sigma}_\tau^2$

Since $MS(\tau)$ and $MS(e)$ are independent and $SS(e)/\sigma_e^2 \sim \chi^2_{n.-v}$, the variance of $\hat{\sigma}_\tau^2$ in formula (3.2.5) is given by

$$Var(\hat{\sigma}_\tau^2) = \frac{1}{d^2}\left\{Var[MS(\tau)] + 2\left(\frac{\sigma_e^2}{n.-v}\right)^2(n.-v)\right\}. \tag{3.2.6}$$

But,

$$\begin{aligned} Var[MS(\tau)] &= \frac{1}{(v-1)^2}\left[2tr(\mathbf{Q\Sigma})^2] + 4E(\mathbf{y})'\mathbf{Q\Sigma Q}E(\mathbf{y})\right] \\ &= \frac{2}{(v-1)^2}tr[(\mathbf{Q\Sigma})^2]. \end{aligned} \tag{3.2.7}$$

This follows from applying, for example, Corollary 1.2 in Searle (1971, p. 57) and the fact that $E(\mathbf{y}) = \mu\mathbf{1}_{n.}$ and $\mathbf{Q1}_{n.} = \mathbf{0}$. Now, from formula (3.2.3) we have

$$tr[(\mathbf{Q\Sigma})^2] = \sigma_\tau^4 tr(\mathbf{M}_1^2) + 2\sigma_\tau^2\sigma_e^2 tr(\mathbf{M}_1\mathbf{M}_2) + \sigma_e^4 tr(\mathbf{M}_2^2), \tag{3.2.8}$$

where

$$\mathbf{M}_1 = \bigoplus_{i=1}^{v}\mathbf{J}_{n_i} - \frac{1}{n.}\mathbf{1}_{n.}\mathbf{a}'$$

$$\mathbf{M}_2 = \bigoplus_{i=1}^{v}\left(\frac{\mathbf{J}_{n_i}}{n_i}\right) - \frac{\mathbf{J}_{n.}}{n.}.$$

From formulas (3.2.6), (3.2.7), and (3.2.8) we obtain

$$Var(\hat{\sigma}_\tau^2) = \frac{2(n.^2g + g^2 - 2n.h)}{(n.^2-g)^2}\sigma_\tau^4 + \frac{4n.}{n.^2-g}\sigma_\tau^2\sigma_e^2 + \frac{2n.^2(n.-1)(v-1)}{(n.-v)(n.^2-g)^2}\sigma_e^4, \tag{3.2.9}$$

where $g = \sum_{i=1}^{v} n_i^2$, $h = \sum_{i=1}^{v} n_i^3$. In particular, if the data set is balanced with $n_i = n$ for $i = 1, 2, \ldots, v$, then (3.2.9) becomes

$$Var(\hat{\sigma}_\tau^2) = \frac{2}{n^2}\left[\frac{(n\sigma_\tau^2 + \sigma_e^2)^2}{v-1} + \frac{\sigma_e^4}{v(n-1)}\right]. \tag{3.2.10}$$

Theorem 3.2.1. For a fixed value of $n.$, $Var(\hat{\sigma}_\tau^2)$ attains a minimum for all $\sigma_\tau^2, \sigma_e^2$ if and only if the data set is balanced.

Proof. From formula (3.2.3) we have that

$$tr(\mathbf{Q\Sigma}) = \left(n. - \frac{g}{n.}\right)\sigma_\tau^2 + (v-1)\sigma_e^2.$$

We also have that

$$d = \frac{1}{v-1}\left(n. - \frac{g}{n.}\right).$$

Thus

$$\begin{aligned} d\sigma_\tau^2 &= \frac{1}{v-1}[tr(\mathbf{Q\Sigma}) - (v-1)\sigma_e^2] \\ &= \frac{tr(\mathbf{Q\Sigma})}{v-1} - \sigma_e^2. \end{aligned}$$

From formulas (3.2.6) and (3.2.7) we can then write

$$Var(\hat{\sigma}_\tau^2) = \frac{2tr[(\mathbf{Q\Sigma})^2]}{d^2(v-1)^2} + \frac{2\sigma_e^4}{d^2(n. - v)}.$$

For a fixed $n.$, $g = \sum_{i=1}^{v} n_i^2$ attains its minimum when $n_i = n$ (or as nearly so as every n_i being an integer will allow) for all $i = 1, 2, \ldots, v$ (this can be easily shown by the method of Lagrange multipliers). Hence, d is at its maximum, namely $d = n$, if and only if $n_i = n$ for all i. The second term in the formula for $Var(\hat{\sigma}_\tau^2)$ has therefore a minimum for all σ_e^2. Now,

$$\begin{aligned} \frac{2tr[(\mathbf{Q\Sigma})^2]}{d^2(v-1)^2} &= \frac{2\sigma_\tau^4 tr[(\mathbf{Q\Sigma})^2]}{[tr(\mathbf{Q\Sigma}) - (v-1)\sigma_e^2]^2} \\ &= \frac{2\sigma_\tau^4 tr[(\mathbf{Q\Sigma})^2]}{[tr(\mathbf{Q\Sigma})]^2} \frac{1}{\left[1 - \frac{(v-1)\sigma_e^2}{tr(\mathbf{Q\Sigma})}\right]^2} \\ &= \frac{2\sigma_\tau^4 tr[(\mathbf{Q\Sigma})^2]}{[tr(\mathbf{Q\Sigma})]^2} \frac{1}{\left[1 - \frac{\sigma_e^2}{d\sigma_\tau^2 + \sigma_e^2}\right]^2} \\ &= \frac{2\sigma_\tau^4 tr[(\mathbf{Q\Sigma})^2]}{[tr(\mathbf{Q\Sigma})]^2} \left(1 + \frac{\sigma_e^2}{d\sigma_\tau^2}\right)^2. \end{aligned}$$

The term $1 + \sigma_e^2/d\sigma_\tau^2$ is at its minimum, namely $1 + \sigma_e^2/n\sigma_\tau^2$, if and only if $n_i = n$ for all i and all σ_τ^2 and σ_e^2. Furthermore, by making use of Theorem 9.1.22 in Graybill (1983, p. 303), it is easy to show that the minimum value of $tr[(\mathbf{Q\Sigma})^2]/[tr(\mathbf{Q\Sigma})]^2$ is equal to the reciprocal of the rank of $\mathbf{Q}$, which is equal to $v - 1$. This minimum is achieved if and only if $n_i = n$ for all i and all σ_τ^2 and τ_e^2. We therefore conclude that $Var(\hat{\sigma}_\tau^2)$ is at its minimum, which is given by formula (3.2.10), if and only if the data set is balanced. This completes the proof of Theorem 3.2.1. □

It should be noted that for a fixed $n.$ and a given v, allocating an equal number, n, of observations to each level of factor A may not be feasible unless

$n.$ is an integer multiple of v. Anderson and Crump (1967, p. 502) showed that for fixed $n.$ and v, the optimal plan is to allocate $t+1$ observations to each of ℓ levels, and t observations to each of $v-\ell$ levels, where t and ℓ are such that

$$n. = vt + \ell, \qquad 0 \le \ell < v.$$

This assignment is as close as possible to equal numbers per level.

Theorem 3.2.1 implies that imbalance causes an increase in the variance of $\hat{\sigma}_\tau^2$ provided that the size of the data remains the same. This result was noted by Singh (1989) who arrived at the same conclusion on the basis of a numerical study in which certain a priori values of σ_τ^2 and σ_e^2, and several selected values of n_i were used (see also Singh, 1992; Caro et al., 1985).

3.2.2. The Probability of a Negative $\hat{\sigma}_\tau^2$

The estimator $\hat{\sigma}_\tau^2$ can be negative. This is an undesirable feature of ANOVA estimation in general since the true value of σ_τ^2 is nonnegative. From formula (3.2.5), the probability of a negative $\hat{\sigma}_\tau^2$ is

$$P[\hat{\sigma}_\tau^2 < 0] = P[MS(\tau) < MS(e)]. \tag{3.2.11}$$

Using (3.2.4) and the fact that $\frac{SS(e)}{\sigma_e^2} \sim \chi^2_{n.-v}$, formula (3.2.11) can be written as

$$P(\hat{\sigma}_\tau^2 < 0) = P\left(\frac{1}{v-1}\sum_{i=1}^{s} \lambda_i \chi^2_{m_i} < \frac{\sigma_e^2}{n.-v}\chi^2_{n.-v}\right). \tag{3.2.12}$$

Since the λ_i's are positive, $\sum_{i=1}^{s} \lambda_i \chi^2_{m_i}$ can be approximately represented as $\lambda\chi^2_m$ according to Satterthwaite's (1941) procedure, where

$$\lambda = \frac{\sum_{i=1}^{s} m_i \lambda_i^2}{\sum_{i=1}^{s} m_i \lambda_i} \tag{3.2.13}$$

$$m = \frac{\left(\sum_{i=1}^{s} m_i \lambda_i\right)^2}{\sum_{i=1}^{s} m_i \lambda_i^2}. \tag{3.2.14}$$

Formulas (3.2.13) and (3.2.14) were obtained by equating the first two moments of $\sum_{i=1}^{s} \lambda_i \chi^2_{m_i}$ to those of $\lambda\chi^2_m$ (see also Section 2.6 of Chapter 2). Using this approximation in (3.2.12), we obtain

$$\begin{aligned} P(\hat{\sigma}_\tau^2 < 0) &\approx P\left[\frac{\lambda}{v-1}\chi^2_m < \frac{\sigma_e^2}{n.-v}\chi^2_{n.-v}\right] \\ &= P\left[\frac{\chi^2_m}{m} < \frac{(v-1)\sigma_e^2}{m\lambda}\,\frac{\chi^2_{n.-v}}{n.-v}\right] \\ &= P\left[F_{m,n.-v} < \frac{(v-1)\sigma_e^2}{m\lambda}\right]. \end{aligned} \tag{3.2.15}$$

On the basis of results given in Khuri (1995), this approximation is adequate if and only if the value of Δ is close to one, where

$$\Delta = \frac{\left(\sum_{i=1}^{s} m_i\right)\left(\sum_{i=1}^{s} m_i\lambda_i^2\right)}{\left(\sum_{i=1}^{s} m_i\lambda_i\right)^2}. \tag{3.2.16}$$

Note that, in general, $1 \leq \Delta \leq \sum_{i=1}^{s} m_i$. If $\Delta = 1$, then the λ_i's must be equal and Satterthwaite's approximation becomes exact, that is, $\sum_{i=1}^{s} \lambda_i \chi^2_{m_i} = \lambda\chi^2_m$. Vice versa, if the λ_i's are equal, then $\Delta = 1$. In particular, for a balanced data set with $n_i = n(i = 1, 2, \ldots, v)$, $\lambda_i = n\sigma_\tau^2 + \sigma_e^2$ for $i = 1, 2, \ldots, s$ and hence Satterthwaite's approximation is exact. In this case, $m = \sum_{i=1}^{s} m_i = v - 1$ and

$$P(\hat{\sigma}_\tau^2 < 0) = P\left[F_{v-1, vn-v} < \frac{\sigma_e^2}{n\sigma_\tau^2 + \sigma_e^2}\right].$$

From formulas (3.2.14) and (3.2.16) we note that

$$m = \frac{\sum_{i=1}^{s} m_i}{\Delta}.$$

Since the m_i's are the multiplicities of the nonzero eigenvalues, $\lambda_1, \lambda_2, \ldots, \lambda_s$, of $\mathbf{Q\Sigma}$ (see formula (3.2.3)), $\sum_{i=1}^{s} m_i = \text{rank}(\mathbf{Q\Sigma}) = \text{rank}(\mathbf{Q})$. But, from formula (3.2.2), $\text{rank}(\mathbf{Q}) = tr(\mathbf{Q}) = v - 1$ (since $\mathbf{Q}$ is idempotent). It follows that

$$m = \frac{v-1}{\Delta}. \tag{3.2.17}$$

Furthermore, we have that (see formulas (3.2.13) and (3.2.14))

$$\begin{aligned} m\lambda &= \sum_{i=1}^{s} m_i\lambda_i \\ &= tr(\mathbf{Q\Sigma}) \\ &= \left(n_. - \frac{1}{n_.}\sum_{i=1}^{v} n_i^2\right)\sigma_\tau^2 + (v-1)\sigma_e^2. \end{aligned} \tag{3.2.18}$$

The last equality results from taking the trace of $\mathbf{Q\Sigma}$ in formula (3.2.3). Since $\Delta > 1$ whenever the λ_i's are unequal, which occurs when the data set is unbalanced, we conclude from formula (3.2.17) that imbalance causes a reduction in the value of m, which is the numerator's number of degrees of freedom of the F-variate in (3.2.15). Also, since

$$n_. - \frac{1}{n_.}\sum_{i=1}^{v} n_i^2 = \frac{1}{n_.}\left[n_.^2 - \frac{n_.^2}{v} - \sum_{i=1}^{v}(n_i - \overline{n_.})^2\right],$$

where $\overline{n_.} = n_./v$, we conclude that imbalance causes a reduction in the value of $n_. - (1/n_.)\sum_{i=1}^{v} n_i^2$ and hence in the value of $m\lambda$, as can be seen from (3.2.18), for a fixed $n_.$.

Another approximate value of $P(\hat{\sigma}_\tau^2 < 0)$ can be obtained by using Hirotsu's (1979, pp. 578–579) approximation (see Appendix 3.1). By applying this approximation to formula (3.2.11), we get

$$\begin{aligned}
P(\hat{\sigma}_\tau^2 < 0) &= P\left[\frac{MS(\tau)}{MS(e)} < 1\right] \\
&= P\left[\frac{SS(\tau)/\xi f}{MS(e)/\sigma_e^2} < \frac{\sigma_e^2(v-1)}{\xi f}\right] \\
&\approx P[F_{f,n.-v} < q] - \left[\delta \Big/ \left\{3(f+2)(f+4)B\left(\frac{f}{2}, \frac{n.-v}{2}\right)\right\}\right] \\
&\quad \times \left[1 + \frac{fq}{n.-v}\right]^{-(f+n.-v)/2} \left[\frac{fq}{n.-v}\right]^{f/2} \\
&\quad \times \left[(f+2)(f+4) - \frac{2(f+n.-v)(f+4)}{1+\frac{n.-v}{fq}}\right. \\
&\quad \left. + \frac{(f+n.-v+2)(f+n.-v)}{\{1+\frac{n.-v}{fq}\}^2}\right],
\end{aligned}$$

where $B(\cdot,\cdot)$ denotes the beta function, and

$$\begin{aligned}
\xi &= \frac{\kappa_2[SS(\tau)]}{2\kappa_1[SS(\tau)]} \\
&= \frac{\sum_{i=1}^{s} m_i\lambda_i^2}{\sum_{i=1}^{s} m_i\lambda_i} = \lambda, \text{ (see formula (3.2.13))} \\
f &= \frac{2\kappa_1^2[SS(\tau)]}{\kappa_2[SS(\tau)]} \\
&= \frac{\left(\sum_{i=1}^{s} m_i\lambda_i\right)^2}{\sum_{i=1}^{s} m_i\lambda_i^2} = m, \text{ (see formula (3.2.14))} \\
q &= \frac{\sigma_e^2(v-1)}{\xi f}, \text{ and} \\
\delta &= \frac{1}{2}\left\{\frac{\kappa_1[SS(\tau)]\kappa_3[SS(\tau)]}{\kappa_2^2[SS(\tau)]}\right\} - 1 \\
&= \frac{\left(\sum_{i=1}^{s} m_i\lambda_i\right)\left(\sum_{i=1}^{s} m_i\lambda_i^3\right)}{\left(\sum_{i=1}^{s} m_i\lambda_i^2\right)^2} - 1,
\end{aligned}$$

where $\kappa_i[SS(\tau)]$ denotes the ith cumulant of $SS(\tau) = \sum_{j=1}^{s} \lambda_j \chi_{m_j}^2 (i = 1, 2, 3)$.

3.2.2.1. Exact Value of $P(\hat{\sigma}_\tau^2 < 0)$. We have seen that $\hat{\sigma}_\tau^2$ is a linear combination of independent central chi-squared variates of the form (see formula (3.2.5))

$$\hat{\sigma}_\tau^2 = \frac{1}{d}\left[\frac{1}{v-1}\sum_{i=1}^{s}\lambda_i\chi_{m_i}^2 - \frac{\sigma_e^2}{n_{.}-v}\chi_{n_{.}-v}^2\right]. \tag{3.2.19}$$

The exact distribution of $\hat{\sigma}_\tau^2$ can be obtained by a numerical inversion of the characteristic function of the right-hand side of (3.2.19), using Imhof's (1961) method. By applying formula (3.2) in Imhof (1961), the exact probability of a negative $\hat{\sigma}_\tau^2$ is given by

$$P(\hat{\sigma}_\tau^2 < 0) = \frac{1}{2} - \frac{1}{\pi}\int_0^\infty \frac{\sin[\theta(u)]}{u\rho(u)}du, \tag{3.2.20}$$

where

$$\begin{aligned}
\theta(u) &= \frac{1}{2}\sum_{i=1}^{s} m_i \arctan\left[\frac{\lambda_i u}{d(v-1)}\right] - \frac{(n_{.}-v)}{2}\arctan\left[\frac{\sigma_e^2 u}{d(n_{.}-v)}\right] \\
\rho(u) &= \left[1 + \frac{\sigma_e^4 u^2}{d^2(n_{.}-v)^2}\right]^{(n_{.}-v)/4}\prod_{i=1}^{s}\left[1 + \frac{\lambda_i^2 u^2}{d^2(v-1)^2}\right]^{m_i/4}.
\end{aligned}$$

Formula (3.2.20) has the advantage of showing the effects of $\lambda_1, \lambda_2, \ldots, \lambda_s$ and $m_1, m_2, \ldots, m_s$ on the probability value in (3.2.20). We recall that the λ_i's and the m_i's are, respectively, the nonzero eigenvalues and corresponding multiplicities of the matrix $\mathbf{Q\Sigma}$ in (3.2.3).

Alternatively, the exact value of $P(\hat{\sigma}_\tau^2 < 0)$ can be obtained by using a computer algorithm given by Davies (1980), which is based on a method proposed by Davies (1973). This algorithm is described as Algorithm AS155 in the journal *Applied Statistics*, and can be easily accessed through STATLIB, which is an e-mail and FTP-based retrieval system for statistical software. A description of how to retrieve software via STATLIB is given in Newton (1993).

An expression of the exact value of $P(\hat{\sigma}_\tau^2 < 0)$ was given by Singh (1989) using an infinite weighted sum of incomplete beta functions. Singh utilized this expression to numerically investigate the effect of imbalance on this probability value. He concluded that imbalance increases the probability of a negative value of $\hat{\sigma}_\tau^2$.

3.2.3. Power of the Test Concerning σ_τ^2

We recall from Table 3.1 that a test statistic for testing the hypothesis H_τ : $\sigma_\tau^2 = 0$ is given by $F = MS(\tau)/MS(e)$, which, under H_τ, has the F-distribution with $v-1$ and $n_{.}-v$ degrees of freedom. This test is uniformly most powerful invariant if the data set is balanced (see Chapter 2). The same is not true, however, in the unbalanced case. As was seen earlier, $SS(\tau)$ is not, in general, expressible as a scaled central chi-squared variate if $\sigma_\tau^2 > 0$. Hence, F does not have the F-distribution when $\sigma_\tau^2 > 0$. Donner and Koval (1989) investigated the effect of imbalance on this F-test by comparing its power to that of the corresponding likelihood ratio test over varying degrees of imbalance. They concluded that the likelihood ratio test was "likely to be better for very unbalanced designs with a large amount of data." We should mention that a locally best invariant (LBI) test of H_τ was derived in Das and Sinha (1987) (see Section 4.2 of Chapter 4).

3.3. MEASURES OF IMBALANCE FOR THE ONE-WAY MODEL

Whenever a data set is unbalanced, a question usually arises as to how unbalanced it is. In order to answer such a question, a certain measure is needed for determining the degree of departure from balance. If this can be accomplished, then it becomes possible to quantify the degree of imbalance instead of using subjective descriptions such as nearly balanced, moderately unbalanced, or highly unbalanced. Moreover, such a measure can be used to compare designs on the basis of their degrees of imbalance. It can also be used to study the effect of imbalance on the properties of tests and parameter estimators associated with a given model, as will be seen in Section 3.6.

In this section, we describe two measures of imbalance, originally introduced by Ahrens and Pincus (1981) for the special case of the one-way model. A general procedure for measuring imbalance will be developed in Section 3.4 (see also Ahrens and Sanchez, 1982, 1988, and Lera, 1994).

Consider again the one-way model given by formula (3.2.1). Let $\mathbf{D} = \{n_1, n_2, \ldots, n_v\}$ denote the associated design. We describe $\mathbf{D}$ as being balanced if the n_i's are equal, otherwise, $\mathbf{D}$ is said to be unbalanced. Ahrens and Pincus (1981) provided general guidelines for the construction of a measure of imbalance for $\mathbf{D}$, or equivalently, model (3.2.1): Such a measure should be

(i) A symmetric function of the n_i's.

(ii) Invariant under k-fold replications of the design, that is, $\mathbf{1}_k' \otimes \mathbf{D}$, as well as under the case in which each of the n_i's is multiplied by a constant k, that is, $k\mathbf{D} = \{kn_1, kn_2, \ldots, kn_v\}$.

Ahrens and Pincus introduced two measures of imbalance, which they denoted by $\gamma(\mathbf{D})$ and $\nu(\mathbf{D})$, that satisfy the above requirements, where

$$\gamma(\mathbf{D}) = \frac{v}{\bar{n}.\sum_{i=1}^{v} \frac{1}{n_i}}, \qquad (3.3.1)$$

$\overline{n.} = n./v$,

$$\nu(\mathbf{D}) = \frac{1}{v \sum_{i=1}^{v} \left(\frac{n_i}{n.}\right)^2}. \tag{3.3.2}$$

It can be verified that

$$\nu(\mathbf{D}) = \frac{1}{1 + (cv)^2}, \tag{3.3.3}$$

where cv is the coefficient of variation of the n_i's, that is,

$$(cv)^2 = \frac{\sum_{i=1}^{v} (n_i - \overline{n.})^2}{v\overline{n.}^2}.$$

Hence, $\nu(\mathbf{D}) \leq 1$. Furthermore, $\nu(\mathbf{D}) > 1/v$. This follows from the fact that $\sum_{i=1}^{v} n_i^2 < n.^2$ for $v > 1$. Note that the first measure, $\gamma(\mathbf{D})$, is the ratio of the harmonic mean of the n_i's to the arithmetic mean $\overline{n.}$. This implies that $0 < \gamma(\mathbf{D}) \leq 1$. Both measures attain the value one if and only if the design $\mathbf{D}$ is balanced. The closer $\gamma(\mathbf{D})$ and $\nu(\mathbf{D})$ are to their corresponding lower bounds, namely 0 and $1/v$, respectively, the more unbalanced $\mathbf{D}$ is.

The measure $\nu(\mathbf{D})$ can be used to evaluate the effect of imbalance on the variance of the ANOVA estimator $\hat{\sigma}_\tau^2$ (see formula (3.2.5)), and on the power of the test concerning σ_τ^2. This will be shown in the next section.

3.3.1. The Effect of Imbalance on $Var(\hat{\sigma}_\tau^2)$

Consider the random one-way model given by formula (3.2.1) and the ANOVA estimator $\hat{\sigma}_\tau^2$ in formula (3.2.5). The variance of $\hat{\sigma}_\tau^2$ is described in formula (3.2.9). Let $\mathbf{N}$ be a $v \times v$ matrix whose $(i,j)^{th}$ element is $(1/n.)\sqrt{n_i n_j}$ $(i,j = 1, 2, \ldots, v)$. Let $\mathbf{T}$ be the matrix

$$\mathbf{T} = (\mathbf{I}_v - \mathbf{N})\mathbf{diag}(n_1, n_2, \cdots, n_v)(\mathbf{I}_v - \mathbf{N}).$$

Formula (3.2.9) can be written as (see Ahrens and Pincus, 1981, p. 230)

$$Var(\hat{\sigma}_\tau^2) = 2\left[\sigma_\tau^4 tr(\mathbf{T}^2) + 2\sigma_\tau^2\sigma_e^2 tr(\mathbf{T}) + \frac{(v-1)(n.-1)}{n.-v}\sigma_e^4\right] / [tr(\mathbf{T})]^2. \tag{3.3.4}$$

It can be verified that $\mathbf{N}$ is idempotent of rank 1. Hence, $\mathbf{I}_v - \mathbf{N}$ is idempotent of rank $v - 1$, which is also the rank of $\mathbf{T}$. Let $r_1, r_2, \ldots, r_{v-1}$ be the nonzero eigenvalues of the latter matrix. These eigenvalues are positive since $\mathbf{T}$ is positive semidefinite. Then, $tr(\mathbf{T}) = \sum_{i=1}^{v-1} r_i$, $tr(\mathbf{T}^2) = \sum_{i=1}^{v-1} r_i^2$, and formula (3.3.4) can be written as

$$Var(\hat{\sigma}_\tau^2) = 2\left[\sigma_\tau^4 \frac{\sum_{i=1}^{v-1} r_i^2}{\left(\sum_{i=1}^{v-1} r_i\right)^2} + 2\sigma_\tau^2\sigma_e^2 \frac{1}{\sum_{i=1}^{v-1} r_i} + \frac{(v-1)(n.-1)}{(n.-v)\left(\sum_{i=1}^{v-1} r_i\right)^2}\sigma_e^4\right]. \tag{3.3.5}$$

Now,

$$\begin{aligned}\sum_{i=1}^{v-1} r_i &= tr[(\mathbf{I}_v - \mathbf{N})\mathbf{diag}(n_1, n_2, \cdots, n_v)] \\ &= n.\left[1 - \sum_{i=1}^{v}\left(\frac{n_i}{n.}\right)^2\right] \\ &= n.\left[1 - \frac{1}{v\nu(\mathbf{D})}\right], \text{ by formula (3.3.2).}\end{aligned}$$

We note that for a fixed $n.$, the coefficients of $\sigma_\tau^2\sigma_e^2$ and σ_e^4 in formula (3.3.5) increase as $\nu(\mathbf{D})$ decreases, that is, as the degree of imbalance increases. On the other hand, $\sum_{i=1}^{v-1} r_i$ attains its maximum value of $n.(1 - 1/v)$ if and only if the design is balanced. Also, since $(v-1)\sum_{i=1}^{v-1} r_i^2 \geq \left(\sum_{i=1}^{v-1} r_i\right)^2$ by the Cauchy–Schwarz inequality, the first term in (3.3.5) attains its minimum value of $\sigma_\tau^4/(v-1)$ if and only if the r_i's are equal, that is, if and only if the design is balanced. Hence, as was seen in Section 3.2.1, the minimum value of $Var(\hat{\sigma}_\tau^2)$ is achieved if and only if the design is balanced. The value of this minimum is equal to

$$\min[Var(\hat{\sigma}_\tau^2)] = 2\left\{\frac{\sigma_\tau^4}{v-1} + \frac{2\sigma_\tau^2\sigma_e^2}{n.(1-\frac{1}{v})} + \frac{(v-1)(n.-1)\sigma_e^4}{(n.-v)[n.(1-\frac{1}{v})]^2}\right\},$$

which is equivalent to the expression in (3.2.10). In general, since $\sum_{i=1}^{v-1} r_i^2 \leq \left(\sum_{i=1}^{v-1} r_i\right)^2$, then from (3.3.5), an upper bound on $Var(\hat{\sigma}_\tau^2)$ is given by

$$Var(\hat{\sigma}_\tau^2) \leq 2\left\{\sigma_\tau^4 + \frac{2\sigma_\tau^2\sigma_e^2}{n.[1-\frac{1}{v\nu(\mathbf{D})}]} + \frac{(v-1)(n.-1)\sigma_e^4}{(n.-v)n.^2[1-\frac{1}{v\nu(\mathbf{D})}]^2}\right\}.$$

This upper bound decreases as $\nu(\mathbf{D})$ increases, that is, as imbalance decreases, for all values of the variance components.

We note from formula (3.3.5) that it is not possible to express $Var(\hat{\sigma}_\tau^2)$ explicitly as a function of only $\nu(\mathbf{D})$ for any given values of $n.$, v, σ_τ^2, and σ_e^2. It is therefore very difficult to study the effects of different patterns of imbalance. This was pointed out by several authors. See, for example, Ahrens and Sanchez (1992), Searle et al. (1992, p. 75 and pp. 222–224), and Singh (1992). In Section 3.6 we show how to resolve this problem by establishing an approximate empirical relationship between $Var(\hat{\sigma}_\tau^2)$ and $\nu(\mathbf{D})$.

3.3.2. The Effect of Imbalance on the Test Concerning σ_τ^2

We recall that a test statistic for testing the null hypothesis $H_\tau : \sigma_\tau^2 = 0$ is given by $F = MS(\tau)/MS(e)$, which, under H_τ, has the F-distribution with

$v-1$ and $n.-v$ degrees of freedom. As was seen earlier in Section 3.2.3, F does not have the F-distribution when $\sigma_\tau^2 > 0$. In this case, $SS(\tau)$ can be expressed as a linear combination $\sum_{i=1}^{s} \lambda_i \chi^2_{m_i}$ of independent central chi-squared variates with positive coefficients (see formula 3.2.4). Using Satterthwaite's (1941) approximation, $SS(\tau)$ can be approximated as a scaled chi-squared variate, $\lambda\chi^2_m$, where λ and m are given in formulas (3.2.13) and (3.2.14), respectively. Thus $MS(\tau)/MS(e)$ is distributed approximately as $\omega F_{m,n.-v}$, where

$$\begin{aligned}
\omega &= \frac{E[MS(\tau)]}{\sigma_e^2} \\
&= \frac{\left(n. - \frac{1}{n.}\sum_{i=1}^{v} n_i^2\right)\sigma_\tau^2 + (v-1)\sigma_e^2}{(v-1)\sigma_e^2} \\
&= 1 + \frac{n.}{v-1}\left[1 - \frac{1}{v\nu(\mathbf{D})}\right]\frac{\sigma_\tau^2}{\sigma_e^2}, \text{ by (3.3.2).}
\end{aligned}$$

Thus the power of the test is approximately given by $P(\omega F_{m,n.-v} > F_{\alpha,v-1,n.-v})$. We note that ω attains its maximum value, namely $1 + n.\sigma_\tau^2/(v\sigma_e^2)$, when $\nu(\mathbf{D}) = 1$, that is, when the data set is balanced. Furthermore, from formula (3.2.14),

$$\begin{aligned}
m &= \frac{\left(\sum_{i=1}^{s} m_i\lambda_i\right)^2}{\sum_{i=1}^{s} m_i\lambda_i^2} \\
&\leq \sum_{i=1}^{s} m_i,
\end{aligned}$$

since

$$\begin{aligned}
\left(\sum_{i=1}^{s} m_i\lambda_i\right)^2 &= \left(\sum_{i=1}^{s} m_i^{1/2} m_i^{1/2}\lambda_i\right)^2 \\
&\leq \left(\sum_{i=1}^{s} m_i\right)\left(\sum_{i=1}^{s} m_i\lambda_i^2\right)
\end{aligned}$$

by the Cauchy–Schwarz inequality. But, $\sum_{i=1}^{s} m_i$ is the rank of the matrix $\mathbf{Q}$ in formula (3.2.2), that is, $\sum_{i=1}^{s} m_i = v-1$. Hence, $m \leq v-1$. Equality is achieved when the λ_i's are equal, or equivalently, when the data set is balanced.

As was mentioned earlier in Section 3.2.3, another test concerning H_τ is the likelihood ratio (LR) test, which involves determining the value of the likelihood function for the complete model and its value under the condition of H_τ. Donner and Koval (1989) studied the effect of varying $\nu(\mathbf{D})$ on the power of the F-test and compared it against the power of the LR-test. They concluded that the F-test is consistently more powerful than the LR-test at all values of v if $\nu(\mathbf{D}) \geq 0.90$, that is, if the design is only mildly unbalanced. However, for extremely unbalanced designs, the LR-test can be appreciably more powerful than the F-test. Finally, we mention that the effect of varying $\nu(\mathbf{D})$ on the power of the LBI test of Das and Sinha (1987) has not been studied in the statistical literature.

3.4. A GENERAL PROCEDURE FOR MEASURING IMBALANCE

In this section we present a general procedure, introduced by Khuri (1987), for measuring imbalance, which can be applied to any linear model. It can also be used to measure departures from certain types of balance, such as proportionality of subclass frequencies, partial balance, and last-stage uniformity. These types will be defined later in this section. The main idea behind this general procedure is the following: The cell frequencies are considered to have a multinomial distribution whose expected value is represented by a loglinear model of the same form as the original model. A null hypothesis H_0 is set up, which holds true if and only if the expected frequencies are equal. A test statistic, namely that of Pearson's goodness of fit, which we denote by X^2, is used to test H_0. The proposed measure of imbalance, denoted by ϕ, is given by

$$\phi = \frac{1}{1+\kappa^2}, \tag{3.4.1}$$

where $\kappa^2 = X^2/n_0$, and n_0 is the sum of all frequencies. The division of X^2 by n_0 causes ϕ to be invariant to any replication of the design, a condition required of any measure of imbalance as was seen earlier in case of Ahrens and Pincus's (1981) measure.

We note that $0 \leq \phi \leq 1$. The upper limit is attained if and only if the data set is balanced. Large values of X^2, and hence small values of ϕ, correspond to high degrees of imbalance.

Let us now apply the aforementioned procedure to some well-known models.

3.4.1. The One-Way Classification Model

Consider again the one-way model given in formula (3.2.1), where τ_i can be either fixed or random. Suppose that the n_i's have a multinomial distribution such that $n_i \sim Bin(n., \pi_i)$, where $Bin(n., \pi_i)$ denotes the binomial distribution associated with $n.$ trials with π_i being the probability of "success" on a single trial. By "success" we mean having an observation that belongs to level $i(=1,2,\ldots,v)$ of factor A. If f_i is the expected frequency $E(n_i)$, then $f_i = n.\pi_i(i =$

$1, 2, \ldots, v$). These expected frequencies can be represented on a logarithmic scale by the loglinear model

$$\log f_i = \overline{\mu} + \overline{\tau}_i, \quad i = 1, 2, \cdots, v, \tag{3.4.2}$$

where

$$\begin{aligned} \overline{\mu} &= \frac{1}{v}\sum_{i=1}^{v} \log f_i \\ &= \log n. + \frac{1}{v}\sum_{i=1}^{v} \log \pi_i \\ \overline{\tau}_i &= \log f_i - \overline{\mu} \\ &= \log \pi_i - \frac{1}{v}\sum_{j=1}^{v} \log \pi_j, \quad i = 1, 2, \cdots, v. \end{aligned}$$

We note that $\sum_{i=1}^{v} \overline{\tau}_i = 0$ and that model (3.4.2) is of the same form as model (3.2.1) except for the error term.

Consider the null hypothesis H_0 that the π_i's are equal, that is, $H_0 : \pi_i = \frac{1}{v}$ for $i = 1, 2, \ldots, v$. If H_0 is true, then the expected frequencies take the value $f_i = \frac{n.}{v} = \overline{n.}$. For the sample of observed frequencies $n_1, n_2, \ldots, n_v$, Pearson's statistic for testing H_0 is given by

$$X^2 = \sum_{i=1}^{v} \frac{(n_i - \overline{n.})^2}{\overline{n.}}.$$

For large samples and under H_0, X^2 has approximately a chi-squared distribution with $v - 1$ degrees of freedom. Large values of X^2 indicate a high degree of imbalance for the design $\mathbf{D} = \{n_1, n_2, \ldots, n_v\}$. We therefore refer to H_0 as the hypothesis of complete balance. As a measure of imbalance we consider ϕ, which is given by formula (3.4.1), where $\kappa^2 = X^2/n.$. Note that $\phi = 1$ if and only if $X^2 = 0$, that is, if and only if the design $\mathbf{D}$ is balanced. We also note that ϕ can be expressed as

$$\begin{aligned} \phi &= \frac{1}{1 + \frac{v}{n_.^2}\sum_{i=1}^{v}(n_i - \overline{n.})^2} \\ &= \frac{1}{v\sum_{i=1}^{v}(n_i/n.)^2}, \end{aligned}$$

which is identical to $\nu(\mathbf{D})$ in formula (3.3.2).

3.4.2. The Two-Way Classification Model

Consider the model

$$y_{ijk} = \mu + \tau_i + \beta_j + (\tau\beta)_{ij} + e_{ijk} \tag{3.4.3}$$

$(i = 1, 2, \ldots, v;\ j = 1, 2, \ldots, b;\ k = 1, 2, \ldots, n_{ij})$, where μ is a fixed unknown parameter and τ_i and β_j can be either fixed or random. As before, we consider that the observed frequencies, n_{ij}, follow a multinomial distribution with cell probabilities denoted by $\pi_{ij}(i = 1, 2, \ldots, v;\ j = 1, 2, \ldots, b)$. Hence, for a fixed $n_{..} = \sum_{i,j} n_{ij}$, $f_{ij} = n_{..}\pi_{ij}$, where $f_{ij} = E(n_{ij})$ is the expected frequency for the $(i, j)^{th}$ cell. Consider next the loglinear model

$$\log f_{ij} = \overline{\mu} + \overline{\tau}_i + \overline{\beta}_j + \overline{(\tau\beta)}_{ij}, \tag{3.4.4}$$

where

$$\begin{aligned} \overline{\mu} &= \frac{1}{vb} \sum_{i=1}^{v} \sum_{j=1}^{b} \log f_{ij} \\ \overline{\tau}_i &= \frac{1}{b} \sum_{j=1}^{b} \log f_{ij} - \overline{\mu} \\ \overline{\beta}_j &= \frac{1}{v} \sum_{i=1}^{v} \log f_{ij} - \overline{\mu} \\ (\overline{\tau\beta})_{ij} &= \log f_{ij} - \frac{1}{v} \sum_{h=1}^{v} \log f_{hj} - \frac{1}{b} \sum_{\ell=1}^{b} \log f_{i\ell} + \overline{\mu}. \end{aligned} \tag{3.4.5}$$

Note that $\sum_{i=1}^{v} \overline{\tau}_i = \sum_{j=1}^{b} \overline{\beta}_j = \sum_{i=1}^{v} (\overline{\tau\beta})_{ij} = \sum_{j=1}^{b} (\overline{\tau\beta})_{ij} = 0$. As in Section 3.4.1, the hypothesis of complete balance in this case is $H_0 : \pi_{ij} = \frac{1}{vb}$ for all i, j. The expected frequencies under this hypothesis are $f_{ij} = \frac{n_{..}}{vb} = \overline{n_{..}}$. Hence, the corresponding test statistic is

$$X^2 = \sum_{i=1}^{v} \sum_{j=1}^{b} \frac{(n_{ij} - \overline{n_{..}})^2}{\overline{n_{..}}},$$

which, under H_0, has an approximate chi-squared distribution with $vb - 1$ degrees of freedom. A measure of departure from complete balance is given by formula (3.4.1) where $\kappa^2 = X^2/n_{..}$. Note that if H_0 is true, model (3.4.4) reduces to $\log f_{ij} = \overline{\mu}$ since the expected frequencies are equal.

In particular, if π_{ij} is of the form $\pi_{ij} = \pi_{i.}\pi_{.j}$, where $\pi_{i.}$ and $\pi_{.j}$ are, respectively, the marginal probabilities for the ith row and j^{th} column, then model (3.4.4) reduces to

$$\log f_{ij} = \overline{\mu} + \overline{\tau}_i + \overline{\beta}_j. \tag{3.4.6}$$

This follows from the fact that, in this case, $\log f_{ij} = \log n_{..} + \log \pi_{i.} + \log \pi_{.j}$. By applying formula (3.4.5), it is easy to show that $(\overline{\tau\beta})_{ij} = 0$. Note that when $\pi_{ij} = \pi_{i.}\pi_{.j}$, the rows and columns of the table of frequencies are said to be independent. Vice versa, if $\log f_{ij}$ is represented by model (3.4.6), then the rows and columns are independent. To show this, we have

$$f_{ij} = \exp(\overline{\mu} + \overline{\tau}_i + \overline{\beta}_j).$$

Hence,

$$\begin{aligned}
\pi_{ij} &= \frac{f_{ij}}{n_{..}} = \frac{1}{n_{..}} e^{\overline{\mu}} e^{\overline{\tau}_i} e^{\overline{\beta}_j} \\
\pi_{i.} &= \frac{1}{n_{..}} e^{\overline{\mu}} e^{\overline{\tau}_i} \sum_{j=1}^{b} e^{\overline{\beta}_j} \\
\pi_{.j} &= \frac{1}{n_{..}} e^{\overline{\mu}} e^{\overline{\beta}_j} \sum_{i=1}^{v} e^{\overline{\tau}_i} \\
1 &= \frac{1}{n_{..}} e^{\overline{\mu}} \sum_{i=1}^{v} e^{\overline{\tau}_i} \sum_{j=1}^{b} e^{\overline{\beta}_j}.
\end{aligned}$$

It follows that

$$\begin{aligned}
\pi_{ij} &= \frac{1}{n_{..}^2} \left(e^{\overline{\mu}}\right)^2 e^{\overline{\tau}_i} e^{\overline{\beta}_j} \sum_{i=1}^{v} e^{\overline{\tau}_i} \sum_{j=1}^{b} e^{\overline{\beta}_j} \\
&= \pi_{i.} \pi_{.j}.
\end{aligned}$$

The maximum likelihood estimates of $\pi_{i.}$ and $\pi_{.j}$ are $\hat{\pi}_{i.} = \frac{n_{i.}}{n_{..}}$, $\hat{\pi}_{.j} = \frac{n_{.j}}{n_{..}}$, respectively, where $n_{i.} = \sum_{j=1}^{b} n_{ij}$, $n_{.j} = \sum_{i=1}^{v} n_{ij}$. Hence, the expected frequency $f_{ij} = n_{..}\pi_{i.}\pi_{.j}$ is estimated by $\hat{f}_{ij} = n_{i.}n_{.j}/n_{..}$. This value corresponds to the case of the so-called proportional subclass frequencies. For this special case, the test statistic concerning the hypothesis $H_0 : \pi_{ij} = \pi_{i.}\pi_{.j}$ is given by

$$X_p^2 = \sum_{i=1}^{v} \sum_{j=1}^{b} \frac{(n_{ij} - \hat{f}_{ij})^2}{\hat{f}_{ij}},$$

which, under H_0, has an asymptotic chi-squared distribution with $(v-1)(b-1)$ degrees of freedom. To measure departure of the n_{ij}'s from their corresponding proportional subclass frequencies we use formula (3.4.1) with $\kappa^2 = X_p^2/n_{..}$. Small values of ϕ indicate a significant departure.

3.4.3. The Three-Way Classification Model

The approach used in Section 3.4.2 can be extended to any n-way classification model. For example, consider the three-way classification model,

$$y_{ijkl} = \mu + \tau_i + \beta_j + \gamma_k + (\tau\beta)_{ij} + (\tau\gamma)_{ik} + (\beta\gamma)_{jk} + (\tau\beta\gamma)_{ijk} + e_{ijkl}$$

($i = 1, 2, \ldots, v$; $j = 1, 2, \ldots, b$; $k = 1, 2, \ldots, p$; $l = 1, 2, \ldots, n_{ijk}$). The corresponding loglinear model for the expected frequency $f_{ijk} = E(n_{ijk})$ is

$$\log f_{ijk} = \overline{\mu} + \overline{\tau}_i + \overline{\beta}_j + \overline{\gamma}_k + (\overline{\tau\beta})_{ij} + (\overline{\tau\gamma})_{ik} + (\overline{\beta\gamma})_{jk} + (\overline{\tau\beta\gamma})_{ijk}, \qquad (3.4.7)$$

where $\sum_{i=1}^{v} \overline{\tau}_i = \sum_{j=1}^{b} \overline{\beta}_j = \sum_{k=1}^{p} \overline{\gamma}_k = \sum_{i=1}^{v} (\overline{\tau\beta})_{ij} = \sum_{j=1}^{b} (\overline{\tau\beta})_{ij} = \cdots = \sum_{k=1}^{p} (\overline{\tau\beta\gamma})_{ijk} = 0$.

Several reduced loglinear models can be obtained from model (3.4.7). For example, the model

$$\log f_{ijk} = \overline{\mu} + \overline{\tau}_i + \overline{\beta}_j + \overline{\gamma}_k$$

corresponds to the case of proportional subclass frequencies for which the estimated expected frequencies are given by

$$\hat{f}_{ijk} = \frac{n_{i..} n_{.j.} n_{..k}}{n_{...}^2},$$

where $n_{i..} = \sum_{j=1}^{b} \sum_{k=1}^{p} n_{ijk}$, $n_{.j.} = \sum_{i=1}^{v} \sum_{k=1}^{p} n_{ijk}$, and $n_{..k} = \sum_{i=1}^{v} \sum_{j=1}^{b} n_{ijk}$. The goodness of fit of each reduced model can be checked by using Pearson's approximate chi-squared statistic

$$X^2 = \sum_{i=1}^{v} \sum_{j=1}^{b} \sum_{k=1}^{p} \frac{(n_{ijk} - \hat{f}_{ijk})^2}{\hat{f}_{ijk}},$$

where here $\hat{f}_{ijk}$ is the appropriate estimated expected frequency for the (ijk)-cell (using maximum likelihood estimation) corresponding to the given reduced model. Departure from each case considered can be measured by using formula (3.4.1) with $\kappa^2 = X^2/n_{...}$. In particular, for the case of complete balance, $\log f_{ijk} = \overline{\mu}$,

$$X^2 = \sum_{i=1}^{v} \sum_{j=1}^{b} \sum_{k=1}^{p} \frac{(n_{ijk} - \overline{n}_{...})^2}{\overline{n}_{...}},$$

where $\overline{n}_{...} = n_{...}/vbp$.

Another statistic that can be used to check goodness of fit is Wilks' likelihood ratio statistic (see, for example, Agresti, 1990, p. 174),

$$G^2 = 2 \sum_{i=1}^{v} \sum_{j=1}^{b} \sum_{k=1}^{p} n_{ijk} \log\left(\frac{n_{ijk}}{\hat{f}_{ijk}}\right).$$

This statistic has the desirable feature of being monotone increasing as terms are deleted from the full model in (3.4.7). It can therefore be used to compare the goodness of fit of two nested models (that is, one model is obtained from the other by deleting one or more terms). Khuri (1987) considered several such models that were derived from model (3.4.7).

3.5. SPECIAL TYPES OF IMBALANCE

In all the models we have considered so far, imbalance occurs only in the last stage of the associated design. In other words, in the set of subscripts that identify the response y, the range of the last subscript is not the same for all combinations of levels of the preceding subscripts. We shall now consider situations in which imbalance involves one or more of the latter subscripts.

3.5.1. The Two-Fold Nested Classification Model

Let us consider the model

$$y_{ijk} = \mu + \tau_i + \beta_{i(j)} + e_{ijk} \tag{3.5.1}$$

($i = 1, 2, \ldots, v$; $j = 1, 2, \ldots, b_i$; $k = 1, 2, \ldots, n_{ij}$). Here, imbalance occurs in the last two stages. Some particular types of balance can be considered. For example, the case in which $n_{ij} = n$ for all i, j is referred to as last-stage uniformity. If, in addition, the values of $b_i (i = 1, 2, \ldots, v)$ are equal, then we have the case of complete balance. If, however, $n_{ij} = n_{ij'}$, for $i = 1, 2, \ldots, v$ and $j \neq j'$, then the associated design is said to be partially balanced. In this case, the numbers of observations for the various levels of the nested factor are the same within each level of the nesting factor.

The corresponding loglinear model for model (3.5.1) is derived in the following manner: Let f_{ij} denote the expected frequency, $E(n_{ij})$. Then, $f_{ij} = n_{..}\pi_{ij} = n_{..}\pi_i \pi_{j|i}$, where π_i is the probability of belonging to level i of the nesting factor and $\pi_{j|i}$ is the conditional probability of belonging to level j of the nested factor given level i of the nesting factor. Thus

$$\log f_{ij} = \overline{\mu} + \overline{\tau}_i + \overline{\beta}_{i(j)}, \tag{3.5.2}$$

where

$$\begin{aligned}
\overline{\mu} &= \log n_{..} + \frac{1}{b_.} \sum_{i=1}^{v} b_i \log \pi_i + \frac{1}{b_.} \sum_{i=1}^{v} \sum_{j=1}^{b_i} \log \pi_{j|i} \\
\overline{\tau}_i &= \log \pi_i + \frac{1}{b_i} \sum_{j=1}^{b_i} \log \pi_{j|i} - \frac{1}{b_.} \sum_{\ell=1}^{v} b_\ell \log \pi_\ell - \frac{1}{b_.} \sum_{\ell=1}^{v} \sum_{j=1}^{b_\ell} \log \pi_{j|\ell} \\
\overline{\beta}_{i(j)} &= \log \pi_{j|i} - \frac{1}{b_i} \sum_{\ell=1}^{b_i} \log \pi_{\ell|i},
\end{aligned}$$

where $b_. = \sum_{i=1}^{v} b_i$. We note that $\sum_{i=1}^{v} b_i \overline{\tau}_i = 0$, $\sum_{j=1}^{b_i} \overline{\beta}_{i(j)} = 0$ for $i = 1, 2, \ldots, v$.

For the particular case of partial balance, $\pi_{j|i} = 1/b_i$ for all i and j. The maximum likelihood estimate of π_i is $n_{i.}/n_{..}$. Hence, f_{ij} is estimated by $\hat{f}_{ij} = n_{i.}/b_i$. A measure of departure from partial balance is then given by ϕ in formula (3.4.1), where

$$\kappa^2 = \frac{1}{n_{..}} \sum_{i=1}^{v} \sum_{j=1}^{b_i} \frac{(n_{ij} - \hat{f}_{ij})^2}{\hat{f}_{ij}}. \tag{3.5.3}$$

In this case, model (3.5.2) reduces to

$$\log f_{ij} = \overline{\mu} + \overline{\tau}_i,$$

since $\overline{\beta}_{i(j)} = 0$ for all i, j.

For the case of last-stage uniformity, $\pi_{ij} = 1/b.$ for all i, j. Correspondingly, model (3.5.2) reduces to

$$\log f_{ij} = \overline{\mu}.$$

A measure of departure from last-stage uniformity is again given by ϕ in formula (3.4.1), where

$$\kappa^2 = \frac{1}{n_{..}} \sum_{i=1}^{v} \sum_{j=1}^{b_i} \frac{(n_{ij} - \overline{n}_{..})^2}{\overline{n}_{..}}. \tag{3.5.4}$$

Now, to measure departure from complete balance we need to consider two types of imbalance: imbalance in the values of b_i and imbalance in the values of n_{ij}. For this purpose, we regard the b_i's as having a multinomial distribution independently of the multinomial distribution of the n_{ij}'s such that $b_i \sim Bin(b., p_i)$. As in Section 3.4.1, the corresponding measure of imbalance concerning the b_i's is given by

$$\phi_1 = \frac{1}{1 + \kappa_1^2},$$

where $\kappa_1^2 = (1/b.)X_1^2$ and

$$X_1^2 = \sum_{i=1}^{v} \frac{(b_i - \overline{b}.)^2}{\overline{b}.},$$

where $\overline{b}. = b./v$. The second type of imbalance is measured by

$$\phi_2 = \frac{1}{1 + \kappa_2^2},$$

where $\kappa_2^2 = (1/n..)X_2^2$ and

$$X_2^2 = \sum_{i=1}^{v}\sum_{j=1}^{b_i} \frac{(n_{ij} - \bar{n}..)^2}{\bar{n}..}.$$

The statistics X_1^2 and X_2^2 are independent due to the independence of the corresponding multinomial distributions. Hence, to measure departure from complete balance we consider the quantity

$$\phi = \frac{1}{1 + \kappa_0^2},$$

where

$$\kappa_0^2 = \frac{b.\kappa_1^2 + n..\kappa_2^2}{b. + n..}. \tag{3.5.5}$$

The measures ϕ_1, ϕ_2, and ϕ were used by Hernandez et al. (1992) and Hernandez and Burdick (1993) to study the effect of imbalance on the coverage probabilities of confidence intervals on the variance components for the two-fold nested model.

Example 3.5.1. Cummings and Gaylor (1974) used several designs to illustrate the combined effects of dependence and nonchi-squaredness of the mean squares for τ_i and $\beta_{i(j)}$, in a random two-fold nested classification model, on the size of Satterthwaite's approximate F-test concerning the hypothesis $H_\tau : \sigma_\tau^2 = 0$. Three of these designs are described in Table 3.2 and are also depicted in Figure 3.1. Design 1 is partially balanced for which the mean

Table 3.2. Designs for a Two-fold Nested Model

	i			
	1	2	3	4
Design 1				
b_i	1	1	4	4
n_{ij}	1	4	1,1,1,1	4,4,4,4
Design 2				
b_i	2	2	2	2
n_{ij}	1,5	1,5	1,5	1,5
Design 3				
b_i	1	2	2	4
n_{ij}	1	1,4	1,8	1,2,3,4

Source: A. I. Khuri (1987). Reproduced with permission of Akademie Verlag.

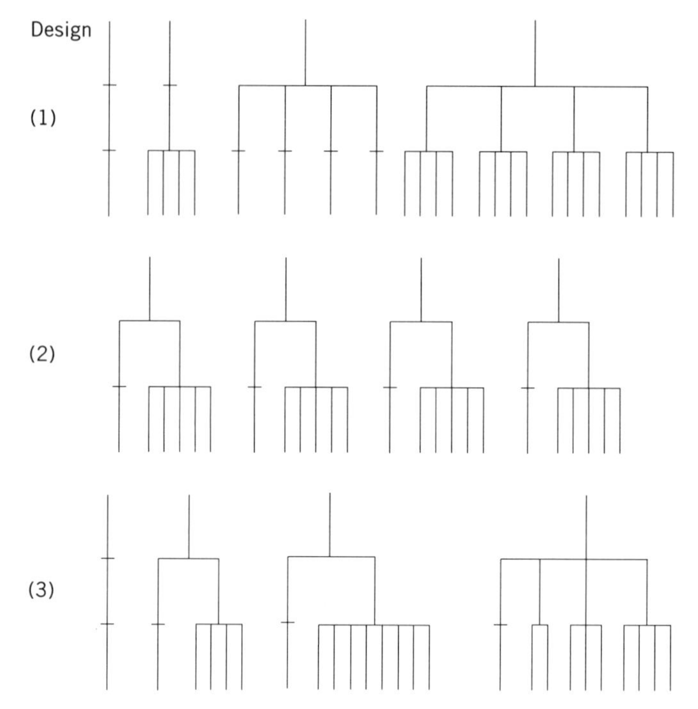

Figure 3.1. Three designs for a two-fold nested model. *Source:* A. I. Khuri (1987). Reproduced with permission of Akademie Verlag.

squares $MS(\tau)$ and $MS(\beta(\tau))$ associated with τ_i and $\beta_{i(j)}$, respectively, are independent, but are not distributed as scaled chi-squared variates. For design 2, the two mean squares are dependent, but have the scaled chi-squared distribution. As for design 3, the mean squares neither are independent nor have the scaled chi-squared distribution.

For each of designs 1, 2, and 3, measures of departures from partial balance, last-stage uniformity, and complete balance are evaluated using formula (3.4.1) with κ^2 given by formulas (3.5.3), (3.5.4), and (3.5.5), respectively. The results are shown in Table 3.3. We note that except for partial balance in case of design 1, none of the three designs has strong balance properties. Design 3 is the most unbalanced with respect to last-stage uniformity and complete balance.

3.5.2. A Model With a Mixture of Cross-Classified and Nested Effects

Let us now consider a model involving three factors, denoted by A, B, and C, with A and C crossed and B nested within A. The model is of the form

$$y_{ijkl} = \mu + \tau_i + \beta_{i(j)} + \gamma_k + (\tau\gamma)_{ik} + (\beta\gamma)_{i(jk)} + e_{ijkl} \qquad (3.5.6)$$

Table 3.3. Measures of Departure from Balance

Design	Partial Balance	Last-Stage Uniformity	Complete Balance
1	1.00	0.735	0.735
2	0.692	0.692	0.750
3	0.730	0.615	0.656

$(i = 1, 2, \ldots, v;\ j = 1, 2, \ldots, b_i;\ k = 1, 2, \ldots, p;\ l = 1, 2, \ldots, n_{ijk})$, where τ_i and γ_k are the effects of levels i and k of A and C, respectively, and $\beta_{i(j)}$ denotes the effect of the jth level of B nested within the ith level of A.

For this model we can have the following types of balance:

1. Proportional subclass frequencies involving the AB subclasses and the levels of factor C.
2. Partial balance with respect to n_{ijk} for fixed levels of i; that is, values of n_{ijk} are equal for the same level i of factor A.
3. Last-stage uniformity with respect to n_{ijk}; that is, the values of n_{ijk} are the same for all i, j, k.
4. Complete balance; that is, the values of b_i and n_{ijk} are constant for all i, j, k.

Departure from each of the above types of balance can be measured by using formula (3.4.1). Table 3.4 provides the values of κ^2 for the above four types of designs.

Table 3.4. Values of κ^2 for the Four Types of Imbalance for Model (3.5.6)

Type	κ^2
(1)	$\dfrac{1}{n_{...}} \sum_{i,j,k} \dfrac{(n_{ijk} - n_{ij.}n_{..k}/n_{...})^2}{n_{ij.}n_{..k}/n_{...}}$
(2)	$\dfrac{1}{n_{...}} \sum_{i,j,k} \dfrac{[n_{ijk} - n_{i..}/(b_i p)]^2}{n_{i..}/(b_i p)}$
(3)	$\dfrac{1}{n_{...}} \sum_{i,j,k} \dfrac{[n_{ijk} - n_{...}/(b_. p)]^2}{n_{...}/(b_. p)}$
(4)	$\dfrac{b_.\kappa_{41}^2 + n_{...}\kappa_{42}^2}{b_. + n_{...}}$, where

$$\kappa_{41}^2 = \frac{1}{b_.} \sum_i \frac{(b_i - \bar{b}_.)^2}{\bar{b}_.}, \quad \kappa_{42}^2 = \frac{1}{n_{...}} \sum_{i,j,k} \frac{[n_{ijk} - n_{...}/(b_. p)]^2}{n_{...}/(b_. p)}$$

3.6. A GENERAL METHOD FOR DETERMINING THE EFFECT OF IMBALANCE

Earlier in this chapter, we examined certain aspects of the effect of imbalance on the analysis concerning the random one-way model. In this respect, it would be desirable to be able to make use of the corresponding measure of imbalance, ϕ, in order to acquire a deeper insight into the effect of imbalance. This was accomplished to a certain extent, as was seen in Sections 3.3.1 and 3.3.2. For example, it was noted in Section 3.3.1 that expressing $Var(\hat{\sigma}_\tau^2)$ explicitly as a function of $\phi = \nu(\mathbf{D})$ was not possible for given values of $n.$, v, σ_τ^2, and σ_e^2. This is attributed to the fact that $Var(\hat{\sigma}_\tau^2)$ depends in a complex way on $n_i (i = 1, 2, \ldots, v)$, as well as on the true values of the variance components (see formula 3.2.9). How then can the effect of varying ϕ on $Var(\hat{\sigma}_\tau^2)$ be determined? One possible solution to this problem is to establish an empirical relationship that provides an approximation of $Var(\hat{\sigma}_\tau^2)$ by a polynomial function that depends on $\phi, n.$, and the true values of the variance components. If such an approximation is adequate for all values of ϕ, then it can be used in lieu of $Var(\hat{\sigma}_\tau^2)$ to assess the effect of imbalance. A demonstration of this approach will be given in Section 3.6.2.

The development of the aforementioned empirical relationship depends on a method for generating designs $\mathbf{D}$ having a specified degree of imbalance and a specified value of $n..$ This method will be described in Section 3.6.1.

It should be noted that quantities other than $Var(\hat{\sigma}_\tau^2)$ can be chosen on which to study the effect of imbalance. For example, the power of a test concerning a variance component and the coverage probability of a corresponding confidence interval are such quantities. Furthermore, since a general measure of imbalance is available, as was seen in Section 3.4, the study of the effect of imbalance can be extended beyond the one-way model to higher-order models.

3.6.1. Generation of Designs Having a Specified Degree of Imbalance for the One-Way Model

Consider the one-way model (3.2.1). Suppose that $n. = \sum_{i=1}^{v} n_i$ and $\phi = \nu(\mathbf{D})$ in formula (3.3.2) have predetermined values. It may be recalled from Section 3.3 that

$$\frac{1}{v} < \phi \leq 1. \tag{3.6.1}$$

In the present section, we show how to determine values of $n_1, n_2, \ldots, n_v$ that correspond to specified values of $\phi, n.$, and v. This is equivalent to solving the equations

$$\sum_{i=1}^{v} n_i = n. \tag{3.6.2}$$

$$\sum_{i=1}^{v} n_i^2 = \frac{n.^2}{v\phi}, \tag{3.6.3}$$

for $n_1, n_2, \ldots, n_v$, where $\phi, n.$, and v are given.

Finding solutions to equations (3.6.2) and (3.6.3) requires using a method given by Khuri (1996). Another related method was presented by Donner and Koval (1987), who generated values of n_i using an underlying probability distribution, namely, the truncated negative binomial distribution. The resulting values, however, have on the average a specified value of ϕ, but do not necessarily satisfy equations (3.6.2) and (3.6.3) for the given values of ϕ, $n.$, and v. We now provide a description of Khuri's (1996) method.

Let $x_i = n_i/n.$, $i = 1, 2, \ldots, v$. Substituting x_i in equations (3.6.2) and (3.6.3), we obtain

$$\sum_{i=1}^{v} x_i = 1 \tag{3.6.4}$$

$$\sum_{i=1}^{v} x_i^2 = \frac{1}{v\phi}. \tag{3.6.5}$$

Equation (3.6.4) represents the interior of a regular $(v-1)$-dimensional simplex, which we denote by T_{v-1}. Note that without the constraint $0 < x_i < 1$, $i = 1, 2, \ldots, v$, equation (3.6.4) represents a hyperplane in a v-dimensional Euclidean space. We denote this hyperplane by P_{v-1}. Furthermore, equation (3.6.5) represents a $(v-1)$-dimensional hypersphere, which we denote by S_{v-1}. It is centered at the origin and has a radius equal to $(v\phi)^{-1/2}$. From the double inequality (3.6.1) we note that

$$\frac{1}{v^{\frac{1}{2}}} \leq \frac{1}{(v\phi)^{\frac{1}{2}}} < 1. \tag{3.6.6}$$

It follows that the set of all feasible solutions of equations (3.6.4) and (3.6.5) consists of the coordinates of points of the form $\mathbf{x} = (x_1, x_2, \ldots, x_v)'$ that belong to the intersection of the hyperplane P_{v-1} with the hypersphere S_{v-1}, provided that $0 < x_i < 1$ for $i = 1, 2, \ldots, v$. This intersection is nonempty because by (3.6.6), the radius of S_{v-1} is larger than or equal to $v^{-1/2}$, which represents the distance of P_{v-1} from the origin. We now proceed to find the solutions of the aforementioned equations.

Let $\mathbf{c}_{v-1}$ denote the centroid of the simplex T_{v-1}, that is, the point $(1/v, 1/v, \ldots, 1/v)$ in the v-dimensional Euclidean space. This point is now considered as the center of a new coordinates system chosen so that its vth axis is orthogonal to the hyperplane P_{v-1}. If $z_1, z_2, \ldots, z_v$ denote the coordinates of a point in the new system, then the relationship between $z_1, z_2, \ldots, z_v$ and $x_1, x_2, \ldots, x_v$ is given by

$$\mathbf{x} = \mathbf{W}(\mathbf{z} + \mathbf{a}_v), \tag{3.6.7}$$

where $\mathbf{x} = (x_1, x_2, \ldots, x_v)'$, $\mathbf{z} = (z_1, z_2, \ldots, z_v)'$, $\mathbf{a}_v = (0, 0, \ldots, 0, 1/\sqrt{v})'$, and $\mathbf{W}$ is an orthogonal matrix. Note that $1/\sqrt{v}$ is the distance of $\mathbf{c}_{v-1}$ from the origin of the x_i-coordinates system. The columns of $\mathbf{W}$ consist of unit vectors in the direction of the v axes of the z_i-coordinates system. Thus the vth column of $\mathbf{W}$ is of the form $(1/\sqrt{v})\ \mathbf{1}_v$, since it is orthogonal to the hyperplane P_{v-1}. It is easy to see that in the z_i-coordinates system, equations (3.6.4) and (3.6.5) can be expressed as

$$\begin{aligned} z_v &= 0 \\ \sum_{i=1}^{v-1} z_i^2 &= \frac{1-\phi}{v\phi}. \end{aligned} \tag{3.6.8}$$

The second equation in (3.6.8) results from the fact that the intersection of the hypersphere S_{v-1} with the hyperplane P_{v-1} is a $(v-2)$-dimensional hypersphere, which we denote by S_{v-2}, whose radius is given by

$$\begin{aligned} \varrho &= \left(\frac{1}{v\phi} - \frac{1}{v}\right)^{1/2} \\ &= \left(\frac{1-\phi}{v\phi}\right)^{1/2}. \end{aligned} \tag{3.6.9}$$

It can be seen that the less unbalanced the data set is (that is, the closer ϕ is to 1), the smaller the radius ϱ.

Any point on S_{v-2} depends on only $v-2$ independent parameters. Using, for example, the spherical coordinates associated with $z_1, z_2, \ldots, z_{v-1}$, we obtain the equations

$$\begin{aligned} z_1 &= \varrho \cos \phi_1 \\ z_2 &= \varrho \sin \phi_1 \cos \phi_2 \\ z_3 &= \varrho \sin \phi_1 \sin \phi_2 \cos \phi_3 \\ &\vdots \\ z_{v-3} &= \varrho \sin \phi_1 \sin \phi_2 \cdots \sin \phi_{v-4} \cos \phi_{v-3} \\ z_{v-2} &= \varrho \sin \phi_1 \sin \phi_2 \cdots \sin \phi_{v-4} \sin \phi_{v-3} \cos \phi_{v-2} \\ z_{v-1} &= \varrho \sin \phi_1 \sin \phi_2 \cdots \sin \phi_{v-4} \sin \phi_{v-3} \sin \phi_{v-2} \\ z_v &= 0, \end{aligned} \tag{3.6.10}$$

where $\phi_1, \phi_2, \ldots, \phi_{v-2}$ are used as independent parameters with $0 \le \phi_1 \le \pi, 0 \le \phi_2 \le \pi, \ldots, 0 \le \phi_{v-3} \le \pi, 0 \le \phi_{v-2} \le 2\pi$ (see, for example, Edwards, 1973, p. 268).

On the basis of the aforementioned geometric ideas, we can now find feasible solutions to equations (3.6.4) and (3.6.5) that correspond to a given value of ϕ. This is done as follows:

1. Compute ϱ for the given ϕ using (3.6.9).
2. Choose values of $\phi_1, \phi_2, \ldots, \phi_{v-2}$ at random from the uniform distribution $\phi_i \sim U(0, \pi)$, $i = 1, 2, \ldots, v-3$, $\phi_{v-2} \sim U(0, 2\pi)$.
3. For each choice of $\phi_1, \phi_2, \ldots, \phi_{v-2}$, compute $z_1, z_2, \ldots, z_{v-1}$ using the spherical coordinates transformation in (3.6.10). Note that $z_v = 0$.
4. Compute $\mathbf{x} = (x_1, x_2, \ldots, x_v)'$ using formula (3.6.7). The resulting values of $x_1, x_2, \ldots, x_v$ represent a feasible solution to equations (3.6.4) and (3.6.5) provided that $x_i > 0$ for $i = 1, 2, \ldots, v$. Any $\mathbf{x}$ for which at least one x_i does not satisfy this condition will be discarded.

Having obtained a solution to equations (3.6.4) and (3.6.5), we can then compute $n.x_i (i = 1, 2, \ldots, v)$. If such quantities are integer valued for all i then we write $n_i = n.x_i (i = 1, 2, \ldots, v)$ and report $\{n_1, n_2, \ldots, n_v\}$ as a design $\mathbf{D}$ with the specified values of $n.$ and ϕ. In general, however, $n.x_i$ may not have an integer value. In this case, equations (3.6.2) and (3.6.3) do not have an exact integer solution. An approximate solution would therefore be needed. We now show how this can be accomplished in an optimal manner.

Let $w_i = n.x_i$, where $(x_1, x_2, \ldots, x_v)'$ is a solution to equations (3.6.4) and (3.6.5) obtained as described earlier $(i = 1, 2, \ldots, v)$. We then have

$$\begin{aligned} \sum_{i=1}^{v} w_i &= n. \\ \sum_{i=1}^{v} w_i^2 &= \frac{n.^2}{v\phi}. \end{aligned} \tag{3.6.11}$$

These equations are identical to (3.6.2) and (3.6.3) except that the w_i's are not necessarily integer valued. Each w_i is rounded off to the nearest integer. Consequently, we obtain the vector $\boldsymbol{\psi} = (\psi_1, \psi_2, \ldots, \psi_v)'$, where ψ_i is either equal to $[w_i]$ or to $[w_i] + 1$ with $[w_i]$ being the greatest integer in w_i. Note that $[w_i] \le w_i < [w_i] + 1 (i = 1, 2, \ldots, v)$. Thus for each value of $\mathbf{w} = (w_1, w_2, \ldots, w_v)'$ there can be 2^v possible values of $\boldsymbol{\psi}$. Let H_w denote the set consisting of all such values. Consider now the set

$$H_w^* = \{\boldsymbol{\psi} \in H_w | \sum_{i=1}^{v} \psi_i = n.\}. \tag{3.6.12}$$

Lemma 3.6.1. The set H_w^* in (3.6.12) is nonempty.
Proof. We have that

$$\begin{aligned} w_i &= \delta_i [w_i] + (1 - \delta_i)(1 + [w_i]), \ 0 < \delta_i \le 1 \\ &= [w_i] + 1 - \delta_i, \ i = 1, 2, \cdots, v. \end{aligned} \tag{3.6.13}$$

Using equation (3.6.11), we get

$$n_{.} = \sum_{i=1}^{v}[w_i] + \sum_{i=1}^{v}(1 - \delta_i).$$

This implies that $\sum_{i=1}^{v}(1 - \delta_i)$ is a positive integer. We denote such an integer by v_0. Note that $0 \le v_0 < v$. Now, ψ_i can be expressed as

$$\begin{aligned} \psi_i &= \delta_i^*[w_i] + (1 - \delta_i^*)([w_i] + 1), \ \ \delta_i^* = 0, 1 \\ &= [w_i] + 1 - \delta_i^* \\ &= w_i - (1 - \delta_i) + (1 - \delta_i^*), \ \ i = 1, 2, \cdots, v. \end{aligned} \tag{3.6.14}$$

From equations (3.6.11) and (3.6.14) we then have

$$\sum_{i=1}^{v} \psi_i = n_{.} - v_0 + \sum_{i=1}^{v}(1 - \delta_i^*).$$

By choosing v_0 of the δ_i^*'s equal to zero and the remaining $v - v_0$ equal to one, we get

$$\sum_{i=1}^{v} \psi_i = n_{.}.$$

Thus for such a choice of the δ_i^*'s, $\boldsymbol{\psi} \in H_w^*$. This concludes the proof of Lemma 3.6.1. □

The coordinates of any point in H_w^* satisfy equation (3.6.2), but may not satisfy equation (3.6.3). We therefore select $\boldsymbol{\psi}^* \in H_w^*$ such that

$$\left| \sum_{i=1}^{v} \psi_i^{*2} - \frac{n_{.}^2}{v\phi} \right| \le \left| \sum_{i=1}^{v} \psi_i^2 - \frac{n_{.}^2}{v\phi} \right|$$

for all $\boldsymbol{\psi} \in H_w^*$, where ψ_i^* is the ith element of $\boldsymbol{\psi}^*$. Thus $\psi_1^*, \psi_2^*, \ldots, \psi_v^*$ represent an approximate integer solution to equations (3.6.2) and (3.6.3) provided that $\psi_i^* > 0$ for all i. The design $\mathbf{D} = \{n_1, n_2, \ldots, n_v\}$ with $n_i = \psi_i^*$, $i = 1, 2, \ldots, v$, is therefore "closest" to satisfying equations (3.6.2) and (3.6.3) for the given values of ϕ and $n_{.}$. Such a design is not unique since any permutation of the elements of $\mathbf{D}$ also satisfies the equations.

Note that if ϕ is close to $1/v$, that is, when the design is severely unbalanced, then some of the ψ_i^*'s may be equal to zero. This causes $\boldsymbol{\psi}^*$ to be an infeasible solution. In this case, the zero elements of $\boldsymbol{\psi}^*$ should be replaced by 1. To maintain equation (3.6.2), a comparable reduction of some of the nonzero ψ_i^*'s should be made. This reduction should be carried out in a manner that keeps $\left|\sum_{i=1}^{v} \psi_i^2 - n_{.}^2/(v\phi)\right|$ as small as possible.

3.6.2. An Example

We now demonstrate the application of the methodology of determining the effect of imbalance using an example given in (Khuri, 1996, Section 3). Consider the random one-way model (3.2.1) with the same assumptions stated earlier in Section 3.2. Let us investigate the effect of imbalance on the quantity

$$\mathcal{F} = \frac{1}{\sigma_e^4} Var(\hat{\sigma}_\tau^2).$$

Using formula (3.2.9), $\mathcal{F}$ can be written as

$$\begin{aligned} \mathcal{F} &= \frac{2(n._{}^2 g + g^2 - 2n.h)}{(n.^2 - g)^2} \left(\frac{\eta}{1-\eta}\right)^2 + \frac{4n.}{n.^2 - g} \frac{\eta}{1-\eta} \\ &\quad + \frac{2n.^2(n. - 1)(v-1)}{(n. - v)(n.^2 - g)^2}, \end{aligned} \tag{3.6.15}$$

where, if we recall, $g = \sum_{i=1}^{v} n_i^2$, $h = \sum_{i=1}^{v} n_i^3$, and

$$\eta = \frac{\sigma_\tau^2}{\sigma_\tau^2 + \sigma_e^2}. \tag{3.6.16}$$

We note that $\mathcal{F}$ depends on the design $\mathbf{D} = \{n_1, n_2, \ldots, n_v\}$ and on the variance components through η. It is not, however, directly related to the measure of imbalance $\phi = \nu(\mathbf{D})$ defined in formula (3.3.2).

In order to determine the effect of imbalance on the precision of estimating σ_τ^2, we need to develop an empirical relationship between $\mathcal{F}$ and the parameters $n.$, ϕ, and η. Response surface techniques can be used for this purpose, where $\mathcal{F}$ is treated as a response variable and $n.$, ϕ, and η as control variables. Let us consider the following second-degree model

$$\begin{aligned} \mathcal{F} &= \alpha_0 + \alpha_1 n_c + \alpha_2 \phi + \alpha_3 \eta + \alpha_{12} n_c \phi + \alpha_{13} n_c \eta \\ &\quad + \alpha_{23} \phi \eta + \alpha_{11} n_c^2 + \alpha_{22} \phi^2 + \alpha_{33} \eta^2 + e, \end{aligned} \tag{3.6.17}$$

where n_c is a coded (scaled) value of $n.$, the α's are unknown constant coefficients, and e is a random error with a zero mean. Fitting this model requires calculating the value of $\mathcal{F}$, using formula (3.6.15), at each of several combinations of values of n_c, ϕ, and η, which can be obtained according to a particular response surface design suitable for this model. We consider, for example, a 3^3 factorial design. This is an efficient design for fitting a second-degree model. It requires three levels of each of n_c, ϕ, and η, giving rise to 27 design combinations. Other second-degree designs could have been considered. For a survey of such designs, see, for example, Khuri and Cornell (1996, Chapter 4).

We recall that $1/v < \phi \leq 1$. Suppose that $v = 5$. Hence, $0.20 < \phi \leq 1$. The selected levels of $n.$, ϕ, and η are

$$\begin{aligned} n.&: 25, 50, 75 \\ \phi&: 0.32, 0.60, 0.85 \\ \eta&: 0.20, 0.40, 0.80. \end{aligned}$$

The coded value n_c is given by

$$n_c = \frac{n. - 50}{50}.$$

Hence, the corresponding levels of n_c are $-0.50, 0, 0.50$.

For each triple $(n., \phi, \eta)$, several values of $\mathbf{D} = \{n_1, n_2, n_3, n_4, n_5\}$ are generated by using the algorithm described in Section 3.6.1, and the corresponding values of $\mathcal{F}$ are obtained from formula (3.6.15). A total of 297 runs were thus generated. The first 20 runs are displayed in Table 3.5. Note that the actual value of ϕ, obtained from a generated design $\mathbf{D}$, may not be exactly identical to the target value of ϕ specified by the 3^3 factorial design. In Table 3.5, ϕ_a denotes an actual value of ϕ computed from formula (3.3.2) using the values of n_1, n_2, n_3, n_4, n_5 from the generated design $\mathbf{D}$.

The data from the 297 runs are used to fit model (3.6.17). On the basis of the previously described levels of n_c, ϕ, and η, the region of interest, denoted by Λ, is determined by the inequalities

$$\begin{aligned} -0.5 &\leq n_c \leq 0.5 \\ 0.32 &\leq \phi \leq 0.85 \\ 0.20 &\leq \eta \leq 0.80. \end{aligned}$$

The variance of the random error e in model (3.6.17) is not constant throughout Λ. Since ϕ_a is close to ϕ, values of $\mathcal{F}$ obtained from each triple $(n., \phi, \eta)$ are considered as near replicates. Therefore, they can be used to estimate the error variances at all the 297 runs. Let ν_i and s_i^2 denote the number of near replicates of $\mathcal{F}$ and the corresponding sample variance at the ith combination of the triple $(n., \phi, \eta)$, $i = 1, 2, \ldots, 27$. The resulting values of s_i^2 are shown in Table 3.6. Consequently, an estimate of the variance–covariance matrix of the random errors in model (3.6.17) is given by $\hat{\mathbf{\Sigma}}_e$, where

$$\hat{\mathbf{\Sigma}}_e = \mathbf{diag}(s_1^2 \mathbf{I}_{v_1}, s_2^2 \mathbf{I}_{v_2}, \cdots, s_{27}^2 \mathbf{I}_{v_{27}}).$$

Estimates of the parameters in model (3.6.17) are obtained by the method of weighted least squares using the estimated variance–covariance matrix $\hat{\mathbf{\Sigma}}_e$. The resulting prediction equation is given by

$$\begin{aligned} \hat{\mathcal{F}} = \; & 2.59 - 0.09 n_c + 0.06\phi - 19.19\eta \\ & +0.07 n_c \phi - 0.19 n_c \eta - 1.95 \phi \eta \\ & +0.07 n_c^2 + 0.16 \phi^2 + 36.95 \eta^2. \end{aligned} \quad (3.6.18)$$

The coefficient of determination for model (3.6.18) is $R^2 = 0.9959$, indicating a good fit for this model. The standard errors of the weighted least-squares estimates of the α's are given in Table 3.7. We note that most of the parameters in the model are significantly different from zero.

Having developed an adequate empirical relationship according to model (3.6.18), we can now utilize $\hat{\mathcal{F}}$ to have a better insight into the effect of ϕ, as well as the effects of $n.$ and η, on the precision of estimating σ_τ^2. By exploring the response surface defined by $\hat{\mathcal{F}}$ over the region Λ, it can be shown that small values of $\hat{\mathcal{F}}$ occur for large values of $n.$ and ϕ and small values of η. Some contour plots of $\hat{\mathcal{F}}$ are shown in Figures 3.2, 3.3, 3.4. In Figure 3.2, $n. = 75$. Similar patterns, but with larger values of $\hat{\mathcal{F}}$, result when $n. = 25, 50$. In Figure 3.3, $\eta = 0.20$. Here also, similar contour plots and larger values of $\hat{\mathcal{F}}$ result when $\eta = 0.40$, 0.80. In Figure 3.4, $\phi = 0.85$. Changing ϕ to 0.32 and 0.60 produce similar contours, but with larger values of $\hat{\mathcal{F}}$.

We note from Tables 3.5 and 3.6 that $\phi = 1$ was not used in building model (3.6.18). The reason for this is that when $\phi = 1$, the design **D** is balanced, giving rise to only one value for each $n_i (i = 1, 2, 3, 4, 5)$, namely $n./5$. In this case, it is not possible to obtain replications on $\mathcal{F}$, and hence, the corresponding sample variance (for the weighted least-squares fit) is equal to zero. To determine if model (3.6.18) provides good predictions when $\phi = 1$, values of $\mathcal{F}$ are obtained at 9 combinations of $n.$ and η using formula (3.6.15) with

Table 3.5. Generated Designs and Corresponding Values of $Var(\hat{\sigma}_\tau^2)/\sigma_e^4$

$n.$	ϕ	η	n_1	n_2	n_3	n_4	n_5	ϕ_a	$Var(\hat{\sigma}_\tau^2)/\sigma_e^4$
25	0.32	0.2	19	3	1	1	1	0.3351	0.24842
25	0.32	0.2	19	2	1	2	1	0.3369	0.24115
25	0.60	0.2	1	5	12	1	6	0.6039	0.14328
25	0.60	0.2	1	2	2	9	11	0.5924	0.14830
25	0.60	0.2	1	3	2	7	12	0.6039	0.14199
25	0.60	0.2	3	13	2	2	5	0.5924	0.13780
25	0.85	0.2	4	8	3	3	7	0.8503	0.11403
25	0.85	0.2	3	7	2	6	7	0.8503	0.11502
25	0.85	0.2	4	9	3	4	5	0.8503	0.11305
25	0.85	0.2	1	7	5	6	6	0.8503	0.11600
50	0.32	0.2	1	2	39	2	6	0.3193	0.13756
50	0.32	0.2	1	39	3	1	6	0.3189	0.13948
50	0.32	0.2	1	39	4	1	5	0.3197	0.13608
50	0.60	0.2	3	25	2	10	10	0.5967	0.08464
50	0.60	0.2	3	24	1	12	10	0.6024	0.08732
50	0.60	0.2	4	25	11	2	8	0.6024	0.08301
50	0.60	0.2	2	24	2	12	10	0.6039	0.08682
50	0.60	0.2	4	26	4	7	9	0.5967	0.07959
50	0.60	0.2	3	25	12	3	7	0.5980	0.08368
50	0.60	0.2	10	2	1	23	14	0.6024	0.09027

Table 3.6. Values of the Sample Variances

$n.$	ϕ	η	s_i^2
25	0.32	0.2	0.0000264
25	0.60	0.2	0.0000187
25	0.85	0.2	0.0000016
50	0.32	0.2	0.0000029
50	0.60	0.2	0.0000161
50	0.85	0.2	0.00000018
75	0.32	0.2	0.0000283
75	0.60	0.2	0.0000145
75	0.85	0.2	0.00000065
25	0.32	0.4	0.0007655
25	0.60	0.4	0.0009481
25	0.85	0.4	0.0000817
50	0.32	0.4	0.0001206
50	0.60	0.4	0.0008264
50	0.85	0.4	0.0000088
75	0.32	0.4	0.0012815
75	0.60	0.4	0.0007419
75	0.85	0.4	0.0000328
25	0.32	0.8	0.8374782
25	0.60	0.8	1.236448
25	0.85	0.8	0.105921
50	0.32	0.8	0.1462931
50	0.60	0.8	1.0760354
50	0.85	0.8	0.0112776
75	0.32	0.8	1.5970851
75	0.60	0.8	0.9648283
75	0.85	0.8	0.0424402

$n_i = n./5$, $i = 1, 2, 3, 4, 5$. These values are then compared against the corresponding values of $\hat{\mathcal{F}}$ using model (3.6.18). The results are given in Table 3.8. We note that the values of $\mathcal{F}$ and $\hat{\mathcal{F}}$ are quite close even though the latter function is evaluated at points outside the region Λ.

As mentioned earlier in this section, the methodology of studying the effect of imbalance can be easily extended to other unbalanced models using the general measure of imbalance in Section 3.4. Furthermore, quantities other than $\mathcal{F}$ in formula (3.6.15) can be considered. For example, the size or power of an approximate test concerning a variance component, the probability that one of its estimators is negative, and the coverage probability of a corresponding confidence interval are such quantities. It is also feasible to use response surface techniques to construct efficient designs for estimating variance components.

Table 3.7. Weighted Least-Squares Estimates for Model (3.6.17)

Parameter	Estimate	Standard Error	P-Value
α_0	2.59	0.0283	0.0001
α_1	−0.09	0.0123	0.0001
α_2	0.06	0.0422	0.1508
α_3	−19.19	0.1644	0.0001
α_{12}	0.07	0.0127	0.0001
α_{13}	−0.19	0.0301	0.0001
α_{23}	−1.95	0.1146	0.0001
α_{11}	0.07	0.0033	0.0001
α_{22}	0.16	0.0282	0.0001
α_{33}	36.95	0.2206	0.0001

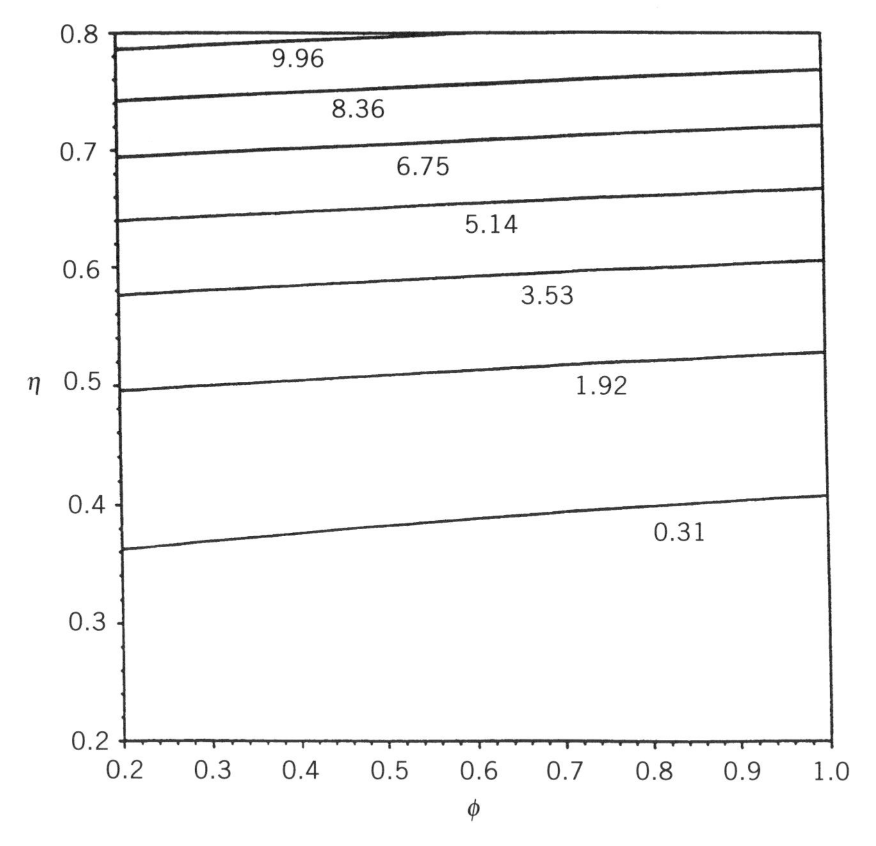

Figure 3.2. Contour plots of $\hat{\mathcal{F}}, n_{.} = 75$. *Source*: A. I. Khuri (1996). Reproduced with permission of Elsevier Science BV.

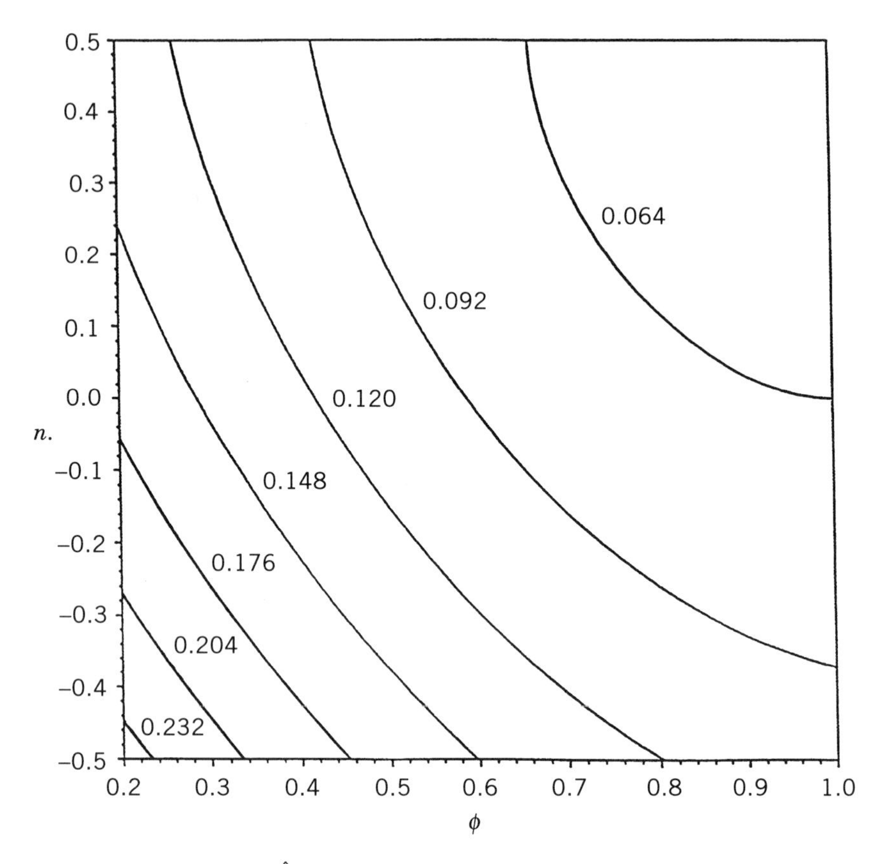

Figure 3.3. Contour plots of $\hat{\mathcal{F}}$, $\eta = 0.20$. *Source*: A. I. Khuri (1996). Reproduced with permission of Elsevier Science BV.

Table 3.8. Values of $\mathcal{F}$ and $\hat{\mathcal{F}}$ When $\phi = 1$

$n.$	η	$\mathcal{F}$	$\hat{\mathcal{F}}$	$\mathcal{F} - \hat{\mathcal{F}}$
25	0.20	0.105	0.106	−0.001
25	0.40	0.379	0.331	0.048
25	0.80	8.824	9.649	−0.825
50	0.20	0.062	0.064	−0.002
50	0.40	0.294	0.270	0.024
50	0.80	8.405	9.551	−1.146
75	0.20	0.050	0.058	−0.008
75	0.40	0.269	0.245	0.024
75	0.80	8.269	9.489	−1.220

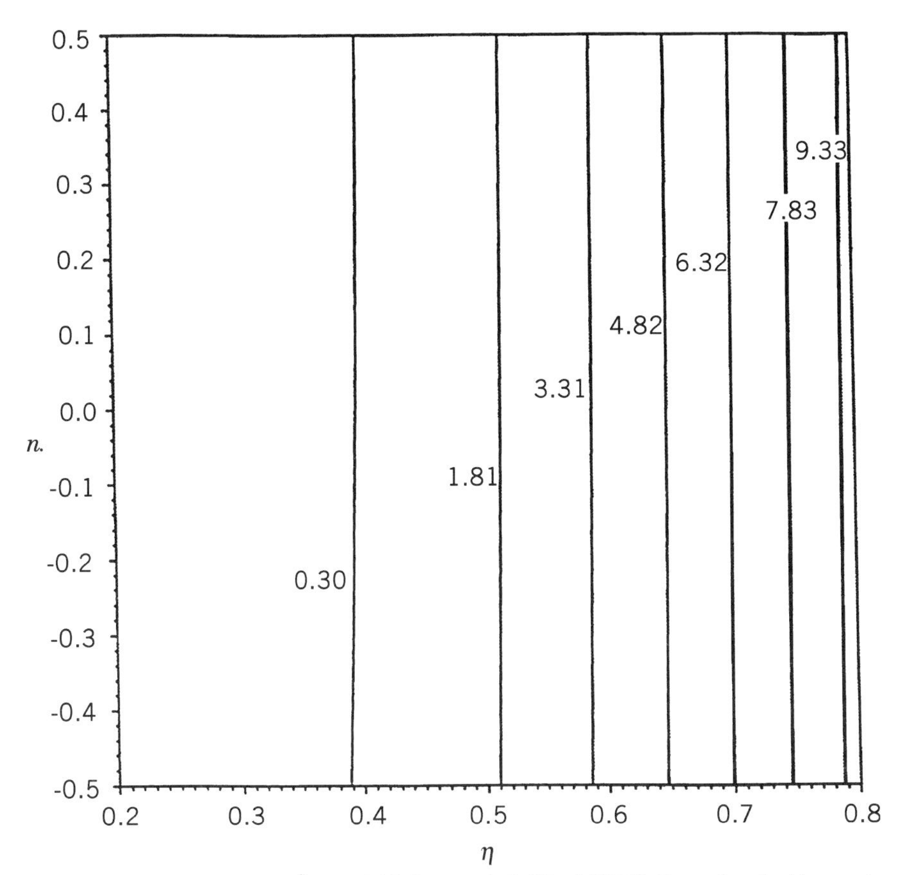

Figure 3.4. Contour plots of $\hat{\mathcal{F}}$, $\phi = 0.85$. *Source*: A. I. Khuri (1996). Reproduced with permission of Elsevier Science BV.

3.7. SUMMARY

Different types of balance can be considered for a given linear model involving crossed and/or nested effects. If imbalance affects only the last stage of the associated design, a certain loglinear model can be used to represent the expected frequencies. This was demonstrated in the special cases of proportionality of subclass frequencies, last-stage uniformity, and complete balance. The loglinear model for each one of these cases is obtained as a particular submodel of the full loglinear model associated with the given model. The goodness of fit of each submodel is determined by an approximate chi-squared statistic. The value of this statistic is then used to measure departure from the corresponding type of balance.

If imbalance occurs in several stages of the associated design, then the corresponding measure of imbalance is obtained as a weighted average of the measures of imbalance associated with the various stages. This was demonstrated in Sections 3.5.1 and 3.5.2.

The immediate advantage of measuring imbalance is the ability to describe in a quantitative manner different degrees of imbalance such as extreme imbalance, moderate imbalance, and near balance. The ultimate goal of such a practice is to have a better understanding of the effects of imbalance on the efficiencies of estimators and tests concerning the parameters of an unbalanced mixed model. A demonstration of how this can be carried out was given in Section 3.6 for the special case of the one-way model using the measure $\phi = \nu(\mathbf{D})$ (see formula 3.3.2). A method for generating designs having a specified value of ϕ was utilized to build an empirical relationship in which ϕ is one of several control variables that affect a quantity of interest such as $\mathcal{F}$ in formula (3.6.15). This relationship was then used to study the effect of ϕ, as well as the effects of other control variables, on the precision of the ANOVA estimator of σ_τ^2. Extensions of the methodology to other unbalanced models and other quantities of interest can be easily developed.

APPENDIX 3.1

Hirotsu's Approximation

Hirotsu's (1979) approximation concerns the upper probability values of a statistic of the form

$$H = \frac{\mathbf{x}'\mathbf{V}\mathbf{x}/(\xi f)}{\hat{\sigma}^2/\sigma^2},$$

where $\mathbf{x}$ is normally distributed with a mean $\boldsymbol{\mu}$ and a variance–covariance matrix $\boldsymbol{\Gamma}$, $\mathbf{V}$ is a nonnegative matrix, $\hat{\sigma}^2/\sigma^2$ is distributed as $(1/f_2)\chi^2_{f_2}$ independently of $\mathbf{x}'\mathbf{V}\mathbf{x}$, and ξ and f are given by

$$\begin{aligned} \xi &= \frac{1}{2}\kappa_2(\mathbf{x}'\mathbf{V}\mathbf{x})/\kappa_1(\mathbf{x}'\mathbf{V}\mathbf{x}), \\ f &= 2\kappa_1^2(\mathbf{x}'\mathbf{V}\mathbf{x})/\kappa_2(\mathbf{x}'\mathbf{V}\mathbf{x}), \end{aligned}$$

where $\kappa_i(\mathbf{x}'\mathbf{V}\mathbf{x})$ denotes the ith cumulant of $\mathbf{x}'\mathbf{V}\mathbf{x}$. According to Hirotsu (1979, formula 2.4), the upper probability value $P(H \geq q)$ for a given q is approximately equal to

$$\begin{aligned} P(H \geq q) \approx\ & P(F_{f,f_2} \geq q) + [\delta/\{3(f+2)(f+4)B(\tfrac{1}{2}f, \tfrac{1}{2}f_2)\}] \\ & \times (1 + fq/f_2)^{-(f+f_2)/2} \times (fq/f_2)^{f/2} \\ & \times \left[(f+2)(f+4) - \frac{2(f+f_2)(f+4)}{1+f_2/(fq)} + \frac{(f+f_2+2)(f+f_2)}{\{1+f_2/(fq)\}^2}\right], \end{aligned}$$

where $B(\cdot,\cdot)$ denotes the beta function, and

$$\delta = \frac{1}{2}\left[\kappa_1(\mathbf{x}'\mathbf{V}\mathbf{x})\kappa_3(\mathbf{x}'\mathbf{V}\mathbf{x})/\kappa_2^2(\mathbf{x}'\mathbf{V}\mathbf{x})\right] - 1.$$

The approximation of $P(H \geq q)$ was developed via a generalized Laguerre polynomial expansion of the true distribution of the statistic $\mathbf{x}'\mathbf{V}\mathbf{x}/(2\xi)$ and was reported in Hirotsu (1979) to be quite satisfactory.

EXERCISES

3.1. Consider the ANOVA table for model (3.2.1) (Table 3.1).

(a) Verify that

$$E[SS(\tau)] = \left(n. - \frac{1}{n.}\sum_{i=1}^{v} n_i^2\right)\sigma_\tau^2 + (v-1)\sigma_e^2.$$

(b) Show that $SS(\tau)$ and $SS(e)$ are independent.

(c) Find the covariance between $\hat{\sigma}_\tau^2$ and $\hat{\sigma}_e^2$, the ANOVA estimators of σ_τ^2 and σ_e^2, respectively.

(d) Give an expression for the moment generating function of $\hat{\sigma}_\tau^2$. [Hint: Use formulas (3.2.4) and (3.2.5).]

3.2. Consider again Table 3.1.

(a) Find the expected value of $SS(\tau)/SS(e)$

(b) Show that

$$\frac{1}{d}\left[\frac{n. - v - 2}{n. - v}\,\frac{MS(\tau)}{MS(e)} - 1\right]$$

is an unbiased estimator of σ_τ^2/σ_e^2, where $MS(\tau) = SS(\tau)/(v-1)$, $MS(e) = SS(e)/(n. - v)$, and $d = 1/(v-1)[n. - (1/n.)\sum_{i=1}^{v} n_i^2]$.

3.3. Consider model (3.2.1) under the following assumptions:

(i) The τ_i's are identically and independently distributed as $N(0, \sigma_\tau^2)$.

(ii) The e_{ij}'s are independently distributed as $N(0, \sigma_i^2)$, $i = 1, 2, \cdots, v$; $j = 1, 2, \cdots, n_i$ (heteroscedastic random errors).

(iii) The τ_i's and e_{ij}'s are independent.

(a) Under assumptions (i)–(iii), what distributions do $SS(\tau)$ and $SS(e)$ have, where $SS(\tau)$ and $SS(e)$ are the sums of squares in Table 3.1?

(b) Are $SS(\tau)$ and $SS(e)$ independent under assumptions (i)–(iii)?

(c) Show that the ANOVA estimator, $\hat{\sigma}_\tau^2$, in formula (3.2.5) is a biased estimator of σ_τ^2 under assumptions (i)–(iii).

3.4. Consider the unbalanced random one-way model

$$y_{ij} = \mu + \tau_i + e_{ij}$$

$i = 1,2,3,4$; $j = 1,2,\cdots,n_i$, where $n_1 = 5$, $n_2 = 4$, $n_3 = 4$, and $n_4 = 6$. The τ_i's and e_{ij}'s are independently distributed as $N(0, \sigma_\tau^2)$ and $N(0, \sigma_e^2)$, respectively. Define the random variables U_1, U_2, U_3, and U_4 as

$$U_i = c_1 \bar{y}_{i.} + c_{2i} \sum_{j=1}^{n_i} \ell_{ij} y_{ij}, \quad i = 1,2,3,4,$$

where c_1, c_{2i}, and ℓ_{ij} are constants such that $\sum_{j=1}^{n_i} \ell_{ij} = 0 \ (i = 1,2,3,4)$, and $\bar{y}_{i.} = \frac{1}{n_i} \sum_{j=1}^{n_i} y_{ij}$.

(a) Show that $E(U_i) = c_1 \mu$, $i = 1,2,3,4$.

(b) Show that

$$Var(U_i) = c_1^2 \sigma_\tau^2 + \left(\frac{c_1^2}{n_i} + c_{2i}^2 \sum_{j=1}^{n_i} \ell_{ij}^2 \right) \sigma_e^2, \quad i = 1,2,3,4.$$

(c) Show that U_1, U_2, U_3, and U_4 are independently distributed and have the normal distribution.

(d) Show that c_1, c_{2i}, and ℓ_{ij} can be determined so that

$$Var(U_i) = \sigma_\tau^2 + 2\sigma_e^2, \quad i = 1,2,3,4.$$

(e) Derive an unbiased estimator of $\sigma_\tau^2 + 2\sigma_e^2$ using U_1, U_2, U_3, and U_4.

(f) Make use of the result in part (e) to obtain an exact $(1-\alpha)100\%$ confidence interval on $\sigma_\tau^2 + 2\sigma_e^2$.
[Note: This problem is based on a method proposed by Burdick and Sielken (1978) for the construction of exact confidence intervals on nonnegative linear combinations of the variance components for the one-way model.]

3.5. Consider the two-fold nested model

$$y_{ijk} = \mu + \tau_i + \beta_{i(j)} + e_{ijk}$$

$i = 1,2,\cdots,v$; $j = 1,2,\cdots,b_i$; $k = 1,2,\cdots,n_{ij}$. It is assumed that $\tau_i \sim N(0, \sigma_\tau^2)$, $\beta_{i(j)} \sim N(0, \sigma_{\beta(\alpha)}^2)$, $e_{ijk} \sim N(0, \sigma_e^2)$, and that the τ_i's, $\beta_{i(j)}$'s, and e_{ijk}'s are independent.
Consider the following ANOVA table:

Source	Sum of Squares	d.f.
τ_i	$\mathbf{y}'\mathbf{Q}_1\mathbf{y}$	$v-1$
$\beta_{i(j)}$	$\mathbf{y}'\mathbf{Q}_2\mathbf{y}$	$b_. - v$
Error	$\mathbf{y}'\mathbf{Q}_3\mathbf{y}$	$n_{..} - b_.$

where $\mathbf{y}$ is the vector of y_{ijk}'s, and

$$\mathbf{Q}_1 = \oplus_{i=1}^{v}\left(\frac{\mathbf{J}_{n_{i.}}}{n_{i.}}\right) - \frac{\mathbf{J}_{n_{..}}}{n_{..}}$$

$$\mathbf{Q}_2 = \oplus_{i=1}^{v}\left[\oplus_{j=1}^{b_i}\left(\frac{\mathbf{J}_{n_{ij}}}{n_{ij}}\right) - \frac{\mathbf{J}_{n_{i.}}}{n_{i.}}\right]$$

$$\mathbf{Q}_3 = \mathbf{I}_{n_{..}} - \oplus_{i=1}^{v}\left[\oplus_{j=1}^{b_i}\left(\frac{\mathbf{J}_{n_{ij}}}{n_{ij}}\right)\right].$$

(a) Verify that for Design 1 in Table 3.2, $\mathbf{y}'\mathbf{Q}_1\mathbf{y}$ and $\mathbf{y}'\mathbf{Q}_2\mathbf{y}$ are independent.

(b) Verify that for Design 2 in Table 3.2, $\mathbf{y}'\mathbf{Q}_1\mathbf{y}$ and $\mathbf{y}'\mathbf{Q}_2\mathbf{y}$ are each distributed as a scalar multiple of a chi-squared variate.

(c) Find the expected values of $\mathbf{y}'\mathbf{Q}_1\mathbf{y}$ and $\mathbf{y}'\mathbf{Q}_2\mathbf{y}$ in general.

(d) Obtain an approximate F statistic for testing the hypothesis $H_\tau: \sigma_\tau^2 = 0$ at the $\alpha = 0.05$ level.

(e) Using Design 1, obtain an approximate value for the true level of significance for the test in part (d) given that $\sigma_{\beta(\alpha)}^2 = 4$, $\sigma_e^2 = 1$.

(f) Redo part (e) using Designs 2 and 3 in Table 3.2. Compare the results with those in part (e).

3.6. The sum of squares $SS(\tau)$ in Table 3.1 is approximately distributed as $\lambda\chi_m^2$, where λ and m are given in formulas (3.2.13) and (3.2.14), respectively. The closeness of this approximation depends on the value of $\Delta - 1$, where Δ is given in formula (3.2.16). Recall that $\Delta \geq 1$, and that equality is achieved if and only if the approximation is exact.

(a) Compute the value of $\Delta - 1$ for the following designs:

(i) $n_1 = 5, n_2 = 3, n_3 = 6, n_4 = 5, n_5 = 3, n_6 = 8$.
(ii) $n_1 = 2, n_2 = 3, n_3 = 2, n_4 = 3, n_5 = 3, n_6 = 17$,
given that $\sigma_\tau^2/\sigma_e^2 = 3.0$.

(b) Redo part (a) using the following sum of squares (instead of $SS(\tau)$):

$$SS^*(\tau) = \tilde{n}\left[\sum_{i=1}^{v}\bar{y}_{i.}^2 - \frac{1}{v}\left(\sum_{i=1}^{v}\bar{y}_{i.}\right)^2\right],$$

where $\bar{y}_{i.} = y_{i.}/n_i$, $\tilde{n} = v/\sum_{i=1}^{v} 1/n_i$.
[Note: $SS^*(\tau)$ is the sum of squares derived by Thomas and Hultquist (1978, pp. 584–585).]

3.7. **(a)** Use the algorithm described in Section 3.6.1 to generate two designs for a one-way model given that $v = 5$, $n_{.} = 50$, $\phi = 0.80$.

(b) If $\eta = \sigma_{\tau}^2/(\sigma_{\tau}^2 + \sigma_e^2) = 0.30$, which of the two designs generated in part (a) is better in terms of $\mathcal{F}$ in formula 3.6.15?

BIBLIOGRAPHY

Agresti, A. (1990). *Categorical Data Analysis.* Wiley, New York.

Ahrens, H. J. and Pincus, R. (1981). "On two measures of unbalancedness in a one-way model and their relation to efficiency." *Biometrical Journal*, 23, 227–235.

Ahrens, H. J. and Sanchez, J. E. (1982). "Unbalancedness and efficiency in estimating components of variance: MINQUE and ANOVA procedure." *Biometrial Journal*, 24, 649–661.

Ahrens, H. J. and Sanchez, J. E. (1988). "Unbalancedness of designs, measures of." In: *Encyclopedia of Statistical Sciences*, Vol. 9 (S. Kotz, N. L. Johnson, Eds.), Wiley, New York, 383–386.

Ahrens, H. J. and Sanchez, J. E. (1992). "Imbalance and its influence on variance component estimation." *Biometrical Journal*, 34, 539–555.

Anderson, R. L. and Crump, P. P. (1967). "Comparisons of designs and estimation procedures for estimating parameters in a two-stage nested process." *Technometrics*, 9, 499–516.

Burdick, R. K. and Sielken, R. L. (1978). "Exact confidence intervals for linear combinations of variance components in nested classifications." *Journal of the American Statistical Association*, 73, 632–635.

Caro, R. F., Grossman, M., and Fernando, R. L. (1985). "Effects of data imbalance on estimation of heritability." *Theoretical and Applied Genetics*, 69, 523–530.

Cummings, W. B. and Gaylor, D. W. (1974). "Variance component testing in unbalanced nested designs." *Journal of the American Statistical Association*, 69, 765–771.

Das, R. and Sinha, B. K. (1987). "Robust optimum invariant unbiased tests for variance components." In: *Proceedings of the Second International Tampere Conference in Statistics* (T. Pukkila, S. Puntanen, Eds.), University of Tampere, Finland, 317–342.

Davies, R. B. (1973). "Numerical inversion of a characteristic function." *Biometrika*, 60, 415–417.

Davies, R. B. (1980). "The distribution of a linear combination of χ^2 random variables." *Applied Statistics*, 29, 323–333.

Donner, A. and Koval, J. J. (1987). "A procedure for generating group sizes from a one-way classification with a specified degree of imbalance." *Biometrical Journal*, 29, 181–187.

Donner, A. and Koval, J. J. (1989). "The effect of imbalance on significance-testing in one-way Model II analysis of variance." *Communications in Statistics—Theory and Methods*, 18, 1239–1250.

Edwards, C. H., Jr. (1973). *Advanced Calculus of Several Variables*. Academic Press, New York.

Graybill, F. A. (1983). *Matrices with Applications in Statistics*, Second Edition. Wadsworth, Belmont, California.

Hernandez, R. P. and Burdick, R. K. (1993). "Confidence intervals on the total variance in an unbalanced two-fold nested design." *Biometrical Journal*, 35, 515–522.

Hernandez, R. P., Burdick, R. K., and Birch, N. J. (1992). "Confidence intervals and tests of hypotheses on variance components in an unbalanced two-fold nested design." *Biometrical Journal*, 34, 387–402.

Hirotsu, C. (1979). "An F approximation and its application." *Biometrika*, 66, 577–584.

Imhof, J. P. (1961). "Computing the distribution of quadratic forms in normal variables." *Biometrika*, 48, 419–426.

Johnson, N. L. and Kotz, S. (1970). *Continuous Univariate Distributions–2*. Wiley, New York.

Khuri, A. I. (1987). "Measures of imbalance for unbalanced models." *Biometrical Journal*, 29, 383–396.

Khuri, A. I. (1995). "A measure to evaluate the closeness of Satterthwaite's approximation." *Biometrical Journal*, 37, 547–563.

Khuri, A. I. (1996). "A method for determining the effect of imbalance." *Journal of Statisitcal Planning and Inference*, 55, 115–129.

Khuri, A. I. and Cornell, J. A. (1996). *Response Surfaces*, Second Edition. Dekker, New York.

Lera Marqués, L. (1994). "Measures of imbalance for higher-order designs." *Biometrical Journal*, 36, 481–490.

Newton, H. J. (1993). "New developments in statistical computing." *The American Statistician*, 47, 146.

Satterthwaite, F. E. (1941). "Synthesis of variance." *Psychometrika*, 6, 309–316.

Searle, S. R. (1971). *Linear Models*. Wiley, New York.

Searle, S. R., Casella, G., and McCulloch, C. E. (1992). *Variance Components*. Wiley, New York.

Singh, B. (1989). "A comparison of variance component estimators under unbalanced situations." *Sankhyā*, Series B, 51, 323–330.

Singh, B. (1992). "On the effect of unbalancedness and heteroscedasticity on the ANOVA estimator of group variance component in one-way random model." *Biometrical Journal*, 34, 91–96.

Thomas, J. D. and Hultquist, R. A. (1978). "Interval estimation for the unbalanced case of the one-way random effects model." *Annals of Statistics*, 6, 582–587.

CHAPTER 4

Unbalanced One-Way and Two-Way Random Models

4.1. INTRODUCTION

This chapter deals with the derivation of various exact and optimum tests for testing the significance of the variance components in the unbalanced one-way random model, the unbalanced two-way crossed classification model, and the unbalanced two-fold nested classification model. Unlike the balanced case, the analysis of such unbalanced models is somewhat complicated. The main difficulty stems from the fact that in the unbalanced situation, the partitioning of the total sum of squares can be done in a variety of ways and hence, there is no unique way to write the ANOVA table. Furthermore, the sums of squares in an ANOVA table corresponding to an unbalanced model are not in general independent or distributed as chi-squared variates. Consequently, until fairly recently, no exact or optimum test procedures were known for testing hypotheses in unbalanced models, except in a few special cases. In this chapter we shall describe the various exact and optimum test procedures that are available for the one-way and two-way random models.

The unbalanced one-way random model is analyzed in Section 4.2. It is shown that the ANOVA-based F-test is valid for testing the significance of the random effect variance component. However, this test is not UMPI, unless we have a balanced model. In fact, a UMPI test does not exist in the unbalanced case. An LBI test exists and it turns out to be different from the ANOVA-based F-test. The LBI test is also derived in Section 4.2.

The two-way crossed classification model with and without interaction is considered in Section 4.3. This section mainly deals with the construction of exact tests for testing the significance of the main effects and interaction variance components. In some instances, the exact tests are obtained using an idea due to Wald (1947), generalized in Seely and El-Bassiouni (1983). (This procedure is explained in Section 1.2.) However, the Wald method is not applicable for testing certain hypotheses, such as hypotheses concerning the main effects variance components when interaction is present. The results due to Khuri and Littell (1987) are useful in such situations. By a suitable linear transformation of the data, these authors have derived sums of squares quite similar to the balanced ANOVA sums of squares. Exact F-tests are then constructed based on such sums of squares. Details appear in Section 4.3. In the same section, we have also given a brief discussion of the optimum test derived in Mathew and Sinha (1988), applicable to models without interaction.

Section 4.4 deals with the unbalanced two-fold nested classification model. The problem is to test the significance of the variance components corresponding to the nesting as well as the nested effects. A Wald-type test is developed for testing a hypothesis regarding the nested effect variance component. For testing the significance of the nesting effect variance component, we have given the results due to Khuri (1987).

Until recently, tests concerning the main effects variance component in a random two-way crossed classification model with interaction, and concerning the nesting effect variance component in a random twofold nested model, were performed using ANOVA-based approximate F-tests and utilizing Satterthwaite's approximation; see Tan and Cheng (1984), Khuri (1987), and Khuri and Littell (1987). The simulation study in Khuri (1987) and Khuri and Littell (1987) show that Satterthwaite's approximation can be highly unreliable for producing critical values of the tests. The exact F-tests developed in this chapter are obviously free of this drawback. The simulation study reported by the above authors also shows that, in terms of power, such exact F-tests are at least as good as the approximate tests in situations where Satterthwaite's approximation is satisfactory. Consequently, for testing the significance of the variance components in unbalanced two-way models, we do not recommend tests based on Satterthwaite's approximation.

4.2. UNBALANCED ONE-WAY RANDOM MODELS

Consider the observations y_{ij} $(i = 1, 2, \ldots, v;\ j = 1, 2, \ldots, n_i)$ following the unbalanced one-way random model given by

$$y_{ij} = \mu + \tau_i + e_{ij}, \tag{4.2.1}$$

where μ is a fixed unknown parameter and the τ_i's and e_{ij}'s are independent random variables with $\tau_i \sim N(0, \sigma_\tau^2)$ and $e_{ij} \sim N(0, \sigma_e^2)$ $(i = 1, 2, \ldots, v;\ j = 1, 2, \ldots, n_i)$. The τ_i's are the random treatment effects and the e_{ij}'s are the experimental error terms. In this section, we shall address the problem of testing

$$H_\tau : \sigma_\tau^2 = 0 \text{ vs } H_1 : \sigma_\tau^2 > 0. \tag{4.2.2}$$

In order to write model (4.2.1) in matrix notation, let $\mathbf{y} = (y_{11}, y_{12}, \ldots, y_{1n_1}, y_{21}, y_{22}, \ldots, y_{2n_2}, \ldots, y_{v1}, y_{v2}, \ldots, y_{vn_v})'$ and $\boldsymbol{\tau} = (\tau_1, \tau_2, \ldots, \tau_v)'$. Furthermore, let $\mathbf{1}_m$ denote the m-component vector of ones. Then (4.2.1) can equivalently be written as

$$\mathbf{y} = \mu\mathbf{1}_{n.} + \mathbf{diag}(\mathbf{1}_{n_1}, \mathbf{1}_{n_2}, \cdots, \mathbf{1}_{n_v})\boldsymbol{\tau} + \mathbf{e}, \tag{4.2.3}$$

where, $n_. = \sum_{i=1}^{v} n_i$, $\mathbf{diag}(\boldsymbol{A}_1, \boldsymbol{A}_2 \ldots, \boldsymbol{A}_v)$ denotes a block-diagonal matrix with $\boldsymbol{A}_1, \boldsymbol{A}_2, \ldots, \boldsymbol{A}_v$ along the blocks and $\mathbf{e}$ is defined in a manner similar to $\mathbf{y}$. The distributional assumptions on the τ_i's and e_{ij}'s immediately give

$$\boldsymbol{\tau} \sim N(\mathbf{0}, \sigma_\tau^2 \boldsymbol{I}_v), \ \mathbf{e} \sim N(\mathbf{0}, \sigma_e^2 \boldsymbol{I}_{n.}). \tag{4.2.4}$$

From (4.2.3) and (4.2.4) we get

$$\mathrm{E}(\mathbf{y}) = \mu\mathbf{1}_{n.}, \ Var(\mathbf{y}) = \sigma_\tau^2 \mathbf{diag}(\boldsymbol{J}_{n_1}, \boldsymbol{J}_{n_2}, \cdots, \boldsymbol{J}_{n_v}) + \sigma_e^2 \boldsymbol{I}_{n.}, \tag{4.2.5}$$

where, $\boldsymbol{J}_m = \mathbf{1}_m \mathbf{1}_m'$. Let $\bar{y}_{i.} = 1/n_i \sum_{j=1}^{n_i} y_{ij}$ and $\bar{y}_{..} = 1/n_. \sum_{i=1}^{v} \sum_{j=1}^{n_i} y_{ij}$. Table 3.1 in Chapter 3 gives the ANOVA table for model (4.2.1), along with the expected values of the two mean squares, namely, the mean square due to the τ_i's (the treatment mean square) and the mean square due to the e_{ij}'s (the error mean square), respectively denoted by $MS(\tau)$ and $MS(e)$. It is well known that $SS(e) \sim \sigma_e^2 \chi_{n_.-v}^2$ and, under $H_\tau : \sigma_\tau^2 = 0$, $SS(\tau) \sim \sigma_e^2 \chi_{v-1}^2$. Hence, from the table of expected values in Table 3.1, it is clear that an F-test for testing (4.2.2) can be obtained using the ratio F_τ given by

$$F_\tau = \frac{SS(\tau)}{(v-1)} \Bigg/ \frac{SS(e)}{(n_. - v)}, \tag{4.2.6}$$

which follows a central F-distribution with $v - 1$ and $n_. - v$ degrees of freedom, when $H_\tau : \sigma_\tau^2 = 0$ is true.

The development given so far is quite standard. We shall now analyze the above testing problem via invariance, and explore the existence of an optimum invariant test. We note that the testing problem (4.2.2) for model (4.2.3) is invariant under the action of the group of transformations $\mathcal{G} = \{(c, \alpha): c > 0, \ \alpha \text{ real}\}$ acting on $\mathbf{y}$ as

$$\mathbf{y} \to c(\mathbf{y} + \alpha\mathbf{1}_{n.}).$$

Note that (4.2.3) is a special case of the general mixed model analyzed in Section 6.2 of Chapter 6. From Theorem 6.2.2, it follows that a UMPI test exists for the testing problem (4.2.2) if and only if the nonzero eigenvalues of the matrix

$$\boldsymbol{C} = \mathbf{diag}(n_1, n_2, \cdots, n_v) - \frac{1}{n_.}\mathbf{n}\mathbf{n}' \tag{4.2.7}$$

are all equal, where $\mathbf{n} = (n_1, n_2, \ldots, n_v)'$. It is shown in Lemma 4.2.1 below that the nonzero eigenvalues of $\boldsymbol{C}$ in (4.2.7) are all equal if and only if the n_i's are all equal, that is, (4.2.3) is a balanced model. Thus, the results in

Chapter 2 are applicable. Hence the F-test based on F_τ in (4.2.6) provides a UMPI test for the testing problem (4.2.2) (see Example 2.2 in Section 3 of Chapter 2).

In the unbalanced case (i.e., when the n_i's are not all equal), the nonzero eigenvalues of $\boldsymbol{C}$ in (4.2.7) are not all equal, and it follows from Theorem 6.2.2 in Chapter 6 that there is no UMPI test for testing (4.2.2). However, the same theorem provides an LBI test in this situation. The LBI test rejects $H_\tau : \sigma_\tau^2 = 0$ for large values of the statistic L given by

$$\begin{aligned} L &= \frac{\mathbf{y}'(\boldsymbol{I}_{n.} - 1/n.\boldsymbol{J}_{n.})\mathbf{diag}(\boldsymbol{J}_{n_1}, \boldsymbol{J}_{n_2}, \cdots, \boldsymbol{J}_{n_v})(\boldsymbol{I}_{n.} - 1/n.\boldsymbol{J}_{n.})\mathbf{y}}{\mathbf{y}'(\boldsymbol{I}_{n.} - 1/n.\boldsymbol{J}_{n.})\mathbf{y}} \\ &= \frac{\sum_{i=1}^{v} n_i^2(\bar{y}_{i.} - \bar{y}_{..})^2}{\sum_{i=1}^{v}\sum_{j=1}^{n_i}(y_{ij} - \bar{y}_{i.})^2 + \sum_{i=1}^{v} n_i(\bar{y}_{i.} - \bar{y}_{..})^2}. \end{aligned} \tag{4.2.8}$$

The LBI test was derived by Das and Sinha (1987). In order to apply the LBI test in practice, one should be able to compute the percentiles of the statistic L in (4.2.8), or be able to compute the observed significance level of the test based on this statistic. One can use an algorithm due to Davies (1980) or an approximation due to Hirotsu (1979) to compute the observed significance level of the LBI test. Hirotsu's approximation is discussed in Appendix 3.1 of Chapter 3.

Note that the F-test based on F_τ in (4.2.6) is valid for testing H_τ: $\sigma_\tau^2 = 0$ in the balanced as well as unbalanced situations. It is also easy to verify that in the balanced case, the test based on L in (4.2.8) reduces to the F-test based on F_τ in (4.2.6). However, in the unbalanced case, the statistic L provides a test that is different from the one based on F_τ. Furthermore, the test based on L is expected to provide a larger local power compared to the F-test based on F_τ.

The following lemma states some properties of the eigenvalues of the matrix $\boldsymbol{C}$ in (4.2.7). The result and its proof are due to LaMotte (1976).

Lemma 4.2.1. Consider the unbalanced one-way random model (4.2.1) and let $\boldsymbol{C}$ be defined as in (4.2.7). Let s denote the number of distinct n_i's and suppose n_i is repeated g_i times $(i = 1, 2, \ldots, s)$. Assume that $n_1 < n_2 < \ldots < n_s$. Then

- **(a)** if $g_i > 1$, n_i is an eigenvalue of $\boldsymbol{C}$ with multiplicity $g_i - 1$,
- **(b)** between each pair of adjacent n_i's, $\boldsymbol{C}$ has an eigenvalue of multiplicity one and,
- **(c)** the nonzero eigenvalues of $\boldsymbol{C}$ are all equal if and only if the n_i's are all equal.

Proof. Write $\boldsymbol{n}^\delta = \mathbf{diag}(n_1, n_2, \ldots, n_v)$. Suppose λ is *not* an eigenvalue of $\boldsymbol{C}$ and also $\lambda \neq n_i$ $(i = 1, 2, \ldots, v)$. Then $\lambda \boldsymbol{I}_v - \boldsymbol{C}$ and $\lambda \boldsymbol{I}_v - \boldsymbol{n}^\delta$ are nonsingular matrices. Thus,

$$
\begin{aligned}
|\lambda \boldsymbol{I}_v - \boldsymbol{C}| &= \left|\lambda \boldsymbol{I}_v - \boldsymbol{n}^\delta + \frac{1}{n_.}\mathbf{n}\mathbf{n}'\right| = |\lambda \boldsymbol{I}_v - \boldsymbol{n}^\delta| \left|\boldsymbol{I}_v + \frac{1}{n_.}(\lambda \boldsymbol{I}_v - \boldsymbol{n}^\delta)^{-1}\mathbf{n}\mathbf{n}'\right| \\
&= \left(\prod_{i=1}^{s}(\lambda - n_i)^{g_i}\right)\left(1 + \frac{1}{n_.}\sum_{i=1}^{s}\frac{g_i n_i^2}{\lambda - n_i}\right) \\
&= \left(\lambda \prod_{i=1}^{s}(\lambda - n_i)^{g_i}\right)\left(\frac{1}{\lambda} + \frac{1}{n_.}\sum_{i=1}^{s}\frac{g_i n_i^2}{\lambda(\lambda - n_i)}\right) \\
&= \left(\lambda \prod_{i=1}^{s}(\lambda - n_i)^{g_i}\right)\left(\frac{1}{\lambda} + \frac{1}{n_.}\sum_{i=1}^{s} g_i n_i \left(\frac{1}{\lambda - n_i} - \frac{1}{\lambda}\right)\right) \\
&= \left(\lambda \prod_{i=1}^{s}(\lambda - n_i)^{g_i}\right)\left(\frac{1}{n_.}\sum_{i=1}^{s}\frac{g_i n_i}{\lambda - n_i}\right) \\
&= \left(\frac{1}{n_.}\lambda \prod_{i=1}^{s}(\lambda - n_i)^{g_i - 1}\right)\left(\sum_{i=1}^{s} g_i n_i \prod_{j\neq i}^{s}(\lambda - n_j)\right).
\end{aligned}
\tag{4.2.9}
$$

In order to arrive at the equality before the last one in (4.2.9), we have used $\sum_{i=1}^{s} g_i n_i = n_{.}$. By continuity, we therefore have

$$
|\lambda \boldsymbol{I}_v - \boldsymbol{C}| = \left(\frac{1}{n_.}\lambda \prod_{i=1}^{s}(\lambda - n_i)^{g_i - 1}\right)\left(\sum_{i=1}^{s} g_i n_i \prod_{j\neq i}^{s}(\lambda - n_j)\right), \tag{4.2.10}
$$

for all λ. Thus $\lambda = 0$ is an eigenvalue of $\boldsymbol{C}$ with multiplicity one and for each $g_i > 1$, n_i is an eigenvalue of multiplicity $g_i - 1$. The remaining eigenvalues of $\boldsymbol{C}$ are the roots of

$$
f(\lambda) = \sum_{i=1}^{s} g_i n_i \prod_{j\neq i}^{s}(\lambda - n_j). \tag{4.2.11}
$$

Note that

$$
\begin{aligned}
f(n_1) &= (-1)^{s-1} g_1 n_1 \prod_{j\neq 1}^{s} |n_1 - n_j| \\
f(n_2) &= (-1)^{s-2} g_2 n_2 \prod_{j\neq 2}^{s} |n_2 - n_j| \\
&\cdots\cdots\cdots \\
f(n_i) &= (-1)^{s-i} g_i n_i \prod_{j\neq i}^{s} |n_i - n_j|.
\end{aligned}
\tag{4.2.12}
$$

Thus $f(\lambda)$ changes sign between each pair of adjacent n_i's. Hence $f(\lambda)$ has exactly one root of multiplicity one between each pair of adjacent n_i's. From (4.2.12), it follows that no n_i is such a root. This completes the proof of (a) and (b). Part (c) follows from (a) and (b). This completes the proof of the lemma. □

4.3. TWO-WAY RANDOM MODELS

Let y_{ijk} be observations following the two-way random model given by

$$\begin{aligned} y_{ijk} &= \mu + \tau_i + \beta_j + (\tau\beta)_{ij} + e_{ijk}, \\ & i = 1, 2, \cdots, v;\; j = 1, 2, \cdots, b;\; k = 1, 2, \cdots, n_{ij}, \end{aligned} \tag{4.3.1}$$

where μ is a fixed unknown parameter, τ_i's, β_j's, $(\tau\beta)_{ij}$'s, and e_{ijk}'s are independent random variables having the distributions $\tau_i \sim N(0, \sigma_\tau^2)$, $\beta_j \sim N(0, \sigma_\beta^2)$, $(\tau\beta)_{ij} \sim N(0, \sigma_{\tau\beta}^2)$, and $e_{ijk} \sim N(0, \sigma_e^2)$. Thus the "$(ij)^{th}$ cell" has n_{ij} observations where $n_{ij} > 0$. The τ_i's and β_j's represent the main effects, $(\tau\beta)_{ij}$'s represent interaction effects, and the e_{ijk}'s represent the error terms. If $\sigma_{\tau\beta}^2 = 0$, then the model (4.3.1) reduces to a model without interaction. In this section, we shall first consider models without interaction and derive tests for testing the significance of the variance components σ_τ^2 and σ_β^2. We shall then consider models with interaction (i.e., the model (4.3.1)), and derive tests for testing the significance of the three variance components $\sigma_{\tau\beta}^2$, σ_τ^2, and σ_β^2. It will be shown that exact F-tests exist in all the cases. However, optimum invariant tests exist only in very special situations, depending on the nature of the n_{ij}'s. When interaction is present, the derivation of the exact tests for testing the significance of σ_τ^2 and σ_β^2 presented in this chapter requires the assumptions $n_{ij} \geq 1$ and $\sum_{i=1}^{v}\sum_{j=1}^{b} n_{ij} > 2bv - 1$. However, when interaction is absent, the derivation of such tests does not require any condition on the n_{ij}'s.

Writing $\mathbf{y}_{ij} = (y_{ij1}, y_{ij2}, \ldots, y_{ijn_{ij}})'$, $\mathbf{y} = (\mathbf{y}_{11}', \mathbf{y}_{12}', \cdots, \mathbf{y}_{1b}', \mathbf{y}_{21}', \mathbf{y}_{22}', \cdots, \mathbf{y}_{2b}', \cdots, \mathbf{y}_{v1}', \mathbf{y}_{v2}', \cdots, \mathbf{y}_{vb}')'$, $\boldsymbol{\tau} = (\tau_1, \tau_2, \ldots, \tau_v)'$, $\boldsymbol{\beta} = (\beta_1, \beta_2, \ldots, \beta_b)'$, and $(\boldsymbol{\tau\beta}) = ((\tau\beta)_{11}, (\tau\beta)_{12}, \ldots, (\tau\beta)_{1b}, \ldots, (\tau\beta)_{21}, (\tau\beta)_{22}, \ldots, (\tau\beta)_{2b}, \ldots, (\tau\beta)_{v1}, (\tau\beta)_{v2}, \ldots, (\tau\beta)_{vb})'$, the model (4.3.1) can be written as

$$\mathbf{y} = \mathbf{1}_{n_{..}}\mu + \boldsymbol{X}_1\boldsymbol{\tau} + \boldsymbol{X}_2\boldsymbol{\beta} + \boldsymbol{X}_3(\boldsymbol{\tau\beta}) + \mathbf{e}, \tag{4.3.2}$$

where $n_{..} = \sum_{i=1}^{v}\sum_{j=1}^{b} n_{ij}$ and $\mathbf{e}$ is defined similarly to $\mathbf{y}$. The design matrices $\boldsymbol{X}_1$, $\boldsymbol{X}_2$ and $\boldsymbol{X}_3$ are given by

$$\begin{aligned} \boldsymbol{X}_1 &= \mathbf{diag}(\mathbf{1}_{n_{1.}}, \cdots, \mathbf{1}_{n_{v.}}), \\ \boldsymbol{X}_2 &= \left[\mathbf{diag}(\mathbf{1}_{n_{11}}', \cdots, \mathbf{1}_{n_{1b}}') : \mathbf{diag}(\mathbf{1}_{n_{21}}', \cdots, \mathbf{1}_{n_{2b}}') : \ldots : \mathbf{diag}(\mathbf{1}_{n_{v1}}', \cdots, \mathbf{1}_{n_{vb}}')\right]' \\ \boldsymbol{X}_3 &= \mathbf{diag}(\mathbf{1}_{n_{11}}, \cdots, \mathbf{1}_{n_{1b}}, \mathbf{1}_{n_{21}}, \cdots, \mathbf{1}_{n_{2b}}, \cdots, \mathbf{1}_{n_{v1}}, \cdots, \mathbf{1}_{n_{vb}}). \end{aligned} \tag{4.3.3}$$

In (4.3.3), $n_{i.} = \sum_{j=1}^{b} n_{ij}$. The properties of $\boldsymbol{X}_1$, $\boldsymbol{X}_2$, and $\boldsymbol{X}_3$ stated in Lemma 4.3.1 below will be used in the sequel. The proof of the lemma is omitted since it follows from the expressions for the matrices $\boldsymbol{X}_1$, $\boldsymbol{X}_2$, and $\boldsymbol{X}_3$ given in (4.3.3). (See Exercise 4.1 and its solution in the Appendix.)

Lemma 4.3.1. Let the matrices $\boldsymbol{X}_1$, $\boldsymbol{X}_2$, and $\boldsymbol{X}_3$ be defined as in (4.3.3), and let n_0 denote the number of nonzero n_{ij}'s. Then

- **(a)** $\text{rank}(\boldsymbol{X}_3) = n_0$,
- **(b)** $\text{rank}(\boldsymbol{X}_1\colon \boldsymbol{X}_3) = \text{rank}(\boldsymbol{X}_2\colon \boldsymbol{X}_3) = \text{rank}(\boldsymbol{X}_1\colon \boldsymbol{X}_2\colon \boldsymbol{X}_3) = \text{rank}(\boldsymbol{X}_3)$,
- **(c)** $\mathcal{C}(\boldsymbol{X}_i) \subset \mathcal{C}(\boldsymbol{X}_3)$, for $i = 1, 2$, where $\mathcal{C}(.)$ denotes range space (column space), and
- **(d)** $\mathbf{1}_{n_{..}} \in \mathcal{C}(\boldsymbol{X}_i)$, $i = 1, 2, 3$. □

4.3.1. Models Without Interaction: Exact Tests

When interaction is absent, the model (4.3.1) reduces to

$$\begin{aligned} y_{ijk} &= \mu + \tau_i + \beta_j + e_{ijk}, \\ &\qquad i = 1, 2, \cdots, v;\ j = 1, 2, \cdots, b;\ k = 1, 2, \cdots, n_{ij}, \end{aligned} \tag{4.3.4}$$

or, equivalently (from (4.3.2))

$$\mathbf{y} = \mathbf{1}_{n_{..}}\mu + \boldsymbol{X}_1\boldsymbol{\tau} + \boldsymbol{X}_2\boldsymbol{\beta} + \mathbf{e}. \tag{4.3.5}$$

In (4.3.5), the random vectors $\boldsymbol{\tau}$, $\boldsymbol{\beta}$ and $\mathbf{e}$ are independently distributed as $\boldsymbol{\tau} \sim N(\mathbf{0}, \sigma_\tau^2 \boldsymbol{I}_v)$, $\boldsymbol{\beta} \sim N(\mathbf{0}, \sigma_\beta^2 \boldsymbol{I}_b)$, and $\mathbf{e} \sim N(\mathbf{0}, \sigma_e^2 \boldsymbol{I}_{n_{..}})$. Model (4.3.4) (or (4.3.5)) is appropriate for analyzing data from a block design involving v treatments in b blocks, when the block effects as well as the treatment effects are random and the blocks and treatments do not interact. The value of n_{ij} is the number of observations for the ith treatment in the jth block, and the jth block has block size $n_{.j}$, given by $n_{.j} = \sum_{i=1}^{v} n_{ij}$. When the τ_i's and β_j's are fixed effects, we have the F-tests based on the ANOVA decomposition for testing the equality of the τ_i's, and also for testing the equality of the β_j's. We shall show that when the τ_i's and β_j's are random, the same F-tests continue to be valid for testing the significance of the variance components σ_τ^2 and σ_β^2. As was pointed out earlier in this section, our development does not require any assumptions on the n_{ij}'s; they can be arbitrary and some of them can even be zeros.

Let $SS(e)$ and $SS(\tau|\beta)$, respectively, denote the sum of squares due to error and the sum of squares due to $\boldsymbol{\tau}$ (adjusted for $\boldsymbol{\beta}$) computed using model (4.3.5). A representation for $SS(\tau|\beta)$ can be obtained by considering the reduced model as follows. Let $\text{rank}(\boldsymbol{X}_2) = r_2$ and let $\boldsymbol{Z}_2$ be an $n_{..} \times (n_{..} -$

r_2) matrix satisfying $\boldsymbol{Z}_2'\boldsymbol{X}_2 = \mathbf{0}$ and $\boldsymbol{Z}_2'\boldsymbol{Z}_2 = \boldsymbol{I}_{n_{..}-r_2}$. Then the model for $\mathbf{u}_2 = \boldsymbol{Z}_2'\mathbf{y}$ is

$$\mathbf{u}_2 = \boldsymbol{Z}_2'\boldsymbol{X}_1\boldsymbol{\tau} + \boldsymbol{Z}_2'\mathbf{e}. \tag{4.3.6}$$

In deriving (4.3.6), we have used the fact that $\boldsymbol{Z}_2'\mathbf{1}_{n_{..}} = \mathbf{0}$. This follows from the fact that $\mathbf{1}_{n_{..}}$ belongs to the range space of $\boldsymbol{X}_2$; see Lemma 4.3.1. Then $SS(\tau|\beta)$ computed from the model (4.3.5) is the same as the sum of squares due to $\boldsymbol{\tau}$ computed using the model (4.3.6). Also, $SS(e)$ computed using (4.3.5) is the same as the sum of squares due to error computed using the model (4.3.6) (i.e., $SS(e) = \mathbf{u}_2'\mathbf{u}_2 - SS(\tau|\beta)$). (See Exercise 4.3.) From (4.3.6), we see that

$$SS(\tau|\beta) = \mathbf{u}_2'\boldsymbol{Z}_2'\boldsymbol{X}_1(\boldsymbol{X}_1'\boldsymbol{Z}_2\boldsymbol{Z}_2'\boldsymbol{X}_1)^{-}\boldsymbol{X}_1'\boldsymbol{Z}_2\mathbf{u}_2, \tag{4.3.7}$$

where the superscript '$-$' denotes generalized inverse. Note that $\boldsymbol{Z}_2\boldsymbol{Z}_2' = \boldsymbol{I} - \boldsymbol{P}_{\boldsymbol{X}_2}$, where $\boldsymbol{P}_{\boldsymbol{X}_2} = \boldsymbol{X}_2(\boldsymbol{X}_2'\boldsymbol{X}_2)^{-}\boldsymbol{X}_2'$ is the orthogonal projection matrix onto the range space of $\boldsymbol{X}_2$. Let m_v denote the degrees of freedom associated with $SS(\tau|\beta)$ and m_e denote the degrees of freedom associated with $SS(e)$. Then

$$m_v = \text{rank}(\boldsymbol{X}_1'\boldsymbol{Z}_2) = \text{rank}[\boldsymbol{X}_1'(\boldsymbol{I} - \boldsymbol{P}_{\boldsymbol{X}_2})\boldsymbol{X}_1],\ m_e = n_{..} - \text{rank}(\boldsymbol{X}_1 : \boldsymbol{X}_2). \tag{4.3.8}$$

Since $Var(\mathbf{u}_2) = \sigma_\tau^2\boldsymbol{Z}_2'\boldsymbol{X}_1\boldsymbol{X}_1'\boldsymbol{Z}_2 + \sigma^2\boldsymbol{I}_{n_{..}-r_2}$, using (4.3.6), (4.3.7), and (4.3.8), we get

$$\begin{aligned} \text{E}(SS(\tau|\beta)) &= m_v\sigma_e^2 + \text{tr}(\boldsymbol{Z}_2'\boldsymbol{X}_1\boldsymbol{X}_1'\boldsymbol{Z}_2)\sigma_\tau^2 = m_v\sigma_e^2 + \text{tr}[\boldsymbol{X}_1'(\boldsymbol{I} - \boldsymbol{P}_{\boldsymbol{X}_2})\boldsymbol{X}_1]\sigma_\tau^2, \\ \text{E}(SS(e)) &= m_e\sigma_e^2. \end{aligned} \tag{4.3.9}$$

It can be readily verified that $SS(e)/\sigma_e^2 \sim \chi^2_{m_e}$ and, under H_τ: $\sigma_\tau^2 = 0$, $SS(\tau|\beta)/\sigma_e^2 \sim \chi^2_{m_v}$. Hence, an F-test for testing $H_\tau : \sigma_\tau^2 = 0$ can be obtained using the F-ratio given by

$$F_\tau = \frac{SS(\tau|\beta)/m_v}{SS(e)/m_e}. \tag{4.3.10}$$

A similar test can be obtained for testing the significance of σ_β^2 also. We note that the F-ratio F_τ in (4.3.10) coincides with the F-ratio that can be obtained based on the ANOVA decomposition in the fixed effects case.

In order to carry out the test based on F_τ in (4.3.10), an explicit computation of the matrix $\boldsymbol{Z}_2$ and the vector $\mathbf{u}_2$ in (4.3.6) is not necessary. This is due to the fact that the sum of squares required to obtain the statistic in (4.3.10) can be obtained from the ANOVA decomposition based on the model (4.3.5). Our development in this section simply shows that the sum of squares so obtained in the fixed effects case can also be used to form F-ratios for testing the significance of the variance components in the random effects case. The idea of using fixed effects sums of squares to draw inferences concerning variance components was exploited by Wald (1947), and is explained in Section 1.2 of Chapter 1. The test based on F_τ in (4.3.10) is in fact a Wald's test.

In the context of block designs with fixed treatment effects and random block effects, F_τ in (4.3.10) is the intra-block F-ratio for testing the equality of the treatment effects. It is well known that in this case, another F-test for testing the same can be obtained using inter-block information. The same is also true when τ_i's and β_j's are random and the problem is to test H_τ: $\sigma_\tau^2 = 0$. That is, inter-block information can indeed be used to test the significance of the treatment variance component. For this problem, an exact test that combines the inter- and intra-block F-tests can be obtained in the special case when the $n_{.j}$'s are all equal (where, $n_{.j} = \sum_{i=1}^{v} n_{ij}$). Details of this testing procedure appear in Chapter 7.

4.3.2. Models Without Interaction: Optimum Tests

Note that the testing problem H_τ: $\sigma_\tau^2 = 0$ in the model (4.3.5) is invariant under the transformation $\mathbf{y} \to c(\mathbf{y} + a\mathbf{1}_{n_{..}})$, for $c > 0$ and a real. In the balanced case, that is, when the n_{ij}'s are all equal in (4.3.4), the F-test based on F_τ in (4.3.10) is UMPIU for testing H_τ: $\sigma_\tau^2 = 0$. This follows from the general results in Chapter 2. When the n_{ij}'s are unequal, a UMPIU test does not exist for testing H_τ: $\sigma_\tau^2 = 0$. An LBIU test was derived by Mathew and Sinha (1988) in the special case when $n_{.j}$'s are all equal and $n_{i.}$'s are all equal $\left(\text{where, } n_{i.} = \sum_{j=1}^{b} n_{ij} \text{ and } n_{.j} = \sum_{i=1}^{v} n_{ij}\right)$. In the context of a block design, this condition is equivalent to the block sizes being equal and the treatments being replicated the same number of times. We shall now describe the LBIU test, omitting the details of the derivation.

Assume that the $n_{i.}$'s are all equal and the $n_{.j}$'s are all equal. Let r and l respectively denote the common value of the $n_{i.}$'s and the common value of the $n_{.j}$'s. Write $y_{i..} = \sum_{j=1}^{b}\sum_{k=1}^{n_{ij}} y_{ijk}$, $y_{.j.} = \sum_{i=1}^{v}\sum_{k=1}^{n_{ij}} y_{ijk}$ and $\bar{y}_{...} = 1/n_{..}\sum_{i=1}^{v}\sum_{j=1}^{b}\sum_{k=1}^{n_{ij}} y_{ijk}$. In the context of block designs, $y_{i..}$ is the sum of the observations corresponding to the ith treatment and $y_{.j.}$ is the sum of the observations in the jth block. The LBIU test for testing H_τ: $\sigma_\tau^2 = 0$ in the model (4.3.5) is given by the following theorem.

Theorem 4.3.1. Assume that the $n_{i.}$'s are all equal and the $n_{.j}$'s are all equal. Let l denote the common value of the $n_{.j}$'s and suppose $y_{i..}$, $y_{.j.}$ and $\bar{y}_{...}$ are as defined above. Let $SS_0(\beta) = \sum_{j=1}^{b}(y_{.j.}/l - \bar{y}_{...})^2$ and $SS_0(e) = \sum_{i=1}^{v}\sum_{j=1}^{b}\sum_{k=1}^{n_{ij}}(y_{ijk} - y_{.j.}/l)^2$. For testing H_τ: $\sigma_\tau^2 = 0$ in the model (4.3.5), (a) the LBIU test rejects H_τ for large values of F_τ in (4.3.10) if $\sigma_\beta^2/\sigma_e^2$ is large, and (*b*) the LBIU test rejects H_τ for large values of $\sum_{i=1}^{v} y_{i..}^2$, conditional on $(\bar{y}_{...}, SS_0(\beta), SS_0(e))$ if $\sigma_\beta^2/\sigma_e^2$ is small. □

We note that $SS_0(\beta)$ and $SS_0(e)$ defined in Theorem 4.3.1 are, respectively, the sum of squares due to the β_j's and the sum of squares due to error in the model (4.3.4), under H_τ: $\sigma_\tau^2 = 0$. An unpleasant feature of the LBIU test in Theorem 4.3.1 is its dependence on the ratio $\sigma_\beta^2/\sigma_e^2$, which is typically unknown. Even if we have some idea about the magnitude of this ratio, application of the LBIU test requires obtaining the conditional distribution of $\sum_{i=1}^{v} y_{i..}^2$, given $(\bar{y}_{...}, SS_0(\beta), SS_0(e))$. These aspects make the test difficult to use in practical applications. Of course, the test based on F_τ in (4.3.10) is an exact test that is easy to use. However, the latter test does not fully utilize the available information in the data. To be specific, it ignores "inter-block information". In Chapter 7, we shall develop exact invariant tests that perform well compared to the test based on F_τ, in terms of power, and which appropriately combine intra-block and inter-block information, while avoiding the above unpleasant features of the LBIU test.

4.3.3. Models With Interaction: Exact Tests

We shall now derive some exact test procedures for testing the significance of the three variance components $\sigma_{\tau\beta}^2$, σ_τ^2, and σ_β^2 in the model (4.3.1). The exact test that we shall derive for testing $H_{\tau\beta}$: $\sigma_{\tau\beta}^2 = 0$ is the ANOVA F-test derived by Thomsen (1975) and is also the Wald's test derived in Seely and El-Bassiouni (1983) (see Section 1.2 in Chapter 1 and Section 6.2 in Chapter 6). For testing H_τ: $\sigma_\tau^2 = 0$ and H_β: $\sigma_\beta^2 = 0$, the exact tests described below are due to Khuri and Littell (1987).

4.3.3.1. Testing $H_{\tau\beta} : \sigma_{\tau\beta}^2 = 0$. We shall show that an F-test based on the ANOVA decomposition in the fixed effects model continues to be valid for testing $H_{\tau\beta}$: $\sigma_{\tau\beta}^2 = 0$ in the random effects model. In order to establish this, we shall use arguments similar to those in Section 4.3.1. Toward this, let n_0 denote the number of nonzero n_{ij}'s. From Lemma 4.3.1, rank$(\mathbf{1}_{n_{..}}, \boldsymbol{X}_1, \boldsymbol{X}_2, \boldsymbol{X}_3)$ $=$ rank$(\boldsymbol{X}_3) = n_0$. Hence, $SS(e)$, the sum of squares due to error, computed based on the model (4.3.1) (or equivalently, the model (4.3.2)) has d.f.$=$ $n_{..} - n_0$. Let $SS(\tau\beta|\tau, \beta)$ denote the sum of squares due to $(\boldsymbol{\tau\beta})$, adjusted for $\boldsymbol{\tau}$ and $\boldsymbol{\beta}$, computed using model (4.3.2). An expression for $SS(\tau\beta|\tau, \beta)$ can be easily obtained as shown below, which is quite similar to the expression given in (4.3.7). For this, let $r_{12} =$ rank$(\boldsymbol{X}_1 : \boldsymbol{X}_2)$ and let $\mathbf{Z}_{12}$ be an $n_{..} \times (n_{..} - r_{12})$ matrix satisfying $\mathbf{Z}_{12}'(\boldsymbol{X}_1 : \boldsymbol{X}_2) = \mathbf{0}$ and $\mathbf{Z}_{12}'\mathbf{Z}_{12} = \boldsymbol{I}_{n_{..}-r_{12}}$. Defining $\mathbf{u}_{12} = \mathbf{Z}_{12}'\mathbf{y}$, we get the following, quite similar to the expression in (4.3.7):

$$SS(\tau\beta|\tau, \beta) = \mathbf{u}_{12}'\mathbf{Z}_{12}'\boldsymbol{X}_3(\boldsymbol{X}_3'\mathbf{Z}_{12}\mathbf{Z}_{12}'\boldsymbol{X}_3)^{-}\boldsymbol{X}_3'\mathbf{Z}_{12}\mathbf{u}_{12}. \tag{4.3.11}$$

Similar to (4.3.9), we also have

$$\begin{aligned} \mathrm{E}[SS(\tau\beta|\tau,\beta)] &= r_3\sigma_e^2 + \mathrm{tr}[\boldsymbol{X}_3'(\boldsymbol{I} - \boldsymbol{P}_{(\boldsymbol{X}_1:\boldsymbol{X}_2)})\boldsymbol{X}_3]\sigma_{\tau\beta}^2, \\ \mathrm{E}[SS(e)] &= (n_{..} - n_0)\sigma_e^2, \end{aligned} \tag{4.3.12}$$

where $\boldsymbol{P}_{(\boldsymbol{X}_1:\boldsymbol{X}_2)} = (\boldsymbol{X}_1 : \boldsymbol{X}_2)[(\boldsymbol{X}_1 : \boldsymbol{X}_2)'(\boldsymbol{X}_1 : \boldsymbol{X}_2)]^-(\boldsymbol{X}_1 : \boldsymbol{X}_2)'$ and $r_3 = \mathrm{rank}(\boldsymbol{X}_3'\boldsymbol{Z}) = \mathrm{rank}[\boldsymbol{X}_3'(\boldsymbol{I} - \boldsymbol{P}_{(\boldsymbol{X}_1,\boldsymbol{X}_2)})\boldsymbol{X}_3]$. Also, $SS(e) \sim \chi^2_{n_{..}-n_0}$, and under $H_{\tau\beta}$: $\sigma^2_{\tau\beta} = 0$, $SS(\tau\beta|\tau,\beta) \sim \chi^2_{r_3}$. Hence an F-test for testing $H_{\tau\beta}$: $\sigma^2_{\tau\beta} = 0$ rejects $H_{\tau\beta}$ for large values of the statistic $F_{\tau\beta}$ given by

$$F_{\tau\beta} = \frac{SS(\tau\beta|\tau,\beta)/r_3}{SS(e)/(n_{..} - n_0)}. \tag{4.3.13}$$

REMARK 4.3.1. When $\boldsymbol{\tau}$, $\boldsymbol{\beta}$, and $(\boldsymbol{\tau\beta})$ are fixed effects, to test $H_{\tau\beta}$: $(\tau\beta)_{ij} = 0$ (for all i and j) we need the condition $r_3 = (v-1)(b-1)$. This is the well-known condition for testability of $H_{\tau\beta}$. However, when the above effects are random, the hypothesis $H_{\tau\beta}$: $\sigma^2_{\tau\beta} = 0$ can be tested using the F-ratio in (4.3.13) even if r_3 is less than $(v-1)(b-1)$, because the hypothesis $H_{\tau\beta}$ deals with just one unknown parameter $\sigma^2_{\tau\beta}$. Note that when $r_3 = (v-1)(b-1)$, $F_{\tau\beta}$ in (4.3.13) is the same as the F-ratio for testing $H_{\tau\beta}$: $(\tau\beta)_{ij} = 0$ (for all i and for all j) in the fixed effects case. The test based on $F_{\tau\beta}$ in (4.3.13) is a Wald's test and is described in Seely and El-Bassiouni (1983).

4.3.3.2. Testing $H_\tau : \sigma^2_\tau = 0$ and $H_\beta : \sigma^2_\beta = 0 (\sigma^2_{\tau\beta} > 0)$. The derivation that leads to the F-test based on $F_{\tau\beta}$ in (4.3.13) will not go through for testing H_τ: $\sigma^2_\tau = 0$ and for testing H_β: $\sigma^2_\beta = 0$. That is, the F-test in the fixed effects case for testing the equality of the τ_i's cannot be used for testing H_τ: $\sigma^2_\tau = 0$, when the τ_i's are random. The reason for this is as follows. In order to arrive at (4.3.11), we used the model for $\mathbf{u}_{12}$ (defined before in (4.3.11)) which does not involve $\boldsymbol{\tau}$ and $\boldsymbol{\beta}$. To apply the same idea for testing H_τ: $\sigma^2_\tau = 0$, we need to arrive at a model that involves $\boldsymbol{\tau}$, but not $\boldsymbol{\beta}$ and $(\boldsymbol{\tau\beta})$. In order to eliminate $\boldsymbol{\beta}$ and $(\boldsymbol{\tau\beta})$ from the model (4.3.2), we need to consider the random variable $\mathbf{u}_{23} = \boldsymbol{Z}_{23}'\mathbf{y}$, where $\boldsymbol{Z}_{23}$ satisfies $\boldsymbol{Z}_{23}'[\boldsymbol{X}_2 : \boldsymbol{X}_3] = \boldsymbol{0}$. However, the model for $\mathbf{u}_{23}$ will not involve $\boldsymbol{\tau}$ either, since $\boldsymbol{Z}_{23}$ will also satisfy $\boldsymbol{Z}_{23}'\boldsymbol{X}_1 = \boldsymbol{0}$, in view of Lemma 4.3.1(c). In other words, we cannot get a reduced model that involves only $\boldsymbol{\tau}$, while simultaneously eliminating $\boldsymbol{\beta}$ and $(\boldsymbol{\tau\beta})$. Consequently, the arguments that lead to the F-tests based on the statistics in (4.3.13) or (4.3.10) are not applicable for testing H_τ: $\sigma^2_\tau = 0$ and H_β: $\sigma^2_\beta = 0$. Even in the balanced case, it is well known that the F-ratio for testing the equality of the τ_i's in the fixed effects case is not applicable for testing H_τ: $\sigma^2_\tau = 0$ in the random effects case, when interaction is present. The former has the error sum of squares in the denominator, while the latter has the interaction sum of squares in the denominator.

However, exact F-tests do exist for testing the hypotheses H_τ: $\sigma^2_\tau = 0$ and H_β: $\sigma^2_\beta = 0$ in the model (4.3.1). In the remainder of this section, we shall derive these tests. Our derivation is based on the results in Khuri and Littell (1987) and assumes that $n_{ij} \geq 1$ and $n_{..} > 2bv - 1$. The idea is to apply appropriate orthogonal transformations to the model for the cell means to arrive at independent sums of squares that are scalar multiples of chi-squared random variables, quite similar to those in the balanced case, so that F-ratios can be constructed based on such sums of squares. We shall thus consider the model for the cell means.

For each i and each j, consider the cell mean $\bar{y}_{ij.} = 1/n_{ij} \sum_{k=1}^{n_{ij}} y_{ijk}$. Note that the $\bar{y}_{ij.}$'s are well defined for each i and j due to the assumption $n_{ij} \geq 1$. From (4.3.1), the model for $\bar{y}_{ij.}$ is

$$\bar{y}_{ij.} = \mu + \tau_i + \beta_j + (\tau\beta)_{ij} + \bar{e}_{ij.}, \tag{4.3.14}$$

where $\bar{e}_{ij.}$ is defined similarly to $\bar{y}_{ij.}$. Writing $\bar{\mathbf{y}} = (\bar{y}_{11.}, \bar{y}_{12.}, \ldots, \bar{y}_{vb.})'$, and defining $\bar{\mathbf{e}}$ similarly, (4.3.14) can equivalently be expressed as

$$\bar{\mathbf{y}} = \mu \mathbf{1}_{bv} + (\boldsymbol{I}_v \otimes \mathbf{1}_v)\boldsymbol{\tau} + (\mathbf{1}_v \otimes \boldsymbol{I}_b)\boldsymbol{\beta} + \boldsymbol{I}_{bv}(\boldsymbol{\tau\beta}) + \bar{\mathbf{e}}, \tag{4.3.15}$$

with

$$\begin{aligned} Var(\bar{\mathbf{y}}) &= \sigma_\tau^2(\boldsymbol{I}_v \otimes \boldsymbol{J}_b) + \sigma_\beta^2(\boldsymbol{J}_v \otimes \boldsymbol{I}_b) + \sigma_{\tau\beta}^2 \boldsymbol{I}_{bv} + \sigma_e^2 \boldsymbol{D} \\ &= b\sigma_\tau^2 \left(\boldsymbol{I}_v \otimes \frac{1}{b}\boldsymbol{J}_b\right) + v\sigma_\beta^2 \left(\frac{1}{v}\boldsymbol{J}_v \otimes \boldsymbol{I}_b\right) + \sigma_{\tau\beta}^2 \boldsymbol{I}_{bv} + \sigma_e^2 \boldsymbol{D}, \end{aligned} \tag{4.3.16}$$

where $\boldsymbol{D} = \mathbf{diag}(1/n_{11}, 1/n_{12}, \ldots, 1/n_{bv})$. Note that the matrices $(\boldsymbol{I}_v \otimes 1/b\boldsymbol{J}_b)$ and $(1/v\boldsymbol{J}_v \otimes \boldsymbol{I}_b)$ in (4.3.16) are projection matrices (i.e., symmetric idempotent matrices) that commute and $(bv)^{-1/2}\mathbf{1}_{bv}$ is an eigenvector of both of them corresponding to the eigenvalue one. Thus, there exists a $bv \times bv$ orthogonal matrix $\boldsymbol{P}$ whose first row is $(bv)^{-1/2}\mathbf{1}'_{bv}$ such that

$$\begin{aligned} \boldsymbol{P}\left(\boldsymbol{I}_v \otimes \frac{1}{b}\boldsymbol{J}_b\right)\boldsymbol{P}' &= \mathbf{diag}(1, \boldsymbol{I}_{v-1}, \mathbf{0}, \mathbf{0}), \\ \text{and } \boldsymbol{P}\left(\frac{1}{v}\boldsymbol{J}_v \otimes \boldsymbol{I}_b\right)\boldsymbol{P}' &= \mathbf{diag}(1, \mathbf{0}, \boldsymbol{I}_{b-1}, \mathbf{0}). \end{aligned} \tag{4.3.17}$$

Note that the intersection of the vector spaces $\mathcal{C}(\boldsymbol{I}_v \otimes 1/b\boldsymbol{J}_b)$ and $\mathcal{C}(1/v\boldsymbol{J}_v \otimes \boldsymbol{I}_v)$ is the vector space consisting of scalar multiples of $\mathbf{1}_{bv}$. Consequently, from the choice of $\boldsymbol{P}$, it follows that the intersection of $\mathcal{C}(\boldsymbol{P}(\boldsymbol{I}_v \otimes 1/b\boldsymbol{J}_b)\boldsymbol{P}')$ and $\mathcal{C}(\boldsymbol{P}(1/v\boldsymbol{J}_v \otimes \boldsymbol{I}_v)\boldsymbol{P}')$ consists of vectors that are scalar multiples of $(1, \mathbf{0})'$ (where $\mathbf{0}$ denotes a $(bv - 1) \times 1$ vector of zeros). This explains the expressions for the diagonal matrices that occur in (4.3.17). Let $\boldsymbol{P}\bar{\mathbf{y}} = \mathbf{z} = (z_1, \mathbf{z}'_\tau, \mathbf{z}'_\beta, \mathbf{z}'_{\tau\beta})'$, where z_1 is the first element of $\mathbf{z}$ and $\mathbf{z}_\tau$, $\mathbf{z}_\beta$, and $\mathbf{z}_{\tau\beta}$ are, respectively,

$(v-1)\times 1$, $(b-1)\times 1$, and $(v-1)(b-1)\times 1$ vectors. Using (4.3.16) and (4.3.17), we get

$$\begin{aligned} Var(\mathbf{z}_\tau) &= (b\sigma_\tau^2+\sigma_{\tau\beta}^2)\boldsymbol{I}_{v-1}+\sigma_e^2\boldsymbol{D}_1, \\ Var(\mathbf{z}_\beta) &= (v\sigma_\beta^2+\sigma_{\tau\beta}^2)\boldsymbol{I}_{b-1}+\sigma_e^2\boldsymbol{D}_2, \\ Var(\mathbf{z}_{\tau\beta}) &= \sigma_{\tau\beta}^2\boldsymbol{I}_{(v-1)(b-1)}+\sigma_e^2\boldsymbol{D}_3, \end{aligned} \tag{4.3.18}$$

where $\boldsymbol{D}_1$, $\boldsymbol{D}_2$, and $\boldsymbol{D}_3$ are appropriate submatrices of $\boldsymbol{PDP}'$. We shall denote by $\boldsymbol{D}_0$ the $(bv-1)\times(bv-1)$ submatrix of $\boldsymbol{PDP}'$ that corresponds to $(\mathbf{z}_\tau',\mathbf{z}_\beta',\mathbf{z}_{\tau\beta}')'$. The matrix $\boldsymbol{D}_0$ may be expressed as

$$\boldsymbol{D}_0=\begin{pmatrix} \boldsymbol{D}_1 & \boldsymbol{D}_{12} & \boldsymbol{D}_{13} \\ \boldsymbol{D}_{12}' & \boldsymbol{D}_2 & \boldsymbol{D}_{23} \\ \boldsymbol{D}_{13}' & \boldsymbol{D}_{23}' & \boldsymbol{D}_3 \end{pmatrix}, \tag{4.3.19}$$

for suitable matrices $\boldsymbol{D}_{12}$, $\boldsymbol{D}_{13}$, and $\boldsymbol{D}_{23}$. The error sum of squares, $SS(e)$, computed based on the model (4.3.1) is clearly given by

$$SS(e)=\sum_{i=1}^{v}\sum_{j=1}^{b}\sum_{k=1}^{n_{ij}}(y_{ijk}-\bar{y}_{ij.})^2. \tag{4.3.20}$$

Because of our assumption $n_{ij}\geq 1$, n_0 in Lemma 4.3.1 is equal to bv and $SS(e)/\sigma_e^2$ has a chi-square distribution with d.f. $= n_{..}-bv$. It is easy to verify that each element of $\bar{\mathbf{y}}$ is uncorrelated with $y_{ijk}-\bar{y}_{ij.}$ (for all i and j). Thus the vector $\bar{\mathbf{y}}$, and hence $(\mathbf{z}_\tau',\mathbf{z}_\beta',\mathbf{z}_{\tau\beta}')'$, is distributed independently of $SS(e)$. Note that we can write

$$\begin{aligned} SS(e) &= \mathbf{y}'\boldsymbol{R}\mathbf{y}, \\ \boldsymbol{R} &= \boldsymbol{I}_{n_{..}}-\mathbf{diag}\left(\frac{1}{n_{11}}\boldsymbol{J}_{n_{11}},\cdots,\frac{1}{n_{bv}}\boldsymbol{J}_{n_{bv}}\right). \end{aligned} \tag{4.3.21}$$

Since $\boldsymbol{R}$ is a symmetric idempotent matrix of rank $(n_{..}-bv)$, there exists an $n_{..}\times n_{..}$ orthogonal matrix $\boldsymbol{G}$ such that

$$\begin{aligned} \boldsymbol{R} &= \boldsymbol{G}\,\mathbf{diag}(\boldsymbol{I}_{n_{..}-bv},\mathbf{0})\boldsymbol{G}' \\ &= (\boldsymbol{G}_1:\boldsymbol{G}_2:\boldsymbol{G}_3)\mathbf{diag}(\boldsymbol{I}_{bv-1},\boldsymbol{I}_{n_{..}-2bv+1},\mathbf{0})(\boldsymbol{G}_1:\boldsymbol{G}_2:\boldsymbol{G}_3)' \\ &= \boldsymbol{G}_1\boldsymbol{G}_1'+\boldsymbol{G}_2\boldsymbol{G}_2'. \end{aligned} \tag{4.3.22}$$

In (4.3.22), we have used the assumption $n_{..}>2bv-1$, so that $n_{..}-2bv+1>0$. Note that the matrices $\boldsymbol{G}_1$, $\boldsymbol{G}_2$, and $\boldsymbol{G}_3$ in (4.3.22) have orders $n_{..}\times(bv-1)$, $n_{..}\times(n_{..}-2bv+1)$, and $n_{..}\times bv$ respectively. Using (4.3.22), $SS(e)$ in (4.3.21) can be written as

$$SS(e)=SS_1(e)+SS_2(e), \tag{4.3.23}$$

where $SS_1(e) = \mathbf{y}'\boldsymbol{G}_1\boldsymbol{G}_1'\mathbf{y}$ and $SS_2(e) = \mathbf{y}'\boldsymbol{G}_2\boldsymbol{G}_2'\mathbf{y}$. It is easy to verify that $SS_1(e)/\sigma_e^2$ and $SS_2(e)/\sigma_e^2$ have independent chi-square distributions with $(bv-1)$ and $(n_{..} - 2bv + 1)$ degrees of freedom, respectively. Let $\lambda_{\max}(\boldsymbol{D}_0)$ denote the largest eigenvalue of the matrix $\boldsymbol{D}_0$ given in (4.3.19). Define the vector $\mathbf{w}$ and partition it as follows:

$$(\mathbf{z}_\tau', \mathbf{z}_\beta', \mathbf{z}_{\tau\beta}')' + [\lambda_{\max}(\boldsymbol{D}_0)\boldsymbol{I}_{bv-1} - \boldsymbol{D}_0]^{1/2}\boldsymbol{G}_1'\mathbf{y} = \mathbf{w} = (\mathbf{w}_\tau', \mathbf{w}_\beta', \mathbf{w}_{\tau\beta}')'. \quad (4.3.24)$$

In (4.3.24), $[\lambda_{\max}(\boldsymbol{D}_0)\boldsymbol{I}_{bv-1} - \boldsymbol{D}_0]^{1/2}$ denotes the positive semidefinite square root of the positive semidefinite matrix $[\lambda_{\max}(\boldsymbol{D}_0)\boldsymbol{I}_{bv-1} - \boldsymbol{D}_0]$. The vectors $\mathbf{w}_\tau$, $\mathbf{w}_\beta$, and $\mathbf{w}_{\tau\beta}$ have respective orders $(v-1)\times 1$, $(b-1)\times 1$, and $(v-1)(b-1)\times 1$. The distributional properties of $\mathbf{w}_\tau$, $\mathbf{w}_\beta$, and $\mathbf{w}_{\tau\beta}$ given in Lemma 4.3.2 below are crucial for developing tests of significance of the variance components σ_τ^2 and σ_β^2.

Lemma 4.3.2. The vectors $\mathbf{w}_\tau$, $\mathbf{w}_\beta$, and $\mathbf{w}_{\tau\beta}$ defined in (4.3.24) are independent and normally distributed random vectors with zero means and variance–covariance matrices

$$\begin{aligned} Var(\mathbf{w}_\tau) &= [b\sigma_\tau^2 + \sigma_{\tau\beta}^2 + \lambda_{\max}(\boldsymbol{D}_0)\sigma_e^2]\boldsymbol{I}_{v-1}, \\ Var(\mathbf{w}_\beta) &= [v\sigma_\beta^2 + \sigma_{\tau\beta}^2 + \lambda_{\max}(\boldsymbol{D}_0)\sigma_e^2]\boldsymbol{I}_{b-1}, \\ Var(\mathbf{w}_{\tau\beta}) &= [\sigma_{\tau\beta}^2 + \lambda_{\max}(\boldsymbol{D}_0)\sigma_e^2]\boldsymbol{I}_{(v-1)(b-1)}. \end{aligned} \quad (4.3.25)$$

Furthermore, $\mathbf{w}_\tau$, $\mathbf{w}_\beta$, and $\mathbf{w}_{\tau\beta}$ are distributed independently of $SS_2(e)$ in (4.3.23).

Proof. By construction, the vectors $\mathbf{z}_\tau$, $\mathbf{z}_\beta$, and $\mathbf{z}_{\tau\beta}$ have zero means. Hence, to show that $\mathbf{w}$ in (4.3.24) has mean zero, it is enough to show that $E(\boldsymbol{G}_1'\mathbf{y}) = \mathbf{0}$. Note that $\boldsymbol{R}$ in (4.3.21) satisfies $\boldsymbol{R}\mathbf{1}_{n_{..}} = \mathbf{0}$. Hence, using (4.3.22), we get $(\boldsymbol{G}_1\boldsymbol{G}_1' + \boldsymbol{G}_2\boldsymbol{G}_2')\mathbf{1}_{n_{..}} = \mathbf{0}$, which gives $\mathbf{1}_{n_{..}}'(\boldsymbol{G}_1\boldsymbol{G}_1' + \boldsymbol{G}_2\boldsymbol{G}_2')\mathbf{1}_{n_{..}} = 0$. This implies $\mathbf{1}_{n_{..}}'\boldsymbol{G}_1\boldsymbol{G}_1'\mathbf{1}_{n_{..}} = 0$, or, equivalently, $\boldsymbol{G}_1'\mathbf{1}_{n_{..}} = \mathbf{0}$. Hence, $E(\boldsymbol{G}_1'\mathbf{y}) = \mu\boldsymbol{G}_1'\mathbf{1}_{n_{..}} = \mathbf{0}$. Thus we have established that the vector $\mathbf{w}$ in (4.3.24) has a zero mean. Normality of $\mathbf{w}$ follows from the fact that $\mathbf{w}$ is obtained by a linear transformation of the normally distributed vector $\mathbf{y}$. We shall now derive the variance–covariance matrix of $\mathbf{w}$ in order to arrive at (4.3.25). Toward this, we shall first show that $(\mathbf{z}_\tau', \mathbf{z}_\beta', \mathbf{z}_{\tau\beta}')'$ and $\boldsymbol{G}_1'\mathbf{y}$ are uncorrelated. For this, it is enough to show that $\bar{\mathbf{y}}$ and $\boldsymbol{G}_1'\mathbf{y}$ are uncorrelated, since $(\mathbf{z}_\tau', \mathbf{z}_\beta', \mathbf{z}_{\tau\beta}')$ is obtained through a linear transformation of $\bar{\mathbf{y}}$. Note that

$$\begin{aligned} \bar{\mathbf{y}} &= \boldsymbol{M}\mathbf{y}, \\ \boldsymbol{M} &= \mathbf{diag}(\mathbf{1}_{n_{11}}'/n_{11}, \mathbf{1}_{n_{12}}'/n_{12}, \cdots, \mathbf{1}_{n_{bv}}'/n_{bv}). \end{aligned} \quad (4.3.26)$$

Also, if $\boldsymbol{\Sigma} = Var(\mathbf{y})$, then

$$\boldsymbol{\Sigma} = \sigma_\tau^2\boldsymbol{X}_1\boldsymbol{X}_1' + \sigma_\beta^2\boldsymbol{X}_2\boldsymbol{X}_2' + \sigma_{\tau\beta}^2\boldsymbol{X}_3\boldsymbol{X}_3' + \sigma_e^2\boldsymbol{I}_{n_{..}}, \quad (4.3.27)$$

where $\boldsymbol{X}_1$, $\boldsymbol{X}_2$, and $\boldsymbol{X}_3$ are the design matrices that appear in (4.3.2). Using the expressions for $\boldsymbol{X}_1$, $\boldsymbol{X}_2$, and $\boldsymbol{X}_3$, and the expressions for $\boldsymbol{M}$ in (4.3.26) and $\boldsymbol{R}$ in (4.3.21), it is easy to verify that

$$\boldsymbol{MR} = \mathbf{0}, \ \boldsymbol{RX}_i = \mathbf{0} \ (i = 1, 2, 3). \tag{4.3.28}$$

Hence $\boldsymbol{M\Sigma R} = \mathbf{0}$ (i.e., $\boldsymbol{M\Sigma}(\boldsymbol{G}_1\boldsymbol{G}_1' + \boldsymbol{G}_2\boldsymbol{G}_2') = \mathbf{0}$). Postmultiplying by $\boldsymbol{G}_1$, and using the fact that $\boldsymbol{G}_1'\boldsymbol{G}_1 = \boldsymbol{I}$ and $\boldsymbol{G}_1'\boldsymbol{G}_2 = \mathbf{0}$ (due to the orthogonality of the matrix $\boldsymbol{G}$), results in $\boldsymbol{M\Sigma G}_1 = \mathbf{0}$ (i.e., $\text{Cov}(\bar{\mathbf{y}}, \boldsymbol{G}_1'\mathbf{y}) = \mathbf{0}$). Thus $\bar{\mathbf{y}}$, and hence $(\mathbf{z}_\tau', \mathbf{z}_\beta', \mathbf{z}_{\tau\beta}')'$, is uncorrelated with $\boldsymbol{G}_1'\mathbf{y}$. Thus the variance–covariance matrix of $\mathbf{w}$ in (4.3.24) is given by

$$\begin{aligned} Var(\mathbf{w}) = {} & Var(\mathbf{z}_\tau', \mathbf{z}_\beta', \mathbf{z}_{\tau\beta}')' + [\lambda_{\max}(\boldsymbol{D}_0)\boldsymbol{I}_{bv-1} - \boldsymbol{D}_0]^{1/2}\boldsymbol{G}_1'\boldsymbol{\Sigma}\boldsymbol{G}_1 \\ & \times [\lambda_{\max}(\boldsymbol{D}_0)\boldsymbol{I}_{bv-1} - \boldsymbol{D}_0]^{1/2}. \end{aligned} \tag{4.3.29}$$

Using $\boldsymbol{RX}_i = \mathbf{0}$ (see (4.3.28)), we get $\boldsymbol{G}_1'\boldsymbol{X}_i = \mathbf{0}$ (i = 1, 2, 3). Hence, using the expression for $\boldsymbol{\Sigma}$ in (4.3.27), we have $\boldsymbol{G}_1'\boldsymbol{\Sigma}\boldsymbol{G}_1 = \sigma_e^2\boldsymbol{G}_1'\boldsymbol{G}_1 = \sigma_e^2\boldsymbol{I}_{n_{..}-bv}$. Formula (4.3.29) thus simplifies to

$$Var(\mathbf{w}) = Var(\mathbf{z}_\tau', \mathbf{z}_\beta', \mathbf{z}_{\tau\beta}')' + \sigma_e^2[\lambda_{\max}(\boldsymbol{D}_0)\boldsymbol{I}_{bv-1} - \boldsymbol{D}_0]. \tag{4.3.30}$$

From (4.3.18), (4.3.19), and (4.3.30), we get the covariance matrices given in (4.3.25). Using the fact that $\boldsymbol{G}_1'\boldsymbol{X}_i = \mathbf{0}$ (i = 1, 2, 3) and $\boldsymbol{G}_1'\boldsymbol{G}_2 = \mathbf{0}$, we get $\boldsymbol{G}_1'\boldsymbol{\Sigma}\boldsymbol{G}_2 = \mathbf{0}$ (i.e., $\boldsymbol{G}_1'\mathbf{y}$ and $\boldsymbol{G}_2'\mathbf{y}$ are uncorrelated). Hence, $SS_2(e) = \mathbf{y}'\boldsymbol{G}_2\boldsymbol{G}_2'\mathbf{y}$ is distributed independently of $\boldsymbol{G}_1'\mathbf{y}$. To show that $SS_2(e)$ is distributed independently of $\mathbf{w}$, it only remains to show that $SS_2(e)$ is distributed independently of $\mathbf{z}_\tau$, $\mathbf{z}_\beta$, and $\mathbf{z}_{\tau\beta}$. This can be established by showing that $\text{Cov}(\bar{\mathbf{y}}, \boldsymbol{G}_2'\mathbf{y}) = \mathbf{0}$. The proof of this is similar to the proof of $\text{Cov}(\bar{\mathbf{y}}, \boldsymbol{G}_1'\mathbf{y}) = \mathbf{0}$ given above. This completes the proof of Lemma 4.3.2. □

The exact tests for testing H_τ: $\sigma_\tau^2 = 0$ and H_β: $\sigma_\beta^2 = 0$ are based on the sums of squares and their distributions given in Theorem 4.3.2 below. The proof of the theorem is immediate from Lemma 4.3.2.

Theorem 4.3.2. Let $\mathbf{w}_\tau$, $\mathbf{w}_\beta$, $\mathbf{w}_{\tau\beta}$, and $SS_2(e)$ be as defined in (4.3.23) and (4.3.24). Define $SS(\tau) = \mathbf{w}_\tau'\mathbf{w}_\tau$, $SS(\beta) = \mathbf{w}_\beta'\mathbf{w}_\beta$, and $SS(\tau\beta) = \mathbf{w}_{\tau\beta}'\mathbf{w}_{\tau\beta}$. Then

$$\begin{aligned} SS(\tau)/(b\sigma_\tau^2 + \sigma_{\tau\beta}^2 + \lambda_{\max}(\boldsymbol{D}_0)\sigma_e^2) &\sim \chi_{v-1}^2, \\ SS(\beta)/(v\sigma_\beta^2 + \sigma_{\tau\beta}^2 + \lambda_{\max}(\boldsymbol{D}_0)\sigma_e^2) &\sim \chi_{b-1}^2, \\ SS(\tau\beta)/(\sigma_{\tau\beta}^2 + \lambda_{\max}(\boldsymbol{D}_0)\sigma_e^2) &\sim \chi_{(v-1)(b-1)}^2, \\ SS_2(e)/\sigma_e^2 &\sim \chi_{n_{..}-2bv+1}^2. \end{aligned}$$

Furthermore, $SS(\tau)$, $SS(\beta)$, $SS(\tau\beta)$, and $SS_2(e)$ are all independently distributed. □

From Theorem 4.3.2, it is clear that an F-test for testing H_τ: $\sigma_\tau^2 = 0$ can be obtained based on the F-ratio $[SS(\tau)/(v-1)]/[SS(\tau\beta)/(b-1)(v-1)]$. Similarly, H_β: $\sigma_\beta^2 = 0$ can be tested using the F-ratio $[SS(\beta)/(b-1)]/[SS(\tau\beta)/(b-1)(v-1)]$. Note that $H_{\tau\beta}$: $\sigma_{\tau\beta}^2 = 0$ can be tested using the F-ratio $[SS(\tau\beta)/(b-1)(v-1)]/[SS_2(e)/(n_{..} - 2bv + 1)]$. However, we do not recommend this test, since the denominator d.f for this test is $(n_{..} - 2bv + 1)$, which can be much smaller than the denominator d.f. of the test based on the F-ratio in (4.3.13). Thus, for testing H_τ: $\sigma_{\tau\beta}^2 = 0$, we recommend the F-test based on $F_{\tau\beta}$ in (4.3.13). Note that under the assumption $n_{ij} \geq 1$, which was used to arrive at Theorem 4.3.2, $n_0 = bv$ in (4.3.13).

The implementation of the F-tests mentioned above for testing H_τ: $\sigma_\tau^2 = 0$ and H_β: $\sigma_\beta^2 = 0$ requires computation of the orthogonal matrices $\boldsymbol{P}$ and $\boldsymbol{G}$ that occur in (4.3.17) and (4.3.22). Note that the columns of $\boldsymbol{G}$ are orthonormal eigenvectors of the matrix $\boldsymbol{R}$ in (4.3.21) and hence $\boldsymbol{G}$ can be easily obtained. The matrix $\boldsymbol{P}$ can be computed using the algorithm in Graybill (1983, p. 406).

REMARK 4.3.2. For testing the hypotheses H_τ: $\sigma_\tau^2 = 0$ and H_β: $\sigma_\beta^2 = 0$, it is possible to construct ANOVA-based approximate F-tests (see Khuri and Littell 1987, Section 6). Such tests are based on Satterthwaite's approximation. Through a simulation study, Khuri and Littell (1987) have noted that the approximation can be highly unreliable for producing the required critical values. They have also noted that in situations where the approximation is satisfactory, the exact F-tests based on the sums of squares in Theorem 4.3.2 are superior (in terms of power) to the approximate tests. Thus, a discussion of the approximate tests is not included here.

REMARK 4.3.3. The derivation of the exact tests for testing H_τ: $\sigma_\tau^2 = 0$ and H_β: $\sigma_\beta^2 = 0$ requires the assumption $n_{..} > 2bv - 1$. This assumption is not unreasonable and can, for example, be satisfied if each cell contains at least two observations (i.e., $n_{ij} \geq 2$ for all i and for all j).

REMARK 4.3.4. In the balanced situation (i.e., when the n_{ij}'s are all equal), the exact tests for testing H_τ: $\sigma_\tau^2 = 0$ and H_β: $\sigma_\beta^2 = 0$ reduce to the F-tests based on the sum of squares in the ANOVA table. Note that when the data are balanced, the matrix $\boldsymbol{D}$ in (4.3.16) and $\boldsymbol{D}_0$ in (4.3.19) are both equal to $1/n\boldsymbol{I}_{bv}$, where n is the common value of the n_{ij}'s. Hence, $\lambda_{\max}(\boldsymbol{D}_0) = 1/n$ and $\mathbf{w} = (\mathbf{z}'_\tau, \mathbf{z}'_\beta, \mathbf{z}'_{\tau\beta})'$ in (4.3.24). It is now easy to verify that $nSS(\tau)$, $nSS(\beta)$, and $nSS(\tau\beta)$ are the same as the balanced ANOVA sums of squares associated with the main effects and interactions. Consequently, the exact tests derived in this section reduce to the ANOVA F-tests in the balanced case. (See the solution to Exercise 4.4, given in the Appendix.)

4.3.4. A Numerical Example

We shall now give an example in order to illustrate the application of Theroem 4.3.2. The example is taken from Khuri and Littell (1987, Section 3), and deals with a study of the variation in fusiform rust in southern pine tree plantations, where the trees came from different families and different test locations. The data given in Table 4.1 represent the proportion of symptomatic trees in each plot for trees coming from five different families and four test locations. Note that the number of plots for the different family-location combinations are not the same, resulting in unbalanced data.

Since the data are proportions, we shall do the analysis after making the arcsin(square root) transformation, which is the variance-stabilizing transformation for proportions. Let y_{ijk} denote the transformed observation from the kth plot at the jth test location for the trees coming from the ith family. We then have the model (4.3.1), where τ_i is the effect due to the ith family, β_j is the effect due to the jth location, and $(\tau\beta)_{ij}$ is the corresponding interaction ($i = 1, 2, 3, 4, 5$; $j = 1, 2, 3, 4$). The above effects are assumed to be independent random variables having the normal distributions given following equation (4.3.1). For testing $H_{\tau\beta}\colon \sigma^2_{\tau\beta} = 0$, we can use the ANOVA F-test based on the statistic $F_{\tau\beta}$ given in (4.3.13), with n_0 = number of nonempty

Table 4.1. Proportions of Symptomatic Trees from Five Families and Four Test Locations

	Family number				
Test Number	1	2	3	4	5
1	0.804 0.967 0.970	0.734 0.817 0.833 0.304	0.967 0.930 0.889	0.917	0.850
2	0.867 0.667 0.793 0.458	0.407 0.511 0.274 0.428	0.896 0.717	0.952	0.486 0.467
3	0.409 0.569 0.715 0.487	0.411 0.646 0.310	0.919 0.669 0.669 0.450	0.408 0.435 0.500	0.275 0.256
4	0.587 0.538 0.961 0.300	0.304 0.428	0.928 0.855 0.655 0.800	0.367	0.525

Source: A. I. Khuri and R. C. Littell (1987). Reproduced with permission of the International Biometric Society.

cells = 20, and r_3 = d.f. associated with $SS(\tau\beta|\tau,\beta) = (v-1)(b-1) = 12$. For testing H_τ: $\sigma_\tau^2 = 0$ or H_β: $\sigma_\beta^2 = 0$, we can use Theorem 4.3.2. Note that the conditions $n_{ij} \geq 1$ and $n_{..} \geq 2bv - 1$ are met in this example. In order to apply Theorem 4.3.2, we need to compute the 20×20 matrix $\boldsymbol{P}$ satisfying (4.3.17), the 19×19 matrix $\boldsymbol{D}_0$ in (4.3.19), and the matrices $\boldsymbol{G}_1$ and $\boldsymbol{G}_2$ in (4.3.22). Furthermore, we need $\lambda_{\max}(\boldsymbol{D}_0)$, the maximum eigenvalue of $\boldsymbol{D}_0$, and the matrix $[\lambda_{\max}(\boldsymbol{D}_0)\boldsymbol{I}_{19} - \boldsymbol{D}_0]^{1/2}$, so that the quantities in (4.3.24) can be evaluated. It is possible to give an explicit expression for the matrix $\boldsymbol{P}$ satisfying (4.3.17). For this, let

$$\boldsymbol{P}_1 = \begin{pmatrix} \frac{1}{\sqrt{5}} & \frac{1}{\sqrt{5}} & \frac{1}{\sqrt{5}} & \frac{1}{\sqrt{5}} & \frac{1}{\sqrt{5}} \\ \frac{1}{\sqrt{2}} & -\frac{1}{\sqrt{2}} & 0 & 0 & 0 \\ \frac{1}{\sqrt{6}} & \frac{1}{\sqrt{6}} & -\frac{2}{\sqrt{6}} & 0 & 0 \\ \frac{1}{\sqrt{12}} & \frac{1}{\sqrt{12}} & \frac{1}{\sqrt{12}} & -\frac{3}{\sqrt{12}} & 0 \\ \frac{1}{\sqrt{20}} & \frac{1}{\sqrt{20}} & \frac{1}{\sqrt{20}} & \frac{1}{\sqrt{20}} & -\frac{4}{\sqrt{20}} \end{pmatrix},$$

$$\boldsymbol{P}_2 = \begin{pmatrix} \frac{1}{2} & \frac{1}{2} & \frac{1}{2} & \frac{1}{2} \\ \frac{1}{\sqrt{2}} & -\frac{1}{\sqrt{2}} & 0 & 0 \\ \frac{1}{\sqrt{6}} & \frac{1}{\sqrt{6}} & -\frac{2}{\sqrt{6}} & 0 \\ \frac{1}{\sqrt{12}} & \frac{1}{\sqrt{12}} & \frac{1}{\sqrt{12}} & -\frac{3}{\sqrt{12}} \end{pmatrix}.$$

Then, with $v = 5$ and $b = 4$, $\boldsymbol{P}_1 \otimes \boldsymbol{P}_2$ satisfies

$$(\boldsymbol{P}_1 \otimes \boldsymbol{P}_2)\left(\boldsymbol{I}_5 \otimes \frac{1}{4}\boldsymbol{J}_4\right)(\boldsymbol{P}_1' \otimes \boldsymbol{P}_2') = \boldsymbol{I}_5 \otimes \begin{pmatrix} 1 & \boldsymbol{0} \\ \boldsymbol{0} & \boldsymbol{0} \end{pmatrix},$$

$$(\boldsymbol{P}_1 \otimes \boldsymbol{P}_2)\left(\frac{1}{5}\boldsymbol{J}_5 \otimes \boldsymbol{I}_4\right)(\boldsymbol{P}_1' \otimes \boldsymbol{P}_2') = \begin{pmatrix} 1 & \boldsymbol{0} \\ \boldsymbol{0} & \boldsymbol{0} \end{pmatrix} \otimes \boldsymbol{I}_4.$$

Hence, by suitably permuting the rows of $(\boldsymbol{P}_1 \otimes \boldsymbol{P}_2)$, we can obtain the orthogonal matrix $\boldsymbol{P}$ satisfying (4.3.17). The computation of $\boldsymbol{D}_0$ in (4.3.19) is striaghtforward. Furthermore, $\boldsymbol{G}$ satisfying (4.3.22) can be easily obtained from the spectral decomposition of the matrix $\boldsymbol{R}$ in (4.3.21). The quantities $\mathbf{w}_\tau$, $\mathbf{w}_\beta$, and $\mathbf{w}_{\tau\beta}$ in (4.3.24), and hence, $SS(\tau) = \mathbf{w}_\tau'\mathbf{w}_\tau$, $SS(\beta) = \mathbf{w}_\beta'\mathbf{w}_\beta$, and $SS(\tau\beta) = \mathbf{w}_{\tau\beta}'\mathbf{w}_{\tau\beta}$, can be obtained by straightforward calculation. The results are $SS(\tau) = 0.6172$, with d.f. $= v - 1 = 4$, and $SS(\tau\beta) = 0.4980$, with d.f. $= (v-1)(b-1) = 12$. Hence the F-ratio for testing H_τ: $\sigma_\tau^2 = 0$ has the value $[SS(\tau)/4]/[SS(\tau\beta)/12] = 3.718$. The corresponding P-value is 0.034. Thus, we reject H_τ: $\sigma_\tau^2 = 0$ at the 5% significance level. The hypothesis H_β: $\sigma_\beta^2 = 0$ can be tested similarly.

4.4. RANDOM TWO-FOLD NESTED MODELS

This section deals with the derivation of exact tests for testing the significance of two variance components corresponding to the nested effect and the nesting effect in an unbalanced random two-fold nested model. For testing the significance of the nested effect variance component, we shall show that an exact F-test can be obtained based on the ANOVA decomposition. For testing the significance of the nesting effect variance component, we shall describe the exact test derived by Khuri (1987). The derivations are quite similar to those in Section 4.3.3. We would like to point out that optimality results, similar to those in Theorem 4.3.1, are available for some nested models with random and mixed effects; see Khattree and Naik (1990) and Naik and Khattree (1992). Such optimality results will not, however, be reviewed here.

Suppose that the nesting effect corresponds to a factor A having v levels, and the nested effect corresponds to a factor B having a total of b levels with b_i levels of B nested within the ith level of A $(i = 1, 2, \ldots, v)$. Let n_{ij} denote the number of observations corresponding to the jth level of B nested within the ith level of A. If y_{ijk} $(k = 1, 2, \ldots, n_{ij})$ denote these observations, then the model is given by

$$\begin{aligned} y_{ijk} &= \mu + \tau_i + \beta_{i(j)} + e_{ijk}, \\ &\quad i = 1, 2, \cdots, v;\ j = 1, 2, \cdots, b_i;\ k = 1, 2, \cdots, n_{ij}, \end{aligned} \tag{4.4.1}$$

where μ is an overall mean, τ_i is the effect due to the ith level of A, $\beta_{i(j)}$ is the effect due to the jth level of B nested within the ith level of A, and e_{ijk} is the experimental error term. We assume that μ is a fixed unknown parameter and τ_i's, $\beta_{i(j)}$'s, and e_{ijk}'s are independent random variables with $\tau_i \sim N(0, \sigma_\tau^2)$, $\beta_{i(j)} \sim N(0, \sigma^2_{\beta(\tau)})$, and $e_{ijk} \sim N(0, \sigma_e^2)$. We are interested in testing the hypotheses $H_{\beta(\tau)}$: $\sigma^2_{\beta(\tau)} = 0$ and H_τ: $\sigma_\tau^2 = 0$. Let $y_{ij} = (y_{ij1}, y_{ij2}, \ldots, y_{ijn_{ij}})'$ and $\mathbf{y} = (\mathbf{y}'_{11}, \mathbf{y}'_{12}, \ldots, \mathbf{y}'_{1b_1}, \ldots, \mathbf{y}'_{v1}, \mathbf{y}'_{v2}, \ldots, \mathbf{y}'_{vb_v})'$. Also write $\boldsymbol{\tau} = (\tau_1, \tau_2, \ldots, \tau_v)'$ and $\boldsymbol{\beta} = (\beta_{1(1)}, \beta_{1(2)}, \ldots, \beta_{1(b_1)}, \ldots, \beta_{v(1)}, \beta_{v(2)}, \ldots, \beta_{v(b_v)})'$. Then model (4.4.1) can be written as

$$\mathbf{y} = \mu \mathbf{1}_{n_{..}} + \mathbf{diag}(\mathbf{1}_{n_{1.}}, \cdots, \mathbf{1}_{n_{v.}})\boldsymbol{\tau} + \mathbf{diag}(\mathbf{1}_{n_{11}}, \cdots, \mathbf{1}_{n_{1b_1}}, \cdots, \mathbf{1}_{n_{v1}}, \cdots, \mathbf{1}_{n_{vb_v}})\boldsymbol{\beta} + \mathbf{e}, \tag{4.4.2}$$

where $n_{i.} = \sum_{j=1}^{b_i} n_{ij}$, $n_{..} = \sum_{i=1}^{v} \sum_{j=1}^{b_i} n_{ij}$, and $\mathbf{e}$ is defined similarly to $\mathbf{y}$. We shall assume without loss of generality that $n_{ij} \geq 1$ for all i and for all j, since if any $n_{ij} = 0$, then the corresponding level of B does not occur in the experiment and hence does not have to be taken into account.

4.4.1. Testing $H_{\beta(\tau)} : \sigma^2_{\beta(\tau)} = 0$

An exact F-test for testing $H_{\beta(\tau)}$: $\sigma^2_{\beta(\tau)} = 0$ can be easily developed, and is quite similar to the test for $H_{\tau\beta}$: $\sigma^2_{\tau\beta} = 0$ discussed in Section 4.3.3. Let $SS(\beta(\tau))$ denote the sum of squares due to $\boldsymbol{\beta}$, adjusted for $\boldsymbol{\tau}$ and $SS(e)$ denote the sum of squares due to error computed based on model (4.4.2). (We use the notation $SS(\beta(\tau))$ rather than $SS(\beta|\tau)$ to emphasize that the levels of B are nested within the levels of A.) Then a test for $H_{\beta(\tau)}$: $\sigma^2_{\beta(\tau)} = 0$ rejects $H_{\beta(\tau)}$ for large values of the F-ratio

$$F_\tau = \frac{SS(\beta(\tau))/r}{SS(e)/(n_{..} - b)}, \tag{4.4.3}$$

where r is the degrees of freedom associated with $SS(\beta(\tau))$ and $b = \sum_{i=1}^{v} b_i$. Note that if $\bar{y}_{ij.} = 1/n_{ij} \sum_{k=1}^{n_{ij}} y_{ijk}$, then

$$SS(e) = \sum_{i=1}^{v} \sum_{j=1}^{b_i} \sum_{k=1}^{n_{ij}} (y_{ijk} - \bar{y}_{ij.})^2. \tag{4.4.4}$$

Derivation of the above test is omitted, since it is quite similar to the derivation of the F-test in Section 4.3.3 for testing $H_{\tau\beta}$: $\sigma^2_{\tau\beta} = 0$. The sum of squares required for computing the F-ratio in (4.4.3) can be obtained from the ANOVA decomposition based on model (4.4.2). Note that in the balanced case (i.e., when the n_{ij}'s are all equal and the b_i's are all equal), then r in (4.4.3) is given by $r = v(b_0 - 1)$, b_0 being the common value of the b_i's.

4.4.2. Testing $H_\tau : \sigma^2_\tau = 0$

We shall derive an F-test for testing H_τ: $\sigma^2_\tau = 0$. The derivation parallels that in Section 4.3.3 for testing the main effects variance components and requires the assumption $n_{..} > 2b - 1$. We shall consider the model for the cell means $\bar{y}_{ij.}$'s and apply a series of orthogonal transformations to arrive at two independent sums of squares that are distributed as scalar multiples of chi-squared random variables under H_τ: $\sigma^2_\tau = 0$. An F-ratio will then be constructed based on these sums of squares to arrive at a test for H_τ: $\sigma^2_\tau = 0$.

Let $\bar{\mathbf{y}} = (\bar{y}_{11.}, \bar{y}_{12.}, \ldots, \bar{y}_{1b_1.}, \ldots, \bar{y}_{v1.}, \bar{y}_{v2.}, \ldots, \bar{y}_{vb_v.})'$ and define $\bar{\mathbf{e}}$ similarly. Then from (4.4.1), we get

$$\bar{\mathbf{y}} = \mu \mathbf{1}_b + \mathbf{diag}(\mathbf{1}_{b_1}, \mathbf{1}_{b_2}, \cdots, \mathbf{1}_{b_v})\boldsymbol{\tau} + \boldsymbol{\beta} + \bar{\mathbf{e}}. \tag{4.4.5}$$

From (4.4.5), we obtain

$$Var(\bar{\mathbf{y}}) = \sigma^2_\tau \,\mathbf{diag}(\boldsymbol{J}_{b_1}, \cdots, \boldsymbol{J}_{b_v}) + \sigma^2_{\beta(\tau)} \boldsymbol{I}_b + \sigma^2_e \boldsymbol{E}, \tag{4.4.6}$$

where $\boldsymbol{E} = \mathbf{diag}(1/n_{11}, 1/n_{12}, \ldots, 1/n_{1b_1}, \ldots, 1/n_{v1}, 1/n_{v2}, \ldots, 1/n_{vb_v})$. The residual sum of squares $SS(e)$ in (4.4.4) can be written as

$$\begin{aligned} SS(e) &= \mathbf{y}'\boldsymbol{R}_1\mathbf{y}, \\ \boldsymbol{R}_1 &= \boldsymbol{I}_{n_{..}} - \mathbf{diag}(\boldsymbol{J}_{n_{11}}/n_{11}, \cdots, \boldsymbol{J}_{n_{vb_v}}/n_{vb_v}). \end{aligned} \tag{4.4.7}$$

Since $\boldsymbol{R}_1$ is a symmetric idempotent matrix of rank $n_{..} - b$, there exists an $n_{..} \times n_{..}$ orthogonal matrix $\boldsymbol{K}$ such that

$$\begin{aligned} \boldsymbol{R}_1 &= \boldsymbol{K}\ \mathbf{diag}(\boldsymbol{I}_{n_{..}-b}, \mathbf{0})\boldsymbol{K}' \\ &= [\boldsymbol{K}_1 : \boldsymbol{K}_2 : \boldsymbol{K}_3]\mathbf{diag}(\boldsymbol{I}_{b-1}, \boldsymbol{I}_{n_{..}-2b+1}, \mathbf{0})[\boldsymbol{K}_1 : \boldsymbol{K}_2 : \boldsymbol{K}_3]' \\ &= \boldsymbol{K}_1\boldsymbol{K}_1' + \boldsymbol{K}_2\boldsymbol{K}_2'. \end{aligned} \tag{4.4.8}$$

The matrices $\boldsymbol{K}_1$, $\boldsymbol{K}_2$ and $\boldsymbol{K}_3$ have respective orders $n_{..} \times (b-1)$, $n_{..} \times (n_{..} - 2b + 1)$, and $n_{..} \times b$ (here we are using the assumption that $n_{..} > 2b - 1$). From (4.4.7) we thus get

$$SS(e) = SS_1(e) + SS_2(e), \tag{4.4.9}$$

where $SS_1(e) = \mathbf{y}'\boldsymbol{K}_1\boldsymbol{K}_1'\mathbf{y}$ and $SS_2(e) = \mathbf{y}'\boldsymbol{K}_2\boldsymbol{K}_2'\mathbf{y}$. Note that the matrix $\mathbf{diag}(\boldsymbol{J}_{b_1}, \ldots, \boldsymbol{J}_{b_v})$ is a $b \times b$ matrix of rank v having nonzero eigenvalues $b_1, b_2, \ldots, b_v$ and zero eigenvalue of multiplicity $(b - v)$. Furthermore, the v orthonormal eigenvectors corresponding to the eigenvalues $b_1, b_2, \ldots, b_v$ can be written as the $b \times v$ matrix $\boldsymbol{O}_1'$, where $\boldsymbol{O}_1 = \mathbf{diag}(\mathbf{1}_{b_1}'/\sqrt{b_1}, \ldots, \mathbf{1}_{b_v}'/\sqrt{b_v})$. Let $\boldsymbol{O}_2$ be a $(b - v) \times b$ matrix so that $\boldsymbol{O} = [\boldsymbol{O}_1' : \boldsymbol{O}_2']'$ is an orthogonal matrix. Then, $\boldsymbol{O}\ \mathbf{diag}(\mathbf{1}_{b_1}\mathbf{1}_{b_1}', \ldots, \mathbf{1}_{b_v}\mathbf{1}_{b_v}')\boldsymbol{O}' = \mathbf{diag}(b_1, b_2, \ldots, b_v, \mathbf{0})$ (where $\mathbf{0}$ is a $(b - v) \times (b - v)$ matrix of zeros). If $\mathbf{z} = \boldsymbol{O}\bar{\mathbf{y}}$, then from (4.4.6),

$$Var(\mathbf{z}) = \sigma_\tau^2\ \mathbf{diag}(b_1, b_2, \cdots, b_v, 0) + \sigma_{\beta(\tau)}^2\boldsymbol{I}_b + \sigma_e^2\boldsymbol{OEO}'. \tag{4.4.10}$$

Also,

$$\mathrm{E}(\mathbf{z}) = \mu\boldsymbol{O}\mathbf{1}_b = \mu[\boldsymbol{O}_1' : \boldsymbol{O}_2']'\mathbf{1}_b = \mu\begin{pmatrix}\boldsymbol{O}_1\mathbf{1}_b \\ \boldsymbol{O}_2\mathbf{1}_b\end{pmatrix} = \mu\begin{pmatrix}\mathbf{f} \\ \mathbf{0}\end{pmatrix}, \tag{4.4.11}$$

where $\mathbf{f} = (\sqrt{b_1}, \sqrt{b_2}, \ldots, \sqrt{b_v})'$. In order to arrive at the last expression in (4.4.11), we have used the expression for $\boldsymbol{O}_1$ and the property that $\boldsymbol{O}_2\mathbf{1}_b = \mathbf{0}$. This follows from the fact that $\boldsymbol{O}_2\boldsymbol{O}_1' = \mathbf{0}$ (due to the orthogonality of $\boldsymbol{O}$) and $\mathbf{1}_b$ can be written as a linear combination of the columns of $\boldsymbol{O}_1'$. Let $\mathbf{g} = 1/\sqrt{b}(\mathbf{f}' : \mathbf{0})'$. Then $\mathbf{g}$ satisfies $\mathbf{g}'\mathbf{g} = 1$. Let $\boldsymbol{Q}_1$ be a $b \times (b - 1)$ matrix such that $\boldsymbol{Q} = [\mathbf{g} : \boldsymbol{Q}_1]'$ is a $b \times b$ orthogonal matrix. Then $\boldsymbol{Q}_1'\mathbf{g} = \mathbf{0}$ and if $\mathbf{u} = \boldsymbol{Q}_1'\mathbf{z}$, then $\mathrm{E}(\mathbf{u}) = \boldsymbol{Q}_1'\mu\boldsymbol{O}\mathbf{1}_b = \sqrt{b}\mu\boldsymbol{Q}_1'\mathbf{g} = \mathbf{0}$. Furthermore,

$$Var(\mathbf{u}) = \sigma_\tau^2\boldsymbol{Q}_1'\ \mathbf{diag}(b_1, b_2, \cdots, b_v, \mathbf{0})\boldsymbol{Q}_1 + \sigma_{\beta(\tau)}^2\boldsymbol{I}_{b-1} + \sigma_e^2\boldsymbol{Q}_1'\boldsymbol{OEO}'\boldsymbol{Q}_1. \tag{4.4.12}$$

We now show that the matrix $\boldsymbol{Q}'_1$ $\mathbf{diag}(b_1, b_2, \ldots, b_v, \mathbf{0})\boldsymbol{Q}_1$ is of rank $(v - 1)$. For this purpose, partition the $(b-1) \times b$ matrix $\boldsymbol{Q}'_1$ as $\boldsymbol{Q}'_1 = [\boldsymbol{Q}'_{11} : \boldsymbol{Q}'_{12}]$, where $\boldsymbol{Q}'_{11}$ and $\boldsymbol{Q}'_{12}$ are respectively $(b-1) \times v$ and $(b-1) \times (b-v)$ matrices. Then $\boldsymbol{Q} = [\mathbf{g} : \boldsymbol{Q}_1]' = \begin{pmatrix} \mathbf{f}'/\sqrt{b} & \mathbf{0} \\ \boldsymbol{Q}'_{11} & \boldsymbol{Q}'_{12} \end{pmatrix}$ and from the orthogonality of $\boldsymbol{Q}$ we get $\mathbf{ff}'/b + \boldsymbol{Q}_{11}\boldsymbol{Q}'_{11} = \boldsymbol{I}_v$. Hence, $\boldsymbol{Q}_{11}\boldsymbol{Q}'_{11} = \boldsymbol{I}_v - \mathbf{ff}'/b$ is a symmetric idempotent matrix of rank $(v-1)$. Hence, $\boldsymbol{Q}_{11}$ has rank $v-1$. Thus, $\text{rank}(\boldsymbol{Q}'_1\ \mathbf{diag}(b_1, b_2, \ldots, b_v, \mathbf{0})\boldsymbol{Q}_1) = \text{rank}(\boldsymbol{Q}'_{11}\ \mathbf{diag}(b_1, b_2, \ldots, b_v)\boldsymbol{Q}_{11}) = \text{rank}(\boldsymbol{Q}_{11}) = v - 1$.

Let $\boldsymbol{S}$ be a $(b-1) \times (b-1)$ orthogonal matrix such that $\boldsymbol{Q}'_1\ \mathbf{diag}(b_1, b_2, \ldots, b_v, \mathbf{0})\boldsymbol{Q}_1 = \boldsymbol{S}\ \mathbf{diag}(\boldsymbol{\Delta}, \mathbf{0})\boldsymbol{S}'$, where $\boldsymbol{\Delta}$ is a $(v-1) \times (v-1)$ diagonal matrix consisting of the nonzero eigenvalues of $\boldsymbol{Q}'_1\ \mathbf{diag}(b_1, b_2, \ldots, b_v, \mathbf{0})\boldsymbol{Q}_1$. We now consider the vector $\boldsymbol{S}'\mathbf{u}$. Since $\mathbf{u}$ has zero mean, so does $\boldsymbol{S}'\mathbf{u}$. Furthermore,

$$Var(\boldsymbol{S}'\mathbf{u}) = \sigma_\tau^2\ \mathbf{diag}(\boldsymbol{\Delta}, \mathbf{0}) + \sigma_{\beta(\tau)}^2 \boldsymbol{I}_{b-1} + \sigma_e^2 \boldsymbol{E}_0, \tag{4.4.13}$$

where $\boldsymbol{E}_0 = \boldsymbol{S}'\boldsymbol{Q}'_1\boldsymbol{O}\boldsymbol{E}\boldsymbol{O}'\boldsymbol{Q}_1\boldsymbol{S}$. Define

$$\mathbf{v} = \boldsymbol{S}'\mathbf{u} + [\lambda_{\max}(\boldsymbol{E}_0)\boldsymbol{I}_{b-1} - \boldsymbol{E}_0]^{1/2}\boldsymbol{K}'_1\mathbf{y}, \tag{4.4.14}$$

where $\lambda_{\max}(\boldsymbol{E}_0)$ is the largest eigenvalue of $\boldsymbol{E}_0$ and $\boldsymbol{K}_1$ is defined in (4.4.8). Let $\mathbf{v}$ be partitioned as $\mathbf{v} = [\mathbf{v}'_\tau : \mathbf{v}'_\beta]'$, where $\mathbf{v}_\tau$ and $\mathbf{v}_\beta$ are, respectively, $(v-1) \times 1$ and $(b-v) \times 1$ vectors. Using arguments similar to those in the proof of Lemma 4.3.2, the following lemma can be established.

Lemma 4.4.1. $\mathbf{v}_\tau$ and $\mathbf{v}_\beta$ are independent normally distributed random vectors with zero means and variance–covariance matrices

$$\begin{aligned} Var(\mathbf{v}_\tau) &= \sigma_\tau^2\boldsymbol{\Delta} + [\sigma_{\beta(\tau)}^2 + \lambda_{\max}(\boldsymbol{E}_0)\sigma_e^2]\boldsymbol{I}_{v-1} \\ Var(\mathbf{v}_\beta) &= [\sigma_{\beta(\tau)}^2 + \lambda_{\max}(\boldsymbol{E}_0)\sigma_e^2]\boldsymbol{I}_{b-v}. \qquad \square \end{aligned} \tag{4.4.15}$$

The following result is now immediate.

Theorem 4.4.1. Let $\mathbf{v}_\tau$ and $\mathbf{v}_\beta$ be as given above. Then

$$\mathbf{v}'_\tau \left[\sigma_\tau^2\boldsymbol{\Delta} + (\sigma_{\beta(\tau)}^2 + \lambda_{\max}(\boldsymbol{E}_0)\sigma_e^2)\boldsymbol{I}_{v-1}\right]^{-1} \mathbf{v}_\tau \sim \chi_{v-1}^2$$

$$\mathbf{v}'_\beta\mathbf{v}_\beta/(\sigma_{\beta(\tau)}^2 + \lambda_{\max}(\boldsymbol{E}_0)\sigma_e^2) \sim \chi_{b-v}^2. \qquad \square$$

From Theorem 4.4.1, it follows that under H_τ: $\sigma_\tau^2 = 0$, $\mathbf{v}'_\tau\mathbf{v}_\tau/(\sigma_{\beta(\tau)}^2 + \lambda_{\max}(\boldsymbol{E}_0)\sigma_e^2) \sim \chi_{v-1}^2$. Thus, the F-ratio $[\mathbf{v}'_\tau\mathbf{v}_\tau/(v-1)]/[\mathbf{v}'_\beta\mathbf{v}_\beta/(b-v)]$ can be used for testing H_τ: $\sigma_\tau^2 = 0$.

Implementation of the above test requires computation of $\mathbf{v}_\tau$ and $\mathbf{v}_\beta$, for which one has to compute the matrices $\boldsymbol{K}$, $\boldsymbol{O}$, $\boldsymbol{Q}_1$, and $\boldsymbol{S}$. Note that $\boldsymbol{K}$, $\boldsymbol{O}$, and $\boldsymbol{S}$ are orthogonal matrices that can be obtained from the spectral decomposition of the relevant matrices. The $b \times (b-1)$ matrix $\boldsymbol{Q}_1$ is constructed in such a way that $\boldsymbol{Q} = [\mathbf{g} : \boldsymbol{Q}_1]'$ is an orthogonal matrix (see the discussion following (4.4.11) for the definition of $\mathbf{g}$). The orthogonality of $\boldsymbol{Q}$ gives $\boldsymbol{Q}_1\boldsymbol{Q}_1' = \boldsymbol{I}_b - \mathbf{g}\mathbf{g}'$. Thus the $(b-1)$ columns of $\boldsymbol{Q}_1$ can be obtained by the Gram–Schmidt orthonormalization of the columns of $\boldsymbol{I}_b - \mathbf{g}\mathbf{g}'$.

Note that in the balanced case (i.e., when the n_{ij}'s are all equal and the b_i's are all equal), $\boldsymbol{E}_0 = 1/n\boldsymbol{I}_{b-1}$, where $\boldsymbol{E}_0$ is given in (4.4.13) and n is the common value of the n_{ij}'s. Hence, $\lambda_{\max}\boldsymbol{I}_{b-1} - \boldsymbol{E}_0 = 0$ and, consequently, $\mathbf{v} = \boldsymbol{S}'\mathbf{u}$ (see (4.4.14)). Furthermore, $\boldsymbol{\Delta}$ appearing in (4.4.13) simplifies to $\boldsymbol{\Delta} = b_0\boldsymbol{I}_{v-1}$, where b_0 is the common value of the b_i's. It is now easy to verify that $n\mathbf{v}_\tau'\mathbf{v}_\tau$ and $n\mathbf{v}_\beta'\mathbf{v}_\beta$ reduce, respectively, to the sum of squares due to the τ_i's and the sum of squares due to the $\beta_{j(i)}$'s computed based on the model (4.4.1). In other words, the exact test derived in this section for testing H_τ: $\sigma_\tau^2 = 0$ reduces to the ANOVA-based F-test in the balanced case. (See also the solution to Exercise 4.6, given in the Appendix.)

EXERCISES

4.1. Prove Lemma 4.3.1.

4.2. Consider a three-way unbalanced random effects model without interactions, where the usual normality assumptions hold. For testing the significance of the variance components, show that exact F-tests can be obtained following the procedure in Section 4.3.1. Also show that such exact F-tests coincide with the Wald's variance component tests.

4.3. **(a)** Show that $SS(\tau|\beta)$ given in (4.3.7) coincides with the sum of squares due to $\boldsymbol{\tau}$, adjusted for $\boldsymbol{\beta}$, obtained from the ANOVA decomposition based on the model (4.3.5).

(b) Show that $SS(\tau\beta|\tau,\beta)$ given in (4.3.11) coincides with the sum of squares due to $\boldsymbol{\tau\beta}$, adjusted for $\boldsymbol{\tau}$ and $\boldsymbol{\beta}$, obtained from the ANOVA decomposition based on model (4.3.2).

4.4. Consider the quantities $SS(\tau)$, $SS(\beta)$, and $SS(\tau\beta)$ given in Theorem 4.3.2. Show that in the balanced case for model (4.3.1), $nSS(\tau)$, $nSS(\beta)$, and $nSS(\tau\beta)$ are the same as the balanced ANOVA sums of squares associated with the main effects and interaction, where n is the common value of the n_{ij}'s.

4.5. Using the data in Table 4.1, carry out F-tests for testing H_0: $\sigma_{\tau\beta}^2 = 0$ and H_0: $\sigma_\beta^2 = 0$.

4.6. Consider the vector $\mathbf{v}$ in (4.4.14) and its partition $\mathbf{v} = [\mathbf{v}'_\tau : \mathbf{v}'_\beta]'$. Show that in the balanced case for model (4.4.1), $n\mathbf{v}'_\tau\mathbf{v}_\tau$ and $n\mathbf{v}'_\beta\mathbf{v}_\beta$ are the same as the ANOVA sums of squares associated with the nesting effect and nested effect, respectively, where n is the common value of the n_{ij}'s.

BIBLIOGRAPHY

Das, R. and Sinha, B. K. (1987). "Robust optimum invariant unbiased tests for variance components." In: *Proceedings of the Second International Tampere Conference in Statistics* (T. Pukkila, S. Puntanen, Eds.), University of Tampere, Finland, 317–342.

Davies, R. B. (1980). "The distribution of a linear combination of chi-square random variables." *Applied Statistics*, 29, 323–333.

Graybill, F. A. (1983). *Matrices with Applications in Statistics*, Second Edition. Wadsworth, California.

Hirotsu, C. (1979). "An F approximation and its application." *Biometrika*, 66, 577–584.

Khattree, R. and Naik, D. N. (1990). "Optimum tests for random effects in unbalanced nested designs." *Statistics*, 21, 163–168.

Khuri, A. I. (1987). "An exact test for the nesting effect's variance component in an unbalanced random two-fold nested model." *Statistics and Probability Letters*, 5, 305–311.

Khuri, A. I. and Littell, R. C. (1987). "Exact tests for the main effects variance components in an unbalanced random two-way model." *Biometrics*, 43, 545–560.

LaMotte, L. R. (1976). "Invariant quadratic estimators in the random one-way ANOVA model." *Biometrics*, 32, 793–804.

Mathew, T. and Sinha, B. K. (1988). "Optimum tests in unbalanced two-way models without interaction." *The Annals of Statistics*, 16, 1727–1740.

Naik, D. N. and Khattree, R. (1992). "Optimum tests for treatments when blocks are nested and random." *Statistics*, 23, 101–108.

Seely, J. F. and El-Bassiouni, Y. (1983). "Applying Wald's variance component test." *The Annals of Statistics*, 11, 197–201.

Tan, W. Y. and Cheng, S. S. (1984). "On testing variance components in three-stage unbalanced nested random effects models." *Sankhyā, Series B*, 46, 188–200.

Thomsen, I. (1975). "Testing hypotheses in unbalanced variance components models for two-way layouts." *The Annals of Statistics*, 3, 257–265.

Wald, A. (1947). "A note on regression analysis." *The Annals of Mathematical Statistics*, 18, 586–589.

CHAPTER 5

Random Models with Unequal Cell Frequencies in the Last Stage

5.1. INTRODUCTION

In the previous chapter, a method for deriving exact tests for random two-way crossed classification and two-fold nested models was presented. We may recall that this method is based on using a particular linear transformation of the data that produces sums of squares having properties similar to those of balanced ANOVA sums of squares. In the present chapter, an extension of this method to more general unbalanced random models will be presented. Such models can have any number of nested and/or crossed effects along with their interactions. The only restriction is that the imbalance occurs in the last stage of the associated design with no missing cells (see Chapter 3 for a definition of this kind of imbalance). This extension was first proposed by Khuri (1990).

The reader is advised to review the properties of balanced models described in Chapter 2, in particular, Sections 2.2, 2.3, and 2.4, before reading the remainder of this chapter.

5.2. UNBALANCED RANDOM MODELS WITH IMBALANCE IN THE LAST STAGE ONLY—NOTATION

The basic definitions, properties, and distributional results for a general balanced model were outlined in Sections 2.2, 2.3, and 2.4. In this section, as well as in Section 5.3, we shall address the main topic of this chapter, namely the analysis of a general unbalanced random model with imbalance occurring only in the last stage of the associated design. Exact tests concerning the model's variance components will be derived in Section 5.3. As before in Chapter 4, the method used in the derivation of these tests is based on a particular linear transformation of the data, which reduces the analysis of the unbalanced model to that of a balanced model.

Let us first present the notation needed for the development of the exact tests in Section 5.3.

We recall from Section 2.2 that $\theta = \{k_1, k_2, \ldots, k_s\}$ is the set of subscripts that identify the response y in model (2.2.1). Let us consider an unbalanced data situation such that $k_i = 1, 2, \ldots, a_i$ for $i = 1, 2, \ldots, s-1$, and the range of $k_s (\geq 1)$ is a function of $k_1, k_2, \ldots, k_{s-1}$, that is,

$$k_j = \begin{matrix} 1, 2, \cdots, a_j, & j = 1, 2, \cdots, s-1 \\ 1, 2, \cdots, n_\omega, & j = s, \end{matrix}$$

where

$$\omega = \{k_1, k_2, \cdots, k_{s-1}\}. \tag{5.2.1}$$

The design is therefore balanced with respect to the first $s-1$ subscripts, but is unbalanced with respect to subscript k_s. The corresponding model can then be written as

$$y_\theta = \sum_{i=0}^{\nu} g^{(i)}_{\theta_i(\bar{\theta}_i)} + e_\theta, \tag{5.2.2}$$

where $g^{(i)}_{\theta_i(\bar{\theta}_i)}$ is the same as in the balanced model (2.2.1) for $i = 0, 1, \ldots, \nu$, and e_θ is the experimental error term. We note that the effects in $\sum_{i=0}^{\nu} g^{(i)}_{\theta_i(\bar{\theta}_i)}$ are indexed by the subscripts in ω. We assume that $g^{(i)}_{\theta_i(\bar{\theta}_i)}$, $i = 1, 2, \ldots, \nu$, and e_θ are independent and normally distributed random variables with zero means and variances $\sigma_1^2, \sigma_2^2, \ldots, \sigma_\nu^2$, and σ_e^2, respectively.

Let T be the set of all $(s-1)$-tuples of the form

$$T = \{\omega = (k_1, k_2, \cdots, k_{s-1}) : k_i = 1, 2, \cdots, a_i;\ i = 1, 2, \cdots, s-1\}. \tag{5.2.3}$$

Let c denote the number of elements in T. Then,

$$c = \prod_{i=1}^{s-1} a_i. \tag{5.2.4}$$

For reasons to be explained later in Section 5.3, we assume that $N > 2c - 1$, where $N = \sum_{\omega \in T} n_\omega$ is the total number of observations. This condition is satisfied if, for example, $n_\omega \geq 2$ for all ω in T. Recall that a similar assumption was also made in Sections 4.3 and 4.4 in Chapter 4 (see Subsections 4.3.3.2 and 4.4.2). Model (5.2.2) can be written in vector form as

$$\mathbf{y} = \sum_{i=0}^{\nu} \mathbf{X}_i \boldsymbol{\beta}_i + \mathbf{e}, \tag{5.2.5}$$

where $\mathbf{X}_i$ is a matrix of zeros and ones of order $N \times c_i$, and c_i $(i = 0, 1, 2, \ldots, \nu)$ is the same as in formula (2.4.4). The matrix $\mathbf{X}_0$ is the $\mathbf{1}_N$ vector of ones. The variance–covariance matrix $\boldsymbol{\Sigma}$ of $\mathbf{y}$ is given by

$$\boldsymbol{\Sigma} = \sum_{i=1}^{\nu} \sigma_i^2 \mathbf{X}_i \mathbf{X}_i' + \sigma_e^2 \mathbf{I}_N. \tag{5.2.6}$$

For a given $\omega = (k_1, k_2, \ldots, k_{s-1})$ in T, let $\bar{y}_\omega$ denote the average of y_θ with respect to subscript k_s. Thus

$$\bar{y}_\omega = \frac{1}{n_\omega} \sum_{k_s=1}^{n_\omega} y_\theta, \quad \omega \in T. \tag{5.2.7}$$

Let $\bar{\mathbf{y}}$ denote the $c \times 1$ vector consisting of the values of $\bar{y}_\omega$ for $\omega \in T$. Then,

$$\bar{\mathbf{y}} = \mathbf{D}\mathbf{y}, \tag{5.2.8}$$

where

$$\mathbf{D} = (\mathbf{W}'\mathbf{W})^{-1}\mathbf{W}' \tag{5.2.9}$$

and $\mathbf{W}$ is an $N \times c$ block–diagonal matrix of the form

$$\mathbf{W} = \mathbf{diag}(\mathbf{1}_{n_\omega})_{\omega \in T}. \tag{5.2.10}$$

Now, from model (5.2.2) we have

$$\bar{y}_\omega = \sum_{i=0}^{\nu} g^{(i)}_{\theta_i(\bar{\theta}_i)} + \bar{e}_\omega, \quad \omega \in T, \tag{5.2.11}$$

where

$$\bar{e}_\omega = \frac{1}{n_\omega} \sum_{k_s=1}^{n_\omega} e_\theta, \quad \omega \in T.$$

Since the portion $\sum_{i=0}^{\nu} g^{(i)}_{\theta_i(\bar{\theta}_i)}$ is the same as in the balanced model (2.2.1), we can write model (5.2.11) as

$$\bar{\mathbf{y}} = \sum_{i=0}^{\nu} \mathbf{H}_i \boldsymbol{\beta}_i + \bar{\mathbf{e}}, \tag{5.2.12}$$

where, as in Section 2.2, the matrix $\mathbf{H}_i$ is expressed as a direct product of the form

$$\mathbf{H}_i = \bigotimes_{j=1}^{s-1} \mathbf{L}_{ij}, \quad i = 0, 1, \cdots, \nu, \tag{5.2.13}$$

where

$$\mathbf{L}_{ij} = \begin{matrix} \mathbf{I}_{a_j}, & k_j \in \psi_i \\ \mathbf{1}_{a_j}, & k_j \notin \psi_i \end{matrix} \qquad i = 0, 1, \cdots, \nu; \; j = 1, 2, \cdots, s-1. \tag{5.2.14}$$

We recall that ψ_i is the set of subscripts associated with the ith effect in the model $(i = 0, 1, \ldots, \nu)$ with ψ_i being the empty set for $i = 0$, and $\otimes$ is the symbol of direct product of matrices.

The variance–covariance matrix of $\bar{\mathbf{y}}$ is of the form

$$Var(\bar{\mathbf{y}}) = \sum_{i=1}^{\nu} \sigma_i^2 \mathbf{A}_i + \sigma_e^2 \mathbf{K}, \tag{5.2.15}$$

where $\mathbf{A}_i = \mathbf{H}_i \mathbf{H}_i'$ $(i = 1, 2, \ldots, \nu)$ and $\mathbf{K}$ is the matrix

$$\mathbf{K} = \mathbf{diag}\left(\frac{1}{n_\omega}\right)_{\omega \in T} \tag{5.2.16}$$

Since $\mathbf{A}_i \mathbf{A}_j = \mathbf{A}_j \mathbf{A}_i$, $i \neq j$, there exists an orthogonal matrix $\mathbf{Q}$ of order $c \times c$ that diagonalizes the $\mathbf{A}_i$'s simultaneously, that is,

$$\mathbf{Q} \mathbf{A}_i \mathbf{Q}' = \mathbf{\Lambda}_i, \; i = 0, 1, \cdots, \nu, \tag{5.2.17}$$

where $\mathbf{\Lambda}_i$ is a diagonal matrix. The construction of the matrix $\mathbf{Q}$ will be described in Section 5.3.

5.3. UNBALANCED RANDOM MODELS WITH IMBALANCE IN THE LAST STAGE ONLY—ANALYSIS

As was noted earlier in Section 5.2, the portion $\sum_{i=0}^{\nu} g^{(i)}_{\theta_i(\bar{\theta}_i)}$ in model (5.2.2) is not affected by the imbalance in the last stage of the design and is therefore the same as in a balanced model. Let us then consider the so-called derived model

$$z_\omega = \sum_{i=0}^{\nu} g^{(i)}_{\theta_i(\bar{\theta}_i)}, \qquad \omega \in T, \tag{5.3.1}$$

which has the form of a balanced model with one observation z_ω for each ω in T. This model can be written as

$$\mathbf{z} = \sum_{i=0}^{\nu} \mathbf{H}_i \boldsymbol{\beta}_i, \tag{5.3.2}$$

where $\mathbf{H}_i$ is defined in formula (5.2.13). Let $\mathbf{P}_i$ be a $c \times c$ matrix associated with the sum of squares for the ith effect in model (5.3.2), $i = 0, 1, \ldots, \nu$. Based on the properties of balanced models in Section 2.3, $\mathbf{P}_i$ is an idempotent matrix $(i = 0, 1, \ldots, \nu)$, $\mathbf{P}_i \mathbf{P}_j = \mathbf{0}$, $i \neq j$, and $\sum_{i=0}^{\nu} \mathbf{P}_i = \mathbf{I}_c$. Furthermore, by Lemmas 2.3.1 and 2.3.4 we have the following results:

(a)

$$\mathbf{P}_i = \sum_{j=0}^{\nu} \frac{\lambda_{ij}}{b_j} \mathbf{A}_j, \quad i = 0, 1, \cdots, \nu, \tag{5.3.3}$$

where $\mathbf{A}_j = \mathbf{H}_j \mathbf{H}_j'$, λ_{ij} is a known constant equal to the coefficient of the jth admissible mean in the ith component for the balanced model (5.3.1), and b_j is given by formula (2.3.4) with θ replaced by ω, that is,

$$b_j = \begin{cases} \prod_{k_\ell \notin \psi_j} a_\ell, & \text{if } \psi_j \neq \omega \\ 1, & \text{if } \psi_j = \omega. \end{cases} \quad j = 0, 1, \cdots, \nu. \tag{5.3.4}$$

Note that ψ_j is a subset of ω for $j = 0, 1, \ldots, \nu$.

(b)

$$\mathbf{A}_j \mathbf{P}_i = \kappa_{ij} \mathbf{P}_i, \tag{5.3.5}$$

$i, j = 0, 1, \ldots, \nu$, where

$$\kappa_{ij} = \begin{cases} 0, & \text{if } \psi_i \not\subset \psi_j \\ b_j, & \text{if } \psi_i \subset \psi_j. \end{cases} \tag{5.3.6}$$

Let m_i be the rank of $\mathbf{P}_i$, and let $\mathbf{Q}_i$ be a matrix of order $m_i \times c$ whose rows are orthonormal and span the row space of $\mathbf{P}_i$ $(i = 0, 1, \ldots, \nu)$. We have that $\sum_{i=0}^{\nu} m_i = c$ since $\sum_{i=0}^{\nu} \mathbf{P}_i = \mathbf{I}_c$. Furthermore, the matrices $\mathbf{Q}_0, \mathbf{Q}_1, \ldots, \mathbf{Q}_\nu$ have the following properties given in Lemma 5.3.1.

Lemma 5.3.1.

(a) $\mathbf{Q}_0 = 1/\sqrt{c}\mathbf{1}_c'$.

(b) $\mathbf{Q}_i \mathbf{Q}_i' = \mathbf{I}_{m_i}$, $i = 0, 1, \ldots, \nu$,
$\mathbf{Q}_i \mathbf{Q}_j' = \mathbf{0}, i \neq j$.

(c) $\mathbf{A}_j \mathbf{Q}_i' = \begin{cases} \mathbf{0}, & \text{if } \psi_i \not\subset \psi_j \\ b_j \mathbf{Q}_i', & \text{if } \psi_i \subset \psi_j, \end{cases} \quad i, j = 0, 1, \ldots, \nu.$

Proof.

(a) From formula (5.3.3), $\mathbf{P}_0 = \mathbf{A}_0 / b_0$, and by formula (5.3.4), $b_0 = \prod_{\ell=1}^{s-1} a_\ell = c$. But,

$$\begin{aligned} \mathbf{A}_0 &= \mathbf{H}_0 \mathbf{H}_0' \\ &= \bigotimes_{i=1}^{s-1} \mathbf{J}_{a_i} \\ &= \mathbf{J}_c. \end{aligned}$$

Hence, $\mathbf{P}_0 = 1/c\, \mathbf{J}_c$. It follows that $\mathbf{Q}_0 = 1/\sqrt{c}\mathbf{1}_c'$.

(b) $\mathbf{Q}_i\mathbf{Q}_i' = \mathbf{I}_{m_i}$ follows directly from the definition of $\mathbf{Q}_i$ $(i = 0, 1, \ldots, \nu)$. Now, $\mathbf{Q}_i$ can be expressed as $\mathbf{Q}_i = \mathbf{V}_i\mathbf{P}_i$ for some matrix $\mathbf{V}_i$ of order $m_i \times c$ $(i = 0, 1, \ldots, \nu)$. Then, for $i \neq j$, $\mathbf{Q}_i\mathbf{Q}_j' = \mathbf{V}_i\mathbf{P}_i\mathbf{P}_j\mathbf{V}_j' = \mathbf{0}$, since $\mathbf{P}_i\mathbf{P}_j = \mathbf{0}$.

(c) By formula (5.3.5),

$$\begin{aligned} \mathbf{A}_j\mathbf{Q}_i' &= \mathbf{A}_j\mathbf{P}_i\mathbf{V}_i' \\ &= \kappa_{ij}\mathbf{P}_i\mathbf{V}_i' \\ &= \kappa_{ij}\mathbf{Q}_i'. \end{aligned}$$

Hence, $\mathbf{A}_j\mathbf{Q}_i' = \mathbf{0}$, if $\psi_i \not\subset \psi_j$, and $\mathbf{A}_j\mathbf{Q}_i' = b_j\mathbf{Q}_i'$, if $\psi_i \subset \psi_j$. This completes the proof of Lemma 5.3.1. □

On the basis of Lemma 5.3.1 it is easy to see that the matrix

$$\mathbf{Q} = [\mathbf{Q}_0' : \mathbf{Q}_1' : \cdots : \mathbf{Q}_\nu']' \tag{5.3.7}$$

is orthogonal and diagonalizes $\mathbf{A}_0, \mathbf{A}_1, \ldots, \mathbf{A}_\nu$ simultaneously because

$$\begin{aligned} \mathbf{Q}\mathbf{A}_i\mathbf{Q}' &= \mathbf{Q}\mathbf{A}_i[\mathbf{Q}_0' : \mathbf{Q}_1' : \cdots : \mathbf{Q}_\nu'] \\ &= \mathbf{Q}[\kappa_{oi}\mathbf{Q}_0' : \kappa_{1i}\mathbf{Q}_1' : \cdots : \kappa_{\nu i}\mathbf{Q}_\nu'] \\ &= \mathbf{diag}(\kappa_{oi}\mathbf{I}_{m_0}, \kappa_{1i}\mathbf{I}_{m_1}, \cdots, \kappa_{\nu i}\mathbf{I}_{m_\nu}),\ i = 0, 1, \cdots, \nu. \end{aligned}$$

From formula (5.2.15) we then have

$$\begin{aligned} Var(\mathbf{Q}_i\bar{\mathbf{y}}) &= \mathbf{Q}_i\left(\sum_{j=1}^{\nu}\sigma_j^2\mathbf{A}_j + \sigma_e^2\mathbf{K}\right)\mathbf{Q}_i' \\ &= \sum_{j=1}^{\nu}\sigma_j^2\mathbf{Q}_i\mathbf{A}_j\mathbf{Q}_i' + \sigma_e^2\mathbf{Q}_i\mathbf{K}\mathbf{Q}_i' \\ &= \sum_{j=1}^{\nu}\sigma_j^2\kappa_{ij}\mathbf{I}_{m_i} + \sigma_e^2\mathbf{Q}_i\mathbf{K}\mathbf{Q}_i',\ \text{by Lemma 5.3.1(c)}, \\ &= \delta_i\mathbf{I}_{m_i} + \sigma_e^2\mathbf{Q}_i\mathbf{K}\mathbf{Q}_i', \end{aligned} \tag{5.3.8}$$

where

$$\begin{aligned} \delta_i &= \sum_{j=1}^{\nu}\sigma_j^2\kappa_{ij} \\ &= \sum_{j\in\varphi_i} b_j\sigma_j^2, \end{aligned} \tag{5.3.9}$$

and

$$\varphi_i = \{j : 1 \leq j \leq \nu | \psi_i \subset \psi_j\}, \quad i = 0, 1, \cdots, \nu. \tag{5.3.10}$$

Furthermore, by Lemma 5.3.1(c),

$$
\begin{aligned}
Cov(\mathbf{Q}_i\bar{\mathbf{y}}, \mathbf{Q}_j\bar{\mathbf{y}}) &= \mathbf{Q}_i\left(\sum_{j=1}^{\nu}\sigma_j^2\mathbf{A}_j + \sigma_e^2\mathbf{K}\right)\mathbf{Q}_j' \\
&= \sigma_e^2\mathbf{Q}_i\mathbf{K}\mathbf{Q}_j', \text{ if } i \neq j.
\end{aligned}
$$

Consider now the vector $\mathbf{u}$ defined by

$$
\mathbf{u} = \tilde{\mathbf{Q}}\bar{\mathbf{y}}, \tag{5.3.11}
$$

where

$$
\tilde{\mathbf{Q}} = [\mathbf{Q}_1' : \mathbf{Q}_2' : \cdots : \mathbf{Q}_\nu']',
$$

which is obtained by deleting the first row of $\mathbf{Q}$ in (5.3.7). It follows that

$$
\begin{aligned}
E(\mathbf{u}) &= \tilde{\mathbf{Q}}E(\bar{\mathbf{y}}) \\
&= \tilde{\mathbf{Q}}(\beta_0\mathbf{1}_c) \\
&= \beta_0\sqrt{c}\tilde{\mathbf{Q}}\mathbf{Q}_0', \text{ by Lemma 5.3.1(a)} \\
&= \mathbf{0}, \text{ by Lemma 5.3.1(b).}
\end{aligned}
$$

By applying formula (5.3.8), the variance of $\mathbf{u}$ is given by

$$
Var(\mathbf{u}) = \mathbf{diag}(\delta_1\mathbf{I}_{m_1}, \delta_2\mathbf{I}_{m_2}, \cdots, \delta_\nu\mathbf{I}_{m_\nu}) + \sigma_e^2\mathbf{G}, \tag{5.3.12}
$$

where

$$
\mathbf{G} = \tilde{\mathbf{Q}}\mathbf{K}\tilde{\mathbf{Q}}'. \tag{5.3.13}
$$

We note that the elements of $\mathbf{u}$ are not independent since $\mathbf{G}$ in (5.3.12) is not necessarily a diagonal matrix.

If $\mathbf{G}$ were not present in formula (5.3.12), it would be possible to use $\mathbf{u}$ to obtain independent sums of squares distributed as scaled chi-squared variates of the form $\delta_1\chi^2_{m_1}, \delta_2\chi^2_{m_2}, \ldots, \delta_\nu\chi^2_{m\nu}$. These sums of squares would then be used to derive exact test statistics concerning the variance components $\sigma_1^2, \sigma_2^2, \ldots, \sigma_\nu^2$. This was demonstrated in Chapter 2. In the next section, we demonstrate that such test statistics can still be found even if $\mathbf{G}$ is not equal to zero.

5.3.1. Derivation of Exact Tests

The residual sum of squares for the original unbalanced model (5.2.2) is of the form

$$
\begin{aligned}
SS(e) &= \sum_{\theta}(y_\theta - \bar{y}_\omega)^2 \\
&= \sum_{\omega\in T}\left[\sum_{k_s=1}^{n_\omega}(y_\theta - \bar{y}_\omega)^2\right],
\end{aligned}
$$

where, if we recall, $\theta = (\omega, k_s)$, and T is the set of $(s-1)$-tuples defined in formula (5.2.3). We thus have

$$SS(e) = \mathbf{y}'\mathbf{R}\mathbf{y}, \tag{5.3.14}$$

where

$$\mathbf{R} = \mathbf{I}_N - \mathbf{W}(\mathbf{W}'\mathbf{W})^{-1}\mathbf{W}', \tag{5.3.15}$$

and $\mathbf{W}$ is the block–diagonal matrix in formula (5.2.10). Properties of the matrix $\mathbf{R}$ are outlined in Lemma 5.3.2.

Lemma 5.3.2.

(a) $\mathbf{R}$ is idempotent of rank $N-c$, where c is defined in (5.2.4).
(b) $\mathbf{DR} = \mathbf{0}$, where $\mathbf{D}$ is the matrix in formula (5.2.9).
(c) $\mathbf{RX}_i = \mathbf{0}$, $i = 1, 2, \ldots, \nu$, where $\mathbf{X}_1, \mathbf{X}_2, \ldots, \mathbf{X}_\nu$ are the matrices shown in formula (5.2.5).

Proof. The proofs of (a) and (b) are straightforward. To prove (c), we note that the matrix $\mathbf{X}_i$ can be partitioned into $c = \prod_{i=1}^{s-1} a_i$ submatrices that correspond to the various values of $\omega = (k_1, k_2, \ldots, k_{s-1}) \in T$. The submatrix corresponding to a particular ω is of oder $n_\omega \times c_i$, where c_i is the number of columns of $\mathbf{X}_i$ $(i = 1, 2, \ldots, \nu)$. Let such a submatrix be denoted by $\mathbf{U}_\omega$. Each column of $\mathbf{U}_\omega$ consists of either n_ω zeros or n_ω ones. Hence, $\mathbf{RX}_i$ can be partitioned into c submatrices of the form $(\mathbf{I}_{n_\omega} - \mathbf{J}_{n_\omega}/n_\omega)\,\mathbf{U}_\omega = \mathbf{U}_\omega - \mathbf{U}_\omega = \mathbf{0}$ for $\omega \in T$. It follows that $\mathbf{RX}_i = \mathbf{0}$ for $i = 1, 2, \ldots, \nu$. This completes the proof of Lemma 5.3.2. □

From Lemma 5.3.2 and formula (5.2.6) we conclude that

$$\mathbf{D\Sigma R} = \mathbf{0}. \tag{5.3.16}$$

Furthermore,

$$\frac{1}{\sigma_e^2}\mathbf{R\Sigma} = \mathbf{R}. \tag{5.3.17}$$

This implies that $SS(e)/\sigma_e^2$ has the central chi-squared distribution with $N-c$ degrees of freedom (since $\mathbf{R}$ is idempotent of rank $N-c$). Also, formula (5.3.16) indicates that the vector $\bar{\mathbf{y}}$ defined in (5.2.8) is independent of $SS(e)$.

Since $\mathbf{R}$ is idempotent of rank $N-c$, it can be expressed as

$$\mathbf{R} = \mathbf{C\Lambda C}', \tag{5.3.18}$$

where $\mathbf{C}$ is an orthogonal matrix and $\mathbf{\Lambda}$ is a diagonal matrix with $N-c$ ones and c zeros. Let us now assume that

$$N > 2c - 1. \tag{5.3.19}$$

This assumption allows us to partition $\boldsymbol{\Lambda}$ as

$$\boldsymbol{\Lambda} = \mathbf{diag}(\mathbf{I}_{\xi_1}, \mathbf{I}_{\xi_2}, \mathbf{0}), \tag{5.3.20}$$

where

$$\xi_1 = c - 1 \tag{5.3.21}$$

$$\xi_2 = N - 2c + 1. \tag{5.3.22}$$

Accordingly, the matrix $\mathbf{C}$ in (5.3.18) can be partitioned as $\mathbf{C} = [\mathbf{C}_1 : \mathbf{C}_2 : \mathbf{C}_3]$, where $\mathbf{C}_1, \mathbf{C}_2$, and $\mathbf{C}_3$ are matrices of orders $N \times \xi_1, N \times \xi_2$, and $N \times c$, respectively. We thus have

$$\mathbf{R} = \mathbf{C}_1\mathbf{C}_1' + \mathbf{C}_2\mathbf{C}_2'. \tag{5.3.23}$$

Note that

$$\begin{aligned} \mathbf{C}_i'\mathbf{C}_i &= \mathbf{I}, \quad i = 1, 2, 3 \\ \mathbf{C}_i'\mathbf{C}_j &= \mathbf{0}, \quad i \neq j. \end{aligned}$$

Let us now define the vector $\boldsymbol{\phi}$ as

$$\boldsymbol{\phi} = \mathbf{u} + (\lambda_{\max}\mathbf{I}_{\xi_1} - \mathbf{G})^{1/2}\mathbf{C}_1'\mathbf{y}, \tag{5.3.24}$$

where $\mathbf{G}$ is defined in formula (5.3.13) and $\lambda_{\max}$ is its largest eigenvalue. Here, $(\lambda_{\max}\mathbf{I}_{\xi_1} - \mathbf{G})^{1/2}$ is a symmetric matrix with eigenvalues equal to the square roots of the eigenvalues of $\lambda_{\max}\mathbf{I}_{\xi_1} - \mathbf{G}$, which are nonnegative. Let $\boldsymbol{\phi}$ be partitioned in a manner similar to that of $\mathbf{u}$ in (5.3.11), that is, $\boldsymbol{\phi} = [\boldsymbol{\phi}_1' : \boldsymbol{\phi}_2' : \ldots : \boldsymbol{\phi}_\nu']'$, where $\boldsymbol{\phi}_i$ is of order $m_i \times 1$ with m_i being the number of rows of $\mathbf{Q}_i$ $(i = 1, 2, \ldots, \nu)$. The vector $\boldsymbol{\phi}$ will be used to derive exact tests concerning the variance components. For this purpose we need to determine the distributional properties of $\boldsymbol{\phi}_i$. These are given in the following lemma, which is a generalization of lemmas 4.3.2 and 4.4.1 in Chapter 4.

Lemma 5.3.3.

(a) $E(\boldsymbol{\phi}_i) = \mathbf{0}$, $i = 1, 2, \ldots, \nu$.

(b) $\boldsymbol{\phi}_1, \boldsymbol{\phi}_2, \ldots, \boldsymbol{\phi}_\nu$ are independently distributed as normal random vectors with $\boldsymbol{\phi}_i$ having the variance–covariance matrix

$$Var(\boldsymbol{\phi}_i) = (\delta_i + \lambda_{\max}\sigma_e^2)\mathbf{I}_{m_i}, \quad i = 1, 2, \cdots, \nu, \tag{5.3.25}$$

where δ_i is defined in formula (5.3.9).

(c) $\boldsymbol{\phi}_1, \boldsymbol{\phi}_2, \ldots, \boldsymbol{\phi}_\nu$ are independent of $SS_2(e) = \mathbf{y}'\mathbf{C}_2\mathbf{C}_2'\mathbf{y}$, which is the portion of the residual sum of squares $SS(e)$ corresponding to the matrix $\mathbf{C}_2$ in formula (5.3.23).

Proof.

(a) From model (5.2.12) and Lemma 5.3.1, we have

$$\begin{aligned} E(\mathbf{Q}_i\bar{\mathbf{y}}) &= \mathbf{Q}_i\mathbf{H}_0\beta_0 \\ &= \mathbf{Q}_i\beta_0\mathbf{1}_c \\ &= \mathbf{0}, \quad i = 1, 2, \cdots, \nu. \end{aligned}$$

Hence, by formula (5.3.11), $E(\mathbf{u}) = \mathbf{0}$. From (5.3.15) we note that

$$\mathbf{R1}_N = \mathbf{0}, \tag{5.3.26}$$

since $\mathbf{1}_N = \mathbf{W1}_c$. Using (5.3.23) in (5.3.26), we obtain,

$$(\mathbf{C}_1\mathbf{C}_1' + \mathbf{C}_2\mathbf{C}_2')\mathbf{1}_N = \mathbf{0}. \tag{5.3.27}$$

Multiplying the two sides of (5.3.27) on the left by $\mathbf{C}_1'$ results in $\mathbf{C}_1'\mathbf{1}_N = \mathbf{0}$. It follows that

$$\begin{aligned} E(\mathbf{C}_1'\mathbf{y}) &= \mathbf{C}_1'\beta_0\mathbf{1}_N \\ &= \mathbf{0}. \end{aligned}$$

By combining this result with the fact that $E(\mathbf{u}) = \mathbf{0}$, we conclude that the mean of $\boldsymbol{\phi}$ in (5.3.24) is zero. Thus $E(\boldsymbol{\phi}_i) = \mathbf{0}$ for $i = 1, 2, \ldots, \nu$.

(b) It is clear that the $\boldsymbol{\phi}_i$'s are normally distributed. Now, the vector $\mathbf{u}$ in (5.3.24) is independent of $\mathbf{C}_1'\mathbf{y}$. To show this, we note from (5.3.16) and (5.3.23) that

$$\mathbf{D\Sigma C}_1 = \mathbf{0}. \tag{5.3.28}$$

Hence, $Cov(\bar{\mathbf{y}}, \mathbf{C}_1'\mathbf{y}) = \mathbf{0}$, since $\bar{\mathbf{y}} = \mathbf{Dy}$. Consequently, $\bar{\mathbf{y}}$, and therefore $\mathbf{u}$ in formula (5.3.11), is independent of $\mathbf{C}_1'\mathbf{y}$. The variance–covariance matrix of $\boldsymbol{\phi}$ can then be expressed as

$$Var(\boldsymbol{\phi}) = Var(\mathbf{u}) + (\lambda_{\max}\mathbf{I}_{\xi_1} - \mathbf{G})^{1/2}\mathbf{C}_1'\mathbf{\Sigma C}_1(\lambda_{\max}\mathbf{I}_{\xi_1} - \mathbf{G})^{1/2}. \tag{5.3.29}$$

But, from (5.3.17),

$$\mathbf{R\Sigma R} = \sigma_e^2\mathbf{R}. \tag{5.3.30}$$

Using (5.3.23) in (5.3.30), it can be verified that

$$\begin{aligned} \mathbf{C}_1'\mathbf{\Sigma C}_1 &= \sigma_e^2\mathbf{C}_1'\mathbf{C}_1 \\ &= \sigma_e^2\mathbf{I}_{\xi_1}. \end{aligned}$$

We then have

$$Var(\boldsymbol{\phi}) = Var(\mathbf{u}) + \sigma_e^2(\lambda_{\max}\mathbf{I}_{\xi_1} - \mathbf{G}). \tag{5.3.31}$$

From (5.3.12) and (5.3.31), we get

$$Var(\boldsymbol{\phi}) = \mathbf{diag}(\delta_1 \mathbf{I}_{m_1}, \delta_2 \mathbf{I}_{m_2}, \cdots, \delta_\nu \mathbf{I}_{m_\nu}) + \sigma_e^2 \lambda_{\max} \mathbf{I}_{\xi_1}, \tag{5.3.32}$$

which implies that $\boldsymbol{\phi}_1, \boldsymbol{\phi}_2, \ldots \boldsymbol{\phi}_\nu$ are independent, and that $\boldsymbol{\phi}_i$ has the variance–covariance matrix described in (5.3.25).

(c) Note first that $SS_2(e)$ is independent of $\mathbf{u}$. This is true because $\mathbf{D\Sigma C}_2 = \mathbf{0}$, which can be shown in a manner similar to that of formula (5.3.28). Hence, $\mathbf{D\Sigma C}_2\mathbf{C}_2' = \mathbf{0}$, which implies that $SS_2(e) = \mathbf{y}'\mathbf{C}_2\mathbf{C}_2'\mathbf{y}$ is independent of $\bar{\mathbf{y}} = \mathbf{Dy}$ and is therefore independent of $\mathbf{u}$. Furthermore, $SS_2(e)$ is independent of $\mathbf{C}_1'\mathbf{y}$ since $\mathbf{C}_1'\mathbf{\Sigma C}_2 = \mathbf{0}$, which follows from multiplying the two sides of formula (5.3.30) on the left by $\mathbf{C}_1'$ and on the right by $\mathbf{C}_2$. Consequently, $SS_2(e)$ is independent of $\boldsymbol{\phi}$. This completes the proof of Lemma 5.3.3. □

Corollary 5.3.1. Let $SS_i = \boldsymbol{\phi}_i'\boldsymbol{\phi}_i$, where $\boldsymbol{\phi} = [\boldsymbol{\phi}_1' : \boldsymbol{\phi}_2' : \ldots : \boldsymbol{\phi}_\nu']'$ is given by formula (5.3.24). Then,

(a) $SS_1, SS_2, \ldots, SS_\nu$ are independent.

(b) $SS_i/(\delta_i + \lambda_{\max}\sigma_e^2)$ is distributed as a central chi-squared variate with m_i degrees of freedom $(i = 1, 2, \ldots, \nu)$.

(c) $SS_1, \ldots, SS_2, \ldots, SS_\nu$ are independent of $SS_2(e)/\sigma_e^2$, which has the central chi-squared distribution with $\xi_2 = N - 2c + 1$ degrees of freedom.

□

From Corollary 5.3.1 we conclude that $SS_1, SS_2, \ldots, SS_\nu$ and $SS_2(e)$ act like sums of squares in a balanced ANOVA table. The analysis concerning the variance components $\sigma_1^2, \sigma_2^2, \ldots, \sigma_\nu^2$ can therefore be carried out using these sums of squares just as in a balanced data situation.

In particular, if the data set is balanced, that is, $n_\omega = n_0$ for all $\omega \in T$ (see formula (5.2.3)), then $\mathbf{K} = 1/n_0\mathbf{I}_c$ (see formula (5.2.16)), and $\mathbf{G} = 1/n_0\mathbf{I}_{\xi_1}$, where $\mathbf{G}$ and ξ_1 are defined in (5.3.13) and (5.3.21), respectively. Hence, $\lambda_{\max}$, the largest eigenvalue of $\mathbf{G}$, is equal to $1/n_0$. The vectors $\boldsymbol{\phi}$ and $\mathbf{u}$ in (5.3.24) are therefore identical. In this case,

$$\begin{aligned} SS_i &= \boldsymbol{\phi}_i'\boldsymbol{\phi}_i \\ &= \mathbf{u}_i'\mathbf{u}_i \\ &= \bar{\mathbf{y}}'\mathbf{Q}_i'\mathbf{Q}_i\bar{\mathbf{y}}. \end{aligned}$$

Thus

$$\begin{aligned} \sum_{i=1}^{\nu} SS_i &= \bar{\mathbf{y}}'\left(\sum_{i=1}^{\nu} \mathbf{Q}_i'\mathbf{Q}_i\right)\bar{\mathbf{y}} \\ &= \bar{\mathbf{y}}'\left(\sum_{i=1}^{\nu} \mathbf{P}_i\right)\bar{\mathbf{y}}, \end{aligned}$$

since

$$\begin{aligned}\sum_{i=1}^{\nu}\mathbf{Q}_i'\mathbf{Q}_i &= \mathbf{Q}'\mathbf{Q}-\mathbf{Q}_0'\mathbf{Q}_0\\ &= \mathbf{I}_c-\frac{1}{c}\mathbf{J}_c, \text{ by Lemma 5.3.1(a)}\\ &= \sum_{i=0}^{\nu}\mathbf{P}_i-\mathbf{P}_0, \text{ since } \mathbf{P}_0=\frac{1}{c}\mathbf{J}_c\\ &= \sum_{i=1}^{\nu}\mathbf{P}_i.\end{aligned}$$

But, from formula (5.3.3)

$$\begin{aligned}\bar{\mathbf{y}}'\mathbf{P}_i\bar{\mathbf{y}} &= \bar{\mathbf{y}}'\left(\sum_{j=0}^{\nu}\frac{\lambda_{ij}}{b_j}\mathbf{A}_j\right)\bar{\mathbf{y}}, \quad i=1,2,\cdots,\nu\\ &= \frac{1}{n_0}\mathbf{y}'\left[\sum_{j=0}^{\nu}\frac{\lambda_{ij}}{n_0 b_j}(\mathbf{A}_j\otimes\mathbf{J}_{n_0})\right]\mathbf{y}, \quad i=1,2,\cdots,\nu, \qquad (5.3.33)\end{aligned}$$

since $\bar{\mathbf{y}}=(\mathbf{I}_c\otimes\mathbf{1}_{n_0}')\mathbf{y}/n_0$. Formula (5.3.33) shows that $n_0\bar{\mathbf{y}}'\mathbf{P}_i\bar{\mathbf{y}}$ is the usual sum of squares for the ith effect in a balanced model of the form given in (5.2.2). In other words, $n_0SS_1, n_0SS_2, \ldots, n_0SS_\nu$ reduce to the usual balanced ANOVA sums of squares associated with the corresponding ν effects in the model whenever the data set is balanced. (See Section 2.3 of Chapter 2).

5.4. MORE ON EXACT TESTS

5.4.1. Power of the Exact Tests

Power values for the exact tests in Section 5.3 can be easily obtained just as in a balanced data situation. For example, a test statistic for testing the hypothesis $H_i: \sigma_i^2=0$ is given by $F=MS_i/MS_i^0$, provided that there exists a mean square MS_i^0 such that $E(MS_i)=E(MS_i^0)$ whenever H_i is true, $i=1,2,\ldots,\nu$. If such a test exists, then, under H_i, F has the F-distribution with m_i and m_i^0 degrees of freedom associated with MS_i and MS_i^0, respectively. Note that

$$\begin{aligned}E(MS_i) &= \delta_i+\lambda_{\max}\sigma_e^2, \quad i=1,2,\cdots,\nu\\ E(MS_i^0) &= \delta_i^0+\lambda_{\max}\sigma_e^2,\end{aligned}$$

where δ_i^0 is equal to δ_i when H_i is true, otherwise, $\delta_i^0<\delta_i$.

Let $\pi_i(\delta_i, \delta_i^0, \sigma_e^2, \lambda_{\max}, \alpha)$ denote the power function for the test concerning H_i corresponding to a level of significance α. Then,

$$\begin{aligned}\pi_i(\delta_i, \delta_i^0, \sigma_e^2, \lambda_{\max}, \alpha) &= P\left[\frac{MS_i}{MS_i^0} \geq F_{\alpha, m_i, m_i^0} | \delta_i > \delta_i^0\right] \\ &= P\left[F_{m_i, m_i^0} \geq \frac{\delta_i^0 + \lambda_{\max}\sigma_e^2}{\delta_i + \lambda_{\max}\sigma_e^2} F_{\alpha, m_i, m_i^0}\right] \\ &= P\left[F_{m_i, m_i^0} \geq \frac{1}{1 + s_i} F_{\alpha, m_i, m_i^0}\right], \end{aligned} \tag{5.4.1}$$

where

$$s_i = \frac{\delta_i - \delta_i^0}{\delta_i^0 + \lambda_{\max}\sigma_e^2}.$$

It is easy to see that π_i is a monotone decreasing function of $\lambda_{\max}$ for given values of δ_i, δ_i^0, and σ_e^2. The quantity $\lambda_{\max}$, being the largest eigenvalue of the matrix $\mathbf{G}$ in (5.3.13), is design dependent. Upper and lower bounds on $\lambda_{\max}$ are given in Lemma 5.4.1.

Lemma 5.4.1.

$$\frac{1}{c} \sum_{\omega \in T} \frac{1}{n_\omega} \leq \lambda_{\max} \leq \frac{1}{\min_{\omega \in T}(n_\omega)}, \tag{5.4.2}$$

where the values of $1/n_\omega$ are the diagonal elements of the matrix $\mathbf{K}$ in formula (5.2.16).

Proof. From formula (5.3.13) it is easy to see that

$$\begin{aligned}\lambda_{\max} &= e_{\max}(\tilde{\mathbf{Q}}\mathbf{K}\tilde{\mathbf{Q}}') \\ &\leq \frac{1}{\min_{\omega \in T}(n_\omega)} e_{\max}(\tilde{\mathbf{Q}}\tilde{\mathbf{Q}}'), \end{aligned} \tag{5.4.3}$$

where $e_{\max}(\cdot)$ denotes the largest eigenvalue of a symmetric matrix. But, by (5.3.7), we have

$$\begin{aligned}\mathbf{I}_c = \mathbf{Q}'\mathbf{Q} &= \sum_{i=0}^{\nu} \mathbf{Q}_i'\mathbf{Q}_i \\ &= \frac{1}{c}\mathbf{J}_c + \sum_{i=1}^{\nu} \mathbf{Q}_i'\mathbf{Q}_i, \quad \text{since } \mathbf{Q}_0 = \frac{1}{\sqrt{c}}\mathbf{1}_c' \\ &= \frac{1}{c}\mathbf{J}_c + \tilde{\mathbf{Q}}'\tilde{\mathbf{Q}}. \end{aligned} \tag{5.4.4}$$

Thus, $\tilde{\mathbf{Q}}'\tilde{\mathbf{Q}} = \mathbf{I}_c - 1/c\mathbf{J}_c$, which is an idempotent matrix. Hence,

$$\begin{aligned}e_{\max}(\tilde{\mathbf{Q}}\tilde{\mathbf{Q}}') &= e_{\max}(\tilde{\mathbf{Q}}'\tilde{\mathbf{Q}}) \\ &= 1.\end{aligned}$$

Using inequality (5.4.3), we conclude that

$$\lambda_{\max} \leq \frac{1}{\min_{\omega \in T}(n_\omega)}.$$

Now, to prove the other half of inequality (5.4.2), we note that

$$\lambda_{\max} \geq \frac{1}{c-1} tr(\tilde{\mathbf{Q}}\mathbf{K}\tilde{\mathbf{Q}}'), \tag{5.4.5}$$

since the right-hand side is the average of the eigenvalues of **G**. But,

$$\begin{aligned}
tr(\tilde{\mathbf{Q}}\mathbf{K}\tilde{\mathbf{Q}}') &= tr(\tilde{\mathbf{Q}}'\tilde{\mathbf{Q}}\mathbf{K}) \\
&= tr\left[(\mathbf{I}_c - \frac{1}{c}\mathbf{J}_c)\mathbf{K}\right], \text{ by (5.4.4)} \\
&= tr(\mathbf{K}) - \frac{1}{c} tr(\mathbf{J}_c\mathbf{K}) \\
&= \sum_{\omega \in T} \frac{1}{n_\omega} - \frac{1}{c}(\mathbf{1}_c'\mathbf{K}\mathbf{1}_c) \\
&= \sum_{\omega \in T} \frac{1}{n_\omega} - \frac{1}{c} \sum_{\omega \in T} \frac{1}{n_\omega} \\
&= \frac{c-1}{c} \sum_{\omega \in T} \frac{1}{n_\omega}.
\end{aligned}$$

Hence,

$$\lambda_{\max} \geq \frac{1}{c} \sum_{\omega \in T} \frac{1}{n_\omega}. \tag{5.4.6}$$

We note that the right-hand side of this inequality is the reciprocal of the harmonic mean of the n_ω's. This completes the proof of Lemma 5.4.1. □

From Lemma 5.4.1 we conclude that $\lambda_{\max} \leq 1$ if at least one n_ω is equal to one, otherwise, $\lambda_{\max} < 1$. The smaller $\lambda_{\max}$ is, the higher the power of the test concerning the hypothesis $H_i : \sigma_i^2 = 0$, $i = 1, 2, \ldots, \nu$.

5.4.2. Sufficient Statistics Associated With the Exact Tests

The sums of squares, $SS_1, SS_2, \ldots, SS_\nu$, and $SS_2(e)$ in Corollary 5.3.1 depend on $\boldsymbol{\phi}$ in formula (5.3.24) and also on $\mathbf{C}_2'\mathbf{y}$, since $SS_2(e) = \mathbf{y}'\mathbf{C}_2\mathbf{C}_2'\mathbf{y}$. Consequently, by formulas (5.3.11) and (5.3.24), the exact test statistics in Section 5.3.1 depend on the vector $\bar{\mathbf{y}}$ of cell means, and on $\mathbf{C}_1'\mathbf{y}$ and $\mathbf{C}_2'\mathbf{y}$. The latter two vectors make up the residual sum of squares $SS(e)$ as can be recalled from formulas (5.3.14) and (5.3.23).

Lemma 5.4.2. The residual sum of squares $SS(e)$ and the elements of the vector $\bar{\mathbf{y}}$ of cell means form a set of minimal sufficient statistics for the parameters of model (5.2.2), namely the grand mean and the variance components $\sigma_1^2, \sigma_2^2, \ldots, \sigma_\nu^2, \sigma_e^2$.

Proof. Let $\mathbf{y}_\omega$ denote the vector of n_ω observations in the ωth cell, $\omega \in T$ (see formula (5.2.3)). From model (5.2.2), we have

$$\mathbf{y}_\omega = \left[\sum_{i=0}^{\nu} g^{(i)}_{\theta_i(\bar{\theta}_i)}\right] \mathbf{1}_{n_\omega} + \boldsymbol{\epsilon}_\omega,$$

where $\boldsymbol{\epsilon}_\omega$ is a vector of random errors associated with $\mathbf{y}_\omega$. The variances and covariances associated with the $\mathbf{y}_\omega$'s are of the form

$$Var(\mathbf{y}_\omega) = \left(\sum_{i=1}^{\nu} \sigma_i^2\right) \mathbf{J}_{n_\omega} + \sigma_e^2 \mathbf{I}_{n_\omega}, \quad \omega \in T, \tag{5.4.7}$$

$$Cov\,(\mathbf{y}_\omega, \mathbf{y}_{\omega'}) = \left(\sum_{\psi_i \subset \omega \cap \omega'} \sigma_i^2\right) \mathbf{J}_{n_\omega \times n_{\omega'}}, \quad \omega \neq \omega', \tag{5.4.8}$$

where $\mathbf{J}_{n_\omega \times n_{\omega'}}$ is a matrix of ones of order $n_\omega \times n_{\omega'}$. The summation in (5.4.8) includes all those variance components whose sets of subscripts, namely ψ_i, are common to both ω and ω', $\omega \neq \omega'$. The variance–covariance matrix $\boldsymbol{\Sigma}$ in formula (5.2.6) can then be expressed as

$$\boldsymbol{\Sigma} = (\gamma_{\omega,\omega'}(\boldsymbol{\sigma}^2)\mathbf{J}_{n_\omega \times n_{\omega'}})_{\omega,\omega' \in T} + \sigma_e^2 \mathbf{I}_N, \tag{5.4.9}$$

where $\boldsymbol{\sigma}^2 = (\sigma_1^2, \sigma_2^2, \ldots, \sigma_\nu^2)'$ and $\gamma_{\omega,\omega'}(\boldsymbol{\sigma}^2)$ is a scalar function of $\boldsymbol{\sigma}^2$. We can also write $\boldsymbol{\Sigma}$ as

$$\boldsymbol{\Sigma} = \sigma_e^2(\mathbf{I}_N + \boldsymbol{\Omega}),$$

where

$$\boldsymbol{\Omega} = (h_{\omega,\omega'}(\boldsymbol{\sigma}^2, \sigma_e^2)\mathbf{J}_{n_\omega \times n_{\omega'}})_{\omega,\omega' \in T}, \tag{5.4.10}$$

and

$$h_{\omega,\omega'}(\boldsymbol{\sigma}^2, \sigma_e^2) = \frac{1}{\sigma_e^2}\gamma_{\omega,\omega'}(\boldsymbol{\sigma}^2), \quad \omega, \omega' \in T.$$

Let $\mathbf{A}$ and $\boldsymbol{\Delta}$ be $c \times c$ matrices defined as

$$\mathbf{A} = \left(h_{\omega,\omega'}(\boldsymbol{\sigma}^2, \sigma_e^2)\right)_{\omega,\omega' \in T}$$

$$\boldsymbol{\Delta} = \mathbf{diag}(n_\omega)_{\omega \in T},$$

where c is the number of elements of T. By a result given in Searle (1982, p. 153), the inverse of $\mathbf{I}_N + \boldsymbol{\Omega}$ can be expressed as

$$(\mathbf{I}_N + \boldsymbol{\Omega})^{-1} = \mathbf{I}_N - (\eta_{\omega,\omega'}(\boldsymbol{\sigma}^2, \sigma_e^2)\mathbf{J}_{n_\omega \times n_{\omega'}})_{\omega,\omega' \in T}. \tag{5.4.11}$$

The second matrix on the right-hand side of (5.4.11) is a partitioned matrix of order $N \times N$ with $\eta_{\omega,\omega'}(\boldsymbol{\sigma}^2, \sigma_e^2)$ being a scalar function of the variance components whose values for $\omega, \omega' \in T$ make up the matrix

$$\mathbf{B} = (\eta_{\omega,\omega'}(\boldsymbol{\sigma}^2, \sigma_e^2))_{\omega,\omega' \in T} = (\mathbf{A}\boldsymbol{\Delta} + \mathbf{I}_c)^{-1}\mathbf{A}. \tag{5.4.12}$$

It follows from (5.4.11) that the inverse of $\boldsymbol{\Sigma}$ can be written as

$$\boldsymbol{\Sigma}^{-1} = \frac{1}{\sigma_e^2}\left[\mathbf{I}_N - \left(\eta_{\omega,\omega'}(\boldsymbol{\sigma}^2, \sigma_e^2)\mathbf{J}_{n_\omega \times n_{\omega'}}\right)_{\omega,\omega' \in T}\right]. \tag{5.4.13}$$

Let us now consider the likelihood function associated with $\mathbf{y}$, namely

$$f(\mathbf{y}, \boldsymbol{\zeta}) = \frac{1}{(2\pi)^{N/2}|\boldsymbol{\Sigma}|^{1/2}} \exp\left[-\frac{1}{2}(\mathbf{y} - \mu\mathbf{1}_N)'\boldsymbol{\Sigma}^{-1}(\mathbf{y} - \mu\mathbf{1}_N)\right], \tag{5.4.14}$$

where μ is the grand mean $g^{(0)}_{\theta_0(\overline{\theta}_0)}$ in model (5.2.2) and $\boldsymbol{\zeta} = (\mu, \boldsymbol{\sigma}^{2\prime}, \sigma_e^2)'$. Using (5.4.13), we find

$$\begin{aligned}
&(\mathbf{y} - \mu\mathbf{1}_N)'\boldsymbol{\Sigma}^{-1}(\mathbf{y} - \mu\mathbf{1}_N) \\
&= \frac{1}{\sigma_e^2}(\mathbf{y} - \mu\mathbf{1}_N)'\left[\mathbf{I}_N - \left(\eta_{\omega,\omega'}(\boldsymbol{\sigma}^2, \sigma_e^2)\mathbf{J}_{n_\omega \times n_{\omega'}}\right)_{\omega,\omega' \in T}\right](\mathbf{y} - \mu\mathbf{1}_N) \\
&= \frac{1}{\sigma_e^2}\left[(\mathbf{y} - \mu\mathbf{1}_N)'(\mathbf{y} - \mu\mathbf{1}_N) - \sum_{\omega,\omega' \in T} \eta_{\omega,\omega'}(\boldsymbol{\sigma}^2, \sigma_e^2)(\mathbf{y}_\omega - \mu\mathbf{1}_{n_\omega})'\right. \\
&\qquad \left. \times \mathbf{J}_{n_\omega \times n_{\omega'}}(\mathbf{y}_{\omega'} - \mu\mathbf{1}_{n_{\omega'}})\right] \\
&= \frac{1}{\sigma_e^2}\left[\sum_\theta (y_\theta - \mu)^2 - \sum_{\omega,\omega' \in T} n_\omega n_{\omega'} \eta_{\omega,\omega'}(\boldsymbol{\sigma}^2, \sigma_e^2)(\overline{y}_\omega - \mu)(\overline{y}_{\omega'} - \mu)\right],
\end{aligned} \tag{5.4.15}$$

where θ is the set of subscripts in model (5.2.2). Note that

$$\begin{aligned}
\sum_\theta (y_\theta - \mu)^2 &= \sum_\theta (y_\theta - \overline{y}_\omega)^2 + \sum_\theta (\overline{y}_\omega - \mu)^2 \\
&= SS(e) + \sum_{\omega \in T} n_\omega (\overline{y}_\omega - \mu)^2.
\end{aligned} \tag{5.4.16}$$

From (5.4.14), (5.4.15), and (5.4.16) we conclude that $SS(e)$ and the $\overline{y}_\omega$'s form a set of sufficient statistics for the elements of $\boldsymbol{\zeta}$ by the Neyman factorization theorem (see, for example, Zacks, 1971, Section 2.3). These statistics are also minimal since the likelihood function is of the same form as formula 9 in Khuri and Ghosh (1990), which produced minimal sufficient statistics for an unbalanced two-fold nested model. This completes the proof of Lemma 5.4.2. □

5.5. A NUMERICAL EXAMPLE

Khuri (1990) reported the results of a study concerning the efficiency of workers in assembly lines at several plants. Three plants were randomly selected and four assembly sites and three workers were randomly selected in each plant. The efficiency scores are given in Table 5.1. The data were originally given in Milliken and Johnson (1984, p. 264).

The model for this experiment is

$$y_{ij\ell s} = g^{(0)} + g^{(1)}_{(i)} + g^{(2)}_{i(j)} + g^{(3)}_{i(\ell)} + g^{(4)}_{i(j\ell)} + e_{ij\ell(s)},$$

which can also be expressed in a more traditional form as

$$y_{ij\ell s} = \mu + \tau_i + \beta_{i(j)} + \delta_{i(\ell)} + (\beta\delta)_{i(j\ell)} + e_{ij\ell(s)}, \tag{5.5.1}$$

where τ_i is the effect of the ith plant, $\beta_{i(j)}$ is the effect of the jth site within the ith plant, $\delta_{i(\ell)}$ is the effect of the ℓth worker within the ith plant, $(\beta\delta)_{i(j\ell)}$ is the interaction effect of sites and workers within the ith plant ($i = 1,2,3$; $j = 1,2,3,4$; $\ell = 1,2,3$), and $e_{ij\ell(s)}$ is the error term for the sth replicate ($s = 1,2,\ldots,n_{ij\ell}$). The corresponding variance components are $\sigma_1^2 = \sigma_\tau^2$, $\sigma_2^2 = \sigma^2_{\beta(\tau)}$, $\sigma_3^2 = \sigma^2_{\delta(\tau)}$, $\sigma_4^2 = \sigma^2_{\beta\delta(\tau)}$, $\sigma_5^2 = \sigma_e^2$.

To facilitate the understanding of the application of the exact testing procedure, the reader is referred to Table 5.2. The expected mean squares of $MS_i = SS_i/m_i$ $(i = 1,2,3,4)$ and $MS_2(e) = SS_2(e)/\xi_2$ are given in Table 5.3.

From Tables 5.2 and 5.3 we derive the following results:

1. The value of the test statistic concerning the hypothesis $H_4 : \sigma^2_{\beta\delta(\tau)} = 0$ is $F = MS_4/MS_2(e) = 7.09$ with 18 and 47 degrees of freedom. The corresponding P-value is 3.4×10^{-8}. This is a highly significant result.
2. The value of the test statistic concerning the hypothesis $H_2 : \sigma^2_{\beta(\tau)} = 0$ is $F = MS_2/MS_4 = 0.994$ with 9 and 18 degrees of freedom. The P-value is 0.478. This is a nonsignificant result.
3. The test statistic value concerning the hypothesis $H_3 : \sigma^2_{\delta(\tau)} = 0$ is $F = MS_3/MS_4 = 3.293$ with 6 and 18 degrees of freedom. The corresponding P-value is 0.023. This is a significant result.
4. Finally, the testing of the hypothesis $H_1 : \sigma_\tau^2 = 0$ requires the use of Satterthwaite's (1941) procedure. This is because Table 5.3 contains no mean square with an expected value equal to that of MS_1 under H_1. In this case, we use the test statistic

$$F = \frac{MS_1}{MS_2 + MS_3 - MS_4} = 5.184,$$

which under H_1 has approximately the F distribution with 2 and m^*

Table 5.1. Efficiency Scores Data

		Site			
Plant	Worker	1	2	3	4
1	1	100.6	110.0	100.0	98.2
		106.8	105.8	102.5	99.5
		100.6		97.6	
				98.7	
				98.7	
	2	92.3	103.2	96.4	108.0
		92.0	100.5		108.9
		97.2	100.2		107.9
		93.9	97.7		
		93.0			
	3	96.9	92.5	86.8	94.4
		96.1	85.9		93.0
		100.8	85.2		91.0
			89.4		
			88.7		
2	1	82.6	96.5	87.9	83.6
			100.1	93.5	82.7
			101.9	88.9	87.7
			97.9	92.8	88.0
			95.9		82.5
	2	72.7	71.7	78.4	82.1
			72.1	80.4	79.9
			72.4	83.8	81.9
			71.4	77.7	82.6
				81.2	78.6
	3	82.5	80.9	96.3	77.7
		82.1	84.0	92.4	78.6
		82.0	82.2	92.0	77.2
			83.4	95.8	78.8
			81.5		80.5
3	1	107.6	96.1	101.1	109.1
		108.8	98.5		
		107.2	97.3		
		104.2	93.5		
		105.4			
	2	97.1	91.9	88.0	89.6
		94.2		91.4	86.0
		91.5		90.3	91.2
		99.2		91.5	87.4
				85.7	
	3	87.1	97.8	95.9	101.4
			95.9	89.7	100.1
					102.1
					98.4

Source: A. I. Khuri(1990). Reproduced with permission of Elsevier Science.

Table 5.2. Quantities Used in the Development of the Exact Tests

Quantity	Formula Cited	Corresponding Value
ω	(5.2.3)	(i, j, ℓ)
c	(5.2.4)	36
b_0	(5.3.4)	36
b_1	(5.3.4)	12
b_2	(5.3.4)	3
b_3	(5.3.4)	4
b_4	(5.3.4)	1
$\mathbf{P}_1$	(5.3.3)	$(\mathbf{I}_3 \otimes \mathbf{J}_{12})/12 - \mathbf{J}_{36}/36$
$\mathbf{P}_2$	(5.3.3)	$(\mathbf{I}_{12} \otimes \mathbf{J}_3)/3 - (\mathbf{I}_3 \otimes \mathbf{J}_{12})/12$
$\mathbf{P}_3$	(5.3.3)	$(\mathbf{I}_3 \otimes \mathbf{J}_4 \otimes \mathbf{I}_3)/4 - (\mathbf{I}_3 \otimes \mathbf{J}_{12})/12$
$\mathbf{P}_4$	(5.3.3)	$\mathbf{I}_{36} - (\mathbf{I}_{12} \otimes \mathbf{J}_3)/3 - (\mathbf{I}_3 \otimes \mathbf{J}_4 \otimes \mathbf{I}_3)/4 + (\mathbf{I}_3 \otimes \mathbf{J}_{12})/12$
m_1	(5.3.25)	2
m_2	(5.3.25)	9
m_3	(5.3.25)	6
m_4	(5.3.25)	18
δ_1	(5.3.9)	$12\sigma_\tau^2 + 3\sigma_{\beta(\tau)}^2 + 4\sigma_{\delta(\tau)}^2 + \sigma_{\beta\delta(\tau)}^2$
δ_2	(5.3.9)	$3\sigma_{\beta(\tau)}^2 + \sigma_{\beta\delta(\tau)}^2$
δ_3	(5.3.9)	$4\sigma_{\delta(\tau)}^2 + \sigma_{\beta\delta(\tau)}^2$
δ_4	(5.3.9)	$\sigma_{\beta\delta(\tau)}^2$
ξ_1	(5.3.21)	35
ξ_2	(5.3.22)	47
$\lambda_{\max}$	(5.3.24)	1
SS_1	Corollary 5.3.1	1265.96
SS_2	Corollary 5.3.1	332.313
SS_3	Corollary 5.3.1	733.949
SS_4	Corollary 5.3.1	668.634
$SS_2(e)$	Corollary 5.3.1	246.245

Table 5.3. Expected Mean Squares

Mean Square	Expected Value
$MS_1 = \frac{1}{2}SS_1$	$12\sigma_\tau^2 + 3\sigma_{\beta(\tau)}^2 + 4\sigma_{\delta(\tau)}^2 + \sigma_{\beta\delta(\tau)}^2 + \sigma_e^2$
$MS_2 = \frac{1}{9}SS_2$	$3\sigma_{\beta(\tau)}^2 + \sigma_{\beta\delta(\tau)}^2 + \sigma_e^2$
$MS_3 = \frac{1}{6}SS_3$	$4\sigma_{\delta(\tau)}^2 + \sigma_{\beta\delta(\tau)}^2 + \sigma_e^2$
$MS_4 = \frac{1}{18}SS_4$	$\sigma_{\beta\delta(\tau)}^2 + \sigma_e^2$
$MS_2(e) = \frac{1}{47}SS_2(e)$	σ_e^2

Source: A. I. Khuri (1990). Reproduced with permission of Elsevier Science.

degrees of freedom, where

$$m^* = \frac{(MS_2 + MS_3 - MS_4)^2}{\frac{(MS_2)^2}{9} + \frac{(MS_3)^2}{6} + \frac{(-MS_4)^2}{18}} = 5.477$$

The P-value of the approximate test is 0.055.

EXERCISES

5.1. Consider the model

$$y_{ijk\ell} = \mu + \tau_i + \beta_{i(j)} + \delta_k + (\tau\delta)_{ik} + (\beta\delta)_{i(jk)} + e_{ijk(\ell)},$$
$$i = 1, 2, \cdots, a_1; \ j = 1, 2, \cdots, a_2; \ k = 1, 2, \cdots, a_3; \ \ell = 1, 2, \cdots, n_{ijk}.$$

Let

$$\bar{y}_{ijk} = \frac{1}{n_{ijk}} \sum_{\ell=1}^{n_{ijk}} y_{ijk\ell}.$$

Applying formula (5.2.12)'s format to this model, we obtain

$$\bar{\mathbf{y}} = \sum_{i=0}^{5} \mathbf{H}_i \boldsymbol{\beta}_i + \bar{\mathbf{e}}.$$

(a) Express $\mathbf{H}_i$ as a direct product $(i = 0, 1, \cdots, 5)$.

(b) Use formula (5.3.3) to obtain expressions for the idempotent matrices, $\mathbf{P}_1, \mathbf{P}_2, \mathbf{P}_3$, which are associated with τ_i, $\beta_{i(j)}$, and δ_k, respectively.

(c) What are the values of $\mathbf{A}_1\mathbf{P}_2, \mathbf{A}_2\mathbf{P}_1$, $\mathbf{A}_2\mathbf{P}_2$, and $\mathbf{A}_3\mathbf{P}_1$, where $\mathbf{A}_i = \mathbf{H}_i\mathbf{H}_i' (i = 1, 2, 3)$?

5.2. Consider the example in Section 5.5.

(a) Obtain a 95% confidence interval on $\sigma^2_{\delta(\tau)}/(\sigma^2_{\beta\delta(\tau)} + \sigma^2_e)$.

(b) Give a test statistic for testing the hypothesis, $H_0 : \sigma^2_{\beta\delta(\tau)} = 3\sigma^2_e$ versus $H_a : \sigma^2_{\beta\delta(\tau)} \neq 3\sigma^2_e$.

(c) Compute the probability $P(\hat{\sigma}^2_{\beta(\tau)} < 0)$ in terms of $\sigma^2_{\beta(\tau)}/(\sigma^2_{\beta\delta(\tau)} + \sigma^2_e)$, where $\hat{\sigma}^2_{\beta(\tau)}$ is the ANOVA estimator of $\sigma^2_{\beta(\tau)}$ obtained from Table 5.3.

5.3. A consumer group was interested in examining consistency of prices of a variety of food items sold in large supermarkets. The study was conducted in a random sample of four standard metropolitan areas. Three supermarkets were randomly selected in each of the four areas. Four food items were randomly chosen for the study. The prices of these items (in dollars) were recorded for a random sample of three months. One record was obtained per month from each supermarket for each food item. However, during the three-month period, some items were not available in some of the supermarkets. The following data were obtained:

		Food item			
Area	Supermarket	1	2	3	4
	1	3.15	5.70	1.30	6.12
		3.15	5.68	1.29	6.14
		3.18	5.70	1.29	
		3.28	5.75	1.27	6.18
1	2	3.24	5.72	1.25	6.16
		3.26		1.26	6.15
		3.19	5.65	1.21	6.10
	3	3.18	5.61	1.21	6.11
		3.16			
	1	3.30	5.80	1.51	6.20
		3.28	5.82	1.51	6.20
			5.80	1.52	6.21
2	2	3.25	5.82	1.49	6.24
		3.23	5.79	1.47	6.22
		3.32	5.72	1.46	6.26
	3	3.30	5.74	1.45	6.23
		3.30			
	1	3.29	5.79	1.57	6.30
		3.28	5.79	1.56	6.28
		3.31	5.78		6.31
3	2	3.35	5.81	1.50	6.29
		3.32	5.80	1.49	6.28
				1.49	
		3.24	5.72	1.58	6.32
	3	3.26	5.69	1.55	6.32
		3.23			6.30
	1	3.14	5.50	1.20	6.08
		3.14	5.49	1.22	6.08
		3.12		1.22	
4	2	3.18	5.55	1.18	6.06
		3.18	5.55	1.18	6.04
			5.53		
		3.20	5.59	1.21	6.12
	3	3.18	5.56	1.22	6.11
		3.16			6.11

(a) Give the corresponding population structure.

(b) Write down the complete model.

(c) Is there a significant variation in the food prices among:

   (i) metropolitan areas?

   (ii) supermarkets within metropolitan areas?

(d) Is the variability in the prices of food items different for different metropolitan areas?

BIBLIOGRAPHY

Khuri, A. I. (1990). "Exact tests for random models with unequal cell frequencies in the last stage." *Journal of Statistical Planning and Inference*, 24, 177–193.

Khuri, A. I. and Ghosh, M. (1990). "Minimal sufficient statistics for the unbalanced two-fold nested model." *Statistics and Probability Letters*, 10, 351–353.

Milliken, G. A. and Johnson, D. E. (1984). *Analysis of Messy Data.* Lifetime Learning Publications, Blemont, California.

Satterthwaite, F. E. (1941). "Synthesis of variance." *Psychometrika*, 6, 309–316.

Searle, S. R. (1982). *Matrix Algebra Useful for Statistics.* Wiley, New York.

Zacks, S. (1971). *The Theory of Statistical Inference.* Wiley, New York.

CHAPTER 6

Tests in Unbalanced Mixed Models

6.1. INTRODUCTION

In this chapter we will discuss various exact, approximate, and optimum tests for variance components as well as fixed effects in unbalanced mixed models. Mixed models with two variance components, which are the easiest to study and often admit exact and optimum tests, are described in Section 6.2. As an example, we consider the analysis of a two-way unbalanced mixed model without interaction. In Section 6.3 we take up the analysis of an unbalanced mixed two-way cross-classification model with interaction, involving three variance components, and describe some exact tests for them. Finally, in Section 6.4, we consider a general mixed model with an arbitrary number of variance components and discuss some exact tests of Bartlett–Scheffé type (which hold under some conditions), the role of MINQUE in deriving some approximate tests, and the relevance of Wald's procedure in this context.

6.2. MIXED MODELS WITH TWO VARIANCE COMPONENTS

The most commonly occurring examples of an unbalanced mixed model with two variance components are analogues of Example 2.2.2 described in Chapter 2, representing two-way cross-classification models without interaction, under two different scenarios: one involving a comparison of v fixed treatment effects $\tau_1, \ldots, \tau_v$ when the b block effects $\beta_1, \ldots, \beta_b$ are random, and a second involving a comparison of v random treatment effects when the b block effects are fixed. For a ready reference, the model is described below.

$$y_{ijk} = \mu + \tau_i + \beta_j + e_{ijk}, \quad i = 1, \cdots, v, \quad j = 1, \cdots, b, \quad k = 1, \cdots, n_{ij}. \tag{6.2.1}$$

Here, the errors e_{ijk}'s are assumed to be independent normal with mean

0 and variance σ^2, and the treatment effects τ_i's and the block effects β_j's, whenever random, are assumed to be independent normal with mean 0 and respective variances σ_τ^2 and σ_β^2. Although the above unbalanced model occurs quite frequently in practice and has been extensively studied in the literature, a derivation of optimum tests for $H_\tau : \tau_1 = \ldots = \tau_v$ when the τ's are fixed, and $H_\tau : \sigma_\tau^2 = 0$ when the τ's are random has been attempted fairly recently (Mathew and Sinha, 1988; Mathew, 1989; Westfall, 1989). It turns out that an optimum test for H_τ in the former case can be derived only under some conditions on the n_{ij}'s (i.e., on the underlying design matrix). The derivation of an optimum test for H_τ in the latter case, however, does not require any such condition on the n_{ij}'s.

The reader may note that exact Wald's F-tests, which are not necessarily optimum, can also be used in the above two cases (see Chapter 1).

6.2.1. Test for $H_\tau : \tau_1 = \ldots = \tau_v$

In this section we describe an optimum test of H_τ. Due to the inherent nature of the multiparameter hypothesis H_τ, reduction through *invariance* is essential in order to derive an optimum test. The problem is clearly invariant under the group of transformations which transform $\mathbf{y}$ to $c(\mathbf{y} + \alpha\mathbf{1})$, where $\mathbf{y}$ is the vector of observations y_{ijk}'s, and α and c are scalars with $c > 0$. This is because under this transformation the mean of $\mathbf{y}$ in model (6.2.1) changes from $\mu\mathbf{1}$ to $c(\mu + \alpha)\mathbf{1}$ and the variance of $\mathbf{y}$ changes from σ^2 to $c^2\sigma^2$, both of which have no influence on the null and alternative hypotheses. However, this is a rather small group, and reducing the class of competing tests through this group alone appears to be inadequate to derive an optimum invariant test. On the other hand, it is also clear that to enlarge the above group to a larger one, we must have a certain pattern of the n_{ij}'s. It turns out that if model (6.2.1) corresponds to a $BIBD$ so that it is possible to have a subgroup of the permutation group leaving the problem invariant, the overall group is large enough to guarantee the existence of an optimum invariant test. This is precisely what is accomplished, among other things, in Mathew and Sinha (1988). Hence, in this section, we consider model (6.2.1) only for a $BIBD$ with the parameters b, v, r, k, λ. Note that for a $BIBD$, $n_{ij} = 0$ *or* 1 in (6.2.1). We now proceed to describe the group of transformations keeping the testing problem invariant, and the test which is optimum invariant under this group.

Note that model (6.2.1), under the usual assumption of normality as well as independence of errors and random block effects, can be summarized as

$$E(\mathbf{y}) = \mu\mathbf{1}_{bk} + \mathbf{F}\boldsymbol{\tau}, \qquad Var(\mathbf{y}) = \sigma^2\mathbf{I} + \sigma_\beta^2\mathbf{J}, \tag{6.2.2}$$

where $\mathbf{y} = (\mathbf{y}'_{.1}, \ldots, \mathbf{y}'_{.b})'$, $\mathbf{y}_{.j}$ is the vector of observations from the jth block, $\boldsymbol{\tau} = (\tau_1, \ldots, \tau_v)'$ is the vector of fixed treatment effects, $\mathbf{F}$ is the $bk \times v$ design matrix for the treatments, σ^2 is the error variance, σ_β^2 is the variance of the random block effects, and $\mathbf{J} = \mathbf{diag}(\mathbf{J}_{n_{.1}}, \ldots, \mathbf{J}_{n_{.b}})$ where $n_{.j} = \sum_{i=1}^{v} n_{ij}$, $j = 1, \ldots, b$. Let $\mathcal{G}$ be the group of $bk \times bk$ permutation matrices which, when applied to the bk plots of the $BIBD$, permutes the blocks as a whole and the plots within each block in such a way that for $\mathbf{\Gamma} \in \mathcal{G}$, $\mathbf{\Gamma}\boldsymbol{F} = \boldsymbol{F}\boldsymbol{P}^*$ for some $v \times v$ permutation matrix $\boldsymbol{P}^*$. We should note that $\mathbf{\Gamma}$ does not permute the plots from two different blocks. Using the symmetry of a $BIBD$, it is easy to check that $\mathcal{G}$ is a finite group with $k!b$ elements. It is interesting to note that such a group has earlier been considered by Sinha (1982) for the fixed effects model in connection with invariant estimation of treatment contrasts. For such a $\mathbf{\Gamma}$, it is easy to see that

$$E(\mathbf{\Gamma y}) = \mathbf{\Gamma}(\mu\mathbf{1} + \mathbf{F}\boldsymbol{\tau}) = \mu\mathbf{1} + \mathbf{F}\boldsymbol{\tau}^*, \; Var(\mathbf{\Gamma y}) = \mathbf{\Gamma}(\sigma^2\boldsymbol{I} + \sigma_\beta^2\mathbf{J})\mathbf{\Gamma}' = \sigma^2\boldsymbol{I} + \sigma_\beta^2\boldsymbol{J}, \tag{6.2.3}$$

where $\boldsymbol{\tau}^* = \boldsymbol{P}^*\boldsymbol{\tau}$ represents a permutation of the elements of $\boldsymbol{\tau}$. We are now in a position to describe an optimum invariant test. It is clear from (6.2.3) and the previous discussion that the basic testing problem H_τ remains invariant under the group G of transformations

$$\mathbf{y} \to c(\mathbf{\Gamma y} + \alpha\mathbf{1}), \quad c > 0, \quad \mathbf{\Gamma} \in \mathcal{G}, \tag{6.2.4}$$

where α is an arbitrary scalar. Following Wijsman's *Representation Theorem* (see Appendix 1.1 in Chapter 1) and using properties of a $BIBD$, Mathew and Sinha (1988) showed that the key element R, which is the ratio of the nonnull to null distributions of a maximal invariant under $\mathcal{G}$ (see Section 1.3 in Chapter 1 for the definition of a maximal invariant), can be expressed as

$$R = (k!b)^{-1} \sum_{\mathbf{\Gamma}} \left[1 - \frac{1}{D_\delta} 2\mathbf{y}'\mathbf{\Gamma}'\left(\boldsymbol{I} - \frac{1}{k}\delta\mathbf{J}\right)\boldsymbol{F}\boldsymbol{\tau} + \frac{1}{D_\delta}\boldsymbol{\tau}'\left(r\boldsymbol{I} - \frac{1}{k}\delta\boldsymbol{N}'\boldsymbol{N}\right)\boldsymbol{\tau}\right]^{-(bk-1)/2}, \tag{6.2.5}$$

where $\boldsymbol{N}$ is the incidence matrix of the $BIBD$, and

$$D_\delta = SS_W + (1 - \delta)SS_B. \tag{6.2.6}$$

In (6.2.6), $SS_B = 1/k\sum_{j=1}^{b} B_j^2 - y_{..}^2/bk$ and and $SS_W = \sum_{i=1}^{v}\sum_{j=1}^{b} y_{ij}^2 - 1/k\sum_{j=1}^{b} B_j^2$, where $B_j = \mathbf{y}'_{.j}\mathbf{1}_{n_{.j}}$ and $y_{..} = \mathbf{1}'_{bk}\mathbf{y}$. Furthermore,

$$\delta = \frac{k\theta_1}{1 + k\theta_1}, \qquad \theta_1 = \sigma_\beta^2/\sigma^2. \tag{6.2.7}$$

The summation in (6.2.5) above is over all $bk \times bk$ permutation matrices $\mathbf{\Gamma}$ in $\mathcal{G}$. After a considerable amount of simplification and algebraic manipulations,

Mathew and Sinha (1988) further proved that R can be expanded locally as

$$R = 1 + \frac{a_0}{D_\delta^2}(\mathbf{q}_\delta'\mathbf{q}_\delta + h(\bar{y}_{..}))(\boldsymbol{\tau}'\boldsymbol{\tau}) - \frac{bk-1}{2D_\delta}\boldsymbol{\tau}'\left(r\boldsymbol{I} - \frac{1}{k}\delta\boldsymbol{N}'\boldsymbol{N}\right)\boldsymbol{\tau} + o(\boldsymbol{\tau}'\boldsymbol{\tau}), \tag{6.2.8}$$

where a_0 is a positive constant, and

$$\begin{aligned} \mathbf{q}_\delta &= \mathbf{T} - \frac{1}{k}\delta\mathbf{NB}, \\ \mathbf{B} &= (B_1, \cdots, B_b)', \quad \mathbf{T} = (T_1, \cdots, T_v)'. \end{aligned} \tag{6.2.9}$$

In the above, T_i denotes the sum of the observations corresponding to the ith treatment appearing in the entire $BIBD$, and $B_j = \mathbf{y}_{.j}'\mathbf{1}_{n.j}$ denotes the sum of the observations appearing in the jth block. Also, $h(\bar{y}_{..})$ is a function of only $\bar{y}_{..}$ whose exact nature is unimportant. Here, $\bar{y}_{..}$ is the overall mean of all the observations. Based on the above expansion, Mathew and Sinha (1988) proved the following result.

Theorem 6.2.1. Suppose (6.2.1) is a mixed effects model corresponding to a $BIBD$ with τ's fixed and β's random. Define F_τ as

$$F_\tau = \frac{\mathbf{q}_1'\mathbf{q}_1/(v-1)}{SS_W/(bk-b-v+1)}, \tag{6.2.10}$$

where $\mathbf{q}_1 (= \mathbf{q}_\delta|\delta = 1) = \mathbf{T} - 1/k\mathbf{NB}$. For testing $H_\tau : \tau_1 = \ldots = \tau_v$, (i) the locally best invariant unbiased test rejects H_τ for large values of F_τ if θ_1 is large and (ii) the locally best invariant unbiased test rejects H_τ for large values of $\mathbf{T'T}$, conditional on $(\bar{y}_{..}, SS_B, SS_W)$, if θ_1 is small. □

Incidentally, note that F_τ is the usual intra-block F-statistic obtained from the ANOVA table.

REMARK 6.2.1. An unpleasant feature of the locally optimum test described above is its dependence on the unknown variance ratio θ_1. In Section 7.3 of Chapter 7 an exact and much simpler analysis of the above problem based on a suitable canonical representation is provided, where the ultimate test procedure appropriately combines what are called the intra-block and inter-block F-tests. As already pointed out, the test based on F_τ defined in (6.2.10) is precisely the intra-block F-test.

6.2.2. Optimum Test for $H_\tau : \sigma_\tau^2 = 0$

In this section we describe an optimum test for H_τ under the basic model (6.2.1). In contrast to the fixed effects hypothesis H_τ discussed in Section 6.2.1, it turns out that for testing $H_\tau : \sigma_\tau^2 = 0$, which involves exactly one parameter, an optimum test exists quite generally, that is, under no condition on the replications n_{ij}'s in model (6.2.1) or, in other words, under no condition on the design matrix $\mathbf{X}$ in the general model (2.2.1) of Chapter 2. An optimum test for H_τ under model (6.2.1) was derived in Mathew and Sinha (1988), while Mathew (1989), Westfall (1989), and Lin and Harville (1991) derived it under the general model (2.2.1) involving two variance components. Below we present details of this derivation for a general design matrix under model (2.2.1).

Under the assumption that model (2.2.1) of Chapter 2 involves exactly two variance components, we see that the $N \times 1$ normally distributed random vector $\mathbf{y}$ has mean and variance given by

$$E(\mathbf{y}) = \boldsymbol{X\beta}, \qquad Var(\mathbf{y}) = \sigma^2(\boldsymbol{I} + \theta \boldsymbol{V}), \tag{6.2.11}$$

where $\boldsymbol{X} : N \times m$ is the design matrix, $\boldsymbol{\beta} : m \times 1$ is a vector of unknown parameters, $\sigma^2 = \sigma_2^2 > 0$ is the error variance, $\theta = \sigma_1^2/\sigma_2^2 \geq 0$ represents the ratio of the two variances, and $\boldsymbol{V}$ is a known nonnegative definite matrix. The problem is to derive an optimum test for $H_0 : \theta = 0$ versus $H_1 : \theta > 0$. Obviously, in the context of model (6.2.1) with fixed block effects and random treatment effects, σ^2 represents the error variance, and $\theta = \sigma_\tau^2/\sigma^2$.

In spite of the fact that we are now concerned with a testing problem involving just one parameter, namely, θ, it turns out that an optimum test can be derived only under invariance considerations (combined with the unbiasedness restriction). In other words, the traditional use of unbiasedness alone does not lead to an optimum test, even under the simpler model (6.2.11). To derive an optimum invariant test, we assume without any loss of generality that the design matrix $\boldsymbol{X}$ has full rank m. It is easy to see that the testing problem $H_0 : \theta = 0$ against the alternative H_1 is invariant under the group $\mathcal{G}$ of transformations

$$\mathbf{y} \to c(\mathbf{y} + \boldsymbol{X\alpha}), \quad c > 0, \quad \boldsymbol{\alpha} : m \times 1 \text{ arbitrary.} \tag{6.2.12}$$

Let $\boldsymbol{P} = \boldsymbol{X}(\boldsymbol{X}'\boldsymbol{X})^{-1}\boldsymbol{X}'$ be the orthogonal projection matrix onto the column space of $\boldsymbol{X}$. Write $\boldsymbol{I} - \boldsymbol{P} = \boldsymbol{ZZ}'$, where $\boldsymbol{Z} : N \times (N - m)$ satisfies $\boldsymbol{Z}'\boldsymbol{Z} = \boldsymbol{I}_{N-m}$. Invariance under the shift part, $\boldsymbol{X\alpha}$, of the group $\mathcal{G}$ means that we should restrict our attention only to $\mathbf{Z}'\mathbf{y}$ (so-called error functions) and this yields a maximal invariant under the entire group $\mathcal{G}$ as $T = \mathbf{Z}'\mathbf{y}/||\mathbf{Z}'\mathbf{y}||$. It is now an easy matter to derive the ratio R of nonnull to null densities of T. Applying Wijsman's (1967) theorem (see Appendix 1.1 in Chapter 1), we then readily get

$$\begin{aligned} R &= |\boldsymbol{I} + \theta \boldsymbol{V}|^{-1/2} |\boldsymbol{X}'\boldsymbol{X}|^{1/2} |\boldsymbol{X}'(\boldsymbol{I} + \theta \boldsymbol{V})\boldsymbol{X}|^{-1/2} \\ &\quad \times \left[\frac{\mathbf{y}'\mathbf{Z}(\boldsymbol{I} + \theta \mathbf{Z}'\boldsymbol{V}\mathbf{Z})^{-1}\mathbf{Z}'\mathbf{y}}{\mathbf{y}'\mathbf{Z}\mathbf{Z}'\mathbf{y}} \right]^{-N-m/2}. \end{aligned} \tag{6.2.13}$$

We now proceed to simplify R. Let $r = \text{rank}(\boldsymbol{Z}'\boldsymbol{V}\boldsymbol{Z}) = \text{rank}((\boldsymbol{I}-\boldsymbol{P})\boldsymbol{V}(\boldsymbol{I}-\boldsymbol{P}))$, since $\boldsymbol{Z}\boldsymbol{Z}' = \boldsymbol{I}-\boldsymbol{P}$. The reader should note that r here has nothing to do with r appearing as a design parameter in the $BIBD$ discussed in Section 6.2.1. To avoid trivialities, we assume that $0 < r < N-m$.

We consider two cases.

Case (i). The nonzero eigenvalues of $(\boldsymbol{I}-\boldsymbol{P})\boldsymbol{V}(\boldsymbol{I}-\boldsymbol{P})$ *are all equal.*

Suppose the r nonzero eigenvalues of $(\boldsymbol{I}-\boldsymbol{P})\boldsymbol{V}(\boldsymbol{I}-\boldsymbol{P})$ are all equal, equal to λ. Again, this λ has nothing to do with λ appearing as a design parameter in the context of the $BIBD$ discussed in the earlier section. Obviously, then, the nonzero eigenvalues of $\boldsymbol{Z}'\boldsymbol{V}\boldsymbol{Z}$ are also equal to λ. In the context of the model (6.2.1), this happens if the nonzero eigenvalues of the associated $\boldsymbol{C}$-matrix are all equal (see Section 4.2 of Chapter 4 for a definition of the $\boldsymbol{C}$-matrix). Note that a design satisfying this condition is known as a *variance-balanced* design. Let $\boldsymbol{C}^* = [\boldsymbol{C}_1 : \boldsymbol{C}_2]$ be an orthogonal matrix such that $\boldsymbol{Z}'\boldsymbol{V}\boldsymbol{Z} = \lambda\boldsymbol{C}_1\boldsymbol{C}_1'$ with $\boldsymbol{C}_1 : (N-m)\times r$. Then $\boldsymbol{Z}(\boldsymbol{I}+\theta\boldsymbol{Z}'\boldsymbol{V}\boldsymbol{Z})^{-1}\boldsymbol{Z}' = (1+\lambda\theta)^{-1}\boldsymbol{Z}\boldsymbol{C}_1\boldsymbol{C}_1'\boldsymbol{Z}' + \boldsymbol{Z}\boldsymbol{C}_2\boldsymbol{C}_2'\boldsymbol{Z}' = (1+\lambda\theta)^{-1}\boldsymbol{Q}_1 + \boldsymbol{Q}_0$, where

$$\boldsymbol{Q}_1 = \boldsymbol{Z}\boldsymbol{C}_1\boldsymbol{C}_1'\boldsymbol{Z}', \qquad \boldsymbol{Q}_0 = \boldsymbol{Z}\boldsymbol{C}_2\boldsymbol{C}_2'\boldsymbol{Z}'. \tag{6.2.14}$$

It is easy to verify that $\boldsymbol{Q}_1$ is the orthogonal projection matrix associated with the equal eigenvalues λ of $(\boldsymbol{I}-\boldsymbol{P})\boldsymbol{V}(\boldsymbol{I}-\boldsymbol{P})$, and that $\boldsymbol{Q}_0+\boldsymbol{Q}_1 = \boldsymbol{I}-\boldsymbol{P}$. Note that $\boldsymbol{Q}_0 \neq \boldsymbol{0}$ since $r < N-m$. Under the above assumptions, we get

$$\begin{aligned}\mathbf{y}'\boldsymbol{Z}(\boldsymbol{I}+\theta\boldsymbol{Z}'\boldsymbol{V}\boldsymbol{Z})^{-1}\boldsymbol{Z}'\mathbf{y} &= (1+\lambda\theta)^{-1}\mathbf{y}'\boldsymbol{Q}_1\mathbf{y} + \mathbf{y}'\boldsymbol{Q}_0\mathbf{y} \\ &= \mathbf{y}'(\boldsymbol{I}-\boldsymbol{P})\mathbf{y} - \frac{\lambda\theta}{1+\lambda\theta}\mathbf{y}'\boldsymbol{Q}_1\mathbf{y}.\end{aligned} \tag{6.2.15}$$

We can now simplify R given in (6.2.13) as

$$\begin{aligned}R &= h(\theta)\left[\frac{\mathbf{y}'(\boldsymbol{I}-\boldsymbol{P})\mathbf{y} - \frac{\lambda\theta}{1+\lambda\theta}\mathbf{y}'\boldsymbol{Q}_1\mathbf{y}}{\mathbf{y}'(\boldsymbol{I}-\boldsymbol{P})\mathbf{y}}\right]^{-N-m/2} \\ &= h(\theta)\left[1 - \frac{\lambda\theta}{1+\lambda\theta}\left(1+\frac{\mathbf{y}'\boldsymbol{Q}_0\mathbf{y}}{\mathbf{y}'\boldsymbol{Q}_1\mathbf{y}}\right)^{-1}\right]^{-N-m/2},\end{aligned} \tag{6.2.16}$$

where

$$h(\theta) = |\boldsymbol{I}+\theta\boldsymbol{V}|^{-1/2}|\boldsymbol{X}'\boldsymbol{X}|^{1/2}|\boldsymbol{X}'(\boldsymbol{I}+\theta\boldsymbol{V})\boldsymbol{X}|^{-1/2} \tag{6.2.17}$$

is independent of $\mathbf{y}$. Now let us consider the testing problem $H_0 : \theta = 0$ versus $H_1 : \theta > 0$, which is nontrivial only when $\lambda > 0$. It is clear from the above expression of R that under this situation R is increasing in $\mathbf{y}'\boldsymbol{Q}_1\mathbf{y}/y'\boldsymbol{Q}_0\mathbf{y}$. Consequently, in this case, the $UMPI$ test for the above hypothesis exists and rejects H_0 for large values of

$$F_0 = \frac{N-m-r}{r} \cdot \frac{\mathbf{y}'\boldsymbol{Q}_1\mathbf{y}}{\mathbf{y}'\boldsymbol{Q}_0\mathbf{y}}. \qquad (6.2.18)$$

Case (ii). The nonzero eigenvalues of $(\boldsymbol{I}-\boldsymbol{P})\boldsymbol{V}(\boldsymbol{I}-\boldsymbol{P})$ *are not all equal.*

In this case an UMPI test does not exist, and an LBI test can be derived by expanding R given by (6.2.16) around $\theta = 0$. Using

$$\begin{aligned} |\boldsymbol{I}+\theta\boldsymbol{V}| &= 1+\theta tr(\boldsymbol{V})+o(\theta) \\ |\boldsymbol{X}'(\boldsymbol{I}+\theta\boldsymbol{V})X| &= |\boldsymbol{X}'\boldsymbol{X}|+\theta tr(\boldsymbol{X}'\boldsymbol{V}\boldsymbol{X})(\boldsymbol{X}'\boldsymbol{X})^{-1}+o(\theta) \qquad (6.2.19) \\ \mathbf{y}'\boldsymbol{Z}(\boldsymbol{I}+\theta\boldsymbol{Z}'\boldsymbol{V}\boldsymbol{Z})^{-1}\boldsymbol{Z}'\mathbf{y} &= \mathbf{y}'\boldsymbol{Z}\boldsymbol{Z}'\mathbf{y}-\theta\mathbf{y}'\boldsymbol{Z}\boldsymbol{Z}'\boldsymbol{V}\boldsymbol{Z}\boldsymbol{Z}'\mathbf{y}+o(\theta), \end{aligned}$$

R can be expanded as

$$R=1+\frac{N-m}{2}\theta\frac{\mathbf{y}'(\boldsymbol{I}-\boldsymbol{P})\boldsymbol{V}(\boldsymbol{I}-\boldsymbol{P})\mathbf{y}}{\mathbf{y}'(\boldsymbol{I}-\boldsymbol{P})\mathbf{y}}-\frac{\theta}{2}\left[tr\boldsymbol{V}+tr(\boldsymbol{X}'\boldsymbol{V}\boldsymbol{X})(\boldsymbol{X}'\boldsymbol{X})^{-1}\right]+o(\theta) \qquad (6.2.20)$$

where the last term $o(\theta)$ is uniformly so in $\mathbf{y}$. Hence, for testing $H_0 : \theta = 0$ versus $H_1 : \theta > 0$, the LBI test rejects H_0 for large values of

$$F^* = \frac{\mathbf{y}'(\boldsymbol{I}-\boldsymbol{P})\boldsymbol{V}(\boldsymbol{I}-\boldsymbol{P})\mathbf{y}}{\mathbf{y}'(\boldsymbol{I}-\boldsymbol{P})\mathbf{y}}. \qquad (6.2.21)$$

We can summarize the above results in the following theorem.

Theorem 6.2.2. In model (6.2.11), let $\boldsymbol{P} = \boldsymbol{X}(\boldsymbol{X}'\boldsymbol{X})^{-1}\boldsymbol{X}'$ and $r = \text{rank}((\boldsymbol{I}-\boldsymbol{P})\boldsymbol{V}(\boldsymbol{I}-\boldsymbol{P}))$ and assume $r > 0$.

(a) Suppose $0 < r < N-m$ and the nonzero eigenvalues of $(\boldsymbol{I}-\boldsymbol{P})\boldsymbol{V}(\boldsymbol{I}-\boldsymbol{P})$ are all equal to λ, where $\lambda > 0$, with associated projection matrix $\boldsymbol{Q}_1$. Then for testing $H_0 : \theta = 0$ versus $H_1 : \theta > 0$, the $UMPI$ test rejects H_0 for large values of

$$F_0 = \frac{N-m-r}{r} \cdot \frac{\mathbf{y}'\boldsymbol{Q}_1\mathbf{y}}{\mathbf{y}'\boldsymbol{Q}_0\mathbf{y}}. \qquad (6.2.22)$$

where $\boldsymbol{Q}_0 = (\boldsymbol{I}-\boldsymbol{P})-\boldsymbol{Q}_1$. Further, under H_0, F_0 is distributed as a central F with $(r, N-m-r)$ degrees of freedom.

(b) If the nonzero eigenvalues of $(\boldsymbol{I} - \boldsymbol{P})\boldsymbol{V}(\boldsymbol{I} - \boldsymbol{P})$ are not all equal, then for testing $H_0 : \theta = 0$ versus $H_1 : \theta > 0$, the *LBI* test rejects H_0 for large values of

$$F^* = \frac{\mathbf{y}'(\boldsymbol{I} - \boldsymbol{P})\boldsymbol{V}(\boldsymbol{I} - \boldsymbol{P})\mathbf{y}}{\mathbf{y}'(\boldsymbol{I} - \boldsymbol{P})\mathbf{y}}. \tag{6.2.23}$$

□

REMARK 6.2.2. The UMPI test in (a) of Theorem 6.2.2 can also be derived from the results in Spjϕtvoll (1967) and El-Bassiouni and Seely (1988). The fact that F_0 has a central F-distribution under H_0 can be proved as follows. From the definition of $\boldsymbol{Q}_1$, it follows that when the eigenvalues of $(\boldsymbol{I} - \boldsymbol{P})\boldsymbol{V}(\boldsymbol{I} - \boldsymbol{P})$ are all equal and equal to λ,

$$\boldsymbol{Q}_1 = \frac{1}{\lambda}(\boldsymbol{I} - \boldsymbol{P})\boldsymbol{V}(\boldsymbol{I} - \boldsymbol{P})$$

is an idempotent matrix. Using the above representation, the following properties are easily verified: $\boldsymbol{Q}_0\boldsymbol{Q}_1 = \mathbf{0}$, $\boldsymbol{Q}_0\boldsymbol{V}\boldsymbol{Q}_1 = \mathbf{0}$, $\boldsymbol{Q}_0\boldsymbol{V}\boldsymbol{Q}_0 = \mathbf{0}$, and hence $\boldsymbol{V}\boldsymbol{Q}_0 = \mathbf{0}$. From standard results on the distribution of quadratic forms, it now follows that $\mathbf{y}'\boldsymbol{Q}_0\mathbf{y}$ has a chi-squared distribution, $\mathbf{y}'\boldsymbol{Q}_0\mathbf{y}$ and $\mathbf{y}'\boldsymbol{Q}_1\mathbf{y}$ are independently distributed, and, under H_0, $\mathbf{y}'\boldsymbol{Q}_1\mathbf{y}$ has a chi-squared distribution. It then immediately follows that F_0 has a central F-distribution under H_0.

REMARK 6.2.3. The distribution of F^* under H_0 can be approximated using Hirotsu's (1979) approximation described in Appendix 3.1 of Chapter 3.

REMARK 6.2.4. It is instructive to compare the two tests described in Theorem 6.2.2 with Wald's variance component test (Wald, 1947). In (6.2.11), suppose θ is nonnegative and $\boldsymbol{V}$ is a nonnegative definite matrix of rank s. Write $\boldsymbol{V} = \boldsymbol{B}\boldsymbol{B}'$ where $\boldsymbol{B}$ is an $N \times s$ matrix of rank s. Then for testing $H_0 : \theta = 0$ versus $H_1 : \theta > 0$, Wald's test rejects H_0 for large values of

$$F_W = \frac{N - m - r}{r} \times \frac{\mathbf{y}'(\boldsymbol{I} - \boldsymbol{P})\boldsymbol{B}[\boldsymbol{B}'(\boldsymbol{I} - \boldsymbol{P})\boldsymbol{B}]^-\boldsymbol{B}'(\boldsymbol{I} - \boldsymbol{P})\mathbf{y}}{\mathbf{y}'(\boldsymbol{I} - \boldsymbol{P})\mathbf{y} - \mathbf{y}'(\boldsymbol{I} - \boldsymbol{P})\boldsymbol{B}[\boldsymbol{B}'(\boldsymbol{I} - \boldsymbol{P})\boldsymbol{B}]^-\boldsymbol{B}'(\boldsymbol{I} - \boldsymbol{P})\mathbf{y}}, \tag{6.2.24}$$

where '-' denotes a generalized inverse, and $r = \text{rank}[(\boldsymbol{I} - \boldsymbol{P})\boldsymbol{B}] = \text{rank}[(\boldsymbol{I} - \boldsymbol{P})\boldsymbol{V}(\boldsymbol{I} - \boldsymbol{P})]$. For a derivation of this test and its properties we refer the reader to Section 1.1 of Chapter 1. Comparing with Theorem 6.2.2(b), it follows that Wald's test is equivalent to the LBI test if and only if $(\boldsymbol{I} - \boldsymbol{P})\boldsymbol{V}(\boldsymbol{I} - \boldsymbol{P})$ is a positive scalar multiple of $(\boldsymbol{I} - \boldsymbol{P})\boldsymbol{B}(\boldsymbol{B}'(\boldsymbol{I} - \boldsymbol{P})\boldsymbol{B})^-\boldsymbol{B}'(\boldsymbol{I} - \boldsymbol{P})$. Using $\boldsymbol{I} - \boldsymbol{P} = \boldsymbol{Z}\boldsymbol{Z}'$, this condition is equivalent to

$$\boldsymbol{Z}'\boldsymbol{V}\boldsymbol{Z} = \delta\boldsymbol{Z}'\boldsymbol{B}(\boldsymbol{B}'\boldsymbol{Z}\boldsymbol{Z}'\boldsymbol{B})^-\boldsymbol{B}'\boldsymbol{Z} \tag{6.2.25}$$

for some $\delta > 0$. Since $V = BB'$, this latter condition holds if and only if the nonzero eigenvalues of $Z'VZ$ are all equal and equal to δ, in which case indeed a UMPI test exists (part (a) of Theorem 6.2.2) and Wald's test coincides with the UMPI test. It is thus clear that if a UMPI test does not exist, Wald's test and the LBI test are entirely *different*.

REMARK 6.2.5. Going back to the specific two-way mixed effects model (6.2.1), the UMPI test and the LBI test can be obtained from Theorem 6.2.2 as follows. Let $\boldsymbol{N}$ denote the $v \times b$ matrix consisting of the n_{ij}'s in (6.2.1). Define $r_i = \sum_{j=1}^{b} n_{ij}$, $k_j = \sum_{i=1}^{v} n_{ij}$ $(i = 1, 2, \ldots, v;\ j = 1, 2, \ldots, b)$, and $\boldsymbol{C} = \mathbf{diag}(r_1, r_2, \ldots, r_v) - \boldsymbol{N}\mathbf{diag}(1/k_1, 1/k_2, \ldots, 1/k_b)\boldsymbol{N}'$. ($\boldsymbol{C}$ is referred to as the '$\boldsymbol{C}$-matrix'; see Section 7.2 in the next chapter for some further details when the k_j's are all equal.) Let $\mathbf{q}_1 = \mathbf{T} - \boldsymbol{N}\mathbf{diag}(1/k_1, \ldots, 1/k_b)\mathbf{B}$, and $\text{rank}(\boldsymbol{C}) = u$. From Theorem 6.2.2, we conclude the following: (i) If the nonzero eigenvalues of $\boldsymbol{C}$ are all equal and equal to λ, the UMPI test rejects H_τ: $\sigma_\tau^2 = 0$ for large values of $F_\tau = [(N - b - u)\mathbf{q}_1'\mathbf{q}_1]/[u\lambda SS(e)]$, where $SS(e)$ denotes the error sum of squares under model (6.2.1). Under H_τ, F_τ has a central F-distribution with degrees of freedom $(u, N - b - u)$. (ii) If the nonzero eigenvalues of $\boldsymbol{C}$ are not all equal, then the LBI test rejects H_τ: $\sigma_\tau^2 = 0$ for large values of the statistic $\mathbf{q}_1'\mathbf{q}_1/SS_W$, where $SS_W = \sum_{i=1}^{v}\sum_{j=1}^{b}\sum_{k=1}^{n_{ij}} y_{ijk}^2 - \sum_{j=1}^{b} B_j^2/k_j$. The above explicit expression for the LBI test statistic is obtained by simplifying (6.2.23).

Here $\boldsymbol{C}^-$ denotes a generalized inverse of $\boldsymbol{C}$ and $SS_W = \sum_{i,j} y_{ij}^2 - \sum_j B_j^2/k_j$. The expression for the *LBI* test statistic is obtained by simplifying (6.2.23).

6.3. MIXED TWO-WAY CROSSED-CLASSIFICATION MODELS WITH INTERACTIONS

In this section we present *exact* tests for variance components and fixed effects in an unbalanced mixed two-way crossed-classification model with interaction when there are no empty cells. Such models in the balanced case have already been introduced in Chapter 2, Section 2.2. In the unbalanced situation, such a model can be expressed as

$$y_{ijk} = \mu + \tau_i + \beta_j + (\tau\beta)_{ij} + e_{ijk},\ i = 1, \cdots, v;\ \ j = 1, \cdots, b;\ \ k = 1, \cdots, n_{ij}, \tag{6.3.1}$$

where μ is an unknown constant, τ's are unknown fixed parameters, β's are assumed to be random effects distributed independently and normally with mean 0 and variance σ_β^2, $(\tau\beta)$'s are the random interaction effects distributed independently and normally with mean 0 and variance $\sigma_{\tau\beta}^2$, and e's are the random error components distributed independently and normally with mean 0 and variance σ_e^2. The reader should note that the above model is similar to Example 2.2.3 in Chapter 2 except that here the cell frequencies n_{ij}'s are not all equal, thus making it an *unbalanced* model.

A proper statistical analysis and derivation of good tests for testing relevant hypotheses of variance components σ_β^2, $\sigma_{\tau\beta}^2$, and fixed effects τ's in the context of the above model is quite difficult. When all the effects in model (6.3.1) are random, exact tests of various variance components have been described in Section 4.3 of Chapter 4.

Under the assumption that there are no empty cells (i.e., $n_{ij} \geq 1$ for all i and j), Gallo and Khuri (1990) derived *exact* tests for the fixed effects and the variance components. We describe below their results, which are in the same spirit as those in Khuri and Littell (1987).

Write $\bar{y}_{ij} = \sum_{l=1}^{n_{ij}} y_{ijl}/n_{ij}$, and note that the model for $\bar{y}_{ij}$ is given by

$$\bar{y}_{ij} = \mu + \tau_i + \beta_j + (\tau\beta)_{ij} + \bar{e}_{ij}, \quad i = 1, \cdots, v; \quad j = 1, \cdots, b, \tag{6.3.2}$$

where $\bar{e}_{ij} = \sum_{l=1}^{n_{ij}} e_{ijl}/n_{ij}$. Our analysis of the unbalanced mixed two-way crossed-classification model essentially depends on exploiting the above structure (6.3.2), which in matrix notation can be written as

$$\bar{\mathbf{y}} = \mu\mathbf{1} + \boldsymbol{B}_1\boldsymbol{\tau} + \boldsymbol{B}_2\boldsymbol{\beta} + \boldsymbol{I}_{vb}(\boldsymbol{\tau\beta}) + \bar{\mathbf{e}} \tag{6.3.3}$$

where $\bar{\mathbf{y}} = (\bar{y}_{11}, \bar{y}_{12}, \ldots, \bar{y}_{vb})'$, $\boldsymbol{B}_1 = \boldsymbol{I}_v \otimes \mathbf{1}_b$, $\boldsymbol{B}_2 = \mathbf{1}_v \otimes \boldsymbol{I}_b$, $\boldsymbol{\tau}$, $\boldsymbol{\beta}$ and $(\boldsymbol{\tau\beta})$ are column vectors whose elements are τ_i, β_j, and $(\tau\beta)_{ij}$ respectively, and $\bar{\mathbf{e}} = (\bar{e}_{11}, \bar{e}_{12}, \ldots, \bar{e}_{vb})'$. Here the symbol $\otimes$ denotes the Kronecker product. Note also that the error sum of squares $SS(e)$ is given by

$$SS(e) = \sum_{ijl}(y_{ijl} - \bar{y}_{ij})^2 = \mathbf{y}'\boldsymbol{R}\mathbf{y}, \tag{6.3.4}$$

where the matrix $\boldsymbol{R}$ is given by

$$\boldsymbol{R} = \boldsymbol{I}_{n_{..}} - \mathbf{diag}(\boldsymbol{J}_{n_{11}}/n_{11}, \boldsymbol{J}_{n_{12}}/n_{12}, \cdots, \boldsymbol{J}_{n_{vb}}/n_{vb}). \tag{6.3.5}$$

In the above, $n_{..} = \sum_{ij} n_{ij}$, and $\boldsymbol{J}_{n_{ij}}$ is the matrix of ones of order $n_{ij} \times n_{ij}$.

6.3.1. Derivation of Exact Tests for Variance Components

From (6.3.3), the expected value and the variance–covariance of $\bar{\mathbf{y}}$ are given by

$$E(\bar{\mathbf{y}}) = \mu\mathbf{1}_{vb} + \boldsymbol{B}_1\boldsymbol{\tau}, \qquad Var(\bar{\mathbf{y}}) = \boldsymbol{A}_2\sigma_\beta^2 + \boldsymbol{I}_{vb}\sigma_{\tau\beta}^2 + \boldsymbol{K}{\sigma_e}^2, \tag{6.3.6}$$

where $\boldsymbol{A}_2 = \boldsymbol{B}_2\boldsymbol{B}_2' = (\mathbf{1_v} \otimes \boldsymbol{I}_b)(\mathbf{1_v}' \otimes \boldsymbol{I}_b) = \boldsymbol{J}_v \otimes \boldsymbol{I}_b$, and $\boldsymbol{K} = \mathbf{diag}(n_{11}^{-1}, n_{12}^{-1}, \ldots, n_{vb}^{-1})$. In order to derive exact tests for the variance components σ_β^2 and

$\sigma^2_{\tau\beta}$, we first need to eliminate the nuisance parameters $\boldsymbol{\tau}$ from $E(\bar{\mathbf{y}})$ by making a suitable transformation. To achieve this, we introduce the following transformation:

$$\mathbf{z} = \boldsymbol{E}\bar{\mathbf{y}}, \qquad \boldsymbol{E} = \mathbf{I}_v \otimes \boldsymbol{T}, \tag{6.3.7}$$

where $\boldsymbol{T}$ is a matrix of order $(b-1) \times b$ defined as

$$\boldsymbol{T} = \begin{bmatrix} 1/\sqrt{2} & -1/\sqrt{2} & 0 & \dots & \dots & 0 \\ 1/\sqrt{6} & 1/\sqrt{6} & -2/\sqrt{6} & 0 & \dots & 0 \\ .. & .. & .. & .. & .. & .. \\ 1/\sqrt{b(b-1)} & 1/\sqrt{b(b-1)} & \dots & \dots & \dots & -(b-1)/\sqrt{b(b-1)} \end{bmatrix}. \tag{6.3.8}$$

Clearly, the matrix $\boldsymbol{E}$ has $v(b-1)$ rows and vb columns, and its rows define orthogonal contrasts in the elements of $\bar{\mathbf{y}}$. From (6.3.6) and (6.3.7), we get

$$\begin{aligned} E(\mathbf{z}) &= \mathbf{0}, \\ Var(\mathbf{z}) &= \boldsymbol{E}\boldsymbol{A}_2\boldsymbol{E}'\sigma^2_\beta + \boldsymbol{E}\boldsymbol{E}'\sigma^2_{\tau\beta} + \boldsymbol{E}\boldsymbol{K}\boldsymbol{E}'{\sigma_e}^2 \\ &= \boldsymbol{E}\boldsymbol{A}_2\boldsymbol{E}'\sigma^2_\beta + \mathbf{I}_{v(b-1)}\sigma^2_{\tau\beta} + \boldsymbol{E}\boldsymbol{K}\boldsymbol{E}'{\sigma_e}^2 \end{aligned} \tag{6.3.9}$$

because $\boldsymbol{E}\boldsymbol{E}' = (\mathbf{I}_v \otimes \boldsymbol{T})(\mathbf{I}_v \otimes \boldsymbol{T})' = \mathbf{I}_v \otimes \boldsymbol{T}\boldsymbol{T}' = \mathbf{I}_{v(b-1)}$, since $\boldsymbol{T}\boldsymbol{T}' = \boldsymbol{I}_{b-1}$. Note that the matrix $\boldsymbol{E}\boldsymbol{A}_2\boldsymbol{E}' = \mathbf{J}_v \otimes \mathbf{I}_{b-1}$ has eigenvalues v and 0 with multiplicities $(b-1)$ and $(v-1)(b-1)$, respectively. Since $\boldsymbol{E}\boldsymbol{A}_2\boldsymbol{E}'$ is symmetric, there exists an orthogonal matrix $\boldsymbol{P}$ of order $(vb-v) \times (vb-v)$ such that $\boldsymbol{P}\boldsymbol{E}\boldsymbol{A}_2\boldsymbol{E}'\boldsymbol{P}' = \boldsymbol{D}$, where $\boldsymbol{D}$ is a diagonal matrix whose diagonal elements are the eigenvalues of $\boldsymbol{E}\boldsymbol{A}_2\boldsymbol{E}'$.

We now define

$$\mathbf{u} = \boldsymbol{P}\mathbf{z} \tag{6.3.10}$$

so that

$$E(\mathbf{u}) = \mathbf{0} \tag{6.3.11}$$

and

$$\begin{aligned} Var(\mathbf{u}) &= \boldsymbol{P}\boldsymbol{E}\boldsymbol{A}_2\boldsymbol{E}'\boldsymbol{P}'\sigma^2_\beta + \mathbf{I}_{v(b-1)}\sigma^2_{\tau\beta} + \boldsymbol{P}\boldsymbol{E}\boldsymbol{K}\boldsymbol{E}'\boldsymbol{P}'\sigma^2_e \\ &= \boldsymbol{D}\sigma^2_\beta + \mathbf{I}_{v(b-1)}\sigma^2_{\tau\beta} + \boldsymbol{L}\sigma^2_e \\ &= \mathbf{diag}[(v\sigma^2_\beta + \sigma^2_{\tau\beta})\mathbf{I}_{b-1}, \sigma^2_{\tau\beta}\mathbf{I}_{(v-1)(b-1)}] + \boldsymbol{L}\sigma^2_e, \end{aligned} \tag{6.3.12}$$

where $\boldsymbol{L} = \boldsymbol{P}\boldsymbol{E}\boldsymbol{K}\boldsymbol{E}'\boldsymbol{P}'$. Referring to the matrix $\boldsymbol{R}$ in (6.3.5), note that $\boldsymbol{R}$ is idempotent and symmetric with rank equal to $n_{..} - vb$ where, as before, $n_{..} = \sum_{i,j} n_{ij}$. Hence, there exists an orthogonal matrix $\boldsymbol{H}$ such that

$$\boldsymbol{R} = \boldsymbol{H}\boldsymbol{\Delta}\boldsymbol{H}', \tag{6.3.13}$$

where $\boldsymbol{\Delta}$ is a diagonal matrix whose diagonal elements are the eigenvalues of $\boldsymbol{R}$. The matrix $\boldsymbol{\Delta}$ has $n_{..} - vb$ diagonal elements equal to unity and vb diagonal elements equal to zero. We can therefore partition $\boldsymbol{\Delta}$ as

$$\boldsymbol{\Delta} = \mathbf{diag}(\mathbf{I}_{v(b-1)}, \mathbf{I}_{n_{..}-2vb+v}, \mathbf{0}) \tag{6.3.14}$$

and, conformably, $\boldsymbol{H}$ as

$$\boldsymbol{H} = [\boldsymbol{H}_1 : \boldsymbol{H}_2 : \boldsymbol{H}_3], \tag{6.3.15}$$

where $\mathbf{0}$ is a zero matrix of order $vb \times vb$, $\boldsymbol{H}_1, \boldsymbol{H}_2, \boldsymbol{H}_3$ are matrices of orders $n_{..} \times v(b-1)$, $n_{..} \times (n_{..} - 2vb + v)$, and $n_{..} \times vb$, respectively. Note that the partitioning of $\boldsymbol{\Delta}$ and hence $\boldsymbol{H}$ is possible provided

$$n_{..} > 2vb - v. \tag{6.3.16}$$

Moreover, note that $\boldsymbol{H}_i'\boldsymbol{H}_i$ is an identity matrix for $i = 1, 2, 3$, and $\boldsymbol{H}_i'\boldsymbol{H}_j = \mathbf{0}$ for $i \neq j$. The error sum of squares $SS(e)$ can then be written as

$$SS(e) = Q_1 + Q_2, \qquad Q_1 = \mathbf{y}'\boldsymbol{H}_1\boldsymbol{H}_1'\mathbf{y}, \qquad Q_2 = \mathbf{y}'\boldsymbol{H}_2\boldsymbol{H}_2'\mathbf{y}. \tag{6.3.17}$$

We are now in a position to derive exact tests for the two variance components σ_β^2 and $\sigma_{\tau\beta}^2$. Define a random vector $\mathbf{w}$ as

$$\mathbf{w} = \mathbf{u} + [\lambda_{max}(\boldsymbol{L})\mathbf{I}_{v(b-1)} - \boldsymbol{L}]^{1/2}\boldsymbol{H}_1'\mathbf{y}, \tag{6.3.18}$$

where $\lambda_{max}(\boldsymbol{L})$ is the largest eigenvalue of $\boldsymbol{L}$, and $[\lambda_{max}(\boldsymbol{L})\mathbf{I}_{v(b-1)} - \boldsymbol{L}]^{1/2}$ is a symmetric matrix with eigenvalues equal to the square roots of those of $\lambda_{max}(\boldsymbol{L})\mathbf{I}_{v(b-1)} - \boldsymbol{L}$. Since $\boldsymbol{L}$ is symmetric, all its eigenvalues are real. Also note that because of this choice of $\lambda_{max}(\boldsymbol{L})$, the matrix $\lambda_{max}(\boldsymbol{L})\mathbf{I}_{v(b-1)} - \boldsymbol{L}$ is positive semidefinite so that its eigenvalues are all nonnegative. We require the following lemma whose proof is similar to that of Lemma 4.3.2 of Chapter 4.

Lemma 6.3.1. Let $\mathbf{w}$ in (6.3.18) be partitioned as $\mathbf{w} = (\mathbf{w}_1', \mathbf{w}_2')'$, where $\mathbf{w}_1$ is of oder $(b-1) \times 1$ and $\mathbf{w}_2$ is of order $(v-1)(b-1) \times 1$. Then

(a) $E(\mathbf{w}_1) = \mathbf{0}$, $E(\mathbf{w}_2) = \mathbf{0}$,

(b) $Var(\mathbf{w}_1) = (v\sigma_\beta^2 + \sigma_{\tau\beta}^2 + \lambda_{max}(\boldsymbol{L})\sigma_e^2)\mathbf{I}_{b-1}$, $Var(\mathbf{w}_2) = (\sigma_{\tau\beta}^2 + \lambda_{max}(\boldsymbol{L})\sigma_e^2)$ $\mathbf{I}_{(v-1)(b-1)}$,

(c) $\mathbf{w}_1$ and $\mathbf{w}_2$ are independent normal vectors, and both are independent of Q_2, where Q_2 is defined in (6.3.17). □

From Lemma 6.3.1, we conclude that if

$$SS(\beta) = \mathbf{w}_1'\mathbf{w}_1, \qquad SS(\tau\beta) = \mathbf{w}_2'\mathbf{w}_2, \tag{6.3.19}$$

then $SS(\beta)$ and $SS(\tau\beta)$ are independently distributed such that

$$\begin{aligned} SS(\beta)/(v\sigma_\beta^2 + \sigma_{\tau\beta}^2 + \lambda_{max}(\boldsymbol{L})\sigma_e^2) &\sim \chi_{b-1}^2, \\ SS(\tau\beta)/(\sigma_{\tau\beta}^2 + \lambda_{max}(\boldsymbol{L})\sigma_e^2) &\sim \chi_{(v-1)(b-1)}^2. \end{aligned} \tag{6.3.20}$$

Furthermore, $SS(\beta)$ and $SS(\tau\beta)$ are independent of Q_2, which is distributed as $\sigma_e^2\chi_{n_{..}-2vb+v}^2$. Therefore, for testing $H_{\tau\beta} : \sigma_{\tau\beta}^2 = 0$, we can use the test statistic

$$F_{\tau\beta} = \frac{SS(\tau\beta)/(v-1)(b-1)}{\lambda Q_2/(n_{..} - 2vb + v)}, \tag{6.3.21}$$

which is distributed as central $F_{(v-1)(b-1),n_{..}-2vb+v}$ under $H_{\tau\beta}$. Similarly, for testing $H_\beta : \sigma_\beta^2 = 0$, we use the test statistic

$$F_\beta = \frac{SS(\beta)/(b-1)}{SS(\tau\beta)/(v-1)(b-1)}, \tag{6.3.22}$$

which is distributed as central $F_{(b-1),(v-1)(b-1)}$ under H_β.

6.3.2. Derivation of an Exact Test for Fixed Effects

In this section we describe an exact test for the hypothesis $H_\tau : \boldsymbol{S\tau} = \boldsymbol{\delta}_0$, where $\boldsymbol{S\tau}$ is estimable under model (6.3.1) and $\boldsymbol{\delta}$ is a specified vector. Let us again consider model (6.3.3) for $\bar{\mathbf{y}}$ with its expectation and variance given in (6.3.6). Let $\boldsymbol{A}_1 = \boldsymbol{B}_1\boldsymbol{B}_1' = \boldsymbol{I}_v \otimes \boldsymbol{J}_b$ so that $\boldsymbol{A}_1$ has eigenvalues equal to b and zero with multiplicities v and $vb - v$, respectively. Recall also that $\boldsymbol{A}_2 = \boldsymbol{J}_v \otimes \boldsymbol{I}_b$, which has eigenvalues equal to v and zero with multiplicities b and $vb - b$, respectively. It is easy to see that the two matrices $\boldsymbol{A}_1$ and $\boldsymbol{A}_2$ commute (i.e., $\boldsymbol{A}_1\boldsymbol{A}_2 = \boldsymbol{A}_2\boldsymbol{A}_1$), so that there exists an orthogonal matrix $\boldsymbol{M}$ which simultaneously diagonalizes both $\boldsymbol{A}_1$ and $\boldsymbol{A}_2$. Lastly, we observe that $\text{rank}(\boldsymbol{A}_1 + \boldsymbol{A}_2) = v + b - 1$, and that the first row of $\boldsymbol{M}$ can be chosen as $(vb)^{-1/2}\mathbf{1}_{vb}'$.

Let us partition the matrix $\boldsymbol{M}$ as

$$\boldsymbol{M} = [(vb)^{-1/2}\mathbf{1}_{vb} : \boldsymbol{M}_v' : \boldsymbol{M}_b' : \boldsymbol{M}_{vb}']', \tag{6.3.23}$$

where $\boldsymbol{M}_v$ consists of the $(v-1)$ rows of $\boldsymbol{M}$ such that $\boldsymbol{M}_v\boldsymbol{A}_1\boldsymbol{M}_v' = b\mathbf{I}_{v-1}$ and $\boldsymbol{M}_v\boldsymbol{A}_2\boldsymbol{M}_v' = \mathbf{0}_{(v-1)\times(v-1)}$; $\boldsymbol{M}_b$ consists of the next $(b-1)$ rows of $\boldsymbol{M}$ such that $\boldsymbol{M}_b\boldsymbol{A}_1\boldsymbol{M}_b' = \mathbf{0}_{(b-1)\times(b-1)}$ and $\boldsymbol{M}_b\boldsymbol{A}_2\boldsymbol{M}_b' = v\mathbf{I}_{b-1}$; and finally $\boldsymbol{M}_{vb}$ consists of the remaining $(v-1)(b-1)$ rows of $\boldsymbol{M}$ such that $\boldsymbol{M}_{vb}\boldsymbol{A}_1\boldsymbol{M}_{vb}' = \boldsymbol{M}_{vb}\boldsymbol{A}_2\boldsymbol{M}_{vb}' = \mathbf{0}_{(v-1)(b-1)\times(v-1)(b-1)}$. Define and partition $\mathbf{v}$ as

$$\mathbf{v} = \boldsymbol{M}\bar{\mathbf{y}} = [v_1 : \mathbf{v}_2' : \mathbf{v}_3' : \mathbf{v}_4']' \tag{6.3.24}$$

so that

$$\begin{aligned} v_1 &\sim N_1(\mu(vb)^{1/2} + (b/v)^{1/2}\mathbf{1}_v'\boldsymbol{\tau}, v\sigma_\beta^2 + \sigma_{\tau\beta}^2 + (vb)^{-1}\left(\sum_{ij} n_{ij}^{-1}\right)\sigma_e^2) \\ \mathbf{v}_2 &\sim N_{v-1}(M_v B_1 \boldsymbol{\tau}, \sigma_{\tau\beta}^2 \mathbf{I}_{v-1} + M_v K M_v' \sigma_e^2) \\ \mathbf{v}_3 &\sim N_{b-1}(\mathbf{0}, (v\sigma_\beta^2 + \sigma_{\tau\beta}^2)\mathbf{I}_{b-1} + M_b K M_b' \sigma_e^2) \\ \mathbf{v}_4 &\sim N_{(v-1)(b-1)}(\mathbf{0}, \sigma_{\tau\beta}^2 \mathbf{I}_{(v-1)(b-1)} + M_{vb} K M_{vb}' \sigma_e^2). \end{aligned} \tag{6.3.25}$$

In what follows we shall use

$$\boldsymbol{\phi} = [\mathbf{v}_2' : \mathbf{v}_4']' \tag{6.3.26}$$

to test hypotheses regarding $\boldsymbol{\tau}$. Referring to $\boldsymbol{R}$ and Q_1, Q_2 defined before in (6.3.13) and (6.3.17), respectively, we now consider a different partition of $\boldsymbol{\Delta}$ and $\boldsymbol{H}$ as

$$\boldsymbol{\Delta} = \mathbf{diag}(\mathbf{I}_{\nu_1}, \mathbf{I}_{\nu_2}, \mathbf{0}_{vb\times vb}), \qquad \boldsymbol{H} = [\boldsymbol{G}_1 : \boldsymbol{G}_2 : \boldsymbol{G}_3], \tag{6.3.27}$$

where $\nu_1 = (v-1) + (v-1)(b-1) = b(v-1)$, $\nu_2 = n_{..} - 2vb + b$, and $\boldsymbol{G}_1$, $\boldsymbol{G}_2$, and $\boldsymbol{G}_3$ are conformable matrices of orders $n_{..} \times b(v-1)$, $n_{..} \times (n_{..} - 2vb + b)$, and $n_{..} \times vb$, respectively. Of course, to define $\boldsymbol{G}_2$, we require that

$$n_{..} > 2vb - b. \tag{6.3.28}$$

Combining (6.3.16) with (6.3.28), we will therefore need

$$n_{..} > max[2vb - v, \ 2vb - b]. \tag{6.3.29}$$

Our exact test concerning the fixed effects $\boldsymbol{\tau}$ is based on the final random vector $\boldsymbol{\psi}$ of order $v(b-1) \times 1$ defined as

$$\boldsymbol{\psi} = \boldsymbol{\phi} + [\zeta_{max}(\boldsymbol{\Gamma})\mathbf{I}_{\nu_1} - \boldsymbol{\Gamma}]^{1/2}\boldsymbol{G}_1'\mathbf{y}, \tag{6.3.30}$$

where

$$\begin{aligned} \boldsymbol{\Gamma} &= [\boldsymbol{M}_v' : \boldsymbol{M}_{vb}']'\boldsymbol{K}[\boldsymbol{M}_v' : \boldsymbol{M}_{vb}'] \\ &= \begin{bmatrix} \boldsymbol{M}_v\boldsymbol{K}\boldsymbol{M}_v' & \boldsymbol{M}_v\boldsymbol{K}\boldsymbol{M}_{vb}' \\ \boldsymbol{M}_{vb}\boldsymbol{K}\boldsymbol{M}_v' & \boldsymbol{M}_{vb}\boldsymbol{K}\boldsymbol{M}_{vb}' \end{bmatrix} \end{aligned} \tag{6.3.31}$$

and $\zeta_{max}(\boldsymbol{\Gamma})$ is the largest eigenvalue of $\boldsymbol{\Gamma}$. The following lemma, whose proof appears in Appendix 6.1, gives the mean and variance of $\boldsymbol{\psi}$.

Lemma 6.3.2. $E(\boldsymbol{\psi}) = [\boldsymbol{\tau}'\boldsymbol{B}_1'\boldsymbol{M}_v' : \mathbf{0}]'$, $Var(\boldsymbol{\psi}) = (\sigma_{\tau\beta}^2 + \zeta_{max}(\boldsymbol{\Gamma})\sigma_e^2)\mathbf{I}_{b(v-1)}$.

It is thus clear that the *observable* random vector $\boldsymbol{\psi}$ provides a standard linear model set up with independent and homoscedastic errors, and with $\boldsymbol{M}_v\boldsymbol{B}_1\boldsymbol{\tau}$ as its fixed effects. Write $\boldsymbol{X} = [\boldsymbol{M}'_v : \boldsymbol{M}'_{vb}]'\boldsymbol{B}_1$, which is a matrix of order $b(v-1) \times v$, and of rank $(v-1)$ because $\boldsymbol{X}\boldsymbol{X}' = \mathbf{diag}(b\mathbf{I}_{v-1}, \mathbf{0})$, where $\mathbf{0}$ is a square matrix of order $(v-1)(b-1) \times (v-1)(b-1)$. Hence, the linear model for $\boldsymbol{\psi}$ can be written as

$$\boldsymbol{\psi} = \boldsymbol{X}\boldsymbol{\tau} + \boldsymbol{\eta}, \qquad \boldsymbol{\eta} \sim N(0, \sigma^2 \boldsymbol{I}_{b(v-1)}), \tag{6.3.32}$$

where $\sigma^2 = \sigma^2_{\tau\beta} + \zeta_{max}(\boldsymbol{\Gamma})\sigma^2_e$. It then follows that for a general linear hypothesis

$$H_0 : \boldsymbol{S}\boldsymbol{\tau} = \boldsymbol{\delta}_0, \tag{6.3.33}$$

where $\boldsymbol{S}$ is a matrix of order $t \times v$ and rank t $(\leq v-1)$ such that $\boldsymbol{S}\boldsymbol{\tau}$ is estimable under (6.3.32), and $\boldsymbol{\delta}_0$ is a specified constant vector, an exact test is provided by

$$F = \frac{(\boldsymbol{S}\boldsymbol{\tau}^0 - \boldsymbol{\delta}_0)'[\boldsymbol{S}(\boldsymbol{X}'\boldsymbol{X})^{-1}\boldsymbol{S}']^{-1}(\boldsymbol{S}\boldsymbol{\tau}^0 - \boldsymbol{\delta}_0)/t}{SS_E/[(v-1)(b-1)]}. \tag{6.3.34}$$

Here $(\boldsymbol{X}'\boldsymbol{X})^{-}$ is a g-inverse of $\boldsymbol{X}'\boldsymbol{X}$, $\boldsymbol{\tau}^0 = (\boldsymbol{X}'\boldsymbol{X})^{-}\boldsymbol{X}'\boldsymbol{\psi}$, and SS_E is the residual sum of squares under model (6.3.32), given by

$$SS_E = \boldsymbol{\psi}'[\mathbf{I}_{v(b-1)} - \boldsymbol{X}(\boldsymbol{X}'\boldsymbol{X})^{-}\boldsymbol{X}']\boldsymbol{\psi}. \tag{6.3.35}$$

Under H_0, F has a central F-distribution with t and $(v-1)(b-1)$ degrees of freedom. In particular, if $\boldsymbol{S}$ is of order $(v-1) \times v$ chosen as

$$\boldsymbol{S} = \begin{bmatrix} 1 & -1 & 0 & \ldots & 0 \\ 1 & 0 & -1 & \ldots & 0 \\ .. & .. & .. & .. & .. \\ 1 & 0 & 0 & \ldots & -1 \end{bmatrix}, \tag{6.3.36}$$

then F serves as a valid exact test statistic for testing the hypothesis $H_\tau : \tau_1 = \ldots = \tau_v$.

For some related recent results, see Öfversten (1993) and Christensen (1996). □

6.3.2.1. A Numerical Example (Gallo and Khuri, 1990). The average daily gains (in pounds) of 65 steers from 9 sires and 3 ages of dam were reported in Damon and Harvey (1987, pp. 131, 140). The data are given in Table 6.1. The actual experiment was conducted at the U.S. Range Livestock Experiment Station in Miles City, Montana, over a 10-year period from 1947 through 1956 (see Shelby et al., 1963). A total of 616 Hereford topcross steers were actually fed in this experiment.

Table 6.1. Average Daily Gain (in Pounds) for 67 Steers

	Age of Dam (years)		
Sire	3	4	5-up
1	2.24	2.41	2.58
	2.65	2.25	2.67
			2.71
			2.47
2	2.15[a]	2.29	1.97
		2.26	2.14
			2.44
			2.52
			1.72
			2.75
3	2.38	2.46	2.29
			2.30
			2.94
4	2.50	2.44	2.54
	2.44	2.15	2.74
			2.50
			2.54
5	2.65	2.52	2.79
		2.67	2.33
			2.67
			2.69
6	2.30[a]	3.00	2.25
		2.49	2.49
			2.02
			2.31
7	2.57	2.64	2.37
	2.37		2.22
			1.90
			2.61
			2.13
8	2.16	2.45	1.44
	2.33		1.72
	2.52		2.17
9	2.68	2.43	2.66
		2.36	2.46
		2.44	2.52
			2.42

[a] Artificially added observation.
Source: J. Gallo and A.I. Khuri (1990). Reproduced with permission of International Biometric Society.

The original data in Damon and Harvey (1987) had two empty cells. Since our procedure allows no empty cells, two observations, marked by footnote in Table 6.1, were artificially added to the data.

In this experiment, the age-of-dam effect (τ_i) is fixed and the sire effect (β_j) is random.

(i) Tests of the Variance Components

These tests are based on model (6.3.18). Here $\lambda_{max}(\boldsymbol{L}) = 1.0$, $\mathbf{u} = \boldsymbol{P}\mathbf{z}$, where $\boldsymbol{P}$ is a 24×24 orthogonal matrix whose rows are eigenvectors of $\boldsymbol{E}\boldsymbol{A}_2\boldsymbol{E}' = (\boldsymbol{I}_3 \otimes \boldsymbol{T})(\boldsymbol{J}_3 \otimes \boldsymbol{I}_9)(\boldsymbol{I}_3 \otimes \boldsymbol{T}')$, and $\mathbf{z} = \boldsymbol{E}\bar{\mathbf{y}} = (\boldsymbol{I}_3 \otimes \boldsymbol{T})\bar{\mathbf{y}}$, where the matrix $\boldsymbol{T}$ is of order 8×9 as in (6.3.8). Also, $\boldsymbol{L} = \boldsymbol{P}\boldsymbol{E}\boldsymbol{K}\boldsymbol{E}'\boldsymbol{P}'$, where $\boldsymbol{K}$ is a diagonal matrix with diagonal elements equal to the reciprocal of the 27 cell frequencies, and $\boldsymbol{H}_1$ consists of the first $v(b-1) = 24$ columns of the 67×67 matrix $\boldsymbol{H}$ given in (6.3.15). The columns of the latter matrix are orthonormal eigenvectors of the matrix $\boldsymbol{R}$ in (6.3.13).

Using model (6.3.18) and the data in Table 6.1, the vector $\mathbf{w}$ is computed and then partitioned into $\mathbf{w}_1$ and $\mathbf{w}_2$ of orders 8×1 and 16×1, respectively, as in Lemma 6.3.1. From (6.3.22), it follows that the test statistic for testing $H_\beta : \sigma_\beta^2 = 0$ has the value $F_\beta = 0.871$ with 8 and 16 degrees of freedom. The level of significance being 0.559, there is no indication of a significant sire effect. Also, from (6.3.21), the value of the test statistic for testing $H_{\tau\beta} : \sigma_{\tau\beta}^2 = 0$ is $F_{\tau\beta} = 1.3398$ with 16 and 16 degrees of freedom. It may be noted that, in (6.3.21), $Q_2 = \mathbf{y}'\boldsymbol{H}_2\boldsymbol{H}_2'\mathbf{y}$, where $\boldsymbol{H}_2$ is a matrix of order 67×16 consisting of columns 25 through 40 of $\boldsymbol{H}$ (see (6.3.15)). The level of significance for the latter test being 0.283, there is no indication of a significant interaction effect.

(ii) Tests of the Fixed Effects

We consider the hypothesis $H_\tau : \tau_1 = \tau_2 = \tau_3$, which is the same as (6.3.33) with $\boldsymbol{\delta}_0 = \mathbf{0}$ and $\boldsymbol{S}$ given by (6.3.36). For this test we use model (6.3.30). We first determine the orthogonal matrix $\boldsymbol{M}$, which simultaneously diagonalizes $\boldsymbol{A}_1 = \boldsymbol{I}_3 \otimes \boldsymbol{J}_9$ and $\boldsymbol{A}_2 = \boldsymbol{J}_3 \otimes \boldsymbol{I}_9$, then compute $\mathbf{v} = \boldsymbol{M}\bar{\mathbf{y}}$, and hence $\mathbf{v}_2$, $\mathbf{v}_4$, using the decomposition (6.3.24). This readily gives $\boldsymbol{\phi}$ from (6.3.26). We next use $\boldsymbol{M}$ and $\boldsymbol{K}$ to compute $\boldsymbol{\Gamma}$ and hence $\zeta_{max}(\boldsymbol{\Gamma})$. It turns out that $\zeta_{max}(\boldsymbol{\Gamma}) = 1.0$. Finally, note that $\boldsymbol{G}_1$ consists of the first $b(v-1) = 18$ columns of $\boldsymbol{H}$ (see (6.3.27)). Using these quantities in (6.3.30), we get the value of $\boldsymbol{\psi}$, which is then used to compute $\boldsymbol{\tau}^0$. Here $\boldsymbol{X} = [\boldsymbol{M}_v' : \boldsymbol{M}_{vb}']'(\boldsymbol{I}_3 \otimes \mathbf{1}_9)$. The value of the corresponding test statistic F from (6.3.34) is $F = 0.3132$ with 2 and 16 degrees of freedom. The level of significance being 0.735, the age-of-dam effect does not seem to exist.

6.3.2.2. Another Numerical Example. We shall now consider a second application of the analysis of an *unbalanced* mixed model to a numerical example taken essentially from Montgomery (1991, p. 201, 206). The example deals with an engineer who is designing a battery for use in a device that will be subjected to some extreme variations in temperature. He has three possible choices of the plate material for the battery (i.e., three *treatments* with *fixed* effects) and decides to test all three plate materials at three temperature levels (selected randomly, knowing that he does not have any control over the temperature), which can be considered as *random* block effects. The data representing the effective life (in hours) of the batteries under the experiment appear below (the data in Table 7-7 in Montgomery, 1991 is balanced; we have deleted some observations to make it unbalanced). Clearly, as with the previous one, this is an example of an *unbalanced* mixed model with interaction. We discuss below exact tests for *random* interaction and *random* block effects as developed by Gallo and Khuri (1990). The exact test for the equality of treatment effects can be similarly described as in the previous example.

To derive the exact tests of $H_{\tau\beta} : \sigma_{\tau\beta}^2 = 0$ and $H_\beta : \sigma_\beta^2 = 0$, we compute all the relevant quantities as described in the text as follows.

Table 6.2. Life Data (in Hours) for the Battery Design Experiment

	Temperature(°F)		
Material type	15	70	125
1	130, 155, 74	34, 40, 80, 75	20, 70
2	150, 188, 159, 126	136, 122, 106	25, 70, 58, 45
3	138, 110	174, 120, 150, 139	96, 104, 82

Source: D.C. Montgomery (1991). Reproduced with permission of John Wiley & Sons, Inc.

$b = 3$, $v = 3$, $n_{..} = 29$, $SS(e) = 12058.5833$,

$$R = \mathbf{I}_{29} - \mathbf{diag}(\mathbf{J}_3/3, \mathbf{J}_4/4, \mathbf{J}_2/2, \mathbf{J}_4/4, \mathbf{J}_3/3, \mathbf{J}_4/4, \mathbf{J}_2/2, \mathbf{J}_4/4, \mathbf{J}_3/3),$$

where $\mathbf{J}_k$ is a matrix of order $k \times k$ with all its elements unity, $\mathbf{H}_1 : 29 \times 6$ with its column vectors as $[\mathbf{h}_1 : \ldots : \mathbf{h}_6]$ where

$$\begin{aligned}
\mathbf{h}_1 &= (0, 0, 0, 0.0634, -0.7148, 0.6950, -0.0436, 0, \cdots, 0)' \\
\mathbf{h}_2 &= (0(9 \text{ times}), 0.0634, -0.7148, 0.6950, -0.0436, 0, \cdots, 0)' \\
\mathbf{h}_3 &= (0(16 \text{ times}), 0.0634, -0.7148, 0.6950, -0.0436, 0, \cdots 0)' \\
\mathbf{h}_4 &= (0(22 \text{ times}), 0.0634, -0.7148, 0.6950, -0.0436, 0, 0, 0)' \\
\mathbf{h}_5 &= (0, 0, 0, -0.8636, 0.2434, 0.3464, 0.2738, 0, \cdots, 0)' \\
\mathbf{h}_6 &= (0(9 \text{ times}), -0.8636, 0.2434, 0.3464, 0.2738, 0, \cdots, 0)'
\end{aligned}$$

$\mathbf{H}_2 : 29 \times 14$ with its column vectors as $[\mathbf{a}_1 : \ldots : \mathbf{a}_{14}]$ where

$$\begin{aligned}
\mathbf{a}_1 &= (0(16 \text{ times}), -0.8636, 0.2434, 0.3464, 0.2738, 0, \cdots, 0)' \\
\mathbf{a}_2 &= (0(22 \text{ times}), -0.8636, 0.2434, 0.3464, 0.2738, 0, 0, 0)' \\
\mathbf{a}_3 &= (0.8165, -0.4082, -0.4082, 0, \cdots, 0)' \\
\mathbf{a}_4 &= (0, 0.7071, -0.7071, 0, \cdots, 0)' \\
\mathbf{a}_5 &= (0(7 \text{ times}), 0.7071, -0.7071, 0, \cdots, 0)' \\
\mathbf{a}_6 &= (0(13 \text{ times}), 0.8165, -0.4082, -0.4082, 0, \cdots, 0)' \\
\mathbf{a}_7 &= (0(14 \text{ times}), 0.7071, -0.7071, 0, \cdots, 0)' \\
\mathbf{a}_8 &= (0(20 \text{ times}), 0.7071, -0.7071, 0, \cdots, 0)' \\
\mathbf{a}_9 &= (0(26 \text{ times}), 0.8165, -0.4082, -0.4082)' \\
\mathbf{a}_{10} &= (0(27 \text{ times}), 0.7071, -0.7071)' \\
\mathbf{a}_{11} &= (0, 0, 0, -0.0131, -0.4240, -0.3834, 0.8204, 0, \cdots, 0)' \\
\mathbf{a}_{12} &= (0(9 \text{ times}), -0.0131, -0.4240, -0.3834, 0.8204, 0, \cdots, 0)' \\
\mathbf{a}_{13} &= (0(16 \text{ times}), -0.0131, -0.4240, -0.3834, 0.8204, 0, \cdots, 0)' \\
\mathbf{a}_{14} &= (0(22 \text{ times}), -0.0131, -0.4240, -0.3834, 0.8204, 0, 0, 0)'
\end{aligned}$$

$Q_1 = 2569.6$, $Q_2 = 9489.0$, $\mathbf{L} : 6 \times 6$ with its column vectors as $[\mathbf{l}_1 : \ldots : \mathbf{l}_6]$ where

$$\begin{aligned}
\mathbf{l}_1 &= (0.4044, 0.0160, 0.0331, 0.0419, -0.0070, -0.0338)' \\
\mathbf{l}_2 &= (0.0160, 0.2917, -0.0029, -0.0098, 0.0110, 0.0098)' \\
\mathbf{l}_3 &= (0.0331, -0.0029, 0.3333, -0.0021, 0.0641, 0.0291)' \\
\mathbf{l}_4 &= (0.0419, -0.0098, -0.0021, 0.3084, -0.0157, -0.0069)' \\
\mathbf{l}_5 &= (-0.0070, 0.0110, 0.0641, -0.0157, 0.3734, 0.0288)' \\
\mathbf{l}_6 &= (-0.0338, 0.0098, 0.0291, -0.0069, 0.0288, 0.0288)'
\end{aligned}$$

$\lambda_{max}(\mathbf{L}) = 0.4346$, $\mathbf{u} = (-13.940, 45.352, 82.893, 35.210, -7.494, -4.118)'$, $\mathbf{z} = (44.1353, 35.4836, 24.3363, 72.7022, -15.3796, 33.3743)'$, the orthogonal matrix $\mathbf{P} : 6 \times 6$ with its column vectors as $[\mathbf{p}_1 : \ldots : \mathbf{p}_6]$ where

$$\begin{aligned}
\mathbf{p}_1 &= (0, 0.5774, 0, 0.5774, 0, 0.5774)' \\
\mathbf{p}_2 &= (0.5744, 0, 0.5744, 0, 0.5744, 0)' \\
\mathbf{p}_3 &= (0, 0.8156, 0, -0.4082, 0, -0.4082)' \\
\mathbf{p}_4 &= (-0.4082, 0, 0.5577, 0.5000, -0.1494, -0.5000)' \\
\mathbf{p}_5 &= (-0.5774, 0, -0.2113, 0, 0.7887, 0)' \\
\mathbf{p}_6 &= (-0.4082, 0, 0.5577, -0.5000, -0.1494, 0.5000)'
\end{aligned}$$

$\mathbf{w} = (-12.2521, 36.6365, 73.4592, 41.3508, 0.1326, -0.8318)'$, $\mathbf{w}_1 = (-12.2521, 36.6365)'$ and $\mathbf{w}_2 = (73.4592, 41.3508, 0.1326, -0.8318)'$.

These calculations finally yield $F_{\tau\beta} = 2.6213$ with 4 and 14 degrees of freedom and $F_\beta = 0.4200$ with 2 and 4 degrees of freedom. The 5% table F-value corresponding to 4 and 14 degrees of freedom is 3.11, and corresponding to 2 and 4 degrees of freedom is 6.94. We therefore do not reject the hypotheses $H_{\tau\beta}$ and H_β at 5% level of significance.

We conclude this section with the observation that some further research dealing with the case of empty cells has been recently carried out in an unpublished *Ph.D.* dissertation by Capen (1991).

6.4. GENERAL UNBALANCED MIXED MODELS: EXACT TESTS

In this section, we describe a new class of exact unbiased tests for variance components and fixed effects in the context of a general unbalanced mixed model. These tests, derived by Seifert (1992), are essentially of Bartlett–Scheffé type (see Chapter 2, Section 2.6.2).

Following formula (2.2.1) in Section 2.2 of Chapter 2, a general mixed model can be written as

$$\mathbf{y} = \boldsymbol{X}_1\boldsymbol{\alpha}_1 + \cdots + \boldsymbol{X}_m\boldsymbol{\alpha}_m + \boldsymbol{Z}_1\mathbf{u}_1 + \cdots + \boldsymbol{Z}_c\mathbf{u}_c, \tag{6.4.1}$$

where, as already explained, $\mathbf{y}: N \times 1$ is the vector of observations, $\boldsymbol{X}$'s and $\boldsymbol{Z}$'s are the design matrices, $\boldsymbol{\alpha}$'s are the vectors of fixed effects, and $\mathbf{u}$'s are the vectors of random effects. As always, we assume that the random effects are independently normally distributed as

$$\mathbf{u}_i \sim N(0, \sigma_i^2 \boldsymbol{I}_{..}), \quad i = 1, \cdots, c \tag{6.4.2}$$

with $\sigma_i^2 \geq 0$ for $i = 1, \ldots, c-1$ and $\sigma_c^2 > 0$. Here $\boldsymbol{I}_{..}$ denotes the identity matrix of an appropriate order.

Let us first consider the problem of testing a variance component

$$H_0: \sigma_1^2 = 0 \textit{ versus } H_1: \sigma_1^2 > 0. \tag{6.4.3}$$

The basic idea for the construction of an exact test is to find a linear transformation to a model with *only* two variance components. Then standard techniques are applied. Suppose we have found a matrix $\boldsymbol{C}^*$ such that

$$\mathbf{w}^* = \boldsymbol{C}^* \mathbf{y} = \tilde{\boldsymbol{U}}_1 \mathbf{u}_1 + \tilde{\boldsymbol{U}}_2 \tilde{\mathbf{u}}_2, \tag{6.4.4}$$

where $\tilde{\boldsymbol{U}}_1$ and $\tilde{\boldsymbol{U}}_1$ depend on the design matrices $\boldsymbol{X}$'s and $\boldsymbol{Z}$'s, $f_1 = \text{rank}(\tilde{\boldsymbol{U}}_1) > 0$, and $f_2 = \text{rank}(\tilde{\boldsymbol{U}}_2) - \text{rank}(\tilde{\boldsymbol{U}}_1) > 0$. f_1 and f_2 are called degrees of freedom of Bartlett–Scheffé tests. Here $\tilde{\mathbf{u}}_2$ is a linear combination of $\mathbf{u}_2, \ldots, \mathbf{u}_c$, and is distributed as

$$\tilde{\mathbf{u}}_2 \sim N(0, \tilde{\sigma}_2^2 \boldsymbol{I}), \tag{6.4.5}$$

where $\tilde{\sigma}_2^2 = d_2\sigma_2^2 + \ldots + d_c\sigma_c^2$ is a nonnegative linear combination of $\sigma_2^2, \ldots, \sigma_c^2$, with the coefficient $d_c > 0$.

Note that $\mathbf{u}_1$ and $\tilde{\mathbf{u}}_2$ are independent. Now since model (6.4.4) involves exactly two variance components, namely, σ_1^2 and $\tilde{\sigma}_2^2$, exact tests for the hypothesis H_0 given in (6.4.3) can be easily carried out (see Section 6.2.2). Thus, the exact Wald-type F-test rejects H_0 specified in (6.4.3) for large values of

$$F = \frac{f_2 \mathbf{w}^{*\prime} \tilde{\boldsymbol{V}}_2^{+} \tilde{\boldsymbol{U}}_1 (\tilde{\boldsymbol{U}}_1' \tilde{\boldsymbol{V}}_2^{+} \tilde{\boldsymbol{U}}_1)^{-1} \tilde{\boldsymbol{U}}_1' \tilde{\boldsymbol{V}}_2^{+} \mathbf{w}^*}{f_1 \mathbf{w}^{*\prime} \tilde{\boldsymbol{V}}_2^{+} (\tilde{\boldsymbol{V}}_2 - \tilde{\boldsymbol{U}}_1 (\tilde{\boldsymbol{U}}_1' \tilde{\boldsymbol{V}}_2^{+} \tilde{\boldsymbol{U}}_1)^{-} \tilde{\boldsymbol{U}}_1') \tilde{\boldsymbol{V}}_2^{+} \mathbf{w}^*}, \tag{6.4.6}$$

where $\tilde{\boldsymbol{V}}_2 = \tilde{\boldsymbol{U}}_2 \tilde{\boldsymbol{U}}_2'$, and $\boldsymbol{A}^+$ and $\boldsymbol{A}^-$ denote the Moore–Penrose inverse and a generalized inverse of $\boldsymbol{A}$, respectively. The above F-statistic has a central F-distribution with f_1 and f_2 degrees of freedom under H_0.

From Theorem (6.2.2)(b), the locally best invariant test based on $\mathbf{w}^*$ rejects H_0 for large values of

$$F^* = \frac{\mathbf{w}^{*\prime} \tilde{\boldsymbol{V}}_2^{+} \tilde{\boldsymbol{V}}_1 \tilde{\boldsymbol{V}}_2^{+} \mathbf{w}^*}{\mathbf{w}^{*\prime} \tilde{\boldsymbol{V}}_2^{+} \mathbf{w}^*}. \tag{6.4.7}$$

On the other hand, for alternatives of the form $H_1: \sigma_1 = \Delta \tilde{\sigma}_2$ where Δ is a given constant, following Kariya and Sinha (1985) and Kleffe and Seifert

(1988), the best invariant test rejects H_0 for large values of

$$F_\Delta = \frac{\mathbf{w}^{*\prime} \tilde{\boldsymbol{V}}_2^{+} \mathbf{w}^*}{\mathbf{w}^{*\prime} (\Delta \tilde{\boldsymbol{V}}_1 + \tilde{\boldsymbol{V}}_2)^{+} \mathbf{w}^*}. \tag{6.4.8}$$

It should be noted that, as with F, F^* and F_Δ have known distributions under the null hypothesis. Although their critical values are not readily available, these can be obtained by simulation, numerical integration, or approximate methods (see Section 3.2 of Chapter 3).

As an example, let us consider a random two-way crossed-classification with interaction model, similar to the one in Gallo and Khuri (1990), given by

$$y_{ijk} = \mu + \tau_i + \beta_j + (\tau\beta)_{ij} + e_{ijk} \tag{6.4.9}$$

with $i = 1, 2, 3$, $j = 1, 2, 3, 4$, and treatment-block replications n_{ij}'s given by

$$\begin{aligned} n_{11} &= 4, n_{12} = n_{13} = n_{14} = 0 \\ n_{21} &= n_{22} = 5, n_{23} = 4, n_{24} = 0 \\ n_{31} &= 6, n_{32} = 5, n_{33} = 4, n_{34} = 3. \end{aligned} \tag{6.4.10}$$

Thus in all we have 36 observations in 8 cells and there are 4 empty cells. The reader should note that because of the existence of some empty cells, the approach of Gallo and Khuri (1990) is not possible.

Consider the problem of testing the null hypothesis $H_0 : \sigma_\tau^2 = 0$. To apply the technique discussed above, we first delete a total of 5 observations to get an incomplete design with 4 observations per cell in every nonempty cell except the cell corresponding to $i = 3$, $j = 4$, which has exactly 3 observations y_{34k}, $k = 1, 2, 3$, as before. Note that this is a linear transformation. We next add a dummy observation y_{344} in the cell corresponding to $i = 3$, $j = 4$, to get the design matrix with all nonempty cells having an equal frequency of 4. Let $\bar{y}_{ij.} = \sum_{l=1}^{4} y_{ijl}/4$ be the cell means for the nonempty cells, resulting in the model

$$y_{ij}^{(2)} = \bar{y}_{ij.} = \mu + \tau_i + \beta_j + \epsilon_{ij}, \tag{6.4.11}$$

where $\epsilon_{ij} = (\tau\beta)_{ij} + \bar{e}_{ij.}$ are independently distributed as $N(0, \sigma_{\tau\beta}^2 + (1/4)\sigma_e^2)$. Let $\boldsymbol{y}^{(2)}$ be the vector of $y_{ij}^{(2)}$'s in lexicographical order, and define

$$\boldsymbol{U}_A = \begin{bmatrix} 1 & 0 & 0 \\ 0 & 1 & 0 \\ 0 & 1 & 0 \\ 0 & 1 & 0 \\ 0 & 0 & 1 \\ 0 & 0 & 1 \\ 0 & 0 & 1 \\ 0 & 0 & 1 \end{bmatrix}, \tag{6.4.12}$$

$$\boldsymbol{U}_B = \begin{bmatrix} 1&0&0&0\\ 1&0&0&0\\ 0&1&0&0\\ 0&0&1&0\\ 1&0&0&0\\ 0&1&0&0\\ 0&0&1&0\\ 0&0&0&1 \end{bmatrix}, \tag{6.4.13}$$

and $\boldsymbol{M}_B = \boldsymbol{I} - \boldsymbol{P}_{\boldsymbol{U}_B}$. Then $\mathbf{w}^* = \boldsymbol{M}_B\boldsymbol{y}^{(2)}$ is of the form (6.4.4) with $\tilde{\boldsymbol{U}}_1 = \boldsymbol{M}_B\boldsymbol{U}_A$, $\tilde{\boldsymbol{U}}_2 = \boldsymbol{M}_B$, $\tilde{\sigma}_2^2 = \sigma_{\tau\beta}^2 + (1/4)\sigma_e^2$, $f_1 = 2$ and $f_2 = 4$. Thus appropriate exact tests based on F, F^*, and F_Δ can be defined. It can be easily seen that the exact tests constructed this way do not depend on the dummy quantity y_{344}.

We now describe a general procedure to derive exact tests. The basic idea is to proceed stepwise where the model is reduced in each step by one variance component. After each step we get a model of the form (6.4.1)–(6.4.3), where all parts of the model depend on the number k of the step (i.e., $\mathbf{y}^{(k)}$, $\boldsymbol{Z}_i^{(k)}$, $\mathbf{u}_i^{(k)}$, $\sigma_i^{(k)}$, and $\boldsymbol{V}_i^{(k)}$ for $i = 1, \ldots, c^{(k)}$). Here $\boldsymbol{V}_i = \boldsymbol{Z}_i\boldsymbol{Z}_i'$, $i = 1, \ldots, c$. This is analogous to $\tilde{\boldsymbol{V}}_i$ based on $\tilde{\boldsymbol{U}}_i$ in the reduced model (6.4.4). For simplicity of notation we omit the superscript k in our subsequent discussion. We continue reducing the model until we end up with a model of the form (6.4.4) involving exactly two variance components, and then a Bartlett–Scheffé-type test as described above is constructed. It is, of course, possible that some situations cannot be handled by this stepwise reduction procedure.

An algorithm to accomplish the stepwise reduction is described below. At every step the algorithm uses the vector of residuals. Let $\boldsymbol{Q} = \boldsymbol{P}_{\boldsymbol{Z}_c} - \boldsymbol{P}_{(\boldsymbol{X},\boldsymbol{Z}_1,\ldots,\boldsymbol{Z}_{c-1})}$ where $\boldsymbol{X} = [\boldsymbol{X}_1 : \ldots : \boldsymbol{X}_m]$ and $f_c = \text{rank}(\boldsymbol{Q})$. Then we define $\boldsymbol{\gamma}$ as

$$\boldsymbol{\gamma} = (\boldsymbol{Q}\,\boldsymbol{V}_c\boldsymbol{Q})^{(-1/2)}\boldsymbol{Q}\,\mathbf{y} \sim N(0, \sigma_c^2\boldsymbol{I}_{f_c}). \tag{6.4.14}$$

Since we are interested in tests invariant under translations of fixed effects, our algorithm always starts with the maximal invariant statistic

$$\boldsymbol{y}^{(1)} = \boldsymbol{M}_X\mathbf{y}. \tag{6.4.15}$$

Algorithm 6.4.1. There are essentially five steps in the algorithm.

1. If $c = 2$, $f_1 = \text{rank}(\boldsymbol{Z}_1) > 0$ and $f_2 = \text{rank}(\boldsymbol{Z}_2) - \text{rank}(\boldsymbol{Z}_1) > 0$, the algorithm succeeded.
2. If $\text{rank}(\boldsymbol{Z}_1) = 0$ or $\text{rank}(\boldsymbol{Z}_2) = \text{rank}(\boldsymbol{Z}_1)$, the algorithm has come to a dead end.
3. If there is a design matrix $\boldsymbol{Z}_{i_0}$ with $R(\boldsymbol{Z}_{i_0})$ contained in $R(\boldsymbol{Z}_1)$, use

$$\boldsymbol{y}^{(k+1)} = \boldsymbol{M}_{\boldsymbol{Z}_{i_0}}\boldsymbol{y}^{(k)} \tag{6.4.16}$$

and restart the algorithm with k replaced by $k+1$. This is because such a design matrix $\boldsymbol{Z}_{i_0}$ must be deleted to ensure the construction of exact tests. The covariance matrix of $\boldsymbol{y}^{(k+1)}$ is reduced by the variance component $\sigma_{i_0}^2$. Here, $R(\boldsymbol{A})$ denotes the range or the column space of $\boldsymbol{A}$.

4. If there is a design matrix $\boldsymbol{Z}_{i_0}$ which is not comparable with $\boldsymbol{Z}_1$ in the sense that the range of one matrix neither contains nor is contained in the range of the other, use

$$\boldsymbol{y}^{(k+1)} = \boldsymbol{M}_{\boldsymbol{Z}_{i_0}} \boldsymbol{y}^{(k)} \tag{6.4.17}$$

and restart the algorithm with k replaced by $k+1$. An alternative procedure is to use

$$\boldsymbol{y}^{(k+1)} = (\boldsymbol{P}_{\boldsymbol{Z}_{i_0}} + \boldsymbol{M}_{(\boldsymbol{Z}_1, \boldsymbol{Z}_{i_0})}) \boldsymbol{y}^{(k)}, \tag{6.4.18}$$

which uses intra-block information about the effect $\mathbf{u}_{i_0}$ and makes that effect nested within $\mathbf{u}_1$.

5. Let $\boldsymbol{Z}_{i_0}$ be a design matrix such that $R(\boldsymbol{Z}_{i_0})$ is contained only in $R(\boldsymbol{Z}_c)$. Define $c_K = \lambda_{max}(\boldsymbol{V}_{i_0}^{+} \boldsymbol{V}_c)$, which is known as Khuri's constant, and let $\boldsymbol{P} = \boldsymbol{P}_{\boldsymbol{Z}_{i_0}}$. Recall that $\boldsymbol{V}_{i_0} = \boldsymbol{U}_{i_0} \boldsymbol{U}_{i_0}'$ and $\boldsymbol{V}_{i_0}^{+}$ denotes the Moore–Penrose inverse of $\boldsymbol{V}_{i_0}$. Suppose $f_c \geq \text{rank}[\boldsymbol{P}(c_K \boldsymbol{V}_{i_0} - \boldsymbol{V}_c)\boldsymbol{P}]$. Then use

$$\boldsymbol{y}^{(k+1)} = \boldsymbol{P}\boldsymbol{y}^{(k)} + [\boldsymbol{P}(c_K \boldsymbol{V}_{i_0} - \boldsymbol{V}_c)\boldsymbol{P}]^{(1/2)} \boldsymbol{\gamma}^{(k)} \tag{6.4.19}$$

and restart the algorithm with k replaced by $k+1$.

The constant c_K above ensures that $\boldsymbol{P}(c_K \boldsymbol{V}_{i_0} - \boldsymbol{V}_c)\boldsymbol{P}$ is nonnegative. As $\boldsymbol{P}\boldsymbol{y}^{(k)}$ and $\boldsymbol{\gamma}^{(k)}$ are independent, the covariance matrix of $\boldsymbol{y}^{(k+1)}$ in the final step is of the form

$$Var(\boldsymbol{y}^{(k+1)}) = \sum_{i=1, i \neq i_0}^{i=c-1} \sigma_i^2 \boldsymbol{P}\boldsymbol{V}_i \boldsymbol{P} + (\sigma_{i_0}^2 + c_K \sigma_c^2) \boldsymbol{P}\boldsymbol{V}_{i_0} \boldsymbol{P}. \tag{6.4.20}$$

The model is thus reduced by one variance component.

It is clear that the variance component $(\sigma_{i_0}^2 + c_K \sigma_c^2)$ is a noise variance, and consequently the power of the resulting tests decreases in the constant c_K. Thus we should choose c_K as small as possible in applications.

The final step can be repeated as long as $c > 2$, ending with a model (6.4.4) and a Bartlett–Scheffé test in (6.4.6). □

The assumption on f_c in the final step is made only for simplicity. Quite generally, let $\boldsymbol{P}$ be any projection matrix with $R(\boldsymbol{P})$ contained in $R(\boldsymbol{Z}_{i_0})$, and c_K any constant such that $\boldsymbol{P}(c_K \boldsymbol{V}_{i_0} - \boldsymbol{V}_c)\boldsymbol{P}$ is nonnegative with a rank not greater than f_c. Then $\boldsymbol{y}^{(k+1)}$ defined in (6.4.19) works. For details, see Seifert (1992).

6.5. GENERAL UNBALANCED MIXED MODELS: APPROXIMATE TESTS

In this section we describe a procedure due to Kleffe and Seifert (1988) to construct an approximate test for a variance component in a general unbalanced mixed model. This is essentially based on exploiting some properties of the minimum norm quadratic estimators (MINQUEs) of the underlying variance components.

We begin with the unbalanced mixed model given by (6.4.1), which is written as

$$\mathbf{y} = \boldsymbol{X}_1\boldsymbol{\alpha}_1 + \cdots + \boldsymbol{X}_m\boldsymbol{\alpha}_m + \boldsymbol{Z}_1\mathbf{u}_1 + \cdots + \boldsymbol{Z}_c\mathbf{u}_c, \tag{6.5.1}$$

where, as already explained, $\mathbf{y} : N \times 1$ is the vector of observations, $\boldsymbol{X}$'s and $\boldsymbol{Z}$'s are the design matrices, $\boldsymbol{\alpha}$'s are the vectors of fixed effects, and $\mathbf{u}$'s are the vectors of random effects. As always, we assume that the random effects are independently normally distributed as

$$\mathbf{u}_i \sim N(0, \sigma_i^2 \boldsymbol{I}_{..}), \quad i = 1, \cdots, c \tag{6.5.2}$$

with $\sigma_i^2 \geq 0$ for $i = 1, \ldots, c-1$ and $\sigma_c^2 > 0$. As before, $\boldsymbol{I}_{..}$ denotes the identity matrix of an appropriate order. For estimation of the variance components $\sigma_1^2, \ldots, \sigma_c^2$, various quadratic forms $\mathbf{y}'\boldsymbol{A}\mathbf{y}$ have been considered in the literature, and in this section we confine our attention to the MINQUE method introduced by Rao (1971). According to this method, the MINQUEs of the variance components $\sigma_1^2, \ldots, \sigma_c^2$ are obtained by solving the MINQUE equations

$$\sum_{i=1}^{c} Z(i,j)\hat{\sigma}_i{}^2 = Q_j, \quad j = 1, \cdots, c, \tag{6.5.3}$$

where $Z(i,j) = tr[\boldsymbol{Z}_i'(\boldsymbol{M}\boldsymbol{V}(\mathbf{t})\boldsymbol{M})^+\boldsymbol{Z}_j\boldsymbol{Z}_j'(\boldsymbol{M}\boldsymbol{V}(\mathbf{t})\boldsymbol{M})^+\boldsymbol{Z}_i]$, $\boldsymbol{M} = \boldsymbol{I} - \boldsymbol{X}(\boldsymbol{X}'\boldsymbol{X})^-\boldsymbol{X}'$, $Q_j = \mathbf{y}'(\boldsymbol{M}\boldsymbol{V}(\mathbf{t})\boldsymbol{M})^+\boldsymbol{Z}_j\boldsymbol{Z}_j'(\boldsymbol{M}\boldsymbol{V}(\mathbf{t})\boldsymbol{M})^+\mathbf{y}$, and $\boldsymbol{V}(\mathbf{t}) = t_1\boldsymbol{Z}_1\boldsymbol{Z}_1' + \ldots + t_c\boldsymbol{Z}_c\boldsymbol{Z}_c'$ is an a priori guess of the covariance matrix $\boldsymbol{\Sigma}$ of $\mathbf{y}$, given by $\boldsymbol{\Sigma} = \sigma_1^2\boldsymbol{Z}_1\boldsymbol{Z}_1' + \ldots + \sigma_c^2\boldsymbol{Z}_c\boldsymbol{Z}_c'$. The vector $\tilde{\boldsymbol{\sigma}}^2 = (\hat{\sigma}_1^2, \ldots, \hat{\sigma}_c^2)'$ is called a MINQUE of $\boldsymbol{\sigma}^2 = (\sigma_1^2, \ldots, \sigma_c^2)'$, which generally depends on the prior vector $\mathbf{t} = (t_1, \ldots, t_c)'$.

The asymptotic distribution of $\tilde{\boldsymbol{\sigma}}^2$ is used to derive an approximate test of the hypothesis $H_0 : \boldsymbol{H}\boldsymbol{\sigma}^2 = \boldsymbol{\delta}$, where $\boldsymbol{H} : h \times c$ is a given matrix of rank $h \leq c$, and the known vector $\boldsymbol{\delta}$ belongs to the column space of $\boldsymbol{H}$. It is clear that the asymptotic distribution of $\tilde{\boldsymbol{\sigma}}^2$ depends on the a priori vector $\mathbf{t}$, and a meaningful inference is achieved if $\mathbf{t}$ is replaced by a consistent estimator of $\boldsymbol{\sigma}^2$. We assume that such a $\mathbf{t}$ is available (based on iterated MINQUE, if necessary) and that this kind of $\mathbf{t}$ is used above. Then, Schmidt and Thrum (1981) showed that, under certain conditions, we can use the test statistic

$$T = (\boldsymbol{H}\tilde{\boldsymbol{\sigma}}^2 - \boldsymbol{\delta})'(\boldsymbol{H}\boldsymbol{Z}(\mathbf{t})\boldsymbol{H}')^{-1}(\boldsymbol{H}\tilde{\boldsymbol{\sigma}}^2 - \boldsymbol{\delta})/2, \tag{6.5.4}$$

which is asymptotically distributed as central χ^2 with h degrees of freedom under H_0, where $Var(\tilde{\boldsymbol{\sigma}}^2|\mathbf{t}) = 2\boldsymbol{Z}(\mathbf{t})$ and $\boldsymbol{H}$ is the coefficient matrix in the above null hypothesis H_0. Here, $\boldsymbol{Z}(\mathbf{t})$ stands for the coefficient matrix appearing on the left hand side of (6.5.3). However, no small sample properties of the test statistic T are known.

The conditions under which the above result holds are clarified in Humak (1984) and also in Kleffe and Rao (1987), and are known to hold for a large class of unbalanced models when $\boldsymbol{\sigma}^2$ has all its components positive in the hypothesis H_0. The restrictions are somewhat severe for testing the hypothesis $\sigma_i^2 = 0$ for $i = 1, \ldots, c-1$. Thus, for example, for the one-way unbalanced random effects model with k treatments and $n_1, \ldots, n_k$ as the treatment replications, the condition requires that the n_i's be bounded while k should approach ∞. For details, we refer to the above papers.

To conclude this section, we finally discuss Seifert's (1985) construction of ANOVA-like tests for $H_i : \sigma_i^2 = 0$ for some $1 \le i \le c-1$ based on the MINQUE. It may be noted that, for balanced models, an ANOVA-type test of H_i is based on

$$T_i = \hat{\sigma}_i^2 / \mathbf{l}'\hat{\boldsymbol{\sigma}}^2, \tag{6.5.5}$$

where $\mathbf{l}'\hat{\boldsymbol{\sigma}}^2$ is a nonnegative linear combination of $\hat{\sigma}_1^2, \ldots, \hat{\sigma}_c^2$, without $\hat{\sigma}_i^2$, and the numerator and the denominator of T_i are independent χ^2 variables. For unbalanced models, however, such a result holds only in some special cases. But in general we can find a linear function $\mathbf{l}'\hat{\boldsymbol{\sigma}}^2$ such that the numerator and the denominator in (6.5.5) are uncorrelated, given $\mathbf{t}$. Recall that the MINQUE $\tilde{\boldsymbol{\sigma}}^2$ depends on the a priori vector $\mathbf{t}$, and its covariance matrix is given by

$$Var(\tilde{\boldsymbol{\sigma}}^2|\mathbf{t}) = 2\boldsymbol{Z}(\mathbf{t}) = \boldsymbol{V}^*, say. \tag{6.5.6}$$

Now choose an upper triangular matrix $\boldsymbol{L}^* = [((l(i,j))]$ with diagonal elements all 1, and a diagonal matrix $\boldsymbol{D} = \mathbf{diag}(d(i))$ such that $\boldsymbol{L}^*\boldsymbol{V}^*\boldsymbol{L}'^* = \boldsymbol{D}$. Hence, given $\mathbf{t}$, the components of $\mathbf{z} = \boldsymbol{L}^*\tilde{\boldsymbol{\sigma}}^2$ are uncorrelated and have variances $d(i)$. It is possible to write down explicitly the expressions for the components of $\mathbf{z}$ as follows.

$$\begin{array}{c} z_1 = \hat{\sigma}_1^2 + l(1,2)\hat{\sigma}_2^2 + \ldots + l(1,c)\hat{\sigma}_c^2 \\ z_2 = \hat{\sigma}_2^2 + l(2,3)\hat{\sigma}_3^2 + \ldots + l(2,c)\hat{\sigma}_c^2 \\ \cdots\cdots\cdots\cdots\cdots\cdots\cdots\cdots\cdots\cdots \\ z_c = \hat{\sigma}_c^2. \end{array} \tag{6.5.7}$$

The expectations of z_i's are obtained by replacing $\tilde{\boldsymbol{\sigma}}^2$ by the true vector $\boldsymbol{\sigma}^2$. Thus, the above equations look very much like an ANOVA table for balanced models. By construction, z_i and $z_i - \hat{\sigma}_i^2$ are locally uncorrelated, given $\mathbf{t}$, and we can test the hypothesis H_i using the statistic T_i^* given by

$$T_i^* = \frac{z_i}{z_i - \hat{\sigma}_i^2}. \tag{6.5.8}$$

This test coincides with the usual ANOVA test for balanced models. The null distribution of T_i^* can be approximated by using Satterthwaite's (1941) approximation of the numerator and the denominator of T_i^* using multiples of central χ^2 variables. Standard arguments (see Section 2.4 of Chapter 2) can be used to verify that T_i^* has an approximate F-distribution with f_{1i} and f_2 degrees of freedom where (see Kleffe and Seifert, 1988 for details)

$$f_{1i} = 2(\mathbf{1}'\mathbf{t})^2/d(i), \qquad f_2 = 2(\mathbf{1}'\mathbf{t})^2/\mathbf{1}'\boldsymbol{V}^*\mathbf{1}. \tag{6.5.9}$$

REMARK 6.5.1. There are other approaches toward the construction of tests for variance components. Gnot and Michalski (1994) have described tests based on admissible estimators of variance components. Michalski and Zmyslony (1996) describe a procedure based on the best quadratic invariant unbiased estimator of variance components, under a commutativity condition. These results are not reviewed here.

APPENDIX 6.1

Proof of Lemma 6.3.2

We first note that the basic model (6.3.1) can be expressed as

$$\mathbf{y} = \mu \mathbf{1} + \boldsymbol{X}_1 \boldsymbol{\tau} + \boldsymbol{X}_2 \boldsymbol{\beta} + \boldsymbol{X}_3 (\boldsymbol{\tau\beta}) + \mathbf{e} \tag{6.1.1}$$

where $\boldsymbol{X}_1$, $\boldsymbol{X}_2$ and $\boldsymbol{X}_3$ are matrices of orders $n_{..} \times v$, $n_{..} \times b$ and $n_{..} \times vb$, respectively, given by

$$\boldsymbol{X}_1 = \mathbf{diag}(\mathbf{1}_{n_{1.}}, \cdots, \mathbf{1}_{n_{v.}}), \tag{6.1.2}$$

$$\boldsymbol{X}_2 = \begin{bmatrix} \mathbf{diag}(\mathbf{1}_{n_{11}}, \mathbf{1}_{n_{12}}, \ldots, \mathbf{1}_{n_{1b}}) \\ \mathbf{diag}(\mathbf{1}_{n_{21}}, \mathbf{1}_{n_{22}}, \ldots, \mathbf{1}_{n_{2b}}) \\ \cdots\cdots\cdots\cdots\cdots\cdots \\ \mathbf{diag}(\mathbf{1}_{n_{v1}}, \mathbf{1}_{n_{v2}}, \ldots, \mathbf{1}_{n_{vb}}) \end{bmatrix}, \tag{6.1.3}$$

and

$$\boldsymbol{X}_3 = \mathbf{diag}(\mathbf{1}_{n_{11}}, \mathbf{1}_{n_{12}}, \cdots, \mathbf{1}_{n_{1b}}, \mathbf{1}_{n_{21}}, \mathbf{1}_{n_{22}}, \cdots, \mathbf{1}_{n_{2b}}, \cdots, \mathbf{1}_{n_{v1}}, \mathbf{1}_{n_{v2}}, \cdots, \mathbf{1}_{n_{vb}}). \tag{6.1.4}$$

To prove the first part, note that

$$\begin{aligned} E(\boldsymbol{\psi}) &= E(\boldsymbol{\phi}) + (\zeta_{max}(\boldsymbol{\Gamma})\boldsymbol{I}_{\nu_1} - \boldsymbol{\Gamma})^{1/2} \boldsymbol{G}_1' E(\mathbf{y}) \\ &= E(\boldsymbol{\phi}) + (\zeta_{max}(\boldsymbol{\Gamma})\boldsymbol{I}_{\nu_1} - \boldsymbol{\Gamma})^{1/2} \boldsymbol{G}_1' (\mu \mathbf{1} + \boldsymbol{X}_1 \boldsymbol{\tau}), \end{aligned} \tag{6.1.5}$$

where $\boldsymbol{G}_1$ is such that

$$\mathbf{y}'\boldsymbol{R}\mathbf{y} = \mathbf{y}'(\boldsymbol{G}_1\boldsymbol{G}_1' + \boldsymbol{G}_2\boldsymbol{G}_2')\mathbf{y} \tag{6.1.6}$$

from (6.3.4), (6.3.13), and (6.3.27). But

$$\boldsymbol{G}_1'\mathbf{1} = \mathbf{0}, \qquad \boldsymbol{G}_1'\boldsymbol{X}_1 = \mathbf{0}, \tag{6.1.7}$$

since by definitions of $\boldsymbol{X}_1$ and $\boldsymbol{R}$,

$$\boldsymbol{R}\boldsymbol{I} = \mathbf{0}, \qquad \boldsymbol{R}\boldsymbol{X}_1 = \mathbf{0}. \tag{6.1.8}$$

Hence,

$$\begin{aligned} E(\boldsymbol{\psi}) &= E(\boldsymbol{\phi}) \\ &= [\boldsymbol{M}'_v : \boldsymbol{M}'_{vb}]'E(\bar{\mathbf{y}}), \text{ from (6.3.24) and (6.3.26)} \\ &= [\boldsymbol{M}'_v : \boldsymbol{M}'_{vb}]'[\mu\mathbf{1} + \boldsymbol{B}_1\boldsymbol{\tau}], \text{ from (6.3.3)} \\ &= [\boldsymbol{\tau}'\boldsymbol{B}'_1\boldsymbol{M}'_v : \mathbf{0}]', \end{aligned} \quad (6.1.9)$$

since $\boldsymbol{M}_{vb}\boldsymbol{A}_1\boldsymbol{M}'_{vb} = \mathbf{0}$ and hence $\boldsymbol{M}_{vb}\boldsymbol{B}_1 = \mathbf{0}$. This completes the proof of the first part of the lemma.

To prove the second part, note that $Cov(\boldsymbol{\phi}, \boldsymbol{G}'_1\mathbf{y}) = \mathbf{0}$ since $\bar{\mathbf{y}}$ and the residual sum of squares $\mathbf{y}'\boldsymbol{R}\mathbf{y}$ in (6.3.4) are statistically independent. Thus,

$$\begin{aligned} Var(\boldsymbol{\psi}) &= Var(\boldsymbol{\phi}) \\ &\quad + [\zeta_{max}(\boldsymbol{\Gamma})\mathbf{I}_{\nu_1} - \boldsymbol{\Gamma}]^{1/2}\boldsymbol{G}'_1 Var(\mathbf{y})\boldsymbol{G}_1[\zeta_{max}(\boldsymbol{\Gamma})\mathbf{I}_{\nu_1} - \boldsymbol{\Gamma}]^{1/2} \quad (6.1.10) \\ &= Var(\boldsymbol{\phi}) \\ &\quad + [\zeta_{max}(\boldsymbol{\Gamma})\mathbf{I}_{\nu_1} - \boldsymbol{\Gamma}]^{1/2}\boldsymbol{G}'_1[\boldsymbol{X}_2\boldsymbol{X}'_2\sigma^2_\beta + \boldsymbol{X}_3\boldsymbol{X}'_3\sigma^2_{\tau\beta} + \mathbf{I}\sigma^2_e] \times \\ &\quad \boldsymbol{G}_1[\zeta_{max}(\boldsymbol{\Gamma})\mathbf{I}_{\nu_1} - \boldsymbol{\Gamma}]^{1/2}. \end{aligned}$$

But $\boldsymbol{G}'_1\boldsymbol{X}_2 = \mathbf{0}$, $\boldsymbol{G}'_1\boldsymbol{X}_3 = \mathbf{0}$, and $\boldsymbol{G}'_1\boldsymbol{G}_1 = \mathbf{I}_{\nu_1}$, since by definitions of $\boldsymbol{X}_2$, $\boldsymbol{X}_3$, and $\boldsymbol{R}$, $\boldsymbol{R}\boldsymbol{X}_2 = \mathbf{0}$, $\boldsymbol{R}\boldsymbol{X}_3 = \mathbf{0}$, and the fact that the columns of $\boldsymbol{G}_1$ are orthonormal. It then follows that

$$Var(\boldsymbol{\psi}) = Var(\boldsymbol{\phi}) + [\zeta_{max}(\boldsymbol{\Gamma})\mathbf{I}_{\nu_1} - \boldsymbol{\Gamma}]\sigma^2_e = (\sigma^2_{\tau\beta} + \zeta_{max}(\boldsymbol{\Gamma})\sigma^2_e)\mathbf{I}_{\nu_1}, \quad (6.1.11)$$

since $\nu_1 = b(v-1)$, and

$$\begin{aligned} Var(\boldsymbol{\phi}) &= [\boldsymbol{M}'_v : \boldsymbol{M}'_{vb}]'Var(\bar{\mathbf{y}})[\boldsymbol{M}'_v : \boldsymbol{M}'_{vb}], by (6.3.23), (6.3.24), (6.3.26) \\ &= [\boldsymbol{M}'_v : \boldsymbol{M}'_{vb}]'(\boldsymbol{A}_2\sigma^2_\beta + \mathbf{I}_{vb}\sigma^2_{\tau\beta} + \boldsymbol{K}\sigma^2_e)[\boldsymbol{M}'_v : \boldsymbol{M}'_{vb}], by (6.3.6) \\ &= \sigma^2_{\tau\beta}\mathbf{I}_{\nu_1} + \boldsymbol{\Gamma}\sigma^2_e, \end{aligned} \quad (6.1.12)$$

by the fact that $\boldsymbol{M}_v\boldsymbol{A}_2\boldsymbol{M}'_v = \mathbf{0}$, $\boldsymbol{M}_v\boldsymbol{A}_2\boldsymbol{M}'_{vb} = \mathbf{0}$, and $\boldsymbol{M}_{vb}\boldsymbol{A}_2\boldsymbol{M}'_{vb} = \mathbf{0}$. This completes the proof of the second part of the lemma.

EXERCISES

6.1. Refer to (6.2.3). Prove that $\boldsymbol{\Gamma}\boldsymbol{J}\boldsymbol{\Gamma}' = \boldsymbol{J}$ where $\boldsymbol{\Gamma}$ is a permutation matrix.

6.2. Prove (6.2.19).

6.3. In the context of model (6.2.11), write down the linear model for $\boldsymbol{Z}'\mathbf{y}$ and prove directly that F_0 is distributed as central F under H_0.

6.4. Refer to Remark 6.2.4 and Wald's test described in Chapter 1. Prove that in the context of model (6.2.11), Wald's test is given by F_W defined in (6.2.24).

6.5. Refer to Remark 6.2.5. Verify conclusions (i) and (ii).

6.6. In the context of Subsection 6.3.1, prove that $\boldsymbol{J}_v \otimes \boldsymbol{I}_{b-1}$ has eigenvalues v and 0 with multiplicities $(b-1)$ and $(b-1)(v-1)$, respectively.

6.7. Prove Lemma 6.3.1.

6.8. In the context of the first numerical example, establish that $\zeta_{max}(\Gamma) = 1.0$.

6.9. In the context of the second numerical example, verify that $F_{\tau\beta} = 2.6213$ and $F_{\beta} = 0.4200$.

BIBLIOGRAPHY

Capen, R. C. (1991). "Exact testing procedures for unbalanced random and mixed linear models." Unpublished Ph.D. dissertation, University of Florida, Gainesville, Florida.

Christensen, R. (1996). "Exact tests for variance components." *Biometrics*, 52, 309–314.

Damon, R. A., Jr. and Harvey, W. R. (1987). *Experimental Design, ANOVA, and Regression*. Harper and Row, New York.

Das, R. and Sinha, B. K. (1987). "Robust optimum invariant unbiased tests for variance components." In: *Proceedings of the Second International Tampere Conference in Statistics* (T. Pukkila, S. Puntanen, Eds.), University of Tampere, Finland, 317–342.

El-Bassiouni, M. Y. and Seely, J. F. (1988). "On the power of Wald's variance component test in the unbalanced random one-way model." In: *Optimal Design and Analysis of Experiments* (Dodge, Y., Federov, V.V., Wynn, H.P., Eds.), North-Holland, Amsterdam, 157–165.

Gallo, J. and Khuri, A. I. (1990). "Exact tests for the random and fixed effects in an unbalanced mixed two-way cross-classification model." *Biometrics*, 46, 1087–1095.

Gnot, S. and Michalski, A. (1994). "Tests based on admissible estimators in two variance components models." *Statistics*, 25, 213–223.

Hirotsu, C. (1979). "An F approximation and its application." *Biometrika*, 66, 577–584.

Humak, K. M. S. (1984). *Statistische Methoden der Modellbildung III. Statistische Inferenz fur Kovarianzparameter*. Akademie-Verlag, Berlin.

Kariya, T. and Sinha, B. K. (1985). "Nonnull and optimality robustness of some multivariate tests." *The Annals of Statistics*, 13, 1182–1197.

Kariya, T. and Sinha, B. K. (1989). *Robustness of Statistical Tests*. Academic Press, Boston.

Khuri, A. I. and Littell, R. C. (1987). "Exact tests for the main effects variance component in an unbalanced random two-way model." *Biometrics*, 43, 545–560.

King, M. L. (1980). "Robust tests for spherical symmetry and their applications to least squares regression." *The Annals of Statistics*, 8, 1265–1271.

Kleffe, J. and Seifert, B. (1988). "On the role of MINQUE in testing of hypotheses under mixed linear models." *Communications in Statistics—Theory and Methods*, 17, 1287–1309.

Lehmann, E. L. (1986). *Testing Statistical Hypotheses*, Second Edition. Wiley, New York.

Lin, T. H. and Harville, D. A. (1991). "Some alternatives to Wald's confidence interval and test." *Journal of the American Statistical Association*, 86, 179–187.

Mathew, T. (1989). "Optimum invariant tests in mixed linear models with two variance components." In: *Statistical Data Analysis and Inference* (Y. Dodge, Ed.), North-Holland, Amsterdam, 381–388.

Mathew, T. and Sinha, B. K. (1988). "Optimum tests in unbalanced two-way models without interaction." *The Annals of Statistics*, 16, 1727–1740.

Michalski, A. and Zmyslony, R. (1996). "Testing hypotheses for variance components in mixed linear models." *Statistics*, 27, 297–310.

Montgomery, D. C. (1991). *Design and Analysis of Experiments*, Third Edition. Wiley, New York.

Öfversten, J. (1993). "Exact tests for variance components in unbalnced mixed linear models." *Biometrics*, 49, 45–57.

Rao, C. R. (1971). "Estimation of variance components—MINQUE Theory." *Journal of Multivariate Analysis*, 1, 257–275.

Rao, C. R. and Kleffe, J. (1988). *Estimation of Variance Components and Applications*. North Holland, Amsterdam.

Satterthwaite, F. E. (1941). "Synthesis of variance." *Psychometrika*, 6, 309–316.

Schmidt, W. H. and Thrum, R. (1981). "Contributions to asymptotic theory in regression models with linear covariance structure." *Math. Operationsforsch. und Statistik, Series Statistics*, 12, 243–269.

Seely, J. F. and El-Bassiouni, Y. (1983). "Applying Wald's variance component test." *The Annals of Statistics*, 11, 197–201.

Seifert, B. (1985). "Estimation and test of variance components using the MINQUE method." *Statistics*, 16, 621–635.

Seifert, B. (1992). "Exact tests in unbalanced mixed analysis of variance." *Journal of Statistical Planning and Inference*, 30, 257–266.

Shelby, C. E., Harvey, W. R., Clark, R. T., Quesenberry, J. R., and Woodward, R. R. (1963). "Estimates of phenotypic and genetic parameters in ten years of Miles City R.O.P. steer data." *Journal of Animal Science*, 22, 346–353.

Sinha, Bikas K. (1982). "On complete classes of experiments for certain invariant problems of linear inference." *Journal of Statistical Planning and Inference*, 7, 171–180.

Spjøtvoll, E. (1967). "Optimum invariant tests in unbalanced variance components models." *The Annals of Mathematical Statistics*, 38, 422–428.

Spjøtvoll, E. (1968). "Confidence intervals and tests for variance ratios in unbalanced variance components models." *Review of International Statistical Institute*, 36, 37–42.

Thomsen, I. (1975). "Testing hypotheses in unbalanced variance components models for two-way layouts." *The Annals of Statistics*, 3, 257–265.

Wald, A. (1947). "A note on regression analysis." *The Annals of Mathematical Statistics*, 18, 586–589.

Westfall, P. H. (1989). "Power comparisons for invariant variance ratio tests in mixed ANOVA models." *The Annals of Statistics*, 17, 318–326.

Wijsman, R. A. (1967). "Cross-section of orbits and their applications to densities of maximal invariants." In: *Fifth Berkeley Symposium on Mathematical Statistics and Probability, I*, University of California, Berkeley, 389–400.

CHAPTER 7

Recovery of Inter-block Information

7.1. INTRODUCTION

In block designs with fixed treatment effects and no interaction, it is a well known and an easily verifiable fact that the usual least squares estimator of the treatment effects, computed assuming that the block effects are fixed, continues to provide an unbiased estimator of the treatment effects even when the block effects are random. This least squares estimator of the treatment effects is referred to as the intra-block estimator since it is based on contrasts among the observations within the blocks, thereby eliminating the block effects. Similarly, the usual F-test (known as the intra-block F-test) can always be used to test the equality of the treatment effects irrespective of whether the block effects are fixed effects or random effects. The derivation of Wald's test in Section 1.2 of Chapter 1 establishes this fact. However, when the block effects are random as opposed to being fixed, and are independently and identically distributed as normal, additional information is available for inference concerning the treatment effects. Such information is referred to as inter-block information and this information is based on the sum of the observations in each block. An important problem in this context is that of suitably combining the intra-block and inter-block information to obtain a combined estimator of the treatment effects or to obtain a combined test for testing the equality of the treatment effects. Such a problem of obtaining a combined estimator was first addressed by Yates (1939, 1940) for certain special designs and later by Rao (1947) for general incomplete block designs. Since then, there has been considerable research activity on the problem of combining the intra-block and inter-block estimators of the treatment effects. We refer to Shah (1975, 1992) for a review of these results.

In spite of the fairly extensive literature on the problem of obtaining a combined estimator, the problem of deriving a combined test has received very little attention. An obvious way of combining the intra-block and inter-block F-tests is by using Fisher's procedure, namely, by combining the two individual P-values via the sum of the log P-values multiplied by the factor -2. The null distribution of the statistic so obtained is a chi-squared distribution with 4 degrees of freedom, since the individual P-values are independent. Thus the combined test is very easy to implement. However, as will be seen in Section 7.3, this procedure ignores the underlying structure of the problem. Some tests that do take into account the underlying structure of the problem are proposed in Feingold (1985, 1988). However, Feingold's tests are not exact and are applicable only for testing whether a single contrast among the treatment effects is zero. The first satisfactory solution to the problem of combining the intra-block and inter-block F-tests appears to be due to Cohen and Sackrowitz (1989). These authors derived a simple and exact combined test for BIBDs. Through simulation, they also showed that their test was superior to the usual intra-block F-test. Their results were generalized and extended by Mathew, Sinha, and Zhou (1993) and Zhou and Mathew (1993) by proposing and comparing a variety of tests and also by considering designs other than BIBDs. For BIBDs, the combined tests and their performance are discussed in detail in Section 7.3. It turns out that the problem of combining inter-block and intra-block F-tests also arises in random effects models (i.e., when, in addition to the block effects, the treatment effects are also independent identically distributed normal random variables, distributed independently of the block effects and the experimental error terms, and the problem is to test the significance of the treatment variance component). For this problem, various combined tests and their performance are studied in Section 7.4. In Section 7.5, we have extended the results to general incomplete block designs having equal block sizes. As will become clear later, the assumption of equal block sizes is crucial for the derivation of the results in this chapter.

7.2. NOTATIONS AND TEST STATISTICS

Consider a block design for comparing v treatments in b blocks of k plots each and let y_{ij} denote the observation from the ith plot in the jth block. When the blocks and treatments do not interact, which is the setup we are considering in this chapter, we have the following two-way classification model for the observations y_{ij}'s:

$$y_{ij} = \mu + \sum_{s=1}^{v} \delta_{ij}^{s} \tau_s + \beta_j + e_{ij}, \; i = 1, 2, \cdots, k; \; j = 1, 2, \cdots, b, \qquad (7.2.1)$$

where μ is a general mean, τ_s is the effect due to the sth treatment, β_j is the effect due to the jth block, e_{ij} is the experimental error term, and δ_{ij}^{s} takes the value 1 if the sth treatment occurs in the ith plot of the jth block and is 0 otherwise. Writing $\mathbf{y} = (y_{11}, \ldots, y_{1k}, y_{21}, \ldots, y_{2k}, \ldots, y_{b1}, \ldots, y_{bk})'$, $\boldsymbol{\tau}$

$= (\tau_1, \ldots, \tau_v)'$, and $\boldsymbol{\beta} = (\beta_1, \ldots, \beta_b)'$, model (7.2.1) can be written as

$$\mathbf{y} = \mu \mathbf{1}_{bk} + \boldsymbol{X}_1 \boldsymbol{\tau} + (\boldsymbol{I}_b \otimes \mathbf{1}_k)\boldsymbol{\beta} + \mathbf{e}, \tag{7.2.2}$$

where $\mathbf{e}$ is defined similarly to $\mathbf{y}$, $\boldsymbol{X}_1$ is an appropriate design matrix (with the δ_{ij}^s's as entries), and $\otimes$ denotes Kronecker product (see Chapter 2). We shall assume that $\boldsymbol{\tau}$ is a vector of fixed effects satisfying $\sum_{s=1}^{v} \tau_s = 0$, $\boldsymbol{\beta}$ is a vector of random effects, and $\boldsymbol{\beta}$ and $\mathbf{e}$ are independently distributed with

$$\boldsymbol{\beta} \sim N(\mathbf{0}, \sigma_\beta^2 \boldsymbol{I}_b) \text{ and } \mathbf{e} \sim N(\mathbf{0}, \sigma_e^2 \boldsymbol{I}_{bk}). \tag{7.2.3}$$

For convenience, and also to make the discussion self-contained, we shall first give the expressions for certain estimators and sums of squares under model (7.2.2), where assumption (7.2.3) is satisfied. Derivations of these are omitted, since they are available in standard books on analysis of variance.

Let $n_{sj} = \sum_{i=1}^{k} \delta_{ij}^s$. Then n_{sj} is clearly the number of times the sth treatment occurs in the jth block and $\boldsymbol{N} = ((n_{sj}))$ is the $v \times b$ incidence matrix of the design. It is readily verified that

$$\boldsymbol{N} = \boldsymbol{X}_1'(\boldsymbol{I}_b \otimes \mathbf{1}_k). \tag{7.2.4}$$

Let T_s and B_j respectively denote the sum of the observations corresponding to the sth treatment and the jth block respectively. Write $\mathbf{T} = (T_1, .., T_v)'$ and $\mathbf{B} = (B_1, \ldots, B_b)'$. Then

$$\mathbf{q}_1 = \mathbf{T} - \frac{1}{k}\boldsymbol{N}\mathbf{B} \tag{7.2.5}$$

is referred to as the vector of adjusted treatment totals. Let r_s be the number of times the sth treatment occurs in the design and define

$$\boldsymbol{C}_1 = \mathbf{diag}(r_1, \cdots, r_v) - \frac{1}{k}\boldsymbol{N}\boldsymbol{N}'. \tag{7.2.6}$$

Then

$$\mathbf{q}_1 \sim N(\boldsymbol{C}_1 \boldsymbol{\tau}, \sigma_e^2 \boldsymbol{C}_1). \tag{7.2.7}$$

The random vector $\mathbf{q}_1$ is an *intra-block* quantity, since each component of $\mathbf{q}_1$ can be written as a linear combination of contrasts among observations within blocks. When the τ_i's are fixed effects, the estimator of $\boldsymbol{\tau}$ based on $\mathbf{q}_1$ is referred to as its intra-block estimator. It is well known that for the estimability of all the treatment contrasts based on intra-block information, we must have $\text{rank}(\boldsymbol{C}_1) = v - 1$. Throughout this chapter, we shall assume that this rank condition holds, since this condition is also necessary in order that we can test the equality of all the treatment effects. If s_1^2 denotes the intra-block error sum of squares (i.e., the error sum of squares based on model (7.2.2) when $\boldsymbol{\beta}$ is a vector of fixed effects), we then have

$$s_1^2 \sim \sigma^2 \chi^2_{bk-b-v+1}. \tag{7.2.8}$$

Let

$$F_1 = \frac{\mathbf{q}_1' \boldsymbol{C}_1^- \mathbf{q}_1/(v-1)}{s_1^2/(bk-b-v+1)}, \tag{7.2.9}$$

then F_1 is the intra-block F-ratio for testing the equality of the treatment effects. Recall that for any matrix $\mathbf{A}$, $\mathbf{A}^-$ denotes a generalized inverse.

From (7.2.2) we get the following model for B_j, the sum of the observations from the jth block:

$$B_j = k\mu + \sum_{s=1}^{v} n_{sj}\tau_s + g_j, \tag{7.2.10}$$

where $g_j = k\beta_j + \sum_{i=1}^{k} e_{ij}$. Writing $\mathbf{g} = (g_1, g_2, \ldots, g_b)'$, (7.2.10) can be expressed as

$$\mathbf{B} = k\mu\mathbf{1}_b + \boldsymbol{N}'\boldsymbol{\tau} + \mathbf{g}, \tag{7.2.11}$$

with

$$\mathbf{g} \sim N[\mathbf{0}, k(k\sigma_\beta^2 + \sigma_e^2)\boldsymbol{I}_b]. \tag{7.2.12}$$

Inter-block information is the information regarding the treatment effects that can be obtained using $\mathbf{B}$, based on model (7.2.11). It is easy to verify that under the distributional assumption (7.2.3), $\mathbf{B}$ is distributed independently of $\mathbf{q}_1$ and s_1^2. Define

$$\mathbf{q}_2 = \boldsymbol{N}\left(\boldsymbol{I}_b - \frac{1}{b}\mathbf{1}_b\mathbf{1}_b'\right)\mathbf{B}, \quad \boldsymbol{C}_2 = \boldsymbol{N}\left(\boldsymbol{I}_b - \frac{1}{b}\mathbf{1}_b\mathbf{1}_b'\right)\boldsymbol{N}'. \tag{7.2.13}$$

Then,

$$\mathbf{q}_2 \sim N[\boldsymbol{C}_2\boldsymbol{\tau}, k(k\sigma_\beta^2 + \sigma_e^2)\boldsymbol{C}_2]. \tag{7.2.14}$$

In order to be able to estimate all the treatment contrasts using $\mathbf{B}$, we need the condition $\text{rank}(\boldsymbol{C}_2) = v - 1$, which will be assumed in this chapter. Note that this condition implies $b \geq v$. If s_2^2 is the error sum of squares based on model (7.2.11), then

$$s_2^2 \sim k(k\sigma_\beta^2 + \sigma_e^2)\chi_{b-v}^2. \tag{7.2.15}$$

Let

$$F_2 = \frac{\mathbf{q}_2' \boldsymbol{C}_2^- \mathbf{q}_2/(v-1)}{s_2^2/(b-v)}. \tag{7.2.16}$$

Then F_2 is referred to as the inter-block F-ratio for testing the equality of the treatment effects. Note that when $b = v$, s_2^2 and the F-ratio F_2 are nonexistent, even though the treatment contrasts can be estimated using $\mathbf{q}_2$ in (7.2.14). Our analysis in this chapter will be based on the quantities $\mathbf{q}_1$, s_1^2, $\mathbf{q}_2$, and s_2^2 having the distributions as specified, respectively, in (7.2.7), (7.2.8), (7.2.14), and (7.2.15). These quantities are all independently distributed. Note that in order to obtain the expressions for the above quantities and to arrive at their distributions, we have explicitly used the assumption of equal block sizes.

Assuming that $\text{rank}(\boldsymbol{C}_1) = \text{rank}(\boldsymbol{C}_2) = v - 1$ and $b > v$, the problem we shall address is that of combining the F-tests based on F_1 and F_2 in order to test the equality of the treatment effects. It will also be of interest to see whether $\mathbf{q}_2$ can be used in this testing problem when $b = v$, that is, when s_2^2 and F_2 are nonexistent. When the block design is a BIBD, this problem is analyzed in detail in Section 7.3 and we have derived and compared several tests. The same problems are discussed in Section 7.5 for a general incomplete block design with equal block sizes. When the treatment effects are random, distributed independently and identically as normal, the F-ratios F_1 and F_2 can be used to test the significance of the treatment variance component. The problem of combining the above F-tests in this context is investigated in Section 7.4 for a BIBD.

7.3. BIBD WITH FIXED TREATMENT EFFECTS

Consider a BIBD with parameters v, b, r, k, and λ, where v, b and k are as in the previous section and r and λ, respectively, denote the number of times each treatment is replicated and the number of blocks in which every pair of treatments occur together. Using the facts that the incidence matrix $\boldsymbol{N}$ of a BIBD satisfies $\boldsymbol{N}\boldsymbol{N}' = (r - \lambda)\boldsymbol{I}_v + \lambda \mathbf{1}_v \mathbf{1}_v'$ and $\boldsymbol{N}\mathbf{1}_b = r\mathbf{1}_v$, the matrices $\boldsymbol{C}_1$ and $\boldsymbol{C}_2$ simplify to

$$\boldsymbol{C}_1 = \frac{\lambda v}{k}\left(\boldsymbol{I}_v - \frac{1}{v}\mathbf{1}_v\mathbf{1}_v'\right), \; \boldsymbol{C}_2 = (r - \lambda)\left(\boldsymbol{I}_v - \frac{1}{v}\mathbf{1}_v\mathbf{1}_v'\right). \tag{7.3.1}$$

Note that $\boldsymbol{C}_2$ is a multiple of $\boldsymbol{C}_1$ and the matrix $(\boldsymbol{I}_v - 1/v\mathbf{1}_v\mathbf{1}_v')$ that occurs in both $\boldsymbol{C}_1$ and $\boldsymbol{C}_2$ is a symmetric idempotent matrix of rank $v - 1$ satisfying $(\boldsymbol{I}_v - 1/v\mathbf{1}_v\mathbf{1}_v')\mathbf{1}_v = \mathbf{0}$. These facts will be used in the analysis that follows. Let $\boldsymbol{O}$ be any $v \times v$ orthogonal matrix whose last column is the vector $1/\sqrt{v}\mathbf{1}_v$. Then we can write $\boldsymbol{O} = [\boldsymbol{O}_1 : 1/\sqrt{v}\mathbf{1}_v]$, where $\boldsymbol{O}_1$ is a $v \times (v - 1)$ matrix denoting the first $(v - 1)$ columns of $\boldsymbol{O}$. The orthogonality of $\boldsymbol{O}$ gives $\boldsymbol{O}_1'\boldsymbol{O}_1 = \boldsymbol{I}_{v-1}$ and $\boldsymbol{O}_1'\mathbf{1}_v = \mathbf{0}$. Consequently, $\boldsymbol{O}'(\boldsymbol{I}_v - 1/v\mathbf{1}_v\mathbf{1}_v')\boldsymbol{O} = \mathbf{diag}(\boldsymbol{I}_{v-1}, 0)$. Hence

$$\boldsymbol{O}'\boldsymbol{C}_1\boldsymbol{O} = \lambda v/k\,\mathbf{diag}(\boldsymbol{I}_{v-1}, 0), \;\; \boldsymbol{O}'\boldsymbol{C}_2\boldsymbol{O} = (r - \lambda)\,\mathbf{diag}(\boldsymbol{I}_{v-1}, 0). \tag{7.3.2}$$

In view of (7.3.2), we have $\boldsymbol{O}'\mathbf{q}_1 \sim N[\lambda v/k\,\mathbf{diag}(\boldsymbol{I}_{v-1}, 0)\boldsymbol{O}'\boldsymbol{\tau}, \sigma_e^2\lambda v/k\;\mathbf{diag}(\boldsymbol{I}_{v-1}, 0)]$ and $\boldsymbol{O}'\mathbf{q}_2 \sim N[(r - \lambda)\,\mathbf{diag}(\boldsymbol{I}_{v-1}, 0)\boldsymbol{O}'\boldsymbol{\tau}, k(k\sigma_\beta^2 + \sigma_e^2)(r - \lambda)\;\mathbf{diag}(\boldsymbol{I}_{v-1}, 0)]$, where $\mathbf{q}_1$ and $\mathbf{q}_2$ are as in (7.2.5) and (7.2.13) respectively. Define

$$\mathbf{x}_1 = \frac{k}{\lambda v}\boldsymbol{O}_1'\mathbf{q}_1, \; \mathbf{x}_2 = \frac{1}{(r - \lambda)}\boldsymbol{O}_1'\mathbf{q}_2, \; \boldsymbol{\tau}^* = \boldsymbol{O}_1'\boldsymbol{\tau}, \tag{7.3.3}$$

and let s_1^2 and s_2^2 be as in (7.2.8) and (7.2.15) respectively. We then have the following canonical form.

$$\begin{aligned} \mathbf{x}_1 \sim N(\boldsymbol{\tau}^*, \tfrac{k}{\lambda v}\sigma_e^2 \boldsymbol{I}_{v-1}),\ \mathbf{x}_2 \sim N[\boldsymbol{\tau}^*, \tfrac{k}{(r-\lambda)}(k\sigma_\beta^2 + \sigma_e^2)\boldsymbol{I}_{v-1}] \\ s_1^2 \sim \sigma_e^2 \chi^2_{bk-b-v+1},\ s_2^2 \sim k(k\sigma_\beta^2 + \sigma_e^2)\chi^2_{b-v}. \end{aligned} \tag{7.3.4}$$

Note that since $\boldsymbol{O}_1'\mathbf{1}_v = \mathbf{0}$, $\boldsymbol{\tau}^*$ in (7.3.3) is a $(v-1)\times 1$ vector of treatment contrasts and the equality of the treatment effects is equivalent to $\boldsymbol{\tau}^* = \mathbf{0}$. The F-ratios F_1 and F_2 in (7.2.9) and (7.2.16) simplify to

$$\begin{aligned} F_1 &= \frac{[k/\lambda v]\mathbf{q}_1'\mathbf{q}_1/(v-1)}{s_1^2/(bk-b-v+1)} = \frac{[\lambda v/k]\mathbf{x}_1'\mathbf{x}_1/(v-1)}{s_1^2/(bk-b-v+1)} \\ F_2 &= \frac{[1/(r-\lambda)]\mathbf{q}_2'\mathbf{q}_2/(v-1)}{s_2^2/(b-v)} = \frac{(r-\lambda)\mathbf{x}_2'\mathbf{x}_2/(v-1)}{s_2^2/(b-v)}. \end{aligned} \tag{7.3.5}$$

Note that if $b = v$ (i.e., for a symmetrical BIBD), s_2^2 and F_2 are nonexistent. We shall first consider the case $b > v$. The situation $b = v$ will be considered later.

7.3.1. Combined Tests When $b > v$

We shall work with the canonical form (7.3.4) since it is easy to identify a group that leaves the testing problem invariant. The tests that we shall derive can be easily expressed in terms of the original variables. In terms of (7.3.4), the testing problem is H_0: $\boldsymbol{\tau}^* = 0$. Consider the group $\mathcal{G} = \{g = (\boldsymbol{\Gamma}, c)$: $\boldsymbol{\Gamma}$ is a $p \times p$ orthogonal matrix and c is a positive scalar$\}$, whose action on $(\mathbf{x}_1, \mathbf{x}_2, s_1^2, s_2^2)$ is given by

$$g(\mathbf{x}_1, \mathbf{x}_2, s_1^2, s_2^2) = (c\boldsymbol{\Gamma}\mathbf{x}_1, c\boldsymbol{\Gamma}\mathbf{x}_2, c^2 s_1^2, c^2 s_2^2). \tag{7.3.6}$$

It is clear that the testing problem H_0: $\boldsymbol{\tau}^* = \mathbf{0}$ in model (7.3.4) is invariant under the above group action. In order to derive invariant tests, we shall first compute a maximal invariant statistic. Recall that a maximal invariant statistic, say D, under the group action (7.3.6) is a statistic satisfying two conditions: (i) D is invariant under the group action in (7.3.6) and, (ii) any statistic that is invariant under the group action in (7.3.6) must be a function of D. A maximal invariant parameter is similarly defined; we refer to Lehmann (1986, Chapter 6) for further details. A maximal invariant statistic and a maximal invariant parameter under the group action (7.3.6) are given in the following lemma.

Lemma 7.3.1. Consider model (7.3.4) and the testing problem H_0: $\boldsymbol{\tau}^* = \mathbf{0}$. Let

$$U_1 = \frac{\lambda v}{k}\mathbf{x}_1'\mathbf{x}_1 + s_1^2,\ U_2 = \frac{1}{k}[(r-\lambda)\mathbf{x}_2'\mathbf{x}_2 + s_2^2],\ U = U_1/U_2,\ R = \mathbf{x}_1'\mathbf{x}_2/\|\mathbf{x}_1\|\|\mathbf{x}_2\|. \tag{7.3.7}$$

Then a maximal invariant statistic, say D, under the group action (7.3.6) is given by

$$D = (F_1, F_2, R, U), \tag{7.3.8}$$

where F_1 and F_2 are the F-ratios given in (7.3.5) and R and U are given in (7.3.7). Furthermore, a maximal invariant parameter is $[\boldsymbol{\tau}^{*'}\boldsymbol{\tau}^*/\sigma_e^2, (\sigma_e^2 + k\sigma_\beta^2)/\sigma_e^2]$.

Proof. The lemma can be easily proved using the definition of a maximal invariant given before the statement of the lemma. □

Obviously, a maximal invariant statistic can be expressed in other forms using functions of $\mathbf{x}_1$, $\mathbf{x}_2$, s_1^2, and s_2^2, different from those in D given in (7.3.8). The reason for expressing the maximal invariant as in (7.3.8) is that the tests we shall describe below are directly in terms of F_1, F_2, R, and U. Note that under H_0: $\boldsymbol{\tau}^* = \mathbf{0}$, $U_1/\{(v-1)+(bk-b-v+1)\}$ and $U_2/\{(v-1)+(b-1)\}$ are unbiased estimators of σ_e^2 and $\sigma_e^2 + k\sigma_\beta^2$. This is a property that we shall eventually use. The following lemma establishes some properties of the quantities F_1, F_2, R, U_1, and U_2.

Lemma 7.3.2. Let F_1, F_2, R, U_1, and U_2 be as in Lemma 7.3.1. Then, under H_0: $\boldsymbol{\tau}^* = \mathbf{0}$, (*a*) U_1 and U_2 are complete and sufficient for σ_e^2 and σ_β^2, and (*b*) F_1, F_2, R, U_1, and U_2 are mutually independent.

Proof. For model (7.3.4), the completeness and sufficiency of U_1 and U_2 under H_0: $\boldsymbol{\tau}^* = \mathbf{0}$ is obvious. The independence of F_1 and F_2 is obvious and so is the independence of U_1 and U_2. The independence of (F_1, F_2) and R under H_0: $\boldsymbol{\tau}^* = \mathbf{0}$ follows by noting that $\mathbf{x}_i/\|\mathbf{x}_i\|$ is independent of $\|\mathbf{x}_i\|$ under H_0, and F_i depends on $\mathbf{x}_i$ only through $\|\mathbf{x}_i\|$ $(i = 1, 2)$. Also, under H_0: $\boldsymbol{\tau}^* = \mathbf{0}$, F_1, F_2, and R are ancillary statistics (i.e., statistics whose distributions do not involve any unknown parameters) and hence are independent of the complete sufficient statistics U_1 and U_2, by Basu's theorem (see Lehmann, 1986, p. 191). This completes the proof of the lemma. □

Lemma 7.3.3. Let $P_{H_0}^D$ and $P_{H_1}^D$ respectively denote the distributions of D in (7.3.8) under H_0: $\boldsymbol{\tau}^* = \mathbf{0}$ and H_1: $\boldsymbol{\tau}^* \neq \mathbf{0}$. Then, the density ratio $dP_{H_1}^D(d)/dP_{H_0}^D(d)$ is a monotone-increasing function of $h^*(\mathbf{x}_1, \mathbf{x}_2, s_1^2, s_2^2, \sigma_e^2, \sigma_\beta^2)$ given by

$$\begin{aligned} h^*(\mathbf{x}_1, \mathbf{x}_2, s_1^2, s_2^2, \sigma_e^2, \sigma_\beta^2) \\ = \left\| \frac{\lambda v}{k}\frac{\mathbf{x}_1}{\sigma_e^2} + \frac{(r-\lambda)}{k}\frac{\mathbf{x}_2}{(k\sigma_\beta^2 + \sigma_e^2)} \right\|^2 \Big/ \left(\frac{U_1}{\sigma_e^2} + \frac{U_2}{(k\sigma_\beta^2 + \sigma_e^2)} \right), \end{aligned} \tag{7.3.9}$$

where, U_1 and U_2 are given in (7.3.7).

Proof. Applying the representation theorem due to Wijsman (see Appendix 1.1 in Chapter 1), it follows that the density ratio, say, $R^* =$

$dP^D_{H_1}(d)/dP^D_{H_0}(d)$, can be obtained as

$$R^* = \left[\int_{O(p)}\int_0^\infty exp\left\{-\frac{1}{2}\left(\frac{\lambda v}{k}\frac{(c\boldsymbol{\Gamma}\mathbf{x}_1-\boldsymbol{\tau}^*)'(c\boldsymbol{\Gamma}\mathbf{x}_1-\boldsymbol{\tau}^*)}{\sigma_e^2}+\frac{c^2s_1^2}{\sigma_e^2}\right.\right.\right.$$
$$\left.\left.+\frac{(r-\lambda)}{k}\frac{(c\boldsymbol{\Gamma}\mathbf{x}_2-\boldsymbol{\tau}^*)'(c\boldsymbol{\Gamma}\mathbf{x}_2-\boldsymbol{\tau}^*)}{(k\sigma_\beta^2+\sigma_e^2)}+\frac{c^2s_2^2}{k(k\sigma_\beta^2+\sigma_e^2)}\right)\right\}\times$$
$$\left.c^{bk-2}dcd\Gamma\right]\Big/\left[\int_{O(p)}\int_0^\infty exp\left\{-\frac{c^2}{2}\left(\frac{U_1}{\sigma_e^2}+\frac{U_2}{(k\sigma_\beta^2+\sigma_e^2)}\right)\right\}c^{bk-2}dcd\boldsymbol{\Gamma}\right]$$
$$=\left[exp\left\{-\frac{\boldsymbol{\tau}^{*'}\boldsymbol{\tau}^*}{2}\left(\frac{\lambda v}{k}\frac{1}{\sigma_e^2}+\frac{(r-\lambda)}{k}\frac{1}{(k\sigma_\beta^2+\sigma_e^2)}\right)\right\}\times\right.$$
$$\int_{O(p)}\int_0^\infty exp\left(-\frac{c^2}{2}\right)exp\left\{\frac{c\boldsymbol{\tau}^{*'}\boldsymbol{\Gamma}\left(\frac{\lambda v}{k}\frac{\mathbf{x}_1}{\sigma_e^2}+\frac{(r-\lambda)}{k}\frac{\mathbf{x}_2}{(k\sigma_\beta^2+\sigma_e^2)}\right)}{(\frac{U_1}{\sigma_e^2}+\frac{U_2}{(k\sigma_\beta^2+\sigma_e^2)})^{\frac{1}{2}}}\right\}\times$$
$$\left.c^{bk-2}dcd\boldsymbol{\Gamma}\right]\Big/\left[\int_0^\infty exp\left(-\frac{c^2}{2}\right)c^{bk-2}dc\right], \tag{7.3.10}$$

where $O(p)$ is the group of $p\times p$ orthogonal matrices and $d\boldsymbol{\Gamma}$ denotes the uniform distribution on $O(p)$ (see Eaton, 1983, Chapter 7). The second equation in (7.3.10) is obtained by making the transformation $c\rightarrow c(U_1/\sigma_e^2+U_2/(k\sigma_\beta^2+\sigma_e^2))^{1/2}$. Let $\boldsymbol{\epsilon}$ denote the $(v-1)\times 1$ vector $(1,0,0,\ldots,0)'$. Then there exists $(v-1)\times(v-1)$ orthogonal matrices $\boldsymbol{\Gamma}_1$ and $\boldsymbol{\Gamma}_2$ such that

$$\boldsymbol{\tau}^*=\|\boldsymbol{\tau}^*\|\boldsymbol{\Gamma}_1\boldsymbol{\epsilon},\quad \frac{\lambda v}{k}\frac{\mathbf{x}_1}{\sigma_e^2}+\frac{(r-\lambda)}{k}\frac{\mathbf{x}_2}{(k\sigma_\beta^2+\sigma_e^2)}=\left\|\frac{\lambda v}{k}\frac{\mathbf{x}_1}{\sigma_e^2}+\frac{(r-\lambda)}{k}\frac{\mathbf{x}_2}{\sigma_e^2}\right\|\boldsymbol{\Gamma}_2\boldsymbol{\epsilon}. \tag{7.3.11}$$

Using (7.3.11) and the invariance of $d\boldsymbol{\Gamma}$, (7.3.10) simplifies to

$$R^*=\left[exp\left\{-\frac{\boldsymbol{\tau}^{*'}\boldsymbol{\tau}^*}{2}\left(\frac{\lambda v}{k}\frac{1}{\sigma_e^2}+\frac{(r-\lambda)}{k}\frac{1}{(k\sigma_\beta^2+\sigma_e^2)}\right)\right\}\times\right.$$
$$\int_{O(p)}\int_0^\infty exp\left(-\frac{c^2}{2}\right)exp\left\{\frac{c\|\boldsymbol{\tau}^*\|\left\|\frac{\lambda v}{k}\frac{\mathbf{x}_1}{\sigma_e^2}+\frac{(r-\lambda)}{k}\frac{\mathbf{x}_2}{(k\sigma_\beta^2+\sigma_e^2)}\right\|\boldsymbol{\epsilon}'\boldsymbol{\Gamma}\boldsymbol{\epsilon}}{\left(\frac{U_1}{\sigma_e^2}+\frac{U_2}{(k\sigma_\beta^2+\sigma_e^2)}\right)^{1/2}}\right\}\times$$
$$\left.c^{bk-2}dcd\boldsymbol{\Gamma}\right]\Big/\left[\int_0^\infty exp\left(-\frac{c^2}{2}\right)c^{bk-2}dc\right]. \tag{7.3.12}$$

From (7.3.12), it is obvious that R^* is a monotone increasing function of

$$\left\| \frac{\lambda v}{k} \frac{\mathbf{x}_1}{\sigma_e^2} + \frac{(r-\lambda)}{k} \frac{\mathbf{x}_2}{(k\sigma_\beta^2 + \sigma_e^2)} \right\| \Big/ \left(\frac{U_1}{\sigma_e^2} + \frac{U_2}{(k\sigma_\beta^2 + \sigma_e^2)} \right)^{1/2},$$

that is, of $h^*(.)$ in (7.3.9). This completes the proof of Lemma 7.3.3. □

It is clear from Lemma 7.3.3 that for testing H_0: $\boldsymbol{\tau}^* = \mathbf{0}$ in model (7.3.4), a best invariant test does not exist. However, the lemma suggests that a test based on a suitable modification of $h^*(.)$ in (7.3.9) after replacing σ_e^2 and σ_β^2 by estimators may perform well. Such a possibility will be explored later in this section.

The statistic R in (7.3.7), which is a part of the maximal invariant statistic D in (7.3.8), plays a major role in some of the tests that we shall propose below. The reason for this is that the test that rejects H_0: $\boldsymbol{\tau}^* = \mathbf{0}$ for large values of R is a valid and meaningful one-sided test. This is implied by the properties $\mathrm{E}(R|H_0 : \boldsymbol{\tau}^* = \mathbf{0}) = 0$ and $\mathrm{E}(R|H_1 : \boldsymbol{\tau}^* \neq \mathbf{0}) > 0$. The first property follows trivially from the definition of R in (7.3.7) since $\mathbf{x}_1$ and $\mathbf{x}_2$, and hence R, are distributed symmetrically around zero under H_0: $\boldsymbol{\tau}^* = \mathbf{0}$. For a proof of the second property, see Lemma 7.5.2 in Section 7.5. From the definition of R, it is also clear that the null distribution of R is the same as that of the ordinary product moment correlation under independence in samples of sizes $v-1$ from a bivariate normal population with mean zero. Recall that under H_0: $\boldsymbol{\tau}^* = \mathbf{0}$, R is independent of F_1 and F_2 (see Lemma 7.3.2). Thus there are three independent tests, one each based on F_1, F_2, and R, for testing H_0: $\boldsymbol{\tau}^* = \mathbf{0}$ in the model (7.3.4). We shall now explore various methods for combining these tests.

Fisher's (1932) idea of combining independent tests was to suitably combine the P-values of the tests. Let P_1, P_2, and P_3 denote the P-values of the tests that reject H_0: $\boldsymbol{\tau}^* = \mathbf{0}$ for large values of F_1, F_2, and R. Define

$$Z_i = -\ln P_i \ (i = 1,2,3), \ Z = \sum_{i=1}^{2} Z_i, \ Z^* = \sum_{i=1}^{3} Z_i. \tag{7.3.13}$$

Since we reject H_0: $\boldsymbol{\tau}^* = \mathbf{0}$ for small values of the P_i's, large values of Z and Z^* indicate evidence against H_0. Furthermore, using the fact that P_i's have independent uniform distributions under H_0, it can be easily shown that, under H_0, $2Z$ has a chi-squared distribution with 4 degrees of freedom and $2Z^*$ has a chi-squared distribution with 6 degrees of freedom. Z and Z^* are the test statistics resulting from Fisher's idea of combining tests. We thus have the following rejection regions Φ_1 and Φ_2 of the size α tests based on Z and Z^*.

$$\Phi_1 : 2Z > \chi^2_{\alpha,4}, \ \Phi_2 : 2Z^* > \chi^2_{\alpha,6}, \tag{7.3.14}$$

where $\chi^2_{\alpha,m}$ is the $100(1-\alpha)^{th}$ percentile of the chi-squared distribution with m d.f. The test based on Φ_1 combines those based on F_1 and F_2; the one based on Φ_2 combines those based on F_1, F_2, and R.

A drawback of the tests based on Φ_1 and Φ_2 is that they put equal weights to the individual tests. From model (7.3.4), it is intuitively clear that a combined test should attach a smaller weight to F_2, as σ^2_β gets larger. One way of achieving this is by taking a weighted combination of the Z_i's in (7.3.13) with weights that suitably reflect the magnitude of σ^2_β relative to σ^2_e. Since these variances are unknown, one has to use estimated weights. Thus, let

$$\theta^2 = \frac{\sigma^2_e}{k\sigma^2_\beta + \sigma^2_e}, \tag{7.3.15}$$

and let $\hat{\theta}^2$ be an estimator of θ^2 (the estimators that we shall use will be discussed later in this section). Define

$$\begin{aligned} \gamma_1 &= \frac{1}{1+\hat{\theta}^2}, \ \gamma_2 = 1-\gamma_1, \\ \delta_1 &= \frac{1}{(1+\hat{\theta}^2)^2}, \ \delta_2 = \frac{\hat{\theta}^4}{(1+\hat{\theta}^2)^2}, \ \delta_3 = \frac{2\hat{\theta}^2}{(1+\hat{\theta}^2)^2}, \\ W &= \sum_{i=1}^{2} \gamma_i Z_i, \ W^* = \sum_{i=1}^{3} \delta_i Z_i. \end{aligned} \tag{7.3.16}$$

Note that θ^2 will become smaller as σ^2_β gets larger, and in this case, γ_2 is expected to be smaller than γ_1. Thus, in the weighted linear combination W, Z_2 is expected to get a smaller weight compared to Z_1 when σ^2_β is large. The same is true in W^* also. However, the weights in W^* are motivated by $h^*(.)$ in (7.3.9). In the numerator of $h^*(.)$, the coefficients of $\mathbf{x}_1'\mathbf{x}_1$, $\mathbf{x}_2'\mathbf{x}_2$, and $\mathbf{x}_1'\mathbf{x}_2$ (essentially the numerators of F_1, F_2, and R) are proportional to $1/(1+\theta^2)^2$, $\theta^4/(1+\theta^2)^2$, and $2\theta^2/(1+\theta^2)^2$. The weighted combination W in (7.3.16) was first considered by Cohen and Sackrowitz (1989).

Several choices of $\hat{\theta}^2$ are possible for computing the γ_i's and the δ_i's in (7.3.16). Cohen and Sackrowitz (1989) recommended the following choice $\hat{\theta}^2$.

$$\hat{\theta}^2_{(1)} = \min\{U_1/U_2, 1\}. \tag{7.3.17}$$

As pointed out earlier, under H_0: $\boldsymbol{\tau}^* = \mathbf{0}$, unbiased estimators of σ^2_e and $(\sigma^2_e + k\sigma^2_\beta)$ are, respectively, given by $\hat{\sigma}^2_e = U_1/(bk-b)$ and $k\hat{\sigma}^2_\beta + \hat{\sigma}^2_e = U_2/(b-1)$. Noting that since $\theta^2 \leq 1$, a natural estimator of θ^2 is $\hat{\theta}^2_{(2)}$ given by

$$\hat{\theta}^2_{(2)} = \min\left\{\hat{\sigma}^2_e/(k\hat{\sigma}^2_\beta + \hat{\sigma}^2_e), 1\right\}. \tag{7.3.18}$$

Other reasonable choices can be obtained by noting that the powers of the F-tests based on F_1 and F_2 in (7.3.5) are functions of the noncentrality parameters $(\lambda v/k)\boldsymbol{\tau}^{*'}\boldsymbol{\tau}^*/\sigma_e^2$ and $[(r-\lambda)/k]\boldsymbol{\tau}^{*'}\boldsymbol{\tau}^*/(k\sigma_\beta^2+\sigma_e^2)$, respectively, where $\boldsymbol{\tau}^*$ is the quantity in (7.3.4). Hence, instead of (7.3.15), it is natural to consider $\theta^2 = [\sigma_e^2/\lambda v]/[(k\sigma_\beta^2+\sigma_e^2)/(r-\lambda)]$ and choose an appropriate estimator $\hat{\theta}^2$. This leads us to the following choice:

$$\hat{\theta}_{(3)}^2 = \min\left\{\frac{r-\lambda}{\lambda v}\frac{\hat{\sigma}_e^2}{k\hat{\sigma}_\beta^2+\hat{\sigma}_e^2}, \frac{r-\lambda}{\lambda v}\right\}, \quad \gamma_1 = \frac{1}{1+\hat{\theta}_{(3)}^2}, \quad \gamma_2 = 1-\gamma_1, \qquad (7.3.19)$$

where $\hat{\sigma}_e^2$ and $k\hat{\sigma}_\beta^2+\hat{\sigma}_e^2$ are the estimators based on U_1 and U_2, mentioned above. A choice similar to that in (7.3.19) is considered in Jordan and Krishnamoorthy (1995). In our simulation results, we have used the estimator $\hat{\theta}_{(1)}^2$ in (7.3.17) suggested by Cohen and Sackrowitz (1989) and $\gamma_1 = 1/1+\hat{\theta}_{(1)}^2$. We note that the estimators of θ^2 given above are functions of only U_1 and U_2. This fact is crucial in the proof of the following theorem.

Theorem 7.3.1. Let W and W^* be as defined in (7.3.16). Furthermore, let $\hat{\theta}^2$ be a function of U_1 and U_2 in (7.3.7). Then, Φ_3 and Φ_4, the critical regions of the size α tests that reject H_0: $\boldsymbol{\tau}^* = \mathbf{0}$ for large values of W and W^*, respectively, are given by

$$\begin{aligned} \Phi_3 &: \frac{\gamma_1 e^{-W/\gamma_1} - \gamma_2 e^{-W/\gamma_2}}{2\gamma_1 - 1} \le \alpha, \\ \Phi_4 &: \sum_{i=1}^{3} \frac{\delta_i e^{-W^*/\delta_i}}{\prod_{j=1;j\neq i}^{3}(\delta_j - \delta_i)} \le \alpha. \end{aligned} \qquad (7.3.20)$$

Proof. Note that under H_0: $\boldsymbol{\tau}^* = \mathbf{0}$, Z_1, Z_2, and Z_3 are independent exponential random variables, that is, the density of Z_i is e^{-z_i} $(z_i > 0)$, $i = 1, 2, 3$. In order to derive Φ_3, we shall show that the size α test which rejects H_0 for large values of W, conditionally given U_1 and U_2, leads to Φ_3. The rejection region of such a test is $W \ge C$, where C is to be determined so that the size of the test is α, conditionally given U_1 and U_2 (i.e., conditionally given γ_1). Using the fact that under H_0, Z_1 and Z_2 are independent exponential random variables distributed independently of U_1 and U_2 (see Lemma 7.3.2), and hence independently of γ_1, we see that C is to be determined from

$$1 - \int\int_{\{0<\gamma_1 z_1+\gamma_2 z_2<C\}} e^{-(z_1+z_2)} dz_1 dz_2 = \alpha. \qquad (7.3.21)$$

Upon integration, the left-hand side (LHS) of (7.3.21) simplifies to $\gamma_1 e^{-C/\gamma_1} - \gamma_2 e^{-C/\gamma_2}/2\gamma_1 - 1$, which is the same statistic involved in the definition of Φ_3

with W replaced by C. Since the LHS of (7.3.21) is a decreasing function of C, we immediately see that $W \geq C$ is equivalent to Φ_3. Furthermore, since Φ_3 has conditional size α, it also has unconditional size α. In order to derive Φ_4, consider the test which rejects H_0 when $W^* > C^*$, where C^* is determined subject to the condition that the test has conditional size α, conditionally given U_1 and U_2, that is, conditionally given the δ_i's in (7.3.16). Similar to (7.3.21), we see that C^* is obtained from

$$1 - \int\int\int_{0<\sum_{i=1}^{3} \delta_i z_i < C^*} e^{-\sum_{i=1}^{3} z_i} dz_1 dz_2 dz_3 = \alpha. \tag{7.3.22}$$

We now show that the LHS of (7.3.22) simplifies to

$$\sum_{i=1}^{3} \frac{\delta_i^2 e^{-C^*/\delta_i}}{\prod_{j=1;j\neq i}^{3} (\delta_i - \delta_j)}. \tag{7.3.23}$$

For this, consider

$$\begin{aligned}
1 - &\int_{\{0<\sum_{i=1}^{3} \delta_i z_i < C^*\}} e^{-\sum_{i=1}^{3} z_i} dz_1 dz_2 dz_3 \\
&= 1 - \int_0^{C^*/\delta_3} e^{-z_3} \left(\int_{\{0<\sum_{i=1}^{2} \delta_i z_i < C^* - \delta_3 z_3\}} e^{-\sum_{i=1}^{2} z_i} dz_1 dz_2 \right) dz_3 \\
&= 1 - \int_0^{C^*/\delta_3} e^{-z_3} \left(1 - \sum_{i=1}^{2} \frac{\delta_i e^{-(C^* - \delta_3 Z_3)/\delta_i}}{\prod_{j=1;j\neq i}^{2} (\delta_i - \delta_j)} \right) dz_3
\end{aligned}$$

(using a simplification similar to that for the LHS of (7.3.21))

$$\begin{aligned}
&= e^{-C^*/\delta_3} + \sum_{i=1}^{2} \frac{\delta_i^2 e^{-C^*/\delta_i}}{\prod_{j=1;j\neq i}^{3} (\delta_i - \delta_j)} - \sum_{i=1}^{2} e^{-C^*/\delta_3} \frac{\delta_i^2}{\prod_{j=1;j\neq i}^{3} (\delta_i - \delta_j)} \\
&= \sum_{i=1}^{2} \frac{\delta_i^2 e^{-C^*/\delta_i}}{\prod_{j=1;j\neq i}^{3} (\delta_i - \delta_j)} + e^{-C^*/\delta_3} \left(1 - \sum_{i=1}^{2} \frac{\delta_i^2}{\prod_{j=1;j\neq i}^{3} (\delta_i - \delta_j)} \right) \\
&= \sum_{i=1}^{3} \frac{\delta_i^2 e^{-C^*/\delta_i}}{\prod_{j=1;j\neq i}^{3} (\delta_i - \delta_j)}
\end{aligned}$$

$$\left(\text{using } 1 - \sum_{i=1}^{2} \frac{\delta_i^2}{\prod_{j=1;j\neq i}^{3} (\delta_i - \delta_j)} = \frac{\delta_3^2}{(\delta_3 - \delta_1)(\delta_3 - \delta_2)} \right).$$

Thus we have shown that the LHS of (7.3.22) is equal to (7.3.23). Derivation of Φ_4 is now similar to that of Φ_3. This completes the proof of Theorem 7.3.1. □

Note that the test based on the critical region Φ_4 combines the tests based on F_1, F_2, and R. Another way of combining these three tests was suggested by Cohen and Sackrowitz (1989). Their test is obtained by modifying Φ_3 and is given in the following theorem.

Theorem 7.3.2. Consider the critical region Φ_5 given by

$$\Phi_5 : \frac{\gamma_1 e^{-W/\gamma_1} - \gamma_2 e^{-W/\gamma_2}}{2\gamma_1 - 1} \leq \alpha(1 + R), \tag{7.3.24}$$

where R is given in (7.3.7) and α satisfies $0 < \alpha < 1/2$. Then Φ_5 is the critical region of a size α test for testing H_0: $\boldsymbol{\tau}^* = \mathbf{0}$ in model (7.3.4).

Proof. In order to prove the theorem, we shall use the fact that under H_0, R is distributed independently of Z_1, Z_2, and γ_1 (see Lemma 7.3.2) and furthermore, R is distributed symmetrically around zero (under H_0). Thus the theorem will be proved, if we can show that the conditional size of the test, conditionally given R, is $\alpha(1 + R)$. However, this follows from the proof of Theorem 7.3.1 with α replaced by $\alpha(1 + R)$ in the derivation of Φ_3, thus completing the proof. □

As noted in Lemma 7.3.3, for testing H_0: $\boldsymbol{\tau}^* = \mathbf{0}$ in model (7.3.4), the density ratio of a maximal invariant is an increasing function of the quantity $h^*(.)$ in (7.3.9). However, $h^*(.)$ involves the unknown variance components σ_e^2 and σ_β^2. The performance of a test based on an estimator of $h^*(.)$ is certainly worth investigating. Note that $h^*(\mathbf{x}_1, \mathbf{x}_2, s_1^2, s_2^2; \sigma_e^2, \sigma_\beta^2)$, apart from the multiplier $1/\sigma_e^2$, is the same as the quantity $h(.)$ given by

$$h(\mathbf{x}_1, \mathbf{x}_2, s_1^2, s_2^2; \sigma_e^2, \sigma_\beta^2) = \frac{\frac{\lambda^2 v^2}{k^2} \frac{\mathbf{x}_1'\mathbf{x}_1}{\sigma_e^2} + \frac{(r-\lambda)^2}{k^2} \theta^2 \frac{\mathbf{x}_2'\mathbf{x}_2}{(k\sigma_\beta^2 + \sigma_e^2)} + 2\frac{\lambda v (r-\lambda)}{k^2} \frac{\mathbf{x}_1'\mathbf{x}_2}{(k\sigma_\beta^2 + \sigma_e^2)}}{\frac{U_1}{\sigma_e^2} + \frac{U_2}{(k\sigma_\beta^2 + \sigma_e^2)}}. \tag{7.3.25}$$

Replacing σ_e^2 and $\sigma_e^2 + k\sigma_\beta^2$ by the estimators $\hat{\sigma}_e^2 = U_1/(bk - b)$ and $k\hat{\sigma}_\beta^2 + \hat{\sigma}_e^2 = U_2/(b - 1)$ (where U_1 and U_2 are given in (7.3.7)) and using an estimator $\hat{\theta}^2$ for θ^2 (where $\hat{\theta}^2$ is a function of U_1 and U_2), straightforward simplifications yield

$$h(\mathbf{x}_1, \mathbf{x}_2, s_1^2, s_2^2; \hat{\sigma}_e{}^2, \hat{\sigma}_\beta^2)$$
$$= \frac{\frac{\lambda v}{k}(bk - b)G_1 + \frac{(r-\lambda)}{k}(b - 1)\hat{\theta}^2 G_2 + 2\sqrt{\frac{\lambda v}{k}\frac{(r-\lambda)}{k}}\,\hat{\theta} R\sqrt{(bk - b)(b - 1)G_1 G_2}}{(bk - 1)}, \tag{7.3.26}$$

where

$$
\begin{aligned}
G_1 &= \frac{\frac{\lambda v}{k}\mathbf{x}_1'\mathbf{x}_1}{\frac{\lambda v}{k}\mathbf{x}_1'\mathbf{x}_1 + s_1^2} \sim B(\frac{v-1}{2}, \frac{bk-b-v+1}{2}) \\
G_2 &= \frac{\frac{(r-\lambda)}{k}\mathbf{x}_2'\mathbf{x}_2}{\frac{1}{k}[(r-\lambda)\mathbf{x}_2'\mathbf{x}_2 + s_2^2]} \sim B(\frac{v-1}{2}, \frac{b-v}{2})
\end{aligned}
\tag{7.3.27}
$$

under H_0: $\boldsymbol{\tau}^* = \mathbf{0}$ ($B(.,.)$ denotes the beta distribution). Note that since G_1, G_2, and R are quantities taking values in the interval [0,1], the right-hand side of (7.3.26) assumes its maximum value when G_1, G_2, and R are all equal to one, and the maximum value is $[1/(bk-1)][\sqrt{(\lambda v/k)(bk-b)} + \hat{\theta}\sqrt{[(r-\lambda)/k](b-1)}]^2$. Let T^* be the random variable obtained by dividing the right-hand side of (7.3.26) by the above maximum value. Then

$$
T^* = \frac{\frac{\lambda v}{k}(bk-b)G_1 + \frac{(r-\lambda)}{k}(b-1)\hat{\theta}^2 G_2 + 2\sqrt{\frac{\lambda v}{k}\frac{(r-\lambda)}{k}}\,\hat{\theta} R\sqrt{(bk-b)(b-1)G_1 G_2}}{(\sqrt{\frac{\lambda v}{k}(bk-b)} + \hat{\theta}\sqrt{\frac{(r-\lambda)}{k}(b-1)})^2}.
\tag{7.3.28}
$$

Our simulation results (see Table 7.1) show that when b is large, the conditional null distribution of T^*, conditionally given $\hat{\theta}^2$, can be approximated as $B(\alpha_{\hat{\theta}}, \beta_{\hat{\theta}})$, where $\alpha_{\hat{\theta}}$ and $\beta_{\hat{\theta}}$ are obtained by equating the first two moments of T^*, given $\hat{\theta}^2$, to the first two moments of $B(\alpha_{\hat{\theta}}, \beta_{\hat{\theta}})$. This gives

$$
\begin{aligned}
E(T^*|\hat{\theta}^2) &= \frac{\frac{(v-1)}{k}\{\lambda v + \hat{\theta}^2(r-\lambda)\}}{\left[\sqrt{\frac{\lambda v}{k}(bk-b)} + \hat{\theta}\sqrt{\frac{(r-\lambda)}{k}(b-1)}\right]^2} = \frac{\alpha_{\hat{\theta}}}{\alpha_{\hat{\theta}} + \beta_{\hat{\theta}}} \\
E(T^{*2}|\hat{\theta}^2) &= \frac{\frac{(v^2-1)}{k^2}\left\{\frac{\lambda^2 v^2(bk-b)}{bk-b+2} + \hat{\theta}^4\frac{(r-\lambda)^2(b-1)}{b+1} + 2\hat{\theta}^2\lambda v(r-\lambda)\right\}}{\left[\sqrt{\frac{\lambda v}{k}(bk-b)} + \hat{\theta}\sqrt{\frac{(r-\lambda)}{k}(b-1)}\right]^4} \\
&= \frac{\alpha_{\hat{\theta}}(\alpha_{\hat{\theta}}+1)}{(\alpha_{\hat{\theta}} + \beta_{\hat{\theta}})(\alpha_{\hat{\theta}} + \beta_{\hat{\theta}} + 1)}.
\end{aligned}
\tag{7.3.29}
$$

The reason for modifying (7.3.26) in order to arrive at (7.3.28) is to use this beta approximation. From (7.3.29), we get

$$
\begin{aligned}
\alpha_{\hat{\theta}} &= E(T^*|\hat{\theta}^2)\frac{E(T^*|\hat{\theta}^2) - E(T^{*2}|\hat{\theta}^2)}{E(T^{*2}|\hat{\theta}^2) - E^2(T^*|\hat{\theta}^2)} \quad \text{and} \\
\beta_{\hat{\theta}} &= [1 - E(T^*|\hat{\theta}^2)]\frac{E(T^*|\hat{\theta}^2) - E(T^{*2}|\hat{\theta}^2)}{E(T^{*2}|\hat{\theta}^2) - E^2(T^*|\hat{\theta}^2)}.
\end{aligned}
\tag{7.3.30}
$$

Table 7.1. Simulated Powers (Based on 100,000 Simulations) of the Tests $\Phi_1 - \Phi_7$ for Testing the Equality of the Treatment Effects in a BIBD With $v = 10$, $b = 30$, $k = 3$ ($\alpha = 0.05$ and $\sigma_e^2 = 1$)

$(\sigma_e^2 + k\sigma_\beta^2)$	$\tau^{*'}\tau^*$	Φ_1	Φ_2	Φ_3	Φ_4	Φ_5	Φ_6	Φ_7
	0.0	0.0504	0.0492	0.0504	0.0489	0.0506	0.0497	0.0509
	0.1	0.0659	0.0715	0.0659	0.0691	0.0688	0.0742	0.0696
1	0.4	0.1274	0.1580	0.1274	0.1506	0.1434	0.1752	0.1392
	1.0	0.3033	0.3960	0.3033	0.3699	0.3482	0.4326	0.3218
	4.0	0.9497	0.9828	0.9497	0.9762	0.9690	0.9860	0.9400
	0.0	0.0504	0.0492	0.0502	0.0491	0.0502	0.0495	0.0509
	0.1	0.0641	0.0670	0.0648	0.0648	0.0660	0.0691	0.0696
2	0.4	0.1174	0.1344	0.1201	0.1254	0.1309	0.1453	0.1392
	1.0	0.2712	0.3227	0.2765	0.2886	0.3075	0.3426	0.3218
	4.0	0.9185	0.9526	0.9205	0.9278	0.9431	0.9831	0.9400
	0.0	0.0504	0.0492	0.0505	0.0499	0.0502	0.0497	0.0509
	0.1	0.0628	0.0645	0.0670	0.0642	0.0677	0.0676	0.0696
4	0.4	0.1127	0.1200	0.1298	0.1187	0.1367	0.1346	0.1392
	1.0	0.2547	0.2776	0.2952	0.2674	0.3167	0.3064	0.3218
	4.0	0.8969	0.9174	0.9219	0.8943	0.9388	0.8968	0.9400

Source: T. Mathew, B. K. Sinha, and L. Zhou (1993). Reproduced with permission of the American Statistical Association.

Hence an approximate optimum invariant size α test for H_0: $\boldsymbol{\tau}^* = \mathbf{0}$ in model (7.3.4) has the rejection region Φ_6 given by

$$\Phi_6 : \quad T^* > B_\alpha(\alpha_{\hat{\theta}}, \beta_{\hat{\theta}}), \tag{7.3.31}$$

where $B_\alpha(\alpha_{\hat{\theta}}, \beta_{\hat{\theta}})$ is the $100(1-\alpha)^{th}$ percentile of the beta distribution $B(\alpha_{\hat{\theta}}, \beta_{\hat{\theta}})$.

Table 7.1 gives the powers of the test based on the critical regions Φ_1–Φ_6, along with the power of the intra-block F-test (i.e., the test based on F_1 in (7.3.5)). Analogous to the critical regions in (7.3.14), the intra-block F-test corresponds to the critical region Φ_7 given by

$$\Phi_7 : 2Z_1 > \chi^2_{\alpha,2}, \tag{7.3.32}$$

where Z_1 is defined in (7.3.13). The critical regions Φ_3–Φ_6 involve $\hat{\theta}^2$, an estimator of θ^2 in (7.3.15). In Table 7.1, we have reported the powers of the tests that use $\hat{\theta}^2_{(1)}$ given in (7.3.17), as done by Cohen and Sackrowitz (1989). The simulation results in Table 7.1 are for a BIBD with $v = 10$, $b = 30$, $r = 9$, $k = 3$, and $\lambda = 2$ and the problem is to test the equality of the treatment effects. All the simulations were carried out using SAS (1989) for the values $\alpha = 0.05$ and $\sigma_e^2 = 1$. The normal and chi-squared variates were generated using the SAS functions RANNOR and RANGAM, respectively. From the table, we see that the simulated sizes are close to 0.05 for all the tests. This, in particular, shows that the beta approximation used to arrive at the test Φ_6 is quite accurate. The simulations clearly indicate that tests Φ_5 and Φ_6 are quite superior to the other tests. Moreover, Φ_6 has some advantage over Φ_5 if the ratio $(k\sigma_\beta^2 + \sigma_e^2)/\sigma_e^2$ is not too large and the converse is true for large values of $(k\sigma_\beta^2 + \sigma_e^2)/\sigma_e^2$.

7.3.2. Combined Tests When $b = v$

In a BIBD with $b = v$ (i.e., a symmetrical BIBD), the inter-block error sum of squares s_2^2 in (7.3.4) is nonexistent and consequently, the inter-block F-ratio F_2 in (7.3.5) is not available. The canonical form (7.3.4) now reduces to

$$\begin{aligned} \mathbf{x}_1 &\sim N\left(\boldsymbol{\tau}^*, \frac{k}{\lambda v}\sigma_e^2 \boldsymbol{I}_{v-1}\right), \\ \mathbf{x}_2 &\sim N\left[\boldsymbol{\tau}^*, \frac{k}{(r-\lambda)}(k\sigma_\beta^2 + \sigma_e^2)\boldsymbol{I}_{v-1}\right], \quad s_1^2 \sim \sigma_e^2 \chi^2_{vk-2v+1}, \end{aligned} \tag{7.3.33}$$

where the above quantities are as defined in (7.3.4), and we have used the fact $b = v$ in order to express the degrees of freedom associated with s_1^2. For testing H_0: $\boldsymbol{\tau}^* = \mathbf{0}$ in (7.3.33), we shall now derive and compare several tests. This is similar to what was done in Section 7.3.1. In this context, an interesting observation is that even though F_2 is nonexistent and an estimator of $(k\sigma_\beta^2 + \sigma_e^2)$ is not available, $\mathbf{x}_2$ can still be used to test H_0 through the quantity R given in (7.3.7). The choice of the following three critical regions should be obvious from our discussion in Section 7.3.1.

$$\begin{aligned} \Psi_1 &: \quad 2Z_1 \geq \chi^2_{\alpha,2} \\ \Psi_2 &: 2(Z_1 + Z_3) \geq \chi^2_{\alpha,4} \\ \Psi_3 &: e^{-Z_1} \leq \alpha(1 + R), \end{aligned} \tag{7.3.34}$$

where Z_1 and Z_3 are as given in (7.3.13). (Ψ_3 is also derived in Zhang, 1992). It is clear that the group $\mathcal{G}$ (see (7.3.6)) is still applicable for testing H_0: $\boldsymbol{\tau}^* = \mathbf{0}$ in (7.3.33). An approximately optimum invariant test can be obtained similar to Φ_6 in (7.3.31). Analogous to T^* given in (7.3.28), define

$$T^{**} = \frac{\frac{\lambda v}{k}(bk-b)G_1 + \frac{(r-\lambda)}{k}(b-1)\hat{\theta}^2 + 2\sqrt{\frac{\lambda v}{k}\frac{(r-\lambda)}{k}}\,\hat{\theta}R\sqrt{(bk-b)(b-1)G_1}}{\left[\sqrt{\frac{\lambda v}{k}(bk-b)} + \hat{\theta}\sqrt{\frac{(r-\lambda)}{k}(b-1)}\right]^2}, \tag{7.3.35}$$

where G_1 is as given in (7.3.27) and

$$\hat{\theta}^2 = \min\left\{\frac{U_1}{\mathbf{x}_2'\mathbf{x}_2}, 1\right\}. \tag{7.3.36}$$

Table 7.2. Simulated Powers (Based on 100,000 Simulations) of the Tests Ψ_1–Ψ_4 for Testing the Equality of the Treatment Effects in a Symmetrical BIBD With $v = b = 16$ and $k = 6$ ($\alpha = 0.05$ and $\sigma_e^2 = 1$)

$(\sigma_e^2 + k\sigma_\beta^2)$	$\tau^{*'}\tau^*$	Ψ_1	Ψ_2	Ψ_3	Ψ_4
	0.0	0.0509	0.0502	0.0509	0.0496
	0.1	0.0623	0.0613	0.0628	0.0609
1	0.4	0.0997	0.1008	0.1030	0.1001
	1.0	0.1984	0.2070	0.2123	0.2030
	4.0	0.7717	0.8073	0.8043	0.7723
	0.0	0.0509	0.0502	0.0509	0.0496
	0.1	0.0623	0.0602	0.0625	0.0596
2	0.4	0.0997	0.0953	0.1019	0.0927
	1.0	0.1984	0.1891	0.2081	0.1783
	4.0	0.7717	0.7617	0.7949	0.6799
	0.0	0.0509	0.0502	0.0509	0.0497
	0.1	0.0623	0.0595	0.0623	0.0587
4	0.4	0.0997	0.0914	0.1011	0.0879
	1.0	0.1984	0.1772	0.2050	0.1621
	4.0	0.7717	0.7269	0.7869	0.6031

similar to the choice in (7.3.17) . Once again, the conditional null distribution of T^{**}, conditionally given $\hat{\theta}^2$, can be approximated by the distribution $B(\alpha_{\hat{\theta}}, \beta_{\hat{\theta}})$, where $\alpha_{\hat{\theta}}$ and $\beta_{\hat{\theta}}$ are obtained from (7.3.30) and (7.3.29) with $b - v = 0$. This leads to the following rejection region :

$$\Psi_4 : T^{**} > B_\alpha(\alpha_{\hat{\theta}}, \beta_{\hat{\theta}}). \tag{7.3.37}$$

Simulated powers of the four tests Ψ_1–Ψ_4 described above are given in Table 7.2 for a symmetrical BIBD with $b = v = 16$, and $k = 6$. The simulations were carried out using $\alpha = 0.05$ and $\sigma_e^2 = 1$. The numerical results in Table 7.2 indicate that we should prefer the test Ψ_3 in this testing problem.

7.3.3. A Numerical Example

We shall now apply the results in Section 7.3.1 to a numerical example taken from Lentner and Bishop (1986, pp. 428–429). The example deals with comparing the effects of six diets on the weight gains of domestic rabbits using litters as blocks. The litter effects are assumed to be random. The data given in Table 7.3 are the weight gains (in ounces) for rabbits from 10 litters. In the table, the six treatments (i.e., the six diets) are denoted by t_i $(i = 1, 2, \ldots, 6)$ and the weight gains are given within brackets. Note that the design is a BIBD with $b = 10$, $v = 6$, $r = 5$, $k = 3$, and $\lambda = 2$.

Since we are recommending the tests Φ_5 and Φ_6 based on the simulation results in Table 7.1, we shall apply these tests to the data in Table 7.3.

Table 7.3. Weight Gains (in Ounces) of Domestic Rabbits from 10 Litters Based on Six Diets t_i ($i = 1, 2, \ldots, 6$)

Litters	Treatments and observations		
1	t_6 (42.2)	t_2 (32.6)	t_3 (35.2)
2	t_3 (40.9)	t_1 (40.1)	t_2 (38.1)
3	t_3 (34.6)	t_6 (34.3)	t_4 (37.5)
4	t_1 (44.9)	t_5 (40.8)	t_3 (43.9)
5	t_5 (32.0)	t_3 (40.9)	t_4 (37.3)
6	t_2 (37.3)	t_6 (42.8)	t_5 (40.5)
7	t_4 (37.9)	t_1 (45.2)	t_2 (40.6)
8	t_1 (44.0)	t_5 (38.5)	t_6 (51.9)
9	t_4 (27.5)	t_2 (30.6)	t_5 (20.6)
10	t_6 (41.7)	t_4 (42.3)	t_1 (37.3)

Source: M. Lentner and T. Bishop (1986). Reproduced with permission of Valley Book Company, Blacksburg, Virginia.

Direct computations using the expressions in Section 7.3.1 give (see equations (7.3.5), (7.3.7), (7.3.13), (7.3.16), and (7.3.17)) $F_1 = 3.14$, $F_2 = 0.3172$, $P_1 = 0.04$, $P_2 = 0.70$, $Z_1 = 3.2189$, $Z_2 = 0.3425$, $U_1 = 309.46$, $U_2 = 2796.90$, $R = 0.25$, $\hat{\theta}^2_{(1)} = \min\{U_1/U_2, 1\} = 0.1106$, $\gamma_1 = 1/(1 + \hat{\theta}^2_{(1)}) = 0.90$, $\gamma_2 = 1\text{-}\gamma_1 = 0.10$, and $W = \gamma_1 Z_1 + \gamma_2 Z_2 = 2.93$. Furthermore, from (7.3.27)–(7.3.30), $G_1 = 0.5125$, $G_2 = 0.2839$, $T^* = 0.4349$, $\mathrm{E}(T^* | \hat{\theta}^2_{(1)}) = 0.2079$, $\mathrm{E}(T^{*^2} | \hat{\theta}^2_{(1)}) = 0.0525$, $\alpha_{\hat{\theta}} = 3.4906$, and $\beta_{\hat{\theta}} = 13.2968$. From (7.3.24), the rejection region Φ_5 simplifies to

$$\Phi_5 : 0.0433 \leq \alpha \times 1.25,$$

which holds for $\alpha = 0.05$. Also, using the values $\alpha_{\hat{\theta}} = 3.4906$ and $\beta_{\hat{\theta}} = 13.2968$ given above, we can get the value of $B_{0.05}(\alpha_{\hat{\theta}}, \beta_{\hat{\theta}})$, namely, the 95th percentile of the beta distribution with parameters $\alpha_{\hat{\theta}}$ and $\beta_{\hat{\theta}}$. This turns out to be $B_{0.05}(3.4906, 13.2968) = 0.3846$. From (7.3.31), the rejection region Φ_7, for $\alpha = 0.05$, is given by

$$\Phi_7 : T^* \geq 0.3846,$$

which holds since $T^* = 0.4349$. Thus, for $\alpha = 0.05$, the tests based on Φ_5 and Φ_7 both reject the null hypothesis of equality of the six treatment effects. Thus we conclude that there are significant differences among the six diets. Cohen and Sackrowitz (1989) have discussed the same numerical example and have applied the test Φ_5.

7.4. BIBD WITH RANDOM EFFECTS

This section deals with the problem of testing the significance of the treatment variance component in a BIBD with both block and treatment effects random. We then have model (7.2.2), where $\boldsymbol{\beta}$ and $\mathbf{e}$ satisfy (7.2.3) and $\boldsymbol{\tau}$

satisfies the usual assumption

$$\boldsymbol{\tau} \sim N(\mathbf{0}, \sigma_\tau^2 \boldsymbol{I}_v), \tag{7.4.1}$$

and $\boldsymbol{\tau}$ is distributed independently of $\boldsymbol{\beta}$ and $\mathbf{e}$. The problem is to test H_0: $\sigma_\tau^2 = 0$. When $\boldsymbol{\beta}$ is vector of fixed effects, the usual intra-block F-test can be used to test H_0: $\sigma_\tau^2 = 0$, and this test is also UMPI (see Theorem 6.2.2 in Chapter 6). When the block effects are random, which is our setup in this section, the intra-block F-test (based on F_1 in (7.3.5)) is still valid for testing H_0: $\sigma_\tau^2 = 0$. This is evident from the results in Section 4.3.1. It is easily verified that under (7.2.3) and (7.4.1), the inter-block F-ratio also provides a valid test for testing H_0: $\sigma_\tau^2 = 0$. This essentially follows from the fact that model (7.2.10) involves only the two variance components σ_τ^2 and $k(k\sigma_\beta^2 + \sigma_e^2)$, similar to (6.2.18) in Chapter 6, and the inter-block F-ratio F_2 in (7.3.5) is simply the F-ratio (6.2.18) derived in Chapter 6. Our problem is to obtain a combined test for testing H_0: $\sigma_\tau^2 = 0$. Note that $\mathbf{q}_1$ and $\mathbf{q}_2$ in (7.2.5) and (7.2.13) now have a joint normal distribution with

$$\begin{aligned} &\mathbf{q}_1 \sim N(\mathbf{0}, \sigma_e^2 \boldsymbol{C}_1 + \sigma_\tau^2 \boldsymbol{C}_1^2),\ \mathbf{q}_2 \sim N(\mathbf{0}, k(k\sigma_\beta^2 + \sigma_e^2)\boldsymbol{C}_2 + \sigma_\tau^2 \boldsymbol{C}_2^2), \\ &Cov(\mathbf{q}_1, \mathbf{q}_2) = \sigma_\tau^2 \boldsymbol{C}_1 \boldsymbol{C}_2, \end{aligned} \tag{7.4.2}$$

where $\boldsymbol{C}_1$ and $\boldsymbol{C}_2$ are defined in (7.2.6) and (7.2.13), respectively. Also, s_1^2 and s_2^2 have the same distributions specified in (7.2.8) and (7.2.15) and are distributed independently of $\mathbf{q}_1$ and $\mathbf{q}_2$ in (7.4.2). Defining $\mathbf{x}_1$ and $\mathbf{x}_2$ as in (7.3.3), (7.4.1) and (7.4.2) give the following canonical form:

$$\begin{aligned} \begin{pmatrix} \mathbf{x}_1 \\ \mathbf{x}_2 \end{pmatrix} \sim N(\mathbf{0}, \boldsymbol{V}),\ \boldsymbol{V} &= \textbf{diag}\left[\frac{k}{\lambda v}\sigma_e^2 \boldsymbol{I}_{v-1}, \frac{k}{(r-\lambda)}(k\sigma_\beta^2 + \sigma_e^2)\boldsymbol{I}_{v-1}\right] \\ &\quad + \sigma_\tau^2(\boldsymbol{J}_2 \otimes \boldsymbol{I}_{v-1}), \end{aligned} \tag{7.4.3}$$

where $\boldsymbol{J}_2$ is a 2×2 matrix of ones, and $\otimes$ denotes Kronecker product (see Chapter 2). In order to develop tests for testing H_0: $\sigma_\tau^2 = 0$, the canonical quantities that we shall work with are $\mathbf{x}_1$ and $\mathbf{x}_2$ in (7.4.3) along with s_1^2 and s_2^2 having the distributions specified in (7.2.8) and (7.2.15). Obviously, the F-ratios F_1 and F_2 can be expressed in terms of these quantities; see (7.3.5). Note also that unlike the setup in Section 7.3, $\mathbf{x}_1$ and $\mathbf{x}_2$ are not independently distributed unless H_0: $\sigma_\tau^2 = 0$ is true. Thus the tests based on F_1 and F_2 are independent only under H_0: $\sigma_\tau^2 = 0$. Recall that most of the combined tests proposed in Section 7.3 combine the tests based on F_1 and F_2 along with a test based on the statistic R defined in (7.3.7).

We shall first consider the case $b > v$ so that the F-ratios F_1 and F_2 are available for testing H_0: $\sigma_\tau^2 = 0$. In terms of the canonical variables mentioned above, the testing problem is clearly invariant under the transformation $(\mathbf{x}_1, \mathbf{x}_2, s_1^2, s_2^2) \longrightarrow (c\mathbf{x}_1, c\mathbf{x}_2, c^2 s_1^2, c^2 s_2^2)$, where c is a positive scalar. The following lemma shows that the density ratio, say R^{**}, of a maximal invariant, is once again a monotone increasing function of $h^*(.)$ given in (7.3.9).

Lemma 7.4.1. Consider the testing problem H_0: $\sigma_\tau^2 = 0$ in model (7.4.3) and let S be a maximal invariant under the action of the group $(\mathbf{x}_1, \mathbf{x}_2, s_1^2, s_2^2) \longrightarrow (c\mathbf{x}_1, c\mathbf{x}_2, c^2 s_1^2, c^2 s_2^2)$, where c is a positive scalar. Then the density ratio of S is a monotone increasing function of $h^*(.)$ given in (7.3.9).

Proof. Let $Q_{H_1}^S(s)$ and $Q_{H_0}^S(s)$ respectively denote the distribution of S under H_1: $\sigma_\tau^2 > 0$ and H_0: $\sigma_\tau^2 = 0$. Applying the representation theorem due to Wijsman (see Appendix 1.1 in Chapter 1), the ratio $R^{**} = dQ_{H_1}^S(s)/dQ_{H_0}^S(s)$ can be evaluated as

$$R^{**} = \frac{\int_0^\infty exp\left[-\frac{c^2}{2}\left\{(\mathbf{x}_1' : \mathbf{x}_2')\boldsymbol{V}^{-1}\begin{pmatrix}\mathbf{x}_1\\ \mathbf{x}_2\end{pmatrix} + \frac{s_1^2}{\sigma_e^2} + \frac{s_2^2}{k(k\sigma_\beta^2 + \sigma_e^2)}\right\}\right] c^{bk-2} dc}{\int_0^\infty exp\left[-\frac{c^2}{2}\left\{\frac{U_1}{\sigma_e^2} + \frac{U_2}{(k\sigma_\beta^2 + \sigma_e^2)}\right\}\right] c^{bk-2} dc}, \tag{7.4.4}$$

where $\boldsymbol{V}$ is given in (7.4.3) and U_1 and U_2 are given in (7.3.7). In order to simplify (7.4.4), let

$$\eta = \frac{\sigma_\tau^4}{1 + \sigma_\tau^2[\frac{\lambda v}{k}\frac{1}{\sigma_e^2} + \frac{(r-\lambda)}{k}\frac{1}{(k\sigma_\beta^2+\sigma_e^2)}]}. \tag{7.4.5}$$

It is readily verified that

$$\begin{aligned}\boldsymbol{V}^{-1} &= \Bigg[\mathbf{diag}\left(\frac{\lambda v}{k}\frac{1}{\sigma_e^2}, \frac{(r-\lambda)}{k}\frac{1}{(k\sigma_\beta^2 + \sigma_e^2)}\right) \\ &\quad -\eta\begin{pmatrix}\frac{\lambda v}{k}\frac{1}{\sigma_e^2}\\ \frac{(r-\lambda)}{k}\frac{1}{(k\sigma_\beta^2+\sigma_e^2)}\end{pmatrix}\left(\frac{\lambda v}{k}\frac{1}{\sigma_e^2}, \frac{(r-\lambda)}{k}\frac{1}{(k\sigma_\beta^2 + \sigma_e^2)}\right)\Bigg] \otimes \boldsymbol{I}_{v-1}.\end{aligned} \tag{7.4.6}$$

Furthermore, our testing problem is equivalent to

$$H_0: \ \eta = 0 \text{ vs } H_1: \ \eta > 0. \tag{7.4.7}$$

Using (7.4.6), (7.4.4) simplifies to

$$\begin{aligned}R^{**} &= \left(\int_0^\infty exp\left[-\frac{c^2}{2}\left\{\frac{U_1}{\sigma_e^2} + \frac{U_2}{(k\sigma_\beta^2 + \sigma_e^2)} - \eta\|\frac{\lambda v}{k}\frac{\mathbf{x}_1}{\sigma_e^2} + \frac{(r-\lambda)}{k}\frac{\mathbf{x}_2}{(k\sigma_\beta^2 + \sigma_e^2)}\|^2\right\}\right]\right. \\ &\quad \left.\times c^{bk-2}dc\right)\Bigg/\left(\int_0^\infty exp\left[-\frac{c^2}{2}\left\{\frac{U_1}{\sigma_e^2} + \frac{U_2}{(k\sigma_\beta^2 + \sigma_e^2)}\right\}\right] c^{bk-2}dc\right) \qquad (7.4.8) \\ &= \int_0^\infty exp(-\frac{c^2}{2})\exp\{c^2\eta h^*(.)\}c^{bk-2}dc\Big/\int_0^\infty exp\{-\frac{c^2}{2}\}c^{bk-2}dc.\end{aligned}$$

Table 7.4. Simulated Powers (Based on 100,000 Simulations) of the Tests Φ_1–Φ_7 for Testing H_0: $\sigma_\tau^2 = 0$ in a BIBD With $v = 10$, $b = 30$, $k = 3$ ($\alpha = 0.05$ and $\sigma_e^2 = 1$)

$(\sigma_e^2 + k\sigma_\beta^2)$	σ_τ^2	Φ_1	Φ_2	Φ_3	Φ_4	Φ_5	Φ_6	Φ_7
	0.0	0.0504	0.0492	0.0504	0.0489	0.0506	0.0497	0.0509
	0.1	0.0893	0.3027	0.0893	0.4013	0.1242	0.1781	0.2283
1	0.4	0.4440	0.9072	0.4440	0.9566	0.5364	0.7466	0.6244
	1.0	0.7946	0.9288	0.7946	0.9434	0.8407	0.9327	0.8369
	4.0	0.9611	0.9786	0.9611	0.9731	0.9707	0.9818	0.9415
	0.0	0.0504	0.0492	0.0502	0.0491	0.0502	0.0495	0.0509
	0.1	0.1181	0.1961	0.1211	0.2174	0.1453	0.1923	0.2283
2	0.4	0.4727	0.6895	0.4742	0.7220	0.5413	0.6680	0.6244
	1.0	0.7837	0.8915	0.7843	0.8965	0.8270	0.9003	0.8369
	4.0	0.9551	0.9738	0.9552	0.9674	0.9660	0.9772	0.9415
	0.0	0.0504	0.0492	0.0505	0.0499	0.0502	0.0497	0.0509
	0.1	0.1415	0.1796	0.1766	0.1941	0.1953	0.2047	0.2283
4	0.4	0.4919	0.5973	0.5264	0.5993	0.5726	0.6100	0.6244
	1.0	0.7748	0.8492	0.7847	0.8394	0.8200	0.8555	0.8369
	4.0	0.9463	0.9659	0.9465	0.9570	0.9583	0.9679	0.9415

Source: T. Mathew, B. K. Sinha, and L. Zhou (1993). Reproduced with permission of the American Statistical Association.

From (7.4.8), it is obvious that R^{**} is a monotone increasing function of $h^*(.)$ given in (7.3.9), thus completing the proof of the lemma. □

Note that under $H_0 : \sigma_\tau^2 = 0$, model (7.4.3) reduces to model (7.3.4) under $H_0 : \boldsymbol{\tau}^* = \mathbf{0}$. Thus the tests with the rejection regions Φ_1–Φ_7 in Section 7.3.1 can be used for testing $H_0 : \sigma_\tau^2 = 0$ in model (7.4.3). A similar argument is also valid for the case $b = v$, and in this case, the tests based on the critical regions Ψ_1–Ψ_4 in Section 7.3.2 can be used to test $H_0 : \sigma_\tau^2 = 0$.

In Table 7.4, we give the simulated powers of the tests Φ_1–Φ_7 for testing $H_0 : \sigma_\tau^2 = 0$ in a BIBD with $v = 10$, $b = 30$, $r = 9$, $k = 3$, and $\lambda = 2$ (i.e., the BIBD considered in Table 7.1). For constructing the critical regions Φ_3–Φ_6, we have once again used $\hat{\theta}_{(1)}^2$ given in (7.3.17). The test based on Φ_4 appears to have a definite edge over the others when σ_β^2 is not very large. For large values of σ_β^2, the test based on Φ_7 dominates the others in terms of power.

7.5. GENERAL INCOMPLETE BLOCK DESIGNS

The results in Sections 7.3 and 7.4 were derived exclusively in the setup of a BIBD. When the block effects are random, the canonical forms (7.3.4) and (7.4.3) correspond to BIBDs with fixed treatment effects and random treatment effects, respectively. For a general incomplete block design having equal block sizes, random block effects, and fixed treatment effects, in order to test the equality of the treatment effects, the relevant quantities to consider are $\mathbf{q}_1$, s_1^2, $\mathbf{q}_2$, and s_2^2 given in (7.2.7), (7.2.8), (7.2.14), and (7.2.15), respectively. The canonical form (7.3.4) and the group, whose action is given in (7.3.6), are applicable only to BIBDs, and not to a general incomplete block design. This will also be obvious from the canonical form (7.5.3) given below. Thus, for a general incomplete block design, even an approximately optimum invariant test, similar to the one based on Φ_6 in (7.3.31), does not exist. However, combined tests based on $\mathbf{q}_1$, s_1^2, $\mathbf{q}_2$, and s_2^2 can still be obtained, as will be seen in this section. Such tests can be obtained by suitably generalizing the rejection regions Φ_1–Φ_5 in Section 7.3.1 (for the case $b > v$) and the rejection regions Ψ_1–Ψ_3 in Section 7.3.2 (for the case $b = v$). We shall consider only the situation of fixed treatment effects and the problem of testing the equality of the treatment effects.

7.5.1. The Combined Test

Let $\boldsymbol{C}_1$, $\boldsymbol{C}_2$, $\mathbf{q}_1$, s_1^2, $\mathbf{q}_2$, and s_2^2 be as given in (7.2.6), (7.2.13), (7.2.7), (7.2.8), (7.2.14), and (7.2.15), respectively. We shall assume that $\text{rank}(\boldsymbol{C}_1) = \text{rank}(\boldsymbol{C}_2) = v - 1$ so that the equality of the treatment effects can be tested using intra-block as well as inter-block information. The above rank condition implies $b \geq v$. Let $\boldsymbol{Q}_1$ and $\boldsymbol{Q}_2$ be $v \times (v-1)$ matrices satisfying $\boldsymbol{Q}_1'\boldsymbol{C}_1\boldsymbol{Q}_1 = \boldsymbol{I}_{v-1}$ and $\boldsymbol{Q}_2'\boldsymbol{C}_2\boldsymbol{Q}_2 = \boldsymbol{I}_{v-1}$. Then, from (7.2.7) and (7.2.14),

$$\boldsymbol{Q}_1'\mathbf{q}_1 \sim N(\boldsymbol{Q}_1'\boldsymbol{C}_1\boldsymbol{O}_1\boldsymbol{\tau}^*, \sigma_e^2\boldsymbol{I}_{v-1}),\ \boldsymbol{Q}_2'\mathbf{q}_2 \sim N[\boldsymbol{Q}_2'\boldsymbol{C}_2\boldsymbol{O}_1\boldsymbol{\tau}^*, k(k\sigma_\beta^2 + \sigma_e^2)\boldsymbol{I}_{v-1}], \tag{7.5.1}$$

where, $\boldsymbol{\tau}^*$ and $\boldsymbol{O}_1$ are as given in (7.3.3). Let

$$\mathbf{x}_1 = \boldsymbol{Q}_1'\mathbf{q}_1,\ \mathbf{x}_2 = \boldsymbol{Q}_2'\mathbf{q}_2,\ \boldsymbol{A}_1 = \boldsymbol{Q}_1'\boldsymbol{C}_1\boldsymbol{O}_1,\ \boldsymbol{A}_2 = \boldsymbol{Q}_2'\boldsymbol{C}_2\boldsymbol{O}_1. \tag{7.5.2}$$

Then, from (7.5.1),

$$\begin{aligned} &\mathbf{x}_1 \sim N(\boldsymbol{A}_1\boldsymbol{\tau}^*, \sigma_e^2\boldsymbol{I}_{v-1}),\ \mathbf{x}_2 \sim N(\boldsymbol{A}_2\boldsymbol{\tau}^*, k[k\sigma_\beta^2 + \sigma_e^2)\boldsymbol{I}_{v-1}] \\ &s_1^2 \sim \sigma_e^2\chi^2_{bk-b-v+1},\ s_2^2 \sim k(k\sigma_\beta^2 + \sigma_e^2)\chi^2_{b-v}, \end{aligned} \tag{7.5.3}$$

where s_1^2 and s_2^2 are the same quantities occurring in (7.2.8) and (7.2.15). For testing the equality of the treatment effects (i.e., the hypothesis H_0: $\boldsymbol{\tau}^* = \mathbf{0}$) in a block design having equal block sizes, fixed treatment effects, and random block effects, (7.5.3) is the canonical form, under the usual normality and independence assumptions on the random quantities in the model. The F-ratios F_1 and F_2 in (7.2.9) and (7.2.16) can be expressed in terms of the quantities in (7.5.3) as

$$F_1 = \frac{\mathbf{x}_1'\mathbf{x}_1/(v-1)}{s_1^2/(bk-b-v+1)},\quad F_2 = \frac{\mathbf{x}_2'\mathbf{x}_2/(v-1)}{s_2^2/(b-v)}. \tag{7.5.4}$$

In order to generalize the critical region Φ_5 (given in Section 7.3.1) to the present context, let P_1 and P_2 denote the observed significance levels of the tests that reject H_0: $\boldsymbol{\tau}^* = \mathbf{0}$ in (7.5.3) for large values of F_1 and F_2, respectively. Define

$$Z_i = -\ln P_i,\ \ U_1 = \mathbf{x}_1'\mathbf{x}_1 + s_1^2,\ \text{ and } U_2 = \frac{1}{k}[\mathbf{x}_2'\mathbf{x}_2 + s_2^2]. \tag{7.5.5}$$

Furthermore, let $\boldsymbol{M}$ be a $(v-1)\times(v-1)$ nonsingular matrix satisfying

$$\boldsymbol{M}'\boldsymbol{A}_1'\boldsymbol{A}_1\boldsymbol{M} = \boldsymbol{I}_{v-1},\ \ \boldsymbol{M}'\boldsymbol{A}_2'\boldsymbol{A}_2\boldsymbol{M} = \boldsymbol{\Lambda} = \mathbf{diag}(\lambda_1,\lambda_2,\cdots,\lambda_{v-1}), \tag{7.5.6}$$

where $\lambda_i > 0$ $(i = 1,2,\ldots,v-1)$, and $\boldsymbol{A}_1$ and $\boldsymbol{A}_2$ are the matrices occurring in (7.5.3). Now define

$$\xi = \max_{1\le i\le(v-1)}\frac{\sqrt{\lambda_i}}{1+\lambda_i},\ \ R = \frac{\mathbf{x}_1'\boldsymbol{A}_1(\boldsymbol{A}_1'\boldsymbol{A}_1 + \boldsymbol{A}_2'\boldsymbol{A}_2)^{-1}\boldsymbol{A}_2'\mathbf{x}_2}{\xi\|\mathbf{x}_1\|\ \|\mathbf{x}_2\|}. \tag{7.5.7}$$

It is easy to check that for a BIBD, R defined in (7.5.7) reduces to that in (7.3.7). We shall now establish certain properties of the quantities ξ and R defined in (7.5.7), which will be used in deriving our test.

Lemma 7.5.1. Let $\boldsymbol{A}_1$ and $\boldsymbol{A}_2$ be the $(v-1)\times(v-1)$ matrices occurring in (7.5.3). Then $\max_{\mathbf{x}_1,\mathbf{x}_2}\{(\mathbf{x}_1'\boldsymbol{A}_1(\boldsymbol{A}_1'\boldsymbol{A}_1 + \boldsymbol{A}_2'\boldsymbol{A}_2)^{-1}\boldsymbol{A}_2'\mathbf{x}_2)/(\|\mathbf{x}_1\|\ \|\mathbf{x}_2\|)\} = \xi$, where ξ is defined in (7.5.7).

Proof. Let $\boldsymbol{E} = \boldsymbol{A}_1(\boldsymbol{A}_1'\boldsymbol{A}_1 + \boldsymbol{A}_2'\boldsymbol{A}_2)^{-1}\boldsymbol{A}_2'$ and let $\sigma_i(\boldsymbol{E})$ denote the singular values of $\boldsymbol{E}$. It can be verified that $\sigma_i(\boldsymbol{E}) = \sqrt{\lambda_i}/(1+\lambda_i)$ for $i \le (v-1)$, where the λ_i's are defined in (7.5.6). From Corollary B.1.a in Marshall and Olkin (1979, p. 515), it follows that $\max_{\mathbf{x}_1,\mathbf{x}_2}\{\mathbf{x}_1'\boldsymbol{E}\mathbf{x}_2/\|\mathbf{x}_1\|\ \|\mathbf{x}_2\|\} =$ the largest singular value of $\boldsymbol{E}$, which is $\max_{1\le i\le(v-1)}\sqrt{\lambda_i}/(1+\lambda_i)$. □

From Lemma 7.5.1, it follows that R defined in (7.5.7) satisfies $|R| \le 1$. From the definition of R, it also follows that under H_0: $\boldsymbol{\tau}^* = \mathbf{0}$, R is distributed symmetrically around zero. In particular, $\mathrm{E}(R) = 0$ under H_0: $\boldsymbol{\tau}^* = \mathbf{0}$. We shall now show that under H_1: $\boldsymbol{\tau}^* \ne \mathbf{0}$, $\mathrm{E}(R) > 0$.

Lemma 7.5.2. Let R be as defined in (7.5.7), where $\mathbf{x}_1$ and $\mathbf{x}_2$ have the distribution specified in (7.5.3). Then $\mathrm{E}(R) > 0$ whenever $\tau^* \ne \mathbf{0}$. □

In order to prove Lemma 7.5.2, we shall use the following result, which may be of independent interest.

Lemma 7.5.3. Let L be a univariate random variable following the normal distribution $N(\mu,\sigma^2)$. Then

$$E\left(\frac{L}{|L|}\,\middle|\,|L|\right) = \frac{exp(\frac{|L|\mu}{\sigma^2}) - exp(-\frac{|L|\mu}{\sigma^2})}{exp(\frac{|L|\mu}{\sigma^2}) + exp(-\frac{|L|\mu}{\sigma^2})}. \tag{7.5.8}$$

Proof. From the univariate normal density of L, it is readily verified that the joint distribution of $X = L/|L|$ and $Y = |L|$, say $g(x, y)$, and the density of Y, say $h(y)$, are respectively given by

$$g(x, y) = \frac{1}{\sqrt{2\pi}\sigma} exp\left\{-\frac{1}{2\sigma^2}(y^2 + \mu^2)\right\} exp\left(\frac{xy\mu}{\sigma^2}\right), \quad x = \pm 1, \quad y > 0,$$

$$h(y) = \frac{1}{\sqrt{2\pi}\sigma} exp\left\{-\frac{1}{2\sigma^2}(y^2 + \mu^2)\right\} \left(exp\left(\frac{y\mu}{\sigma^2}\right) + exp\left(-\frac{y\mu}{\sigma^2}\right)\right), \quad y > 0.$$

Thus, the conditional distribution of X given Y is given by

$$\Pr\{X = 1|Y\} = \frac{exp(\frac{y\mu}{\sigma^2})}{(exp(\frac{y\mu}{\sigma^2}) + exp(-\frac{y\mu}{\sigma^2}))},$$

$$\Pr\{X = -1|Y\} = \frac{exp(-\frac{y\mu}{\sigma^2})}{(exp(\frac{y\mu}{\sigma^2}) + exp(-\frac{y\mu}{\sigma^2}))}.$$

The expression (7.5.8) for $E(X|Y)$ (i.e., $E([L/|L|]||L|)$) is now immediate. This completes the proof of Lemma 7.5.3. □

Proof of Lemma 7.5.2: Let $\boldsymbol{M}$ be the $(v - 1) \times (v - 1)$ nonsingular matrix satisfying (7.5.6). Define

$$\begin{aligned} \mathbf{u} &= (u_1, \cdots, u_m)' = \boldsymbol{M}^{-1}\mathbf{A}_1^{-1}\mathbf{x}_1, \quad \mathbf{v} = (v_1, \cdots, v_m)' = \boldsymbol{M}^{-1}\mathbf{A}_2^{-1}\mathbf{x}_2 \\ \mathbf{m} &= \boldsymbol{M}^{-1}\boldsymbol{\tau}^*. \end{aligned} \tag{7.5.9}$$

Then

$$\mathbf{u} \sim N(\mathbf{m}, \sigma_e^2 \boldsymbol{I}_{v-1}) \qquad \mathbf{v} \sim N(\mathbf{m}, k(k\sigma_\beta^2 + \sigma_e^2)\mathbf{\Lambda}^{-1}), \tag{7.5.10}$$

where $\mathbf{\Lambda}$ is given in (7.5.6). Furthermore,

$$\begin{aligned} \frac{\mathbf{x}_1'\boldsymbol{A}_1(\boldsymbol{A}_1'\boldsymbol{A}_1 + \boldsymbol{A}_2'\boldsymbol{A}_2)^{-1}\boldsymbol{A}_2'\mathbf{x}_2}{\|\mathbf{x}_1\|\|\mathbf{x}_2\|} &= \frac{\mathbf{u}'(\boldsymbol{I}_{v-1} + \mathbf{\Lambda})^{-1}\mathbf{\Lambda}\mathbf{v}}{\{(\mathbf{u}'\mathbf{u})(\mathbf{v}'\mathbf{\Lambda}\mathbf{v})\}^{1/2}} \\ &= \frac{\sum_{i=1}^m \frac{\lambda_i}{1+\lambda_i} u_i v_i}{\{(\sum_{j=1}^m u_j^2)(\sum_{j=1}^m \lambda_j v_j^2)\}^{1/2}}. \end{aligned} \tag{7.5.11}$$

The proof is complete if we can show that

$$E\left[\frac{u_i v_i}{\{(\sum_{j=1}^m u_j^2)(\sum_{j=1}^m \lambda_j v_j^2)\}^{1/2}}\right] > 0, \tag{7.5.12}$$

whenever $\mathbf{m} \neq 0$, where $\mathbf{m}$ is given in (7.5.9)). To establish this, note that

$$E\left[\frac{u_i v_i}{\{(\sum_{j=1}^m u_j^2)(\sum_{j=1}^m \lambda_j v_j^2)\}^{1/2}}\right]$$
$$= E\left[\frac{|u_i||v_i|}{\{(\sum_{j=1}^m u_j^2)(\sum_{j=1}^m \lambda_j v_j^2)\}^{1/2}} E\left(\frac{u_i}{|u_i|}\frac{v_i}{|v_i|}\,\middle|\, |u_1|, \cdots, |u_m|, |v_1|, \cdots, |v_m|\right)\right], \tag{7.5.13}$$

where the inner expectation in (7.5.13) is conditionally given $|u_1|$, ..., $|u_m|$, $|v_1|$, ..., $|v_m|$ and the outer expectation is with respect to the distribution of $|u_1|$, ..., $|u_m|$, $|v_1|$, ..., $|v_m|$. From (7.5.10) it is clear that the u_i's and v_i's are independently distributed. Thus, from (7.5.13), it follows that in order to prove (7.5.12)), it is enough to show that

$$E\left(\frac{u_i}{|u_i|}\,\middle|\, |u_i|\right)E\left(\frac{v_i}{|v_i|}\,\middle|\, |v_i|\right) > 0, \tag{7.5.14}$$

whenever $m_i \neq 0$, m_i being the ith component of $\mathbf{m}$. From Lemma 7.5.3, it is clear that $E([u_i/|u_i|]\big||u_i|)$ and $E([v_i/|v_i|]\big||v_i|)$ are both positive or both negative depending on $m_i > 0$ or $m_i < 0$. This establishes (7.5.14) and the proof of Lemma 7.5.2 is complete. □

From Lemma 7.5.2, it follows that R in (7.5.7) can be used to test H_0: $\boldsymbol{\tau}^* = \mathbf{0}$ in model (7.5.3), with large values of R providing the rejection region. Also, the conclusions in Lemma 7.3.2 are valid for model (7.5.3) as well. Thus, U_1 and U_2 in (7.5.5) are complete and sufficient for σ_e^2 and σ_β^2 under H_0: $\boldsymbol{\tau}^* = \mathbf{0}$ and, furthermore, F_1 and F_2 in (7.5.4), R in (7.5.7), and U_1 and U_2 are independently distributed under H_0. Thus, the rejection regions Φ_1–Φ_5 can be defined using the quantities F_1, F_2, U_1, U_2, and R given in this section, and these rejection regions provide combined tests for testing H_0: $\boldsymbol{\tau}^* = \mathbf{0}$ in model (7.5.3) when $b > v$. Furthermore, when $b = v$, combined tests can be obtained based on the rejection regions Ψ_1–Ψ_3 in Section 7.3.2, where these rejection regions are now defined in terms of the quantities given in this section. Note that the above tests involve an estimator of θ^2 and one can use the estimators given in (7.3.17) or (7.3.18). It is also possible to use other choices of $\hat{\theta}^2$ similar to that given in (7.3.19). We have indeed used a choice similar to that in (7.3.19) in the computations that follow.

We shall now give some simulated powers of the combined test Φ_5 in (7.3.24) and compare it with the power of the intra-block F-test for testing the equality of the treatment effects in the following block design with random block effects. We have considered only Φ_5 for the purpose of comparison since, at least for the BIBD, the numerical results indicate that this is the test to be preferred among the tests based on Φ_1–Φ_5. The following design has 8 blocks of 3 plots each and 5 treatments. Let t_1, t_2, t_3, t_4, and t_5 denote the 5 treatments. The design is given by

Block 1	Block 2	Block 3	Block 4	Block 5	Block 6	Block 7	Block 8
t_1	t_1	t_1	t_1	t_2	t_2	t_2	t_3
t_2	t_2	t_3	t_4	t_3	t_3	t_4	t_4
t_3	t_5	t_4	t_5	t_4	t_5	t_5	t_5

(7.5.15)

Note that the treatment t_1 occurs 4 times in the design while the other treatments occur 5 times each. For the above design, the matrices $\boldsymbol{C}_1$ and $\boldsymbol{C}_2$ in (7.2.6) and (7.2.13) are given by

$$\begin{aligned} 3\boldsymbol{C}_1 &= \begin{pmatrix} 8 & -2 & -2 & -2 & -2 \\ -2 & 10 & -3 & -2 & -3 \\ -2 & -3 & 10 & -3 & -2 \\ -2 & -2 & -3 & 10 & -3 \\ -2 & -3 & -2 & -3 & 10 \end{pmatrix} \\ 8\boldsymbol{C}_2 &= \begin{pmatrix} 16 & -4 & -4 & -4 & -4 \\ -4 & 15 & -1 & -9 & -1 \\ -4 & -1 & 15 & -1 & -9 \\ -4 & -9 & -1 & 15 & -1 \\ -4 & -1 & -9 & -1 & 15 \end{pmatrix}. \end{aligned} \tag{7.5.16}$$

It can be verified that $\boldsymbol{C}_1$ and $\boldsymbol{C}_2$ commute (i.e., $\boldsymbol{C}_1\boldsymbol{C}_2 = \boldsymbol{C}_2\boldsymbol{C}_1$). Consequently, we can simultaneously diagonalize them using the same 5×5 orthogonal matrix, say $\boldsymbol{O}$. The matrix $\boldsymbol{O}$ can be chosen as $\boldsymbol{O} = [\boldsymbol{O}_1, 1/\sqrt{5}\mathbf{1}_5]$; see below equation (7.3.1) where $\boldsymbol{O}_1$ is defined and the choice of $\boldsymbol{O}$ is discussed. The corresponding diagonal matrices are

$$\boldsymbol{O}'\boldsymbol{C}_1\boldsymbol{O} = \mathbf{diag}(14/3, 4, 4, 10/3, 0) \text{ and } \boldsymbol{O}'\boldsymbol{C}_2\boldsymbol{O} = \mathbf{diag}(1, 3, 3, 2.5, 0). \tag{7.5.17}$$

Hence the matrices $\boldsymbol{Q}_1$ and $\boldsymbol{Q}_2$ used in (7.5.1) can be taken as

$$\begin{aligned} \boldsymbol{Q}_1 &= \boldsymbol{O}_1\mathbf{diag}(\sqrt{3}/\sqrt{14}, 1/2, 1/2, \sqrt{3}/\sqrt{10}) \\ \boldsymbol{Q}_2 &= \boldsymbol{O}_1\mathbf{diag}(1, 1/\sqrt{3}, 1/\sqrt{3}, 1/\sqrt{2.5}), \end{aligned}$$

so that $\boldsymbol{Q}_1'\boldsymbol{C}_1\boldsymbol{Q}_1 = \boldsymbol{Q}_2'\boldsymbol{C}_2\boldsymbol{Q}_2 = \boldsymbol{I}_4$. The canonical form (7.5.3) can now be written as

$$\begin{aligned} \mathbf{x}_1 &\sim N[\mathbf{diag}(\sqrt{14}/\sqrt{3}, 2, 2, \sqrt{10}/\sqrt{3})\boldsymbol{\tau}^*, \sigma_e^2\boldsymbol{I}_4],\ s_1^2 \sim \sigma_e^2\chi_{12}^2 \\ \mathbf{x}_2 &\sim N[\mathbf{diag}(1, \sqrt{3}, \sqrt{3}, \sqrt{2.5})\boldsymbol{\tau}^*, 3(3\sigma_\beta^2 + \sigma_e^2)\boldsymbol{I}_4],\ s_2^2 \sim 3(3\sigma_\beta^2 + \sigma_e^2)\chi_3^2. \end{aligned} \tag{7.5.18}$$

Thus the matrices $\boldsymbol{A}_1$ and $\boldsymbol{A}_2$ in (7.5.3) are given by $\boldsymbol{A}_1 = \mathbf{diag}(\sqrt{14}/\sqrt{3}, 2, 2, \sqrt{10}/\sqrt{3})$ and $\boldsymbol{A}_2 = \mathbf{diag}(1, \sqrt{3}, \sqrt{3}, \sqrt{2.5})$. Furthermore, the F-ratios in (7.5.4) are given by

$$F_1 = \frac{\mathbf{x}_1'\mathbf{x}_1/4}{s_1^2/12}, \quad F_2 = \frac{\mathbf{x}_2'\mathbf{x}_2/4}{s_2^2/3}. \tag{7.5.19}$$

Also, since $\boldsymbol{A}_1$ and $\boldsymbol{A}_2$ are diagonal matrices, we can take $\boldsymbol{M} = (\boldsymbol{A}_1'\boldsymbol{A}_1)^{-1/2}$ in (7.5.6). The λ_i's in (7.5.6) are then the eigenvalues of $(\boldsymbol{A}_1'\boldsymbol{A}_1)^{-1/2}\boldsymbol{A}_2'\boldsymbol{A}_2 \times \boldsymbol{A}_1'\boldsymbol{A}_1)^{-1/2}$, which are also the eigenvalues of $(\boldsymbol{A}_1'\boldsymbol{A}_1)^{-1}\boldsymbol{A}_2'\boldsymbol{A}_2$. Using these observations and (7.5.7), we get

$$\begin{aligned} \xi &= 0.49487, \\ R &= 2.02073 \times \mathbf{x}_1'\mathbf{diag}(0.38122, 0.49487, 0.49487, 0.49487)\mathbf{x}_2/\|\mathbf{x}_1\|\ \|\mathbf{x}_2\|. \end{aligned} \tag{7.5.20}$$

In what follows, we shall give the simulated power of the test based on Φ_5 in (7.3.24) for the choice of γ_1 discussed below. It turned out that the choice $\gamma_1 = 1/(1 + \hat{\theta}^2)$ as given in (7.3.16), with $\hat{\theta}^2 = \hat{\theta}_{(1)}^2$ or $\hat{\theta}_{(2)}^2$ in (7.3.17) and (7.3.18), resulted in a combined test whose performance was only marginally better or slightly worse compared to the intra-block F-test (based on F_1 in (7.5.19)) for the parameter values considered for simulation. A choice of γ_1 that we shall use is motivated by the arguments that lead to the choice in (7.3.19). The powers of the tests based on F_1 and F_2 in (7.5.19) are functions of $\boldsymbol{\tau}^{*'}A_1'A_1\boldsymbol{\tau}^*/\sigma_e^2$ and $\boldsymbol{\tau}^{*'}A_2'A_2\boldsymbol{\tau}^*/3(3\sigma_\beta^2 + \sigma_e^2)$ respectively. Note that the matrices $\boldsymbol{A}_1$ and $\boldsymbol{A}_2$ satisfy $\boldsymbol{A}_1'\boldsymbol{A}_1 \geq 4\boldsymbol{A}_2'\boldsymbol{A}_2/3$ (i.e., $\boldsymbol{A}_1'\boldsymbol{A}_1 - 4\boldsymbol{A}_2'\boldsymbol{A}_2/3$ is a nonnegative definite matrix). Hence it appears reasonable to choose γ_1 to reflect the magnitude of $\sigma_e^2/4$ relative to that of $(3\sigma_\beta^2 + \sigma_e^2)$. Noting that $\sigma_e^2/4(3\sigma_\beta^2 + \sigma_e^2) \leq 1/4$, we shall thus make the following choice, similar to (7.3.19).

$$\hat{\theta}^2 = \min\left\{\frac{\hat{\sigma}_e^2}{4(3\hat{\sigma}_\beta^2 + \hat{\sigma}_e^2)}, \frac{1}{4}\right\}, \quad \gamma_1 = \frac{1}{1 + \hat{\theta}^2}, \quad \gamma_2 = 1 - \gamma_1, \tag{7.5.21}$$

where $\hat{\sigma}_e^2 = U_1/16$ and $3\hat{\sigma}_\beta^2 + \hat{\sigma}_e^2 = U_2/7$ (U_1 and U_2 are given in (7.5.5)). Thus if P_1 and P_2 are the P-values of the tests that reject H_0: $\boldsymbol{\tau}^* = \mathbf{0}$ for large values of F_1 and F_2 in (7.5.19) and if $Z_i = -\ln P_i$ $(i = 1, 2)$, then our combined test is based on the critical region Φ_5 in (7.3.24), where $W = \gamma_1 Z_1 + \gamma_2 Z_2$, γ_1 and γ_2 are given in (7.5.21) and R is given in (7.5.20).

Table 7.5 gives the simulated powers of the intra-block F-test (based on F_1 in (7.5.19)) and the combined test based on Φ_5 mentioned above for the following alternatives

$$\boldsymbol{\tau}^* = \lambda(1, 1, 1, 1)' \tag{7.5.22}$$

for $\lambda = 0.1, 0.2, 0.3, 0.4, 0.5, 0.8, 1$ and 1.5. We chose $\sigma_e^2 = 1$ for the simulation and $3\sigma_\beta^2 + \sigma_e^2 = 1, 4, 9, 16,$ and 25. In order to numerically compute the power, we used 100,000 simulations. For the same setup, some simulated powers

Table 7.5. Simulated Powers (Based on 100,000 Simulations) of the Intra-block F-Test Based on F_1 in (7.5.19) and the Combined Test Based on Φ_5 in (7.5.5) for Testing H_0: $\tau^* = 0$ in Model (7.5.18) for the Alternatives in (7.5.22) for $\alpha = 0.05$ and $\sigma_e^2 = 1$

		$k\sigma_\beta^2 + \sigma_e^2$				
λ		1	4	9	16	25
0.1	F_1	0.0561	0.0561	0.0561	0.0561	0.0561
	Φ_5	0.0570	0.0562	0.0560	0.0558	0.0556
0.2	F_1	0.0744	0.0744	0.0744	0.0744	0.0744
	Φ_5	0.0783	0.0757	0.0747	0.0744	0.0741
0.3	F_1	0.1066	0.1066	0.1066	0.1066	0.1066
	Φ_5	0.1178	0.1104	0.1086	0.1078	0.1071
0.4	F_1	0.1586	0.1586	0.1586	0.1586	0.1586
	Φ_5	0.1796	0.1655	0.1614	0.1593	0.1581
0.5	F_1	0.2282	0.2282	0.2282	0.2282	0.2282
	Φ_5	0.2659	0.2427	0.2352	0.2314	0.2290
0.8	F_1	0.5399	0.5399	0.5399	0.5399	0.5399
	Φ_5	0.6230	0.5737	0.5566	0.5470	0.5408
1.0	F_1	0.7552	0.7552	0.7552	0.7552	0.7552
	Φ_5	0.8323	0.7870	0.7691	0.7589	0.7525
1.5	F_1	0.9852	0.9852	0.9852	0.9852	0.9852
	Φ_5	0.9958	0.9905	0.9871	0.9848	0.9831

are also reported in Table 2 in Zhou and Mathew (1993). However, their numerical results and some of the expressions in their paper are incorrect.

We note from Table 7.5 that in terms of power, Φ_5 dominates F_1 except for large values of $k\sigma_\beta^2 + \sigma_e^2$. When $k\sigma_\beta^2 + \sigma_e^2$ is large, the test based on F_1 has a slightly larger power compared to the test based on Φ_5; however, the difference is rather insignificant.

The choice of $\hat{\theta}^2$ and γ_1 in (7.5.21) is based on the fact that $\boldsymbol{A}_1'\boldsymbol{A}_1 \geq 4\boldsymbol{A}_2'\boldsymbol{A}_2/3$. Since $\boldsymbol{A}_1'\boldsymbol{A}_1 = \boldsymbol{O}'\boldsymbol{C}_1\boldsymbol{O}$ and $\boldsymbol{A}_2'\boldsymbol{A}_2 = \boldsymbol{O}'\boldsymbol{C}_2\boldsymbol{O}$ (see (7.5.17)), the above condition is equivalent to $\boldsymbol{C}_1 \geq 4\boldsymbol{C}_2/3$. From the expressions (7.2.9) and (7.2.16) for the F-ratios, it is clear that the power functions of the tests based on F_1 and F_2 are functions of the noncentrality parameters $\boldsymbol{\tau}'\boldsymbol{C}_1\boldsymbol{\tau}/\sigma_e^2$ and $\boldsymbol{\tau}'\boldsymbol{C}_2\boldsymbol{\tau}/k(k\sigma_\beta^2 + \sigma_e^2)$, respectively. Following the arguments that lead to (7.5.21), it is clear that if we can find a constant c such that $\boldsymbol{C}_1 \geq c/k\boldsymbol{C}_2$, then one can choose

$$\hat{\theta}^2 = \min\left\{\frac{\hat{\sigma}_e^2}{c(k\hat{\sigma}_\beta^2 + \hat{\sigma}_e^2)}, \frac{1}{c}\right\}, \quad \gamma_1 = \frac{1}{1+\hat{\theta}^2}, \quad \gamma_2 = 1 - \gamma_1, \tag{7.5.23}$$

where $\hat{\sigma}_e^2 = U_1/(bk - b)$ and $k\hat{\sigma}_\beta^2 + \hat{\sigma}_e^2 = U_2/(b-1)$. The choice in (7.5.23) is, of course, similar to that in (7.5.21). In order to be able to use (7.5.23) in practice to arrive at the combined test Φ_5, one should be able to compute the constant c satisfying $\boldsymbol{C}_1 \geq c/k\boldsymbol{C}_2$. Since $\boldsymbol{C}_1$ and $\boldsymbol{C}_2$ are $v \times v$ nonnegative definite matrices of rank $(v-1)$ satisfying $\boldsymbol{C}_1\mathbf{1}_v = \boldsymbol{C}_2\mathbf{1}_v = \mathbf{0}$, we can find a $v \times v$ nonsingular matrix $\boldsymbol{S}$ such that

$$\boldsymbol{S}'\boldsymbol{C}_1\boldsymbol{S} = \mathbf{diag}(\boldsymbol{I}_{v-1}, 0) \text{ and } \boldsymbol{S}'\boldsymbol{C}_2\boldsymbol{S} = \mathbf{diag}(\boldsymbol{\Lambda}, 0), \tag{7.5.24}$$

where $\boldsymbol{\Lambda}$ is a $(v-1) \times (v-1)$ diagonal matrix with positive diagonal elements. (The notation $\boldsymbol{\Lambda}$ was also used in (7.5.6); as we shall see in the next section, $\boldsymbol{\Lambda}$ satisfying (7.5.6) also satisfies (7.5.24)). If $\lambda_{\max}$ denotes the largest diagonal element of $\boldsymbol{\Lambda}$, then $c = k/\lambda_{\max}$ obviously satisfies $\boldsymbol{I}_{v-1} \geq c/k\boldsymbol{\Lambda}$, that is, $\mathbf{diag}(\boldsymbol{I}_{v-1}, 0) \geq c/k\,\mathbf{diag}(\boldsymbol{\Lambda}, 0)$, or equivalently, $\boldsymbol{C}_1 \geq c/k\boldsymbol{C}_2$. In other words, there always exists a positive constant c satisfying $\boldsymbol{C}_1 \geq c/k\boldsymbol{C}_2$. In order to compute such a c, it is not necessary to explicitly obtain the representation (7.5.24). This can be seen as follows. First of all, using the representation (7.5.24), we see that a generalized inverse of $\boldsymbol{C}_1$, say $\boldsymbol{C}_1^-$, is necessarily of the form

$$\boldsymbol{C}_1^- = \boldsymbol{S}\begin{pmatrix} \boldsymbol{I}_{v-1} & \mathbf{a}_1 \\ \mathbf{a}_2' & a_3 \end{pmatrix}\boldsymbol{S}',$$

where $\mathbf{a}_1$ and $\mathbf{a}_2$ are arbitrary $(v-1) \times 1$ vectors and a_3 is an arbitrary scalar. (For the above expression for the generalized inverse, we refer to Rao and Mitra, 1971.) It can be verified that $\lambda_{\max}$, the largest diagonal element of $\boldsymbol{\Lambda}$ in (7.5.24), is also the largest eigenvalue of $\boldsymbol{C}_1^-\boldsymbol{C}_2$. Thus in order to compute the constant c satisfying $\boldsymbol{C}_1 \geq c/k\boldsymbol{C}_2$, first compute $\lambda_{\max}$, the largest eigenvalue of $\boldsymbol{C}_1^-\boldsymbol{C}_2$. Then $c = k/\lambda_{\max}$.

Once c is computed, we can obtain the quantities in (7.5.23) and use them in order to arrive at the critical region Φ_5. When $\boldsymbol{C}_1$ and $\boldsymbol{C}_2$ commute, one can simultaneously diagonalize these matrices as in (7.5.17) and get the constant c satisfying $\boldsymbol{C}_1 \geq c/k\boldsymbol{C}_2$ by comparing the diagonal matrices, as we have done for the design (7.5.15). It is rather easy to verify that $\boldsymbol{C}_1$ and $\boldsymbol{C}_2$ always commute for any design where the treatments are replicated the same number of times (i.e., the r_i's are equal in (7.2.6)), in addition to the block sizes being equal. This can be checked directly by using the expressions for $\boldsymbol{C}_1$ and $\boldsymbol{C}_2$. However, $\boldsymbol{C}_1$ and $\boldsymbol{C}_2$ can commute even if the treatments are not replicated the same number of times in the design. This is the case for the design in (7.5.15).

7.5.2. Some Computational Formulas

In order to carry out the test Φ_5 in the context of the general block design considered in Section 7.5.1, we now give some formulas that express the relevant quantities in terms of $\mathbf{q}_1$, $\mathbf{q}_2$, $\boldsymbol{C}_1$, $\boldsymbol{C}_2$, s_1^2, and s_2^2. In other words, in order to carry out the test, it is not necessary to compute the matrices $\boldsymbol{Q}_1$,

$\boldsymbol{Q}_2$, $\boldsymbol{O}_1$, and $\boldsymbol{M}$ that occur in (7.5.1)–(7.5.6). Note that we already have the expressions for the F-ratios F_1 and F_2 in terms of $\mathbf{q}_1$, $\mathbf{q}_2$, $\boldsymbol{C}_1$, $\boldsymbol{C}_2$, s_1^2, and s_2^2; see (7.2.9) and (7.2.16). Since $\mathbf{x}_1'\mathbf{x}_1 = \mathbf{q}_1'\boldsymbol{C}_1^-\mathbf{q}_1$ and $\mathbf{x}_2'\mathbf{x}_2 = \mathbf{q}_2'\boldsymbol{C}_2^-\mathbf{q}_2$, it follows that (see (7.5.5))

$$U_1 = \mathbf{q}_1'\boldsymbol{C}_1^-\mathbf{q}_1 + s_1^2, \quad U_2 = \frac{1}{k}[\mathbf{q}_1'\boldsymbol{C}_1^-\mathbf{q}_1 + s_2^2]. \tag{7.5.25}$$

Also, from the definitions of $\boldsymbol{A}_1$, $\boldsymbol{A}_2$, $\mathbf{x}_1$, and $\mathbf{x}_2$, it follows that $\boldsymbol{A}_1'\boldsymbol{A}_1 = \boldsymbol{O}_1'\boldsymbol{C}_1\boldsymbol{O}_1$, $\boldsymbol{A}_2'\boldsymbol{A}_2 = \boldsymbol{O}_1'\boldsymbol{C}_2\boldsymbol{O}_1$, $\boldsymbol{A}_1'\mathbf{x}_1 = \boldsymbol{O}_1'\mathbf{q}_1$, and $\boldsymbol{A}_2'\mathbf{x}_2 = \boldsymbol{O}_1'\mathbf{q}_2$. Using arguments similar to those in the discussion following (7.5.24), it follows that $\lambda_1, \lambda_2, \ldots, \lambda_{v-1}$ in (7.5.6) are the eigenvalues of $\boldsymbol{C}_1^-\boldsymbol{C}_2$. In other words, in order to compute ξ in (7.5.7), we only need to compute the eigenvalues of $\boldsymbol{C}_1^-\boldsymbol{C}_2$. As already pointed out earlier, once these eigenvalues are computed, the constant c satisfying $\boldsymbol{C}_1 \geq c/k\boldsymbol{C}_2$ is given by $c = k/\lambda_{\max}$, where $\lambda_{\max}$ is the largest eigenvalue of $\boldsymbol{C}_1^-\boldsymbol{C}_2$. Furthermore, R in (7.5.7) can be expressed as

$$R = \frac{1}{\xi}\frac{\mathbf{q}_1'(\boldsymbol{C}_1 + \boldsymbol{C}_2)^-\mathbf{q}_2}{\sqrt{(\mathbf{q}_1'\boldsymbol{C}_1^-\mathbf{q}_1)(\mathbf{q}_2'\boldsymbol{C}_2^-\mathbf{q}_2)}}. \tag{7.5.26}$$

In the numerical example that follows, we have carried out the computations based on the expressions given above.

7.5.3. A Numerical Example

The block design given in Table 7.6 is for comparing 8 air filters regarding their ability to collect air pollutants. The 8 filters are denoted by t_i $(i = 1, 2, \ldots, 8)$. Since the filtering ability will be affected by environmental conditions, days were taken as blocks with random effects. On each day, three filters were tested. The data obtained were the increase in weight (in grams) of the filter after 40 minutes of operation. The data given in Table 7.6 are taken from Lentner and Bishop (1986, pp. 458–460). (In Table 7.6, we have reproduced only part of the data; for the full data, see Lenter and Bishop, 1986.)

The design has $b = 11$, $v = 8$, and $k = 3$. Direct computations give

$$\boldsymbol{C}_1 = \frac{1}{3}\begin{pmatrix} 10 & -1 & -1 & -2 & -2 & -2 & -1 & -1 \\ -1 & 8 & -1 & -2 & -2 & -1 & -1 & 0 \\ -1 & -1 & 10 & -2 & -2 & -2 & -1 & -1 \\ -2 & -2 & -2 & 10 & -1 & -1 & -1 & -1 \\ -2 & -2 & -2 & -1 & 10 & -1 & -1 & -1 \\ -2 & -1 & -2 & -1 & -1 & 8 & -1 & 0 \\ -1 & -1 & -1 & -1 & -1 & -1 & 6 & 0 \\ -1 & 0 & -1 & -1 & -1 & 0 & 0 & 4 \end{pmatrix},$$

Table 7.6. The Air Filter Data

Block	Treatments and observations		
1	t_1 (18.3)	t_2 (12.2)	t_5 (21.9)
2	t_1 (11.9)	t_3 (22.3)	t_6 (14.5)
3	t_1 (11.4)	t_4 (11.2)	t_7 (15.8)
4	t_1 (25.7)	t_4 (8.9)	t_8 (21.7)
5	t_1 (15.3)	t_5 (18.8)	t_6 (8.4)
6	t_2 (8.8)	t_3 (12.2)	t_4 (22.8)
7	t_2 (14.1)	t_4 (19.0)	t_5 (8.5)
8	t_2 (13.7)	t_6 (10.1)	t_7 (9.6)
9	t_3 (14.8)	t_4 (7.6)	t_6 (12.6)
10	t_3 (18.5)	t_5 (15.4)	t_7 (18.6)
11	t_3 (7.7)	t_5 (13.6)	t_8 (12.1)

Source: M. Lentner and T. Bishop (1986). Reproduced with permission of Valley Book Company, Blacksburg, Virginia.

$$C_2 = \frac{1}{11}\begin{pmatrix} 30 & -9 & -14 & -3 & -3 & 2 & -4 & 1 \\ -9 & 28 & -9 & 2 & 2 & -5 & -1 & -8 \\ -14 & -9 & 30 & -3 & -3 & 2 & -4 & 1 \\ -3 & 2 & -3 & 30 & -14 & -9 & -4 & 1 \\ -3 & 2 & -3 & -14 & 30 & -9 & -4 & 1 \\ 2 & -5 & 2 & -9 & -928 & -1 & -8 & \\ -4 & -1 & -4 & -4 & -4 & -1 & 24 & -6 \\ 1 & -8 & 1 & 1 & 1 & -8 & -6 & 18 \end{pmatrix},$$

$$\mathbf{q}_1 = \frac{1}{3}(9.5, -24.8, 13.1, -6.6, 12.2, -22.8, 7.7, 11.7)'$$

$$\mathbf{q}_2 = \frac{1}{11}(231.3, -28.8, -42.6, -23.9, 56.4, -156.4, -66.7, 30.7)'$$

$$s_1^2 = 481.7217 \quad (\text{d.f.} = 15), \quad s_2^2 = 346.52 \quad (\text{d.f.} = 3).$$

We note that $\text{rank}(C_1) = \text{rank}(C_2) = v - 1 = 7$. Using (7.2.9) and (7.2.16) and the computational formulas given in Section 7.5.2, we get $F_1 = 0.32478$, $F_2 = 0.38529$. From (7.5.5), $Z_1 = 0.07171$, $Z_2 = 0.14479$, $U_1 = 554.7337$, and $U_2 = 219.349$. The eigenvalues of $C_1^- C_2$ are 1.9465, 0, 0.23, 0.4044, 1.3613, 0.9355, 1.0909, 1.0909. These are the λ_i's used in the computation of ξ in (7.5.7). We thus get $\xi = 0.49972$, and, from (7.5.26), $R = 0.42$. Also, $\hat{\sigma}_e^2 = U_1/(bk - b)$ $= 25.215168$ and $k\hat{\sigma}_\beta^2 + \hat{\sigma}_e^2 = U_2/(b-1) = 21.9349$. Also, $c = k/\lambda_{\max} = 3/1.9465 = 1.5412$. This value of c satisfies $C_1 \geq c/kC_2$. From (7.5.23), we thus get $\hat{\theta}^2 = 0.64884$. From (7.3.16), $\gamma_1 = 0.6065$, $\gamma_2 = 0.3935$, and $W = 0.10046$. Consequently, from (7.3.24), the rejection region Φ_5 simplifies to $0.98159 \leq \alpha \times 1.42$. Since $0.98159 \geq \alpha \times 1.42$ holds for $\alpha \leq 0.69$, we accept the null hypothesis that the treatment effects are equal. In other words, we conclude that the 8 filters are equally efficient in terms of their ability to collect air pollutants.

EXERCISES

7.1. Show that for a BIBD, R in (7.5.7) simplifies to that in (7.3.7).

7.2. Show that the expected value of the numerator of R in (7.5.7) is non-negative.

7.3. Show that the quantities $\mathbf{q}_1$, s_1^2, $\mathbf{q}_2$, and s_2^2, defined in Section 7.2, are independently distributed under model (7.2.2), when the distributional assumptions (7.2.3) hold.

7.4. Consider a Youden square design (i.e., a row-column design for comparing v treatments using v rows and k columns $(k < v)$ such that if the rows are treated as blocks, the design is a BIBD). Let y_{ij} denote the observation from the plot in the ith row and j^{th} column $(i = 1, 2, \ldots, v;$ $j = 1, 2, \ldots, k)$. Assume the model

$$y_{ij} = \mu + \alpha_i + \beta_j + \sum_{s=1}^{v} \delta_{ij}^s \tau_s + e_{ij};\ i = 1, 2, \cdots, v;\ j = 1, 2, \cdots, k,$$

where α_i, β_j and τ_s, respectively, denote the effects due to the ith row, j^{th} column and s^{th} treatment, δ_{ij}^s is 1 or 0, depending on whether or not the sth treatment occurs in the ith row and j^{th} column, and the e_{ij}'s are independently distributed as $N(0, \sigma_e^2)$.

(a) Suppose the row effects are independently distributed as $\alpha_i \sim N(0, \sigma_\alpha^2)$, and are independent of the e_{ij}'s. Following the procedures developed in Section 7.3.2, derive tests that recover "inter-row information" for testing H_0: $\tau_1 = \tau_2 = \ldots, = \tau_v$.

(b) Suppose $\alpha_i \sim N(0, \sigma_\alpha^2)$, $\tau_i \sim N(0, \sigma_\tau^2)$, where the α_i's, τ_i's, and e_{ij}'s are all independently distributed. Following the procedures developed in Section 7.4, derive tests for testing H_0: $\sigma_\tau^2 = 0$.

7.5. Explain why the procedure in Section 7.5 for deriving a combined test cannot be applied to a block design having unequal block sizes.

7.6. Let $\mathbf{y}_1 \sim N(\boldsymbol{X}_1\boldsymbol{\beta}, \sigma_1^2\boldsymbol{I}_{n_1})$ and $\mathbf{y}_2 \sim N(\boldsymbol{X}_2\boldsymbol{\beta}, \sigma_2^2\boldsymbol{I}_{n_2})$ be two independent linear models where $\mathbf{y}_i$: $n_i \times 1$ are vectors of observations, X_i: $n_i \times m$ are known design matrices, and $\boldsymbol{\beta}$: $m \times 1$ and $\sigma_i^2 > 0$ are unknown parameters $(i = 1, 2)$. Let $\boldsymbol{K\beta}$ be an estimable linear parametric function under both the models. Following the procedure in Section 7.5, derive a test, combining information from both the models, for testing H_0: $\boldsymbol{K\beta} = 0$.

7.7. Consider the independent linear models $\mathbf{y}_i \sim N(\boldsymbol{X}_i\boldsymbol{\beta}, \sigma_i^2\boldsymbol{I}_{n_i})$ $(i = 1, 2, \ldots, a)$ where $\mathbf{y}_i$: $n_i \times 1$ are vectors of observations, $\boldsymbol{X}_i$: $n_i \times m$ are known design matrices, and $\boldsymbol{\beta}$: $m \times 1$ and $\sigma_i^2 > 0$ are unknown parameters $(i = 1, 2, \ldots, a)$. Let $\boldsymbol{K\beta}$ be an estimable linear parametric function under all the models. Derive a test, combining information from all the models, for testing H_0: $\boldsymbol{K\beta} = 0$.

BIBLIOGRAPHY

Cohen, A. and Sackrowitz, H. B. (1989). "Exact tests that recover inter-block information in balanced incomplete block designs." *Journal of the American Statistical Association*, 84, 556–559.

Eaton, M. L. (1983). *Multivariate Statistics: A Vector Space Approach*. Wiley, New York.

Feingold, M. (1985). "A test statistic for combined intra- and inter-block estimates." *Journal of Statistical Planning and Inference*, 12, 103–114.

Feingold, M. (1988). "A more powerful test for incomplete block designs." *Communications in Statistics—Theory and Methods*, 17, 3107–3119.

Fisher, R. A. (1932). *Statistical Methods for Research Workers*. Oliver and Boyd, London.

Jordan, S. M. and Krishnamoorthy, K. (1995). "On combining independent tests in linear models." *Statistics and Probability Letters*, 23, 117–122.

Lehmann, E. L. (1986). *Testing Statistical Hypotheses*, Second Edition. Wiley, New York.

Lentner, M. and Bishop, T. (1986). *Experimental Design and Analysis*. Valley Book Company, Blacksburg, Virginia.

Marshall, A. W. and Olkin, I. (1979). *Inequalities: Theory of Majorization and Its Applications*. Academic Press, New York.

Mathew, T., Sinha, B. K., and Zhou, L. (1993). "Some statistical procedures for combining independent tests." *Journal of the American Statistical Association*, 88, 912–919.

Rao, C. R. (1947). "General methods of analysis for incomplete block designs." *Journal of the American Statistical Association*, 42, 541–561.

Rao, C. R. and Mitra, S. K. (1971). *Generalized Inverse of Matrices and its Applications*. Wiley, New York.

SAS User's Guide: Statistics, 1989 Edition, SAS Institute, Inc., Cary, North Carolina.

Shah, K. R. (1975). "Analysis of block designs." *Gujarat Statistical Review*, 2, 1–11.

Shah, K. R. (1992). "Recovery of inter-block information: An update." *Journal of Statistical Planning and Inference*, 30, 163–172.

Yates, F. (1939). "The recovery of inter-block information in varietal trials arranged in three dimensional lattice." *Annals of Eugenics*, 9, 136–156.

Yates, F. (1940). "The recovery of inter-block information in balanced incomplete block designs." *Annals of Eugenics*, 10, 317–325.

Zhang, Z. (1992). "Recovery tests in BIBD's with very small degrees of freedom for inter-block errors." *Statistics and Probability Letters*, 15, 197–202.

Zhou, L. and Mathew, T. (1993). "Combining independent tests in linear models." *Journal of the American Statistical Association*, 88, 650–655.

CHAPTER 8

Split-Plot Designs Under Mixed and Random Models

8.1. INTRODUCTION

In factorial experiments, it often happens that certain factors require bigger plots than others for convenience in organizing the experiment. The simple split-plot design is one such experimental design where v levels of a factor, say A, are randomly assigned to the plots (to be referred to as whole plots) in b blocks, and s levels of a second factor, say B, are randomly assigned to subplots (or split-plots) within each whole plot. An example of a split-plot design reported in Yates (1937) (see also John, 1971, Section 5.7) involves three varieties of oats and four levels of manure (nitrogen). Six blocks of three plots were taken, and one plot in each block was sown with each of the three varieties of oats. Each plot was divided into four split-plots, one of which was assigned at random to each level of the nitrogen factor. This is an example of a *randomized block* balanced split-plot design, namely, when each block has v whole plots to which the v levels of A are randomly assigned and each whole plot has s split-plots to which the s levels of B are randomly assigned. The analysis of variance of such a balanced split-plot design is well-known (see Milliken and Johnson, 1984, Chapter 24; Hinkelmann and Kempthorne, 1994, Chapter 13.) However, it can very often happen that the split-plot design is *unbalanced* due to incomplete blocks or incomplete whole plots, or due to missing data, and furthermore, some of the effects could be random. For example, in the experiment involving oats and nitrogen mentioned before, the different blocks may correspond to different locations, and the number of whole plots available at each location need not be the same. Moreover, if a large number of levels of the nitrogen factor is available to the experimenter, usually a few levels will have to be selected randomly to include in the experiment, due to the limitations on the experimental resources. This will obviously result in a mixed effects model for the data. Thus, a proper statistical analysis of such unbalanced split-plot designs is called for. In a somewhat simpler setup, when the whole plots form a completely randomized design, Milliken and Johnson (1984, Chapter 28) considered the analysis of unbalanced split-plot designs assuming a fixed effects model. However, their analysis does not cover the situation when some of the effects are random.

In this chapter, which is primarily based on Mathew and Sinha (1992), we consider the first type of unbalanced split-plot designs in the context of block designs (i.e., the situation where each whole plot has s split-plots to which the s levels of B are assigned, but the v levels of A may not be replicated the same number of times in the design). Throughout this chapter, we provide suitable test procedures for the various hypotheses of interest in this setup assuming the usual linear model for the data. We have also provided the optimum tests whenever they exist. The testing problems are addressed in contexts where the block effects, effects due to the v levels of A, and/or effects due to the s levels of B are either fixed or random. It is hoped that the analysis of the unbalanced split-plot design that we have developed in this chapter will be useful to practitioners who carry out experiments using such designs.

In Section 8.2, for the sake of completeness, we provide a detailed analysis of a *balanced* split-plot design. This section also includes results on the optimality of F-tests under different scenarios of fixed, random, and mixed models. Section 8.3 is devoted to the analysis of an *unbalanced* split-plot design of the type mentioned above.

The framework of a general split-plot design is as follows. We first describe it in the unbalanced case. Let k_i denote the number of whole plots in the ith block $(i = 1, 2, \ldots b)$ and let y_{ijl} denote the response from the split-plot of the jth whole plot in the ith block, where the lth level of B is occurring $(l = 1, 2, \ldots s;\ j = 1, 2, \ldots k_i;\ i = 1, 2, \ldots b)$. Note that the suffix j is used to denote the whole plots within a block rather than the levels of A, since the blocks could be incomplete and certain levels of A may not occur in every block (this may not be a standard notation, but we find it convenient to use). Let $\boldsymbol{\tau} = (\tau_1, \tau_2, \ldots \tau_v)'$ be the vector of main effects due to the v levels of A, $\boldsymbol{\delta} = (\delta_1, \delta_2, \ldots \delta_s)'$ be the vector of main effects due to the s levels of B, $\boldsymbol{\beta} = (\beta_1, \beta_2, \ldots \beta_b)'$ be the vector of block effects, γ_{ul} denote the interaction effect between u^{th} level of A and l^{th} level of B, and $\boldsymbol{\gamma} = (\gamma_{11}, \ldots \gamma_{1s}, \gamma_{21}, \ldots \gamma_{2s}, \ldots \gamma_{v1}, \ldots \gamma_{vs})'$. Whenever these effects are fixed, we assume that $\sum_{u=1}^{v} \tau_u = \sum_{l=1}^{s} \delta_l = \sum_{i=1}^{b} \beta_i = 0$, $\sum_{l=1}^{s} \gamma_{ul} = 0$ for every u, and $\sum_{u=1}^{v} \gamma_{ul} = 0$ for every l. Whenever the effects are random, we assume that $\boldsymbol{\tau} \sim N(\mathbf{0}, \sigma_\tau^2 I_v)$, $\boldsymbol{\delta} \sim N(\mathbf{0}, \sigma_\delta^2 I_s)$, $\boldsymbol{\beta} \sim N(\mathbf{0}, \sigma_\beta^2 I_b)$, and $\boldsymbol{\gamma} \sim N(\mathbf{0}, \sigma_\gamma^2 I_{vs})$, and that $\tau, \delta, \beta, \gamma$ are independent. In order to identify the levels of A that occur in a block, let

$$
\begin{aligned}
f_{ij}^{u} &= 1 \text{ if the } u^{th} \text{ level of } A \text{ occurs in the } j^{th} \text{ whole plot} \\
&\quad\ \text{of the } i^{th} \text{ block} \\
&= 0 \text{ otherwise.}
\end{aligned}
$$

The linear model for analyzing the above split-plot design is then given by

$$\begin{aligned} y_{ijl} &= \mu + \beta_i + \sum_{u=1}^{v} f_{ij}^u \tau_u + \delta_l + \sum_{u=1}^{v} f_{ij}^u \gamma_{ul} + e_{ijl}, \\ l &= 1,2,\cdots s,\ j = 1,2,\cdots k_i,\ i = 1,2,\cdots b, \end{aligned} \tag{8.1.1}$$

where the e_{ijl}'s are normally distributed random error terms, distributed independently of any other random effect in model (8.1.1), and satisfy

$$\begin{aligned} \mathrm{E}(e_{ijl}) &= 0, \\ Cov(e_{ijl}, e_{i'j'l'}) &= \sigma_e^2 \quad \text{for } i = i',\ j = j',\ l = l' \\ &= \rho\sigma_e^2 \quad \text{for } i = i',\ j = j',\ l \neq l' \\ &= 0 \quad \text{for } i \neq i' \quad \text{or } j \neq j'. \end{aligned}$$

It may be noted that the standard notation in the context of split-plot designs is to write $e_{ijl} = g_{ij} + h_{ijl}$ where the "whole plot error" g_{ij} and the "split-plot error" h_{ijl} are independent random variables with $g_{ij} \sim N(0, \sigma_1^2)$ and $h_{ijl} \sim N(0, \sigma_2^2)$ (see Milliken and Johnson, 1984, Chapter 24.) Clearly, this results in the same covariance structure as mentioned above with $\rho = \sigma_1^2/(\sigma_1^2 + \sigma_2^2)$ and $\sigma_e^2 = \sigma_1^2 + \sigma_2^2$.

For $j = 1, 2, \ldots, k_i$ and $i = 1, 2, \ldots, b$, write $\mathbf{f}_{ij} = (f_{ij}^1, f_{ij}^2, \ldots, f_{ij}^v)'$, $\boldsymbol{F}_i = [\mathbf{f}_{i1} : \mathbf{f}_{i2} : \ldots : \mathbf{f}_{ik_i}]'$, and $\mathbf{y}_{ij\cdot} = (y_{ij1}, \ldots, y_{ijs})'$. Also, let $\boldsymbol{F} = [\boldsymbol{F}_1' : \boldsymbol{F}_2' : \ldots : \boldsymbol{F}_b']'$ and $\mathbf{y} = (\mathbf{y}_{11\cdot}', \ldots, \mathbf{y}_{1k_1\cdot}', \mathbf{y}_{21\cdot}', \ldots, \mathbf{y}_{2k_2\cdot}', \ldots,\ \mathbf{y}_{b1\cdot}', \ldots, \mathbf{y}_{bk_b\cdot}')'$. Then, model (8.1.1) can be written as

$$\begin{aligned} \mathbf{y}_{ij\cdot} &= (\mu + \beta_i + \mathbf{f}_{ij}'\boldsymbol{\tau})\mathbf{1}_s + \boldsymbol{\delta} + (\mathbf{f}_{ij}' \otimes I_s)\boldsymbol{\gamma} + \mathbf{e}_{ij\cdot} \\ & j = 1, 2, \cdots, k_i,\ i = 1, 2, \cdots, b, \end{aligned} \tag{8.1.2}$$

or, equivalently,

$$\mathbf{y} = (\mathbf{1}_n \otimes \mathbf{1}_s)\mu + (\boldsymbol{E} \otimes \mathbf{1}_s)\boldsymbol{\beta} + (\boldsymbol{F} \otimes \mathbf{1}_s)\boldsymbol{\tau} + (\mathbf{1}_n \otimes I_s)\boldsymbol{\delta} + (\boldsymbol{F} \otimes I_s)\boldsymbol{\gamma} + \mathbf{e} \tag{8.1.3}$$

where $n = \sum_{i=1}^{b} k_i$, $\mathbf{1}_m$ denotes the $m \times 1$ vector of ones, $\mathbf{e}_{ij\cdot}$ and $\mathbf{e}$ are defined similarly to $\mathbf{y}_{ij\cdot}$ and $\mathbf{y}$, and

$$\boldsymbol{E} = \mathbf{diag}(\mathbf{1}_{k_1}, \cdots, \mathbf{1}_{k_b}). \tag{8.1.4}$$

Note that

$$Var(\mathbf{e}_{ij\cdot}) = \sigma_e^2 \left[(1 - \rho)\boldsymbol{I}_s + \rho\boldsymbol{J}_s\right], \tag{8.1.5}$$

$$Var(\mathbf{e}) = \sigma_e^2 \boldsymbol{I}_n \otimes \left[(1 - \rho)\boldsymbol{I}_s + \rho\boldsymbol{J}_s\right], \tag{8.1.6}$$

where $\boldsymbol{J}_s = \mathbf{1}_s\mathbf{1}_s'$.

Define $\bar{y}_{ij.} = 1/s\mathbf{y}'_{ij.}\mathbf{1}_s$ and $\bar{\delta} = 1/s\boldsymbol{\delta}'\mathbf{1}_s$. Let $\boldsymbol{P} = [1/\sqrt{s}\mathbf{1}_s : \boldsymbol{P}'_1]'$ be an $s \times s$ orthogonal matrix. Obviously, $\boldsymbol{P}_1\boldsymbol{P}'_1 = \boldsymbol{I}_{s-1}$, $\boldsymbol{P}_1\mathbf{1}_s = \mathbf{0}$, and $\boldsymbol{P}'_1\boldsymbol{P}_1 = \boldsymbol{I}_s - \frac{1}{s}\boldsymbol{J}_s$. Using (8.1.2), we now consider the model for $\boldsymbol{P}\mathbf{y}_{ij\cdot}$, which decomposes into models (8.1.7) and (8.1.8) given below:

$$\sqrt{s}\bar{y}_{ij.} = \sqrt{s}\left(\mu + \beta_i + \mathbf{f}'_{ij}\boldsymbol{\tau}\right) + \sqrt{s}\bar{\delta} + \left(\mathbf{f}'_{ij} \otimes \frac{1}{\sqrt{s}}\mathbf{1}'_s\right)\boldsymbol{\gamma} + \sqrt{s}\bar{e}_{ij.} \tag{8.1.7}$$

$$\boldsymbol{P}_1\mathbf{y}_{ij\cdot} = \boldsymbol{P}_1\boldsymbol{\delta} + (\mathbf{f}'_{ij} \otimes \boldsymbol{P}_1)\boldsymbol{\gamma} + \boldsymbol{P}_1\mathbf{e}_{ij\cdot} \tag{8.1.8}$$

where $\bar{e}_{ij.}$ is defined similar to $\bar{y}_{ij.}$. We note that in the fixed effects case, $\bar{\delta} = 0$ and $(\mathbf{f}'_{ij} \otimes \mathbf{1}'_s)\boldsymbol{\gamma} = \mathbf{0}$, for all i and j, by our assumption. We also note that $\sqrt{s}\bar{y}_{ij.}$ and $\boldsymbol{P}_1\mathbf{y}_{ij\cdot}$ are independent irrespective of whether the various effects are fixed or random. Model (8.1.8) is obviously the linear model for the $(s-1)$ orthogonal contrasts among the components of $\mathbf{y}_{ij\cdot}$. It can be verified that $\bar{e}_{ij.}$'s are uncorrelated and

$$Var(\sqrt{s}\bar{e}_{ij.}) = \sigma_e^2\{1 + (s-1)\rho\} = \sigma_*^2 \tag{8.1.9}$$

and

$$Var(\boldsymbol{P}_1\mathbf{e}_{ij.}) = \sigma_e^2(1-\rho)\boldsymbol{I}_{s-1} = \sigma_{**}^2\boldsymbol{I}_{s-1}, \tag{8.1.10}$$

where $\sigma_*^2 = \sigma_e^2\{1 + (s-1)\rho\}$ and $\sigma_{**}^2 = \sigma_e^2(1-\rho)$. The variances of $\bar{y}_{ij.}$'s and $\boldsymbol{P}_1\mathbf{y}_{ij.}$'s and the covariances among them obviously depend on whether the various effects are fixed or random (see cases 8.2(ii)–8.2(viii) in Section 8.2 and cases 8.3(ii)–8.3(viii) in Section 8.3). The decomposition (8.1.7)–(8.1.10) is fundamental to our subsequent development of tests for the various hypotheses in model (8.1.2) or model (8.1.3). It turns out that the test procedures for testing the significance of the block effects and the main effects due to A are based only on model (8.1.7), and those for testing the significance of the main effects due to B and the interaction between A and B are based only on model (8.1.8). As a consequence, the significance of the main effects due to B and the interaction between A and B can always be tested. However, the significance of the main effects due to A (or the significance of the block effects) cannot be tested in the fixed effects case unless the appropriate estimability condition is met. In other words, contrasts among the τ_i's must be estimable based on model (8.1.7) (where $\bar{\delta} = 0$ and $(\mathbf{f}'_{ij} \otimes \mathbf{1}'_s)\boldsymbol{\gamma} = \mathbf{0}$, for all i and j). A condition for this estimability is that the design consisting of the blocks and whole plots is connected (Hinkelmann and Kempthorne, 1994, Chapter 9). An equivalent condition is that the corresponding "$\boldsymbol{C}$ matrix" has rank $(v-1)$. (We shall denote this $\boldsymbol{C}$ matrix by $\boldsymbol{C}_\tau$ and define it following equation (8.3.4).)

We now turn our attention to the case of a *balanced* split-plot design. Here the number of whole plots appearing in each block is the same (i.e.,

$k_1 = \ldots = k_b$), and we assume it to be equal to v, the number of levels of factor A so that each level of A appears in every block exactly once. Without any loss of generality we can take

$$\mathbf{f}_{ij} = (0, \cdots, 1, \cdots, 0)', \quad i = 1, \cdots, b. \tag{8.1.11}$$

The basic equation (8.1.1) in this case takes the form

$$y_{ijl} = \mu + \beta_i + \tau_j + \delta_l + \gamma_{jl} + e_{ijl}, \tag{8.1.12}$$

and equation (8.1.2) reduces to

$$\mathbf{y}_{ij\cdot} = (\mu + \beta_i + \tau_j)\mathbf{1}_s + \boldsymbol{\delta} + \boldsymbol{\gamma}_{j\cdot} + \boldsymbol{e}_{ij\cdot} \tag{8.1.13}$$

where $\boldsymbol{\gamma}_{j\cdot} = (\gamma_{j1}, \ldots, \gamma_{js})'$. Defining $\boldsymbol{P}$ and $\boldsymbol{P}_1$ as before, and writing $\bar{\gamma}_{j.} = (\gamma_{j1} + \ldots + \gamma_{js})/s$, the fundamental equations (8.1.7) and (8.1.8) leading to the relevant inferences can be written as

$$\sqrt{s}\bar{y}_{ij.} = \sqrt{s}(\mu + \beta_i + \tau_j) + \sqrt{s}\bar{\delta} + \sqrt{s}\bar{\gamma}_{j.} + \sqrt{s}\bar{e}_{ij.} \tag{8.1.14}$$

$$\boldsymbol{P}_1\mathbf{y}_{ij\cdot} = \boldsymbol{P}_1\boldsymbol{\delta} + \boldsymbol{P}_1\boldsymbol{\gamma}_{j\cdot} + \boldsymbol{P}_1\mathbf{e}_{ij\cdot}. \tag{8.1.15}$$

Derivation of exact and optimum tests for various fixed effects and variance components are described in Section 8.2. Due to the balanced nature of the underlying linear model, it turns out that the well-known F-tests for block effects, main effects due to the levels of A and B, and the interaction effects between the levels of A and B, continue to be both exact and optimum ($UMPI$), irrespective of the fixed or random nature of the various effects involved in the model (see Chapter 2). To put this in proper perspective, we give below the formulas for these F-statistics denoted by F_β, F_τ, F_δ, F_γ, respectively. Obviously, these F-statistics depend, in turn, on the various sums of squares, which are given below. For this, let $\bar{y}_{i..} = 1/vs\sum_{j=1}^{v}\sum_{l=1}^{s} y_{ijl}$, $\bar{y}_{.j.} = 1/bs\sum_{i=1}^{b}\sum_{l=1}^{s} y_{ijl}$, $\bar{y}_{..l} = 1/bv\sum_{i=1}^{b}\sum_{j=1}^{v} y_{ijl}$, and $\bar{y}_{...} = 1/bvs\sum_{i=1}^{b}\sum_{j=1}^{v}\sum_{l=1}^{s} y_{ijl}$. The various sums of squares and their degrees of freedoms are as follows:

$$SS(\beta) = vs\sum_{i=1}^{b}(\bar{y}_{i..} - \bar{y}_{...})^2, \quad \text{d.f.} = (b-1)$$

$$SS(\tau) = bs\sum_{j=1}^{v}(\bar{y}_{.j.} - \bar{y}_{...})^2, \quad \text{d.f.} = (v-1)$$

$$SS(e_1) = s\sum_{i=1}^{b}\sum_{j=1}^{v}(\bar{y}_{ij.} - \bar{y}_{i..} - \bar{y}_{.j.} + \bar{y}_{...})^2, \quad \text{d.f.} = (b-1)(v-1)$$

$$
\begin{aligned}
SS(\delta) &= bv\sum_{l=1}^{s}(\bar{y}_{..l}-\bar{y}_{...})^2, \quad \text{d.f.} = (s-1)\\
SS(\gamma) &= b\sum_{j=1}^{v}\sum_{l=1}^{s}(\bar{y}_{.jl}-\bar{y}_{.j.}-\bar{y}_{..l}+\bar{y}_{...})^2, \quad \text{d.f.} = (v-1)(s-1)\\
SS(e_2) &= \sum_{i=1}^{b}\sum_{j=1}^{v}\sum_{l=1}^{s}(y_{ijl}-\bar{y}_{ij.}-\bar{y}_{.jl}+\bar{y}_{.j.})^2, \quad \text{d.f.} = v(b-1)(l-1),
\end{aligned}
\tag{8.1.16}
$$

where, $SS(e_1)$ and $SS(e_2)$ are the error sum of squares from the whole plot ANOVA and the split-plot ANOVA, respectively.

As in the previous chapters, H_β will denote the hypothesis $\beta_1 = \ldots = \beta_b$, or, $\sigma_\beta^2=0$, depending on whether the β_i's are fixed effects or random effects; H_τ, H_δ, and H_γ are similarly defined. The F-ratios F_β, F_τ, and so on, for testing the respective hypotheses H_β, H_τ, and so on. are given below. Of course, F_δ is used to test H_δ only when H_γ is accepted.

$$
\begin{aligned}
F_\beta &= \frac{SS(\beta)/(b-1)}{SS(e_1)/(b-1)(v-1)}\\
F_\tau &= \frac{SS(\tau)/(v-1)}{SS(e_1)/(b-1)(v-1)}\\
F_\delta &= \frac{SS(\delta)/(s-1)}{[SS(\gamma)+SS(e_2)]/(bv-1)(s-1)}\\
F_\gamma &= \frac{SS(\gamma)/(v-1)(s-1)}{SS(e_2)/v(b-1)(s-1)}.
\end{aligned}
\tag{8.1.17}
$$

Incidentally, in the case of a balanced split-plot design, $n = bv$, $\boldsymbol{F}_j = \mathbf{I}_v$, $j = 1, \ldots, v$, and $\boldsymbol{F} = \mathbf{1}_b \otimes \mathbf{I}_v$. In Section 8.2, we have briefly demonstrated the optimality of the above F-tests on a case-by-case basis, depending on whether the various effects are fixed or random. These optimality results follow directly from Chapter 2, since we are dealing with a balanced model.

In Section 8.3 we have derived suitable test procedures for various hypotheses concerning the fixed effects and variance components in model (8.1.3), when the effects are fixed, mixed, or random, and the model is *unbalanced*. The exact and optimum test procedures that we have derived are mostly based on invariance of the underlying testing problem with respect to a suitable group of transformations. As already noted, when the levels of A and B are fixed, $\bar{\delta} = 0$, $(\mathbf{f}'_{ij} \otimes \mathbf{1}'_s)\boldsymbol{\gamma} = \mathbf{0}$ for all i and j, and in this case, (8.1.7) reduces to an unbalanced two-way model without interaction involving the main effects due to A which are fixed, and the block effects which can be either fixed or random. In the former case, (8.1.7) is a fixed effects model and in the latter case it is a mixed effects model. On the other hand, if the levels of A and/or B are random, the variance components σ_γ^2, σ_τ^2 and/or σ_δ^2 will obviously appear in the variance–covariance matrix of the $\bar{y}_{ij.}$'s. In the next section, it will be shown that the variance components σ_δ^2 and σ_γ^2 can always be ignored for deriving invariant tests to test the significance of the block effects. However, the significance of the main effects due to A can be tested only when σ_γ^2 is zero (see cases 8.3(iii)–8.3(viii) in Section 8.3) and when this is the case, σ_δ^2 can be ignored for carrying out the test. Thus, if we can conclude that the interaction is not significant (using model (8.1.8)), then the significance of the main effects due to A can be tested using model (8.1.7). Consequently, Wald's test procedure (see Chapter 1) and the test procedures that are available for an unbalanced two-way model without interaction (see Chapters 4, 6, and 7) are immediately applicable for testing the significance of the β_i's and the τ_u's, irrespective of whether the δ_l's and γ_{ul}'s are fixed or random. It should be noted that (8.1.8) continues to be a fixed effects model irrespective of whether the block effects are fixed or random, as long as the levels due to A and B are fixed. In case the levels due to A and/or B are random, the exact and optimum tests that we have presented for testing the significance of the main effects due to A and those due to B assume that σ_γ^2 is zero. It turns out that the standard F-tests derived under the fixed effects model continue to provide valid exact tests in the mixed and random effects situations as well (see equations (8.3.6)–(8.3.11) in Section 8.3 for the expressions of the test statistics). Of course, as noted before, $\boldsymbol{\gamma} = \mathbf{0}$ or $\sigma_\gamma^2 = 0$, as appropriate, is assumed for testing the significance of the main effects due to A and B.

The decomposition (8.1.7)–(8.1.10) also shows that when all the effects are fixed, the F-tests (obtained from the ANOVA table) are optimum invariant for testing the significance of the various effects in (8.1.7) and (8.1.8). This follows from standard linear model theory; see Lehmann (1986, Chapter 7). However, in the mixed or random effects case, optimum invariant tests for various hypotheses exist only under certain restrictions on the design. In Section 8.3 we have exhibited these restrictions explicitly on a case-by-case basis, along with the optimum invariant tests. Some preliminary additional results, which are used in the sequel, are given in Appendix 8.1. As a ready reference, Tables 8.1 and 8.2 at the end of Section 8.3 provide a summary of our results, and should prove to be quite handy and useful to the users of such models. Finally, a numerical example in the unbalanced case is carried out in detail to demonstrate the application of the techniques developed in Section 8.3.

Returning to the setup of Milliken and Johnson (1984, Section 28.1), when the whole plots form a completely randomized design in which the v levels of A need not be replicated the same number of times, but each whole plot has exactly s split-plots, appropriate test procedures can be obtained as special cases of our results by taking the number of blocks to be one in cases 8.3(i), 8.3(iii), 8.3(iv), and 8.3(v) in Section 8.3. It may be noted that for the same design, Milliken and Johnson (1984, Section 28.1) have derived F-tests for testing the significance of the various effects, assuming a fixed effects model.

Our results thus provide a complete solution to the various testing problems that can arise in this context when the effects are fixed, mixed, or random.

8.2. DERIVATION OF EXACT AND OPTIMUM TESTS: BALANCED CASE

For the balanced split-plot design introduced in Section 8.1, we now demonstrate that the F-tests defined in (8.1.17) for various hypotheses of interest are optimum invariant tests when model (8.1.13) is fixed, mixed, or random. In case 8.2(i) below, we have considered the fixed effects ANOVA. In cases 8.2(ii)–8.2(viii), we have specified the effects that are random. It will be understood that the remaining effects are fixed. As mentioned earlier, we have used the decomposition displayed in (8.1.14) and (8.1.15) to derive tests for (H_β, H_τ) and (H_δ, H_γ) respectively.

Case 8.2(i). Fixed Effects ANOVA

In the fixed effects case, since $\bar{\delta} = 0$ and $\bar{\gamma}_{j.} = 0$ for all j, model (8.1.14) becomes

$$\sqrt{s}\bar{y}_{ij.} = \sqrt{s}(\mu + \beta_i + \tau_j) + \sqrt{s}\bar{e}_{ij.}, \tag{8.2.1}$$

which is the standard *balanced* two-way classification model without interaction. Validity and optimality of the exact tests for H_β and H_τ, as given by F_β and F_τ defined in (8.1.17), follow immediately (see Chapter 2).

The split-plot analysis of variance can be carried out in a straightforward manner using model (8.1.15), and the optimality of the other two tests, F_δ and F_γ defined in (8.1.17) for testing the two hypotheses H_δ and H_γ, again follows directly (see Chapter 2).

Case 8.2(ii). Random β_i's

In this case, we again have $\bar{\delta} = 0, \bar{\gamma}_{j.} = 0$, for all j, so that model (8.1.14) reduces to

$$E(\sqrt{s}\bar{y}_{ij.}) = \sqrt{s}(\mu + \tau_j)$$

and the covariance structure is given by

$$\begin{aligned} Cov(\sqrt{s}\bar{y}_{ij.}, \sqrt{s}\bar{y}_{i'j'.}) &= \sigma_*^2 + s\sigma_\beta^2, \text{ if } i = i',\ j = j' \\ &= s\sigma_\beta^2, \quad \text{if } i = i',\ j \neq j' \\ &= 0 \text{ otherwise,} \end{aligned}$$

where σ_*^2 is defined in (8.1.9). Thus the model for $\sqrt{s}\bar{y}_{ij.}$ is a two-way *balanced* mixed model with random block effects and fixed treatment effects, and the optimality of the F-tests based on F_β and F_τ for testing H_β and H_τ follows easily (see Chapter 2).

Returning to tests for H_δ and H_γ, we note that since (8.1.15) is still a fixed effects model, the F-tests based on F_γ and F_δ defined in (8.1.17) continue to be UMPI for H_γ and H_δ, respectively.

Case 8.2(iii). Random τ_u's

Here again $\bar{\delta} = 0$, although the interaction effects γ's are random because τ's are random. Model (8.1.14) reduces to

$$E(\sqrt{s}\bar{y}_{ij.}) = \sqrt{s}(\mu + \beta_i)$$

and

$$\begin{aligned} Cov(\sqrt{s}\bar{y}_{ij.}, \sqrt{s}\bar{y}_{i'j'.}) &= \sigma_*^2 + (s\sigma_\tau^2 + \sigma_\gamma^2), \quad \text{if } i = i',\ j = j' \\ &= s\sigma_\tau^2 + \sigma_\gamma^2, \quad \text{if } i \neq i',\ j = j' \\ &= 0, \text{ otherwise.} \end{aligned}$$

We note that this model involves the two variance components σ_τ^2 and σ_γ^2 only through their sum, which implies that H_τ can be tested only if we assume that $\sigma_\gamma^2 = 0$ (of course, as discussed below, the hypothesis H_γ can be tested based on $\boldsymbol{P}_1\mathbf{y}_{ij.}$). Clearly, then, when $\sigma_\gamma^2 = 0$, the above model reduces to a two-way *balanced* mixed model with fixed block effects and random treatment effects. This is similar to the previous case except that the treatments and the blocks are interchanged, and the optimality of the F-test based on F_τ defined in (8.1.17) for testing H_τ easily follows (see Chapter 2). On the other hand, for testing H_β, we note that $(s\sigma_\tau^2 + \sigma_\gamma^2)$ can be treated as a single variance component, and hence an exact optimum test can then be based on F_β whatever σ_τ^2 and σ_γ^2 are.

Returning to model (8.1.15), since the γ_{jl}'s are random, the model reduces to

$$E(\boldsymbol{P}_1\mathbf{y}_{ij\cdot}) = \boldsymbol{P}_1\boldsymbol{\delta}$$

and

$$\begin{aligned} Cov(\boldsymbol{P}_1\mathbf{y}_{ij\cdot}, \boldsymbol{P}_1\mathbf{y}_{i'j'\cdot}) &= (\sigma_{**}^2 + \sigma_\gamma^2)\boldsymbol{I}, \quad \text{if } i = i',\ j = j' \\ &= \sigma_\gamma^2\boldsymbol{I}, \quad \text{if } i \neq i',\ j = j' \\ &= \mathbf{0}, \text{ otherwise,} \end{aligned}$$

where σ_{**}^2 is as in (8.1.10). Again, if $\sigma_\gamma^2 = 0$, H_δ can be easily tested using F_δ as in the fixed effects case and the resultant F-test based on F_δ defined in (8.1.17) is UMPI. To test H_γ, we note that the model for $\boldsymbol{P}_1\mathbf{y}_{ij\cdot}$ involves two variance components and is similar to (6.2.11). It follows from Theorem 6.2.2(a) that the test based on F_γ defined in (8.1.17) is UMPI.

Case 8.2(iv). Random δ_l's

In this case, model (8.1.14) is written as

$$\begin{aligned} E(\sqrt{s}\bar{y}_{ij.}) &= \sqrt{s}(\mu + \beta_i + \tau_j) \\ Cov(\sqrt{s}\bar{y}_{ij.}, \sqrt{s}\bar{y}_{i'j'.}) &= \sigma_*^2 + \sigma_\delta^2 + \sigma_\gamma^2, \quad \text{if } i = i',\ j = j' \\ &= \sigma_\delta^2 + \sigma_\gamma^2, \quad \text{if } i \neq i',\ j = j' \\ &= 0, \text{ otherwise.} \end{aligned}$$

We note that in the covariance matrix of the vector consisting of the $\sqrt{s}\bar{y}_{ij.}$'s, the coefficient matrix of σ_δ^2 is $\mathbf{1}_n\mathbf{1}_n'$. Hence, if $\sigma_\gamma^2 = 0$, then Lemma A.8.1(*iii*) in Appendix 8.1 is applicable for testing H_τ, and it follows that the F-test based on F_τ defined in (8.1.17) is UMPI. For testing H_β, it can once again be verified that Lemma A.8.1(*iii*) is applicable (even if $\sigma_\gamma^2 \neq 0$) and the test based on F_β is UMPI.

Model (8.1.15), on the other hand, can be written as

$$\begin{aligned} E[(\boldsymbol{I}_n \otimes \boldsymbol{P}_1)\mathbf{y}] &= \mathbf{0} \\ Var[(\boldsymbol{I}_n \otimes \boldsymbol{P}_1)\mathbf{y}] &= \sigma_\delta^2(\boldsymbol{J}_n \otimes \boldsymbol{I}_{s-1}) + \sigma_\gamma^2[\{(\mathbf{1}_b \otimes \boldsymbol{I}_v) \\ &\qquad (\mathbf{1}_b' \otimes \boldsymbol{I}_v)\} \otimes \boldsymbol{I}_{s-1}] + \sigma_{**}^2\boldsymbol{I}_{n(s-1)}, \end{aligned} \tag{8.2.2}$$

where $\boldsymbol{J}_n = \mathbf{1}_n\mathbf{1}_n'$. If $\sigma_\gamma^2 = 0$, the above model involves only two variance components and a UMPI test for H_δ can be obtained by applying Theorem 6.2.2. The UMPI test always exists and coincides with the test based on F_δ defined in (8.1.17). Testing H_γ requires a special attention since model (8.2.2) involves three variance components. Lemma A.8.3 given in Appendix 8.1 is suitable in this situation. To apply Lemma A.8.3, let $\boldsymbol{M} = [\boldsymbol{F}b^{-1/2} : \boldsymbol{M}_1]'$ be an $n \times n$ orthogonal matrix (recall that $\boldsymbol{F} = \mathbf{1}_b \otimes \boldsymbol{I}_v$ and $\boldsymbol{F}'\boldsymbol{F} = b\boldsymbol{I}_v$). Thus $\boldsymbol{M}_1'\boldsymbol{F} = 0$ and $\boldsymbol{M}_1'\mathbf{1}_n = 0$, since $\boldsymbol{F}\mathbf{1}_v = \mathbf{1}_n$. Hence $\boldsymbol{M}\boldsymbol{F}\boldsymbol{F}'\boldsymbol{M}' = \mathbf{diag}(b\boldsymbol{I}_v, \mathbf{0})$ and $\boldsymbol{M}\mathbf{1}_n = (b^{1/2}\mathbf{1}_v', 0)'$. Consider the random vector $(\boldsymbol{M} \otimes \boldsymbol{I}_{s-1})(\boldsymbol{I}_n \otimes \boldsymbol{P}_1)\mathbf{y} = (\boldsymbol{M} \otimes \boldsymbol{P}_1)\mathbf{y}$ which has mean $\mathbf{0}$ and a variance–covariance matrix (from (8.2.2)):

$$\begin{aligned} Var[(\boldsymbol{M} \otimes \boldsymbol{P}_1)\mathbf{y}] = \mathbf{diag}[&\sigma_{**}^2\boldsymbol{I}_{v(s-1)} + b\sigma_\delta^2((\mathbf{1}_v\mathbf{1}_v') \otimes \boldsymbol{I}_{s-1}) \\ &+ b\sigma_\gamma^2(\boldsymbol{I}_v \otimes \boldsymbol{I}_{s-1}),\ \sigma_{**}^2\boldsymbol{I}_{(n-v)(s-1)}]. \end{aligned} \tag{8.2.3}$$

This, however, is similar to (A.8.2) in Appendix 8.1. Applying Lemma A.8.3, we therefore conclude that a UMPU test exists for testing H_γ, and it indeed coincides with the F-test based on F_γ. Furthermore, the same test is also UMPI.

Case 8.2(v). Random τ_j's and δ_l's

Here, model (8.1.14) yields

$$E(\sqrt{s}\bar{y}_{ij.}) = \sqrt{s}(\mu + \beta_i)$$

and

$$\begin{aligned} Cov(\sqrt{s}\bar{y}_{ij.}, \sqrt{s}\bar{y}_{i'j'.}) &= \sigma_*^2 + \sigma_\delta^2 + (s\sigma_\tau^2 + \sigma_\gamma^2), \quad \text{if } i = i', \ j = j' \\ &= \sigma_\delta^2 + (s\sigma_\tau^2 + \sigma_\gamma^2), \quad \text{if } i \neq i', \ j = j' \\ &= \sigma_\delta^2, \ \text{otherwise.} \end{aligned}$$

This model is similar to the model under random τ_j's discussed in case 8.2(iii), except for the extra term σ_δ^2 in the expression for the covariance. However, in view of Lemmas A.8.2 and A.8.1(iii) of Appendix 8.1, the class of invariant tests for testing H_τ and H_β does not depend on σ_δ^2, and hence σ_δ^2 can be ignored while deriving invariant tests for these hypotheses. Thus an exactly similar analysis as in case 8.2(iii) obtains, and the invariant tests described there continue to hold. In particular, F_β defined in (8.1.17) can be used to test H_β, and when $\sigma_\gamma^2 = 0$, F_τ defined in (8.1.17) can be used to test H_τ.

Returning to model (8.1.15) for $\boldsymbol{P}_1\mathbf{y}_{ij\cdot}$, it is easily seen to be the same as the corresponding model discussed in case 8.2(iv). Therefore, tests for H_δ and H_γ remain the same.

Case 8.2(vi). Random β_i's and τ_u's

Since the split-plot effects, δ_l's, are fixed, we get $\bar{\delta} = 0$, and model (8.1.14) reduces to

$$E(\sqrt{s}\bar{y}_{ij.}) = \sqrt{s}\mu$$

and

$$\begin{aligned} Cov(\sqrt{s}\bar{y}_{ij.}, \sqrt{s}\bar{y}_{i'j'.}) &= \sigma_*^2 + s\sigma_\beta^2 + (s\sigma_\tau^2 + \sigma_\gamma^2), \quad \text{if } i = i', \ j = j' \\ &= s\sigma_\beta^2, \quad \text{if } i = i', \ j \neq j' \\ &= s\sigma_\tau^2 + \sigma_\gamma^2, \quad \text{if } i \neq i', \ j = j' \\ &= 0, \ \text{otherwise.} \end{aligned}$$

We first note that for testing H_β, the model can be treated as a two-way *balanced* random effects model involving three variance components, namely, σ_*^2, σ_β^2, and $(s\sigma_\tau^2 + \sigma_\gamma^2)$. By following the derivation of Wald's test given in Chapter 1 (see Section 1.2), it can be verified that the test based on F_β is a Wald's test and hence is a valid exact test for H_β. Next, for testing H_τ, we note that when $\sigma_\gamma^2 = 0$, the model for $\sqrt{s}\bar{y}_{ij.}$ is a two-way *balanced* random effects model without interaction and the F-test based on F_τ is a Wald's test and hence continues to be a valid exact test.

Finally, model (8.1.15) for $\boldsymbol{P}_1\mathbf{y}_{ij.}$ under random β_i's and τ_u's is easily seen to be the same as the corresponding model in case 8.2(iii). Hence the tests based on F_δ and F_γ defined in (8.1.17) for H_δ and H_γ continue to hold in this case.

Case 8.2(vii). Random β_i's and δ_l's

In this case model (8.1.14) reduces to

$$E(\sqrt{s}\bar{y}_{ij.}) = \sqrt{s}(\mu + \tau_j)$$

and

$$\begin{aligned} Cov(\sqrt{s}\bar{y}_{ij.}, \sqrt{s}\bar{y}_{i'j'.}) &= \sigma_*^2 + s\sigma_\beta^2 + \sigma_\delta^2 + \sigma_\gamma^2, \quad \text{if } i = i',\ j = j' \\ &= s\sigma_\beta^2 + \sigma_\delta^2, \quad \text{if } i = i',\ j \neq j' \\ &= \sigma_\delta^2 + \sigma_\gamma^2, \quad \text{if } i \neq i',\ j = j' \\ &= \sigma_\delta^2, \text{ otherwise.} \end{aligned}$$

This model involves the fixed effects μ and τ_u's and four variance components. However, using Lemma A.8.2 in Appendix 8.1, we see that an invariant test for H_β will not depend on the two variance components σ_δ^2 and σ_γ^2, and hence these can be ignored for deriving such a test. This results in a two-way *balanced* mixed effects model with fixed treatment effects and random block effects. Hence, Theorem 6.2.2 can be used to obtain an optimum invariant test for H_β and the test is as given in case 8.2(ii), namely, the test based on F_β.

For testing H_τ, if $\sigma_\gamma^2 = 0$, an application of Lemma A.8.1(*iii*) in Appendix 8.1 shows that invariant tests for H_τ will not depend on σ_δ^2, and hence this variance component can be ignored for deriving such a test. Thus the above model once again reduces to a two-way *balanced* mixed effects model with fixed treatment effects and random block effects, and hence the test based on F_τ continues to be valid since it is a Wald's test.

Model (8.1.15) for $\boldsymbol{P}_1\mathbf{y}_{ij.}$ is easily seen to remain the same as that discussed under case 8.2(iv). Therefore, the same tests for H_δ and H_γ as described before continue to hold in this case.

Case 8.2(viii). Random β_i's, τ_u's and δ_l's

In this case, model (8.1.14) reduces to

$$E(\sqrt{s}\bar{y}_{ij.}) = \sqrt{s}\mu$$

and

$$\begin{aligned} Cov(\sqrt{s}\bar{y}_{ij.}, \sqrt{s}\bar{y}_{i'j'.}) &= \sigma_*^2 + s\sigma_\beta^2 + \sigma_\delta^2 + (s\sigma_\tau^2 + \sigma_\gamma^2), \quad \text{if } i = i',\ j = j' \\ &= s\sigma_\beta^2 + \sigma_\delta^2, \quad \text{if } i = i',\ j \neq j' \\ &= \sigma_\delta^2 + (s\sigma_\tau^2 + \sigma_\gamma^2), \quad \text{if } i \neq i',\ j = j' \\ &= \sigma_\delta^2, \text{ otherwise.} \end{aligned}$$

We first note that in view of Lemma A.8.2 in Appendix 8.1, the variance component σ_δ^2 can be ignored for deriving invariant tests for H_β and H_τ. This results in the model given under case 8.2(vi) and hence all the conclusions in case 8.2(vi) are valid for testing H_β and H_τ. The model for $\boldsymbol{P}_1\mathbf{y}_{ij.}$ once again reduces to the corresponding model under case 8.2(iv), and tests based on F_δ and F_γ continue to be $UMPI$.

REMARK 8.2.1. We have demonstrated above that, as expected, the familiar tests based on F_β, F_τ, F_δ, and F_γ continue to be valid and optimum invariant irrespective of the nature (i.e., fixed, mixed, random) of the model whenever the split-plot design is *balanced*.

8.3. DERIVATION OF EXACT AND OPTIMUM TESTS: UNBALANCED CASE

For the unbalanced split-plot design introduced in Section 8.1, we now provide optimum tests for various hypotheses of interest under (8.1.3), when the model is fixed, mixed, or random. Throughout this section, if the β's or τ's are fixed, then for testing H_β or H_τ, we will assume that the design consisting of the blocks and whole plots is *connected*. In other words, we assume that the corresponding $\boldsymbol{C}$ matrix $\boldsymbol{C}_\tau$ has rank $(v-1)$, where $\boldsymbol{C}_\tau$ is defined following equation (8.3.4). This assumption is necessary for the testability of H_β or H_τ. In case 8.3(i) below, we have considered the fixed effects ANOVA. In cases 8.3(ii)–8.3(viii) we have specified the effects that are random. It will be understood that the remaining effects are fixed. As mentioned before, we have used the fundamental decomposition displayed in (8.1.7) and (8.1.8) to derive tests for (H_β, H_τ) and (H_δ, H_γ), respectively.

Case 8.3(i). Fixed Effects ANOVA

In the fixed effects case, since $\bar{\delta} = 0$ and $(\mathbf{f_{ij}} \otimes \mathbf{1}_s')\boldsymbol{\gamma} = \mathbf{0}$, model (8.1.7) becomes

$$\sqrt{s}\bar{y}_{ij.} = \sqrt{s}(\mu + \beta_i + \mathbf{f}_{ij}'\boldsymbol{\tau}) + \sqrt{s}\bar{e}_{ij.}, \tag{8.3.1}$$

which is the standard unbalanced two-way classification model without interaction. Let n_{ui} denote the number of times the uth level of A occurs in the ith block and $\boldsymbol{N}$ denote the $v \times b$ incidence matrix whose $(u,i)^{th}$ element is n_{ui}. Clearly, $\sum_{i=1}^b n_{ui} = r_u$, the number of whole plots in the design where the uth level of A is occurring, and $\sum_{u=1}^v n_{ui} = k_i$, the number of whole plots in the ith block. Let

$$\mathbf{R} = \mathbf{diag}(r_1, r_2, \cdots, r_v) \tag{8.3.2}$$

$$\text{and } \mathbf{K} = \mathbf{diag}(k_1, k_2, \cdots, k_b). \tag{8.3.3}$$

If $\mathbf{1}_m$ denotes the m-component vector of 1's, it is readily verified that (see the definitions of $\boldsymbol{F}_i$'s just preceding (8.1.2))

$$\begin{aligned}\boldsymbol{F}_i\mathbf{1}_v &= \mathbf{1}_{k_i},\\ \boldsymbol{F}_i'\mathbf{1}_{k_i} &= (n_{1i},\cdots,n_{vi})'\\ \boldsymbol{F}_i'\boldsymbol{F}_i &= \mathbf{diag}(n_{1i},\cdots,n_{vi})\\ \boldsymbol{F}'\boldsymbol{F} &= \sum_{i=1}^{b}\boldsymbol{F}_i'\boldsymbol{F}_i = \boldsymbol{R}.\end{aligned} \tag{8.3.4}$$

Let $B_i = s\sum_{j=1}^{k_i}\bar{y}_{ij.}$, $\mathbf{B} = (B_1,\ldots,B_b)'$, $T_u = s\sum_{i=1}^{b}\sum_{j=1}^{k_i} f_{ij}^u\bar{y}_{ij.}$, $\mathbf{T} = (T_1,\ldots,T_v)'$, $\mathbf{Q}_\tau = \mathbf{T} - \boldsymbol{N}\boldsymbol{K}^{-1}\mathbf{B}$ (vector of the treatment totals adjusted for the blocks), $\mathbf{Q}_\beta = \mathbf{B} - \boldsymbol{N}'\boldsymbol{R}^{-1}\mathbf{T}$ (vector of the block totals adjusted for the treatments), $\boldsymbol{C}_\tau = \boldsymbol{R} - \boldsymbol{N}\boldsymbol{K}^{-1}\boldsymbol{N}'$, and $C_\beta = \boldsymbol{K} - \boldsymbol{N}'\boldsymbol{R}^{-1}\boldsymbol{N}$. Define SSE_1 as

$$SSE_1 = s^2\sum_{i=1}^{b}\sum_{j=1}^{k_i}\bar{y}_{ij.}^2 - \sum_{i=1}^{b}\frac{B_i^2}{k_i} - \mathbf{Q}_\tau'\boldsymbol{C}_\tau^-\mathbf{Q}_\tau, \tag{8.3.5}$$

where $\boldsymbol{C}_\tau^-$ denotes a generalized inverse of $\boldsymbol{C}_\tau$. It may be noted that SSE_1/s is the usual error sum of squares for model (8.3.1). Write $n = \sum_{i=1}^{b}k_i = \sum_{u=1}^{v}r_u$. Assuming that the block design consisting of the blocks and whole plots is connected, the hypothesis H_τ: $\tau_1 = \tau_2 = \ldots = \tau_v$ can be tested using

$$F_\tau = \frac{(n-b-v+1)}{(v-1)}\frac{\mathbf{Q}_\tau'\boldsymbol{C}_\tau^-\mathbf{Q}_\tau}{SSE_1}, \tag{8.3.6}$$

which has a central F-distribution with $(v-1)$ and $(n-b-v+1)$ degrees of freedom, under H_τ. Similarly, H_β: $\beta_1 = \beta_2 = \ldots = \beta_b$ can be tested using

$$F_\beta = \frac{(n-b-v+1)}{(b-1)}\frac{\mathbf{Q}_\beta'\boldsymbol{C}_\beta^-\mathbf{Q}_\beta}{SSE_1}, \tag{8.3.7}$$

which has a central F-distribution with $(b-1)$ and $(n-b-v+1)$ degrees of freedom, under H_β. This completes the whole plot ANOVA using model (8.3.1).

The split-plot analysis of variance can be carried out in a straightforward manner using model (8.1.8). Let $\mathbf{r} = (r_1, r_2, \ldots, r_v)'$. Write $\mathbf{Q}_\gamma = [\boldsymbol{F}'(\boldsymbol{I}_n - 1/n\boldsymbol{J}_n)\otimes(\boldsymbol{I}_s - 1/s\boldsymbol{J}_s)]\,\mathbf{y}$, $\boldsymbol{C}_\gamma = (\boldsymbol{R} - 1/n\mathbf{r}\mathbf{r}')\otimes\boldsymbol{I}_s$, $SS(\gamma) = \mathbf{Q}_\gamma'\boldsymbol{C}_\gamma^-\mathbf{Q}_\gamma = \mathbf{Q}'_\gamma(\boldsymbol{R}^{-1}\otimes\boldsymbol{I}_s)\mathbf{Q}_\gamma$, $SS(\delta) = \mathbf{y}'\,[1/n\boldsymbol{J}_n\otimes(\boldsymbol{I}_s - 1/s\boldsymbol{J}_s)]\,\mathbf{y}$ and $SSE_2 = \mathbf{y}'\,[\boldsymbol{I}_n\otimes(\boldsymbol{I}_s - 1/s\boldsymbol{J}_s)]\,\mathbf{y} - SS(\delta) - SS(\gamma)$. Then $H_\gamma : \boldsymbol{\gamma} = \mathbf{0}$ can be tested using

$$F_\gamma = \frac{(n-v)(s-1)}{(v-1)(s-1)}\frac{SS(\gamma)}{SSE_2}, \tag{8.3.8}$$

which has a central F-distribution with $(v-1)(s-1)$ and $(n-v)(s-1)$ degrees of freedom under H_γ. If $\boldsymbol{\gamma} = \mathbf{0}$, then $H_\delta : \boldsymbol{\delta} = \mathbf{0}$ can be tested using

$$F_\delta = \frac{(n-1)(s-1)}{(s-1)} \frac{SS(\delta)}{SSE_2 + SS(\gamma)}, \tag{8.3.9}$$

which has a central F distribution with $(s-1)$ and $(n-1)(s-1)$ degrees of freedom under H_δ. From standard linear model theory, it also follows that the tests based on $F_\tau, F_\beta, F_\gamma$, and F_δ are UMPI for testing the respective hypotheses under appropriate groups of transformation.

As already pointed out in the beginning of this section, the tests based on F_τ and F_β in (8.3.6) and (8.3.7), respectively, are applicable when the underlying block design is connected. When connectedness is violated, suppose that $\text{rank}(\boldsymbol{C}_\tau) = v_1 < (v-1)$ and $\text{rank}(\boldsymbol{C}_\beta) = b_1 < (b-1)$. In this case, the hypothesis $H_\tau^* : \boldsymbol{C}_\tau \boldsymbol{\tau} = \mathbf{0}$ can be tested using the statistic

$$F_\tau^* = \frac{(n-b-v_1)}{v_1} \frac{\mathbf{Q}_\tau' \boldsymbol{C}_\tau^- \mathbf{Q}_\tau}{SSE_1}, \tag{8.3.10}$$

which has a central F-distribution with v_1 and $(n-b-v_1)$ degrees of freedom under H_τ^*. Similarly, the hypothesis $H_\beta^* : \boldsymbol{C}_\beta \boldsymbol{\beta} = \mathbf{0}$ can be tested using the statistic

$$F_\beta^* = \frac{(n-v-b_1)}{b_1} \frac{\mathbf{Q}_\beta' \boldsymbol{C}_\beta^- \mathbf{Q}_\beta}{SSE_1}, \tag{8.3.11}$$

which has a central F-distribution with b_1 and $(n-v-b_1)$ degrees of freedom under H_β^*. While the tests based on (8.3.10) and (8.3.11) may not be of interest in the fixed effects case, they provide valid exact F-tests for testing $H_\tau : \sigma_\tau^2 = 0$ and $H_\beta : \sigma_\beta^2 = 0$ (under appropriate conditions), when the τ_u's and β_j's are random variables as in some of the cases discussed below.

Case 8.3(ii). Random β_i's

In this case, we obviously have $\bar{\delta} = 0, (\mathbf{f_{ij}} \otimes \mathbf{1}_s')\boldsymbol{\gamma} = \mathbf{0}$, so that model (8.1.7) reduces to

$$E(\sqrt{s}\bar{y}_{ij.}) = \sqrt{s}(\mu + \mathbf{f}_{ij}'\boldsymbol{\tau})$$

and the covariance structure is given by

$$\begin{aligned} Cov(\sqrt{s}\bar{y}_{ij.}, \sqrt{s}\bar{y}_{i'j'.}) &= \sigma_*^2 + s\sigma_\beta^2, \; if \; i = i', \; j = j' \\ &= s\sigma_\beta^2, \quad \text{if } i = i', \; j \neq j' \\ &= 0 \text{ otherwise,} \end{aligned}$$

where σ_*^2 is as in (8.1.9). Thus model for $\sqrt{s}\bar{y}_{ij.}$ is a two-way unbalanced mixed model with random block effects and fixed treatment effects, and it can be used for testing H_β and H_τ. Appropriate exact (not necessarily optimum) and optimum invariant tests for these hypotheses are available in the literature (see Chapter 6). Of course, existence of optimum tests does require some conditions on the design (i.e., on the incidence matrix $\mathbf{N}$). One can continue to use F_τ as a valid exact F-test without any optimality property for testing H_τ if the underlying block design is connected. This is so because the test based on F_τ is a Wald's test. Similarly, F_β^* given in (8.3.11) provides a valid exact F-test (once again without any optimality property) for testing H_β even if the underlying block design is not connected. Theorems 6.2.1 and 6.2.2 can be used to get optimum tests and the conditions under which such tests will exist. For example, for testing H_τ, if the blocks and whole plots form a BIBD with parameters v, b, r (common value of the r_u's), k (common value of the k_i's), and λ, the test based on F_τ is LBIU if $s\sigma_\beta^2/\sigma_*^2$ is large. On the other hand, if $s\sigma_\beta^2/\sigma_*^2$ is small, the LBIU test rejects H_τ for large values of $\mathbf{T}'\mathbf{T}$, conditionally given $(\bar{y}_{...}, \mathbf{B}'\mathbf{B}, SSE_1)$, where $\bar{y}_{...} = 1/vrs\sum_{i,j,l} y_{ijl}$. We also note that for testing H_τ, we are in the situation where the recovery of inter-block information is possible. Thus, if the blocks and the whole plots form a BIBD, all the results given in Section 7.3 are applicable. If the design is not a BIBD, but the k_i's are all equal, the results in Section 7.5 are applicable. Since all the results in the above sections of Chapter 7 are applicable, we skip the details. It should be noted that the tests that recover inter-block information are preferable to the test based only on F_τ, since additional information is being used, as pointed out in Chapter 7.

For testing H_β, though, there always exists an optimum invariant test without any condition on the design. Applying Theorem 6.2.2, we conclude that if the nonzero eigenvalues of $\boldsymbol{C}_\beta$ are all equal, and equal to ϵ, a UMPI test exists and rejects H_β for large values of F_β^*, which now simplifies to $\{(n - v - b_1)/b_1\epsilon\} \cdot \{\mathbf{Q}_\beta'\mathbf{Q}_\beta/SSE_1\}$, where $b_1 = \text{rank}(\boldsymbol{C}_\beta)$. As already noted, this test statistic is distributed as central F with b_1 and $n - v - b_1$ degrees of freedom under H_β. On the other hand, if the nonzero eigenvalues of $\boldsymbol{C}_\beta$ are not all equal, then an LBI test exists and rejects H_β for large values of U_β where

$$U_\beta = \frac{\mathbf{Q}_\beta'\mathbf{Q}_\beta}{SSE_1 + \mathbf{Q}_\beta'\boldsymbol{C}_\beta^-\mathbf{Q}_\beta}. \tag{8.3.12}$$

Returning to tests for H_δ and H_γ, we note that since (8.1.8) is still a fixed effects model, the F-tests based on F_γ and F_δ defined earlier in this section under case 8.3(i) for the fixed effects case continue to be UMPI for H_γ and H_δ, respectively.

Case 8.3(iii). Random τ_u's

Here again $\bar{\delta} = 0$, although the interaction effects γ's are random so that the model (8.1.7) reduces to

$$E(\sqrt{s}\bar{y}_{ij.}) = \sqrt{s}(\mu + \beta_i)$$

and

$$\begin{aligned} Cov(\sqrt{s}\bar{y}_{ij.}, \sqrt{s}\bar{y}_{i'j'.}) &= \sigma_*^2 + (\mathbf{f}_{ij}'\mathbf{f}_{i'j'})(s\sigma_\tau^2 + \sigma_\gamma^2), \quad \text{if } i = i',\ j = j' \\ &= (\mathbf{f}_{ij}'\mathbf{f}_{i'j'})(s\sigma_\tau^2 + \sigma_\gamma^2), \ \text{otherwise.} \end{aligned}$$

We note that this model involves the two variance components σ_τ^2 and σ_γ^2 only through their sum, which implies that H_τ can be tested only if we assume that $\sigma_\gamma^2 = 0$ (of course, as discussed below, the hypothesis H_γ can be tested based on $\boldsymbol{P}_1\mathbf{y}_{ij.}$). Clearly, then, when $\sigma_\gamma^2 = 0$, the above model reduces to a two-way unbalanced mixed model with fixed block effects and random treatment effects. This is similar to the previous case except that the treatments and the blocks are interchanged. An exact test for H_τ based on F_τ^* continues to hold without any condition on the design. On the other hand, an optimum test for H_τ always exists when $\sigma_\gamma^2 = 0$. This test can be described as follows. If the nonzero eigenvalues of $\boldsymbol{C}_\tau$ are all equal, and equal to η, applying Theorem 6.2.2, we see that a UMPI test exists and rejects H_τ for large values of F_τ^*, which now simplifies to $\{(n - b - v_1)/v_1\eta\} \cdot \{\mathbf{Q}_\tau'\mathbf{Q}_\tau/SSE_1\}$, where v_1=rank($\boldsymbol{C}_\tau$). This test statistic is distributed as central F with v_1 and $n - b - v_1$ degrees of freedom under H_τ. On the other hand, if the nonzero eigenvalues of $\boldsymbol{C}_\tau$ are not all equal, then an LBI test exists and rejects H_τ for large values of U_τ, where

$$U_\tau = \frac{\mathbf{Q}_\tau'\mathbf{Q}_\tau}{SSE_1 + \mathbf{Q}_\tau'\boldsymbol{C}_\tau^-\mathbf{Q}_\tau}. \tag{8.3.13}$$

For testing H_β, we note that $(s\sigma_\tau^2 + \sigma_\gamma^2)$ can be treated as a single variance component. An exact test without any optimality can then be based on F_β whatever σ_γ^2 and σ_τ^2 may be, provided that the design is connected. An LBIU test for H_β exists if the underlying block design is the dual of a BIBD (i.e., $\boldsymbol{N}'$ is the incidence matrix of a BIBD). This follows from Theorem 6.2.1. Thus, if $(s\sigma_\tau^2 + \sigma_\gamma^2)/\sigma_*^2$ is large, such an LBIU test rejects H_β for large values of F_β. On the other hand, if $(s\sigma_\tau^2 + \sigma_\gamma^2)/\sigma_*^2$ is small, the LBIU test rejects H_β for large values of $\mathbf{B}'\mathbf{B}$, conditionally given $(\bar{y}_{....}, \mathbf{T}'\mathbf{T}, SSE_1)$. Since the block effects are fixed and the treatment effects are random, an exact test that recovers "intertreatment information" can be obtained for testing H_β, using the results in Chapter 7. For example, if the underlying block design is the dual of a BIBD, the results in Section 7.3 are applicable.

Returning to model (8.1.8), since the γ_{lu}'s are random, the model reduces to

$$E(\boldsymbol{P}_1\mathbf{y}_{ij\cdot}) = \boldsymbol{P}_1\boldsymbol{\delta}$$

and

$$\begin{aligned} Cov(\boldsymbol{P}_1\mathbf{y}_{ij\cdot}, \boldsymbol{P}_1\mathbf{y}_{i'j'\cdot}) &= (\sigma_{**}^2 + \sigma_\gamma^2 \mathbf{f}_{ij}'\mathbf{f}_{i'j'})\boldsymbol{I}_{s-1}, \quad \text{if } i = i',\ j = j' \\ &= \sigma_\gamma^2(\mathbf{f}_{ij}'\mathbf{f}_{i'j'})\boldsymbol{I}_{s-1}, \text{ otherwise,} \end{aligned}$$

where σ_{**}^2 is as in (8.1.10). Again, if $\sigma_\gamma^2 = 0$, H_δ can be easily tested using F_δ as in the fixed effects case and the resultant test is UMPI. To test H_γ, we note that the model for $\boldsymbol{P}_1\mathbf{y}_{ij\cdot}$ involves two variance components and hence Theorem 6.2.2 is applicable. Routine computations then yield that, if $\boldsymbol{R} - 1/n\ \mathbf{rr}'$ has equal nonzero eigenvalues, or equivalently, if the r_i's are all equal, the test based on F_γ is UMPI. If the r_i's are not all equal, then the test which rejects H_γ for large values of U_γ where

$$U_\gamma = \frac{\mathbf{Q}_\gamma'\mathbf{Q}_\gamma}{SSE_2 + \mathbf{Q}_\gamma'\boldsymbol{C}_\gamma^-\mathbf{Q}_\gamma}, \tag{8.3.14}$$

is LBI. Of course, in any event, the test based on F_γ continues to be an exact test without any condition on the r_i's.

Case 8.3(iv). Random δ_l's

In this case, we have

$$\begin{aligned} E(\sqrt{s}\bar{y}_{ij.}) &= \sqrt{s}(\mu + \beta_i + \mathbf{f}_{ij}'\boldsymbol{\tau}) \\ Cov(\sqrt{s}\bar{y}_{ij.}, \sqrt{s}\bar{y}_{i'j'.}) &= \sigma_*^2 + \sigma_\delta^2 + (\mathbf{f}_{ij}'\mathbf{f}_{i'j'})\sigma_\gamma^2, \quad \text{if } i = i',\ j = j' \\ &= \sigma_\delta^2 + (\mathbf{f}_{ij}'\mathbf{f}_{i'j'})\sigma_\gamma^2, \text{ otherwise.} \end{aligned}$$

We note that in the variance–covariance matrix of the vector consisting of the $\sqrt{s}\bar{y}_{ij.}$'s, the coefficient matrix of σ_δ^2 is $\mathbf{1}_n\mathbf{1}_n'$. Hence, if $\sigma_\gamma^2 = 0$, then Lemma A.8.1(*iii*) in Appendix 8.1 is applicable for testing H_τ, and it follows that the F-test based on F_τ is UMPI. For testing H_β, it can once again be verified that Lemma A.8.1(*iii*) is applicable (even if $\sigma_\gamma^2 \neq 0$) and the test based on F_β is UMPI.

Model (8.1.8), on the other hand, can be written as

$$\begin{aligned} E[(\boldsymbol{I}_n \otimes \boldsymbol{P}_1)\mathbf{y}] &= \mathbf{0} \\ Var[(\boldsymbol{I}_n \otimes \boldsymbol{P}_1)\mathbf{y}] &= \sigma_\delta^2(\boldsymbol{J}_n \otimes \boldsymbol{I}_{s-1}) + \sigma_\gamma^2(\boldsymbol{F}\boldsymbol{F}' \otimes \boldsymbol{I}_{s-1}) + \sigma_{**}^2\boldsymbol{I}_{n(s-1)}, \end{aligned} \tag{8.3.15}$$

where $\boldsymbol{J}_n = \mathbf{1}_n\mathbf{1}_n'$. If $\sigma_\gamma^2 = 0$, the above model involves only two variance components and a UMPI test for H_δ can be obtained by applying Theorem 6.2.2. The UMPI test always exists and coincides with the test based on F_δ. Testing H_γ requires special attention since model (8.3.15) involves three variance components. Lemma A.8.3 in Appendix 8.1 is suitable in this situation. To apply Lemma A.8.3, let $\boldsymbol{M} = [\boldsymbol{F}\boldsymbol{R}^{-1/2} : \boldsymbol{M}_1]'$ be an $n \times n$ orthogonal matrix (recall that $\boldsymbol{F}'\boldsymbol{F} = \boldsymbol{R}$, as mentioned in formula (8.3.4)). Thus $\boldsymbol{M}_1'\boldsymbol{F} = 0$ and $\boldsymbol{M}_1'\mathbf{1}_n = 0$, since $\boldsymbol{F}\mathbf{1}_v = \mathbf{1}_n$. Hence $\boldsymbol{M}\boldsymbol{F}\boldsymbol{F}'\boldsymbol{M}' = \mathbf{diag}(\boldsymbol{R}, 0)$ and

$$\boldsymbol{M}\mathbf{1}_n = [\mathbf{r}_0' : 0]', \quad \text{where } \mathbf{r}_0 = (\sqrt{r_1}, \sqrt{r_2}, \cdots \sqrt{r_v})'.$$

Consider the random vector $(\boldsymbol{M} \otimes \boldsymbol{I}_{s-1})(\boldsymbol{I}_n \otimes \boldsymbol{P}_1)\mathbf{y} = (\boldsymbol{M} \otimes \boldsymbol{P}_1)\mathbf{y}$, which has a mean $\mathbf{0}$ and a variance–covariance matrix (from (8.3.15))

$$\begin{aligned} Var[(\boldsymbol{M} \otimes \boldsymbol{P}_1)\mathbf{y}] &= \mathbf{diag}[\sigma_{**}^2\boldsymbol{I}_{v(s-1)} + \sigma_\delta^2(\boldsymbol{r}_0\boldsymbol{r}_0' \otimes \boldsymbol{I}_{s-1}) \\ &\quad + \sigma_\gamma^2(\boldsymbol{R} \otimes \boldsymbol{I}_{s-1}),\ \sigma_{**}^2\boldsymbol{I}_{(n-v)(s-1)}]. \end{aligned} \tag{8.3.16}$$

We now note that if the r_i's are all equal, then (8.3.16) is similar to (A.8.2) in Appendix 8.1. Applying Lemma A.8.3, we therefore conclude that when the r_i's are equal, a UMPU test exists for testing H_γ, and it indeed coincides with the F-test based on F_γ. Furthermore, the same test is also UMPI. When the r_i's are unequal, Lemma A.8.3 can be suitably modified to yield an LBI test for H_γ. Details are omitted.

Case 8.3(v). Random τ_j's and δ_l's

Here, the model in (8.1.7) yields

$$E(\sqrt{s}\bar{y}_{ij.}) = \sqrt{s}(\mu + \beta_i)$$

and

$$\begin{aligned} Cov(\sqrt{s}\bar{y}_{ij.}, \sqrt{s}\bar{y}_{i'j'.}) &= \sigma_*^2 + \sigma_\delta^2 + (\mathbf{f}_{ij}'\mathbf{f}_{i'j'})(s\sigma_\tau^2 + \sigma_\gamma^2), \quad \text{if } i = i',\ j = j' \\ &= \sigma_\delta^2 + (\mathbf{f}_{ij}'\mathbf{f}_{i'j'})(s\sigma_\tau^2 + \sigma_\gamma^2), \text{ otherwise.} \end{aligned}$$

This model is similar to the model under random τ_j's discussed in case 8.3(iii), except for the extra term σ_δ^2 in the expression for the covariance. However, in view of Lemmas A.8.2 and A.8.1(*iii*) of Appendix 8.1, the class of invariant tests for testing H_τ and H_β does not depend on σ_δ^2 and hence σ_δ^2 can be ignored while deriving invariant tests for these hypotheses. Thus an exactly similar analysis as in case 8.3(iii) obtains, and the invariant tests described there continue to hold. In particular, F_β can be used to test H_β for connected designs without any condition on σ_γ^2 and σ_τ^2, and when $\sigma_\gamma^2 = 0$, F_τ^* can be used to test H_τ without any condition on the design. Also, tests that recover inter-block information can be obtained using the results in Chapter 7 under appropriate conditions. Returning to model (8.1.8) for $\boldsymbol{P}_1\mathbf{y}_{ij\cdot}$, it is easily seen to be the same as the corresponding model discussed in case 8.3(iv). Therefore, tests for H_δ and H_γ remain the same.

Case 8.3(vi). Random β_i's and τ_u's

Since the split-plot effects δ_l's are fixed, we get $\bar{\delta} = 0$, and model (8.1.7) reduces to

$$E(\sqrt{s}\bar{y}_{ij.}) = \sqrt{s}\mu$$

and

$$\begin{aligned} Cov(\sqrt{s}\bar{y}_{ij.}, \sqrt{s}\bar{y}_{i'j'.}) &= \sigma_*^2 + s\sigma_\beta^2 + (\mathbf{f}_{ij}'\mathbf{f}_{i'j'})(s\sigma_\tau^2 + \sigma_\gamma^2), \quad \text{if } i = i',\ j = j' \\ &= s\sigma_\beta^2 + (\mathbf{f}_{ij}'\mathbf{f}_{i'j'})(s\sigma_\tau^2 + \sigma_\gamma^2), \quad \text{if } i = i', j \neq j' \\ &= (\mathbf{f}_{ij}'\mathbf{f}_{i'j'})(s\sigma_\tau^2 + \sigma_\gamma^2), \text{ otherwise.} \end{aligned}$$

We first note that for testing H_β, the model can be treated as a two-way unbalanced random effects model involving three variance components, namely, σ_*^2, σ_β^2, and $(s\sigma_\tau^2 + \sigma_\gamma^2)$. Hence it follows that the test based on F_β^* continues to be a valid test for H_β. Theorem 4.3.1 can be used to obtain an LBIU test in the equiblock (i.e., $k_1 = \ldots = k_b$) and equireplicate (i.e., $r_1 = \ldots = r_v$) case. The LBIU test in this case rejects H_β for large values of $\mathbf{Q}_\beta'\mathbf{Q}_\beta/SSE_1$ if $(s\sigma_\tau^2 + \sigma_\gamma^2)/\sigma_*^2$ is large, and for large values of $\mathbf{B}'\mathbf{B}$, conditionally given $(\bar{y}_{...}, \mathbf{T}'\mathbf{T}, SSE_1)$, if $(s\sigma_\tau^2 + \sigma_\gamma^2)/\sigma_*^2$ is small.

Next, for testing H_τ, we note that when $\sigma_\gamma^2 = 0$, the model for $\sqrt{s}\bar{y}_{ij.}$ is a two-way unbalanced random effects model without interaction, and therefore the F-test based on F_τ^* continues to be a valid test. Theorem 4.3.1 can again be applied to yield an LBIU test if the resultant design is equiblock and equireplicate. Specifically, the LBIU test rejects H_τ for large values of $\mathbf{Q}_\tau'\mathbf{Q}_\tau/SSE_1$ if $\sigma_\beta^2/\sigma_*^2$ is large, and for large values of $\mathbf{T}'\mathbf{T}$, conditionally given $(\bar{y}_{...}, \mathbf{B}'\mathbf{B}, SSE_1)$, if $\sigma_\beta^2/\sigma_*^2$ is small.

Model (8.1.8) for $\boldsymbol{P}_1\mathbf{y}_{ij.}$ under random β_i's and τ_u's is easily seen to be the same as the corresponding model in case 8.3(iii). Hence the tests for H_δ and H_γ discussed there continue to hold in this case.

Case 8.3(vii). Random β_i's and δ_l's

In this case, model (8.1.7) reduces to

$$E(\sqrt{s}\bar{y}_{ij.}) = \sqrt{s}(\mu + \mathbf{f}_{ij}'\boldsymbol{\tau})$$

and

$$\begin{aligned} Cov(\sqrt{s}\bar{y}_{ij.}, \sqrt{s}\bar{y}_{i'j'.}) &= \sigma_*^2 + s\sigma_\beta^2 + \sigma_\delta^2 + (\mathbf{f}_{ij}'\mathbf{f}_{i'j'})\sigma_\gamma^2, \quad \text{if } i = i',\ j = j' \\ &= s\sigma_\beta^2 + \sigma_\delta^2 + (\mathbf{f}_{ij}'\mathbf{f}_{i'j'})\sigma_\gamma^2, \quad \text{if } i = i',\ j \neq j' \\ &= \sigma_\delta^2 + (\mathbf{f}_{ij}'\mathbf{f}_{i'j'})\sigma_\gamma^2, \text{ otherwise.} \end{aligned}$$

This model involves the fixed effects, μ and τ_u's, and four variance components. However, using Lemma A.8.2 in Appendix 8.1, we see that an invariant test for H_β will not depend on the two variance components σ_δ^2 and σ_γ^2 and hence these can be ignored for deriving such a test. This results in a two-way unbalanced mixed effects model with fixed treatment effects and random block effects. Hence Theorem 6.2.2 can be used to derive an optimum invariant test for H_β and the UMPI or LBI test is as given in case 8.3(ii). Of course, the invariant test based on F_β^* continues to be valid.

For testing H_τ, if $\sigma_\gamma^2 = 0$, an application of Lemma A.8.1(*iii*) in Appendix 8.1 shows that invariant tests for H_τ will not depend on σ_δ^2, and hence this variance component can be ignored for deriving such a test. Thus the above model once again reduces to a two-way unbalanced mixed effects model with fixed treatment effects and random block effects, and hence the test based on F_τ continues to be valid. Moreover, when the underlying block design is a BIBD, an LBIU test can be obtained as in case 8.3(ii), and exact tests that recover inter-block information can be constructed as already pointed out.

Model (8.1.8) for $\boldsymbol{P}_1\mathbf{y}_{ij.}$ is easily seen to remain the same as that discussed under case 8.3(iv). Therefore, the same tests for H_δ and H_γ as described before continue to hold in this case.

Case 8.3(viii). Random β_i's, τ_u's and δ_l's

In this case, model (8.1.7) reduces to

$$E(\sqrt{s}\bar{y}_{ij.}) = \sqrt{s}\mu$$

and

$$\begin{aligned} Cov(\sqrt{s}\bar{y}_{ij.}, \sqrt{s}\bar{y}_{i'j'.}) &= \sigma_*^2 + s\sigma_\beta^2 + \sigma_\delta^2 + (\mathbf{f}_{ij}'\mathbf{f}_{i'j'})(s\sigma_\tau^2 + \sigma_\gamma^2), \quad \text{if } i = i',\ j = j' \\ &= s\sigma_\beta^2 + \sigma_\delta^2 + (\mathbf{f}_{ij}'\mathbf{f}_{i'j'})(s\sigma_\tau^2 + \sigma_\gamma^2), \quad \text{if } i = i', j \neq j' \\ &= \sigma_\delta^2 + (\mathbf{f}_{ij}'\mathbf{f}_{i'j'})(s\sigma_\tau^2 + \sigma_\gamma^2), \ \text{otherwise.} \end{aligned}$$

We first note that in view of Lemma A.8.2 given in Appendix 8.1, the variance component σ_δ^2 can be ignored for deriving invariant tests for H_β and H_τ. This results in the model given under case 8.3(vi), and hence all the conclusions in case 8.3(vi) are valid for testing H_β and H_τ. The model for $\boldsymbol{P}_1\mathbf{y}_{ij.}$ once again reduces to the corresponding model under case 8.3(iv).

We summarize below all the results on exact tests in Table 8.1 and those on optimum tests in Table 8.2. We also point out that the null distributions of the test statistics U_β, U_τ, and U_γ, given respectively in (8.3.12), (8.3.13), and (8.3.14), can be approximated using Hirotsu's (1979) approximation (see Appendix 3.1 in Chapter 3).

Table 8.1. Test Statistics for Exact Tests in an Unbalanced Split-Plot Design with Conditions for Their Validity

Nature of Effects	$H_{0\beta}$	$H_{0\tau}$	$H_{0\delta}$	$H_{0\gamma}$
1. All fixed	F_β Condition: block design is connected	F_τ Condition: block design is connected	F_δ Condition: $\gamma = 0$	$F\gamma$ Condition: none
2. β random, others fixed	F_β^* Condition: none	F_τ Condition: block design is connected	Same as in 1	Same as in 1
3. τ, γ random, others fixed	F_β Condition: block design is connected	F_τ^* Condition: $\sigma_\gamma^2 = 0$	F_δ Condition: $\sigma_\gamma^2 = 0$	Same as in 1
4. δ, γ random, others fixed	Same as in 1	F_τ Condition: block design is connected and $\sigma_\gamma^2 = 0$	Same as in 3	Same as in 1
5. τ, δ, γ random, others fixed	Same as in 3	Same as in 3	Same as in 3	Same as in 1
6. β, τ, γ random, others fixed	Same as in 2	Same as in 3	Same as in 3	Same as in 1
7. β, δ, γ random, others fixed	Same as in 2	Same as in 4	Same as in 3	Same as in 1
8. All random	Same as in 2	Same as in 3	Same as in 3	Same as in 1

Source: T. Mathew and B.K. Sinha (1992). Reproduced with permission of the American Statistical Association.

Table 8.2. Test Statistics for Optimum Tests in an Unbalanced Split-Plot Design with Nature of Optimality and Conditions for Their Validity

Nature of Effects	$H_{0\beta}$	$H_{0\tau}$	$H_{0\delta}$	$H_{0\gamma}$
1. All fixed	F_β Nature of optimality: UMPI Condition: block design is connected	F_τ Nature of optimality: UMPI Condition: block design is connected	F_δ Nature of optimality: UMPI Condition: $\gamma = 0$	$F\gamma$ Nature of optimality: UMPI Condition: none
2. β random, others fixed	(i) F_β^* Nature of optimality: UMPI Condition: $\boldsymbol{C}_\beta$ has equal nonzero eigenvalues (ii) U_β Nature of optimality: LBI Condition: None	(i) F_τ Nature of optimality: LBIU Condition: $s\sigma_\beta^2/\sigma_*^2$ is large and block design is BIBD (ii) $\mathbf{T'T}$[†] Nature of optimality: LBIU Condition: $s\sigma_\beta^2/\sigma_*^2$ is small and block design is BIBD	Same as in 1	Same as in 1
3. τ, γ random, others fixed	(i) F_β Nature of optimality: LBIU Condition: $(s\sigma_\tau^2 + \sigma_\gamma^2)/\sigma_*^2$ is large and block design is the dual of a BIBD (ii) $\mathbf{B'B}$[‡] Nature of optimality: LBIU Condition: $(s\sigma_\tau^2 + \sigma_\gamma^2)/\sigma_*^2$ is small and block design is dual of a BIBD	(i) F_τ^* Nature of optimality: UMPI Condition: $\boldsymbol{C}_\tau$ has equal nonzero eigenvalues and $\sigma_\gamma^2 = 0$ (ii) U_τ Nature of optimality: LBI Condition: $\sigma_\gamma^2 = 0$	F_δ Nature of optimality: UMPI Condition: $\sigma_\gamma^2 = 0$	(i) F_γ Nature of optimality: UMPI Condition: $\boldsymbol{C}_\gamma$ has equal nonzero eigenvalues (ii) U_γ Nature of optimality: LBI Condition: none

4. δ, γ random, others fixed	Same as in 1	F_τ Nature of optimality: UMPI Condition: block design is connected and $\sigma_\gamma^2 = 0$	Same as in 3	F_γ Nature of optimality: LBI Condition: r_j's are equal
5. τ, δ, γ random, others fixed	Same as in 3	Same as in 3	Same as in 3	Same as in 4
6. β, τ, γ random, others fixed	(i) $\mathbf{Q}'_\beta \mathbf{Q}_\beta / SS_{E_1}$ Nature of optimality: LBIU Condition: $(s\sigma_\tau^2 + \sigma_\gamma^2)/\sigma_*^2$ is large, k_i's are equal and r_j's are equal (ii) $\mathbf{B}'\mathbf{B}$‡ Nature of optimality: LBIU Condition: $(s\sigma_\tau^2 + \sigma_\gamma^2)/\sigma_*^2$ is small, k_i's are equal and r_j's are equal	(i) $\mathbf{Q}'_\tau \mathbf{Q}_\tau / SS_{E_1}$ Nature of optimality: LBIU Condition: $s\sigma_\beta^2/\sigma_*^2$ is large, k_i's are equal and r_j's are equal (ii) $\mathbf{T}'\mathbf{T}$† Nature of optimality: LBIU Condition: $s\sigma_\beta^2/\sigma_*^2$ is small, k_i's are equal and r_j's are equal	Same as in 3	Same as in 3
7. β, δ, γ random, others fixed	Same as in 2	Same as in 2, with the additional condition $\sigma_\gamma^2 = 0$	Same as in 3	Same as in 4
8. All random	Same as in 6	Same as in 6	Same as in 3	Same as in 4

†Conditionally given $(\bar{y}_{...}, \mathbf{B}'\mathbf{B}, SS_{E_1})$.
‡Conditionally given $(\bar{y}_{...}, \mathbf{T}'\mathbf{T}, SS_{E_1})$.
Source: T. Mathew and B.K. Sinha (1992). Reproduced with permission of the American Statistical Association.

8.4. A NUMERICAL EXAMPLE

We shall now apply the results of the analysis of an *unbalanced* split-plot model to a numerical example taken essentially again from Montgomery (1991, p. 468). The example deals with a paper manufacturer who is interested in three different pulp preparation methods and four different cooking temperatures for the pulp and wishes to study their effects on the tensile strength of the paper. The design for this study is as follows. On any day, a batch of pulp is produced by one of the three methods under study. Then this batch is divided into four samples, and each sample is cooked at one of the four temperatures. Then a second batch of pulp is made by another of the three methods, and it is also divided into four samples that are tested at the four temperatures, and so on. The entire data for this study are shown in Table 8.3. We mention that this is a modification of the data on page 468 of Montgomery (1991).

Clearly, we have an example of an *unbalanced* split-plot design of the type discussed in the text in Section 8.3. Moreover, this is indeed a *fixed* effects model with three *block* effects (days), three *whole plot* effects (pulp preparation methods), and four *split-plot* effects (temperatures). To derive the exact tests of various fixed effects, we follow the derivations under case 8.3(i), and compute the relevant quantities as follows.

$b = 3$, $v = 3$, $s = 4$, $n = 7$, $r_1 = 2$, $r_2 = 3$, $r_3 = 2$, $k_1 = 2$, $k_2 = 3$, $k_3 = 2$, $\mathbf{R} = \mathbf{diag}(2,3,2)$, $\mathbf{K} = \mathbf{diag}(2,3,2)$, $\mathbf{B} = (293, 423, 308)'$, $\mathbf{T} = (279, 462, 283)'$, $\mathbf{N}: 3 \times 3$ with its column vectors as $[\mathbf{N}_1, \mathbf{N}_2, \mathbf{N}_3]$ where $\mathbf{N}_1 = (1,1,0)'$, $\mathbf{N}_2 = (1,1,1)'$, $\mathbf{N}_3 = (0,1,1)'$, $\mathbf{F}: 7 \times 3$ with its column vectors as $[\mathbf{F}_1, \mathbf{F}_2, \mathbf{F}_3]$ where $\mathbf{F}_1 = (1,0,1,0,0,0,0)'$, $\mathbf{F}_2 = (0,1,0,1,0,1,0)'$, $\mathbf{F}_3 = (0,0,0,0,1,0,1)'$, $\mathbf{Q}_\tau = (-8.5, 20.5, -12.0)'$, $\mathbf{Q}_\beta = (-0.5, -12.0, 12.5)'$, $\mathbf{C}_\tau: 3 \times 3$ with its column vectors as $[\mathbf{C}_1, \mathbf{C}_2, \mathbf{C}_3]$ where $\mathbf{C}_1 = (1.1667, -0.8333, -0.3333)'$, $\mathbf{C}_2 = (-0.8333, 1.6667, -0.8333)'$, $\mathbf{C}_3 = (-0.3333, -0.8333, 1.1667)'$, $\mathbf{C}_\beta: 3 \times 3$ with its column vectors as $[\mathbf{C}_1^*, \mathbf{C}_2^*, \mathbf{C}_3^*]$ where $\mathbf{C}_1^* = (1.1667, -0.8333, -0.3333)'$, $\mathbf{C}_2^* = (-0.8333, 1.6667, -0.8333)'$, $\mathbf{C}_3^* = (-0.3333, -0.8333, 1.1667)'$, $\mathbf{C}_\tau^- = \mathbf{C}_\beta^-: 3 \times 3$ with its three column vectors as $(0.4000, -0.1333, -0.3333)'$, $(-0.1333, 0.2667, -0.1333)'$, $(-0.2667, -0.1333, 0.4000)'$, $\mathbf{Q}_\tau'\mathbf{C}_\tau^-\mathbf{Q}_\tau = 256.2333$, $\mathbf{Q}_\beta'\mathbf{C}_\beta^-\mathbf{Q}_\beta$

Table 8.3. Data on Tensile Strength of the Paper

Pulp Preparation Method	Block 1		Block 2			Block 3	
Temperature	1	2	1	2	3	2	3
200	30	34	28	31	31	35	32
225	35	41	32	36	30	40	34
250	37	38	40	42	32	30	39
275	36	42	41	40	40	44	45

Source: D.C. Montgomery (1991). Reproduced with permission of John Wiley & Sons, Inc.

puted under $Var(\mathbf{y}) = \sigma^2 \boldsymbol{V}$ for any given $\boldsymbol{V}$ satisfying the condition in (i), coincides with the F-statistic computed under $Var(\mathbf{y}) = \sigma^2 \boldsymbol{I}$. Furthermore, the F-statistic computed under $Var(\mathbf{y}) = \sigma^2 \boldsymbol{V}$ has the same null and nonnull distributions as in the case $\boldsymbol{V} = \boldsymbol{I}$.

Proof of Lemma A.8.1: For a proof of (i), see Corollary 2.1 in Mathew and Bhimasankaram (1983). To prove (ii) and (iii), we assume that the problem is in canonical form (see Lehmann, 1986, Chapter 7). Thus, let $\mathbf{y} = (\mathbf{y}_1', \mathbf{y}_2', \mathbf{y}_3')'$, $E(\mathbf{y}) = (\boldsymbol{\beta}_1', \boldsymbol{\beta}_2', \mathbf{0})'$, and suppose the problem is to test $H_0 : \boldsymbol{\beta}_1 = \mathbf{0}$. If $\boldsymbol{V}$ satisfies the condition $(\boldsymbol{I} - \boldsymbol{P}_0)\boldsymbol{V}(\boldsymbol{I} - \boldsymbol{P}_0) = a(\boldsymbol{I} - \boldsymbol{P}_0)$ for some $a > 0$, then it is easy to see that

$$\boldsymbol{V} = a \begin{pmatrix} \boldsymbol{I} & \boldsymbol{V}_{12} & \mathbf{0} \\ \boldsymbol{V}_{21} & \boldsymbol{V}_{22} & \boldsymbol{V}_{23} \\ \mathbf{0} & \boldsymbol{V}_{32} & \boldsymbol{I} \end{pmatrix} \tag{A.8.1}$$

where $\boldsymbol{V}_{12}$, $\boldsymbol{V}_{22}$, and $\boldsymbol{V}_{23}$ are arbitrary matrices subject only to the condition that $\boldsymbol{V}$ is positive definite. When $Var(\mathbf{y}) = \sigma^2 \boldsymbol{I}$, the above testing problem is invariant under the group of transformations $(\mathbf{y}_1, \mathbf{y}_2, \mathbf{y}_3) \to c(\boldsymbol{Q}_1\mathbf{y}_1, \mathbf{y}_2 + \boldsymbol{\alpha}_2, \boldsymbol{Q}_3\mathbf{y}_3)$, where $c > 0$, $\boldsymbol{\alpha}_2$ is an arbitrary vector and $\boldsymbol{Q}_1$ and $\boldsymbol{Q}_3$ are orthogonal matrices of appropriate dimensions. It is obvious that when $Var(\mathbf{y}) = \sigma^2 \boldsymbol{V}$, where $\boldsymbol{V}$ is given by (A.8.1), then the same group leaves the testing problem invariant and the usual UMPI F-test obtained using $Var(\mathbf{y}) = \sigma^2 \boldsymbol{I}$ is also UMPI when $Var(\mathbf{y}) = \sigma^2 \boldsymbol{V}$.

The proof of (iii) is immediate once we note that the condition on $\boldsymbol{V}$ given in (iii) is equivalent to

$$\boldsymbol{V} = a \begin{pmatrix} \boldsymbol{I} & \mathbf{0} & \mathbf{0} \\ \mathbf{0} & \boldsymbol{I} + \boldsymbol{V}_{22} & \mathbf{0} \\ \mathbf{0} & \mathbf{0} & \boldsymbol{I} \end{pmatrix}.$$

This completes the proof of Lemma A.8.1. □

Lemma A.8.2. Let $\mathbf{y}$ be a normally distributed random vector with $E(\mathbf{y}) = \boldsymbol{X\beta}$, where these quantities are as defined in Lemma A.8.1, and $Var(\mathbf{y}) = \boldsymbol{V}(\boldsymbol{\theta})$, where $\boldsymbol{V}(\boldsymbol{\theta})$ is a function of a vector $\boldsymbol{\theta}$ consisting of unknown parameters. For testing hypotheses concerning $\boldsymbol{\theta}$, the class of tests that are invariant under the group of transformations $\mathbf{y} \to \mathbf{y} + \boldsymbol{X\alpha}$ (where $\boldsymbol{\alpha}$ is an arbitrary vector) coincides with the same class when $Var(\mathbf{y}) = \boldsymbol{V}(\boldsymbol{\theta}) + \boldsymbol{V}_1$ when range $(\boldsymbol{V}_1) \subset$ range $(\boldsymbol{X})$.

Proof. Suppose $\boldsymbol{X}$ is an $n \times m$ matrix of rank r $(r < n)$. Let $\boldsymbol{Z}$ be an $n \times (n - r)$ matrix satisfying $\boldsymbol{Z}'\boldsymbol{X} = 0$ and $\boldsymbol{Z}'\boldsymbol{Z} = \boldsymbol{I}_{n-r}$. Then it is readily verified that a maximal invariant under the group of transformations $\mathbf{y} \to \mathbf{y} + \boldsymbol{X\alpha}$ is the vector $\boldsymbol{Z}'\mathbf{y}$. When range $(\boldsymbol{V}_1) \subset$ range $(\boldsymbol{X})$, $\boldsymbol{Z}'\boldsymbol{V}_1 = 0$. Hence the distribution of $\boldsymbol{Z}'\mathbf{y}$ under $Var(\mathbf{y}) = \boldsymbol{V}(\boldsymbol{\theta}) + \boldsymbol{V}_1$ is the same as its distribution under $Var(\mathbf{y}) = \boldsymbol{V}(\boldsymbol{\theta})$, from which the lemma follows. □

Lemma A.8.3, stated below, deals with a model involving three variance components. Let $\mathbf{y}$ be a $p \times 1$ random vector with $E(\mathbf{y}) = \mathbf{0}$ and

$$Var(\mathbf{y}) = \mathbf{diag}(\sigma^2 \boldsymbol{I}_{vm} + \sigma_1^2(\mathbf{1}_v\mathbf{1}_v' \otimes \boldsymbol{I}_m) + \sigma_2^2 \boldsymbol{I}_{vm},\ \sigma^2 \boldsymbol{I}_{p-vm}) \qquad \text{(A.8.2)}$$

where m is a positive integer satisfying $vm < p$. Consider the testing problem $H_0 : \sigma_2^2 = 0$ versus $H_1 : \sigma_2^2 > 0$.

In Lemma A.8.3, we derive a test that is both UMPU and UMPI for testing H_0. The proof of Lemma A.8.3 is based on the following canonical reduction of the problem.

Suppose $\mathbf{y}$ is partitioned as $\mathbf{y} = [\mathbf{y}_1' : \mathbf{y}_2']'$ where $\mathbf{y}_1$ is $vm \times 1$. Let $\boldsymbol{H} = [1/\sqrt{v}\mathbf{1}_v : \boldsymbol{H}_1{}']'$ be a $v \times v$ orthogonal matrix. Consider

$$(\boldsymbol{H} \otimes \boldsymbol{I}_m)\mathbf{y}_1 = \mathbf{w}_1 = [\mathbf{w}_{11}' : \mathbf{w}_{12}']' \qquad \text{(A.8.3)}$$

where

$$\mathbf{w}_{11} = \left(\frac{1}{\sqrt{v}}\mathbf{1}_s' \otimes \boldsymbol{I}_m\right)\mathbf{y}_1 \qquad \text{(A.8.4)}$$

and

$$\mathbf{w}_{12} = (\boldsymbol{H}_1 \otimes \boldsymbol{I}_m)\mathbf{y}_1. \qquad \text{(A.8.5)}$$

From (A.8.2) and (A.8.3), and using $\boldsymbol{H}\mathbf{1}_v\mathbf{1}_v'\boldsymbol{H}' = \mathbf{diag}(v, 0, \ldots 0)$, we get

$$Cov(\mathbf{w}_1', \mathbf{y}_2')' = \mathbf{diag}[(\sigma^2 + v\sigma_1^2 + \sigma_2^2)\boldsymbol{I}_m,\ (\sigma^2 + \sigma_2^2)\boldsymbol{I}_{(v-1)m},\ \sigma^2 \boldsymbol{I}_{p-vm}]. \qquad \text{(A.8.6)}$$

If $\boldsymbol{P}_s$ denotes an $s \times s$ orthogonal matrix and c is a positive scalar, it is readily verified that the above testing problem is left invariant under the group of transformations $(c, \boldsymbol{P}_m, \boldsymbol{P}_{(v-1)m}, \boldsymbol{P}_{(p-vm)})$ acting as $(\mathbf{w_{11}}, \mathbf{w_{12}}, \mathbf{y_2}) \to c(\boldsymbol{P}_m\mathbf{w_{11}}, \boldsymbol{P}_{(v-1)m}\mathbf{w_{12}}, \boldsymbol{P}_{(p-vm)}\mathbf{y_2})$.

Lemma A.8.3. Consider a $p \times 1$ normally distributed random vector $\mathbf{y}$ with mean $\mathbf{0}$ and variance–covariance matrix given by (A.8.2). Suppose $\mathbf{y}$ is partitioned as $\mathbf{y} = [\mathbf{y}_1' : \mathbf{y}_2']'$, where $\mathbf{y}_1$ is $vm \times 1$. For testing $H_0 : \sigma_2^2 = 0$, the UMPU test rejects H_0 for large values of $\{(p - vm)/(v-1)m\}(\mathbf{y}_1'[(\boldsymbol{I} - (1/v)\mathbf{1}_v\mathbf{1}_v') \otimes \boldsymbol{I}_m]\mathbf{y}_1)/(\mathbf{y}_2'\mathbf{y}_2)$, which is distributed as central F with $((v-1)m, (p - vm))$ degrees of freedom under H_0. Furthermore, the same test is also UMPI.

Proof. Using (A.8.6), it is readily verified that the exponent of the normal density of $[\mathbf{w}_1' : \mathbf{y}_2']'$ is

$$-\frac{1}{2}\left[(\sigma^2 + v\sigma_1^2 + \sigma_2^2)^{-1}\mathbf{w}_{11}'\mathbf{w}_{11} + (\sigma^2 + \sigma_2^2)^{-1}\mathbf{w}_{12}'\mathbf{w}_{12} + (\sigma^2)^{-1}\mathbf{y}_2'\mathbf{y}_2\right].$$

Hence, from Lehmann (1986, Section 4.4), the F-test based on $\mathbf{w}_{12}'\mathbf{w}_{12}/\mathbf{y}_2'\mathbf{y}_2$ is UMPU for testing $H_0 : \sigma_2^2 = 0$ versus $H_1 : \sigma_2^2 > 0$. To show that the same test is also UMPI, note that a maximal invariant statistic under the transformation described above is $(\mathbf{w}_{11}'\mathbf{w}_{11}/\mathbf{y}_2'\mathbf{y}_2,\ \mathbf{w}_{12}'\mathbf{w}_{12}/\mathbf{y}_2'\mathbf{y}_2)$. From standard arguments (see Lehmann 1986, Chapter 6) it follows that the test based on $\mathbf{w}_{12}'\mathbf{w}_{12}/\mathbf{y}_2'\mathbf{y}_2$ is UMPI. The proof is complete in view of the fact that since $\boldsymbol{H}$ is orthogonal, $\boldsymbol{H}_1'\boldsymbol{H}_1 = \boldsymbol{I}_v - 1/v\mathbf{1}_v\mathbf{1}_v'$ and hence from (A.8.5), $\mathbf{w}_{12}'\mathbf{w}_{12} = \mathbf{y}_1'[(\boldsymbol{I}_v - 1/v\mathbf{1}_v\mathbf{1}_v') \otimes \boldsymbol{I}_m]\mathbf{y}_1$. □

EXERCISES

8.1. Consider case 8.2(*iv*). Prove that for testing H_β, Lemma A.8.1(*iii*) is applicable, and hence that the test based on F_β is UMPI.

8.2. Verify (8.2.3).

8.3. Refer to case 8.2(*vi*). Prove that the tests based on F_β and F_τ are Wald's tests.

8.4. Verify (8.3.4).

8.5. Prove that F_τ defined in (8.3.6) has a central F-distribution under H_τ.

8.6. Verify (8.3.16).

BIBLIOGRAPHY

Hinkelmann, K. and Kempthorne, O. (1994). *Design and Analysis of Experiments*, Volume I. Wiley, New York.

Hirotsu, C. (1979). "An F approximation and its application." *Biometrika*, 66, 577–584.

John, P. W. M. (1971). *Statistical Design and Analysis of Experiments*. Macmillan, New York.

Lehmann, E. L. (1986). *Testing Statistical Hypotheses*, Second Edition. Wiley, New York.

Mathew, T. and Bhimasankaram, P. (1983). "On the robustness of the LRT with respect to specification errors in a linear model." *Sankhyā, Series A*, 45, 212–225.

Mathew, T. and Sinha, B. K. (1992). "Exact and optimum tests in unbalanced split-plot designs under mixed and random models." *Journal of the American Statistical Association*, 87, 192–200.

Milliken, G. A. and Johnson, D. E. (1984). *Analysis of Messy Data*. Lifetime Learning Publications, California.

Montgomery, D. C. (1991). *Design and Analysis of Experiments*, Third Edition. Wiley, New York.

Yates, F. (1937), "The design and analysis of factorial experiments." Imperial Bureau of Soil Science, Harpenden, England.

CHAPTER 9

Tests Using Generalized P-Values

9.1. INTRODUCTION

It is well known that in mixed and random models, exact F-tests do not always exist for testing the significance of all the variance components. The standard practice in such situations is to use approximate F-tests based on Satterthwaite's approximation (see Chapter 6, Section 2.6). Usually the hypotheses concerning the variance components that are frequently tested are whether a variance component is zero or not. However, there are applications that call for testing other types of hypotheses involving variance components. For example, one can test if a variance component has a specified nonzero value, or if two variance components, occurring in two different models, are equal. As we shall see, for such testing problems, exact F-tests do not exist. In this chapter, we shall investigate an alternative approach for testing such hypotheses. The approach is based on the concept of a *generalized P-value*, introduced and investigated by Tsui and Weerahandi (1989). The concept is defined and explained in the next section. We shall first give some specific examples of situations where the concept will be applied.

Our first example is a three-way random model with crossed and nested effects, given earlier in Section 5.5. The example deals with a study of the efficiency of workers in assembly lines in several plants, and is described in Milliken and Johnson (1984, pp. 264–266). Even though this unbalanced data is analyzed in Milliken and Johnson, and also in Section 5.5, we shall present the balanced case here for simplicity (the unbalanced case discussed in Section 5.5 is considered later in Section 9.5). Suppose a plants are randomly selected for the study, and within each plant b assembly sites and c workers are randomly selected. The data consist of the efficiency scores and assume that n efficiency scores are available on the lth worker at the jth site in the ith plant, so that we have balanced data. If y_{ijls} denotes the sth efficiency score for the lth worker at the jth site in the ith plant, the model to be used to analyze the data is model (5.5.1), which is

$$y_{ijls} = \mu + \tau_i + \beta_{i(j)} + \delta_{i(l)} + (\beta\delta)_{i(jl)} + e_{ijl(s)}, \tag{9.1.1}$$

Table 9.1. Expected Mean Squares for the Random Model (9.1.1) with Balanced Data

Sum of Squares	d.f.	Expected Mean Squares
$SS(\tau)$	$a-1$	$bcn\sigma_\tau^2 + cn\sigma_{\beta(\tau)}^2 + bn\sigma_{\delta(\tau)}^2 + n\sigma_{\beta\delta(\tau)}^2 + \sigma_e^2$
$SS(\beta(\tau))$	$a(b-1)$	$cn\sigma_{\beta(\tau)}^2 + n\sigma_{\beta\delta(\tau)}^2 + \sigma_e^2$
$SS(\delta(\tau))$	$a(c-1)$	$bn\sigma_{\delta(\tau)}^2 + n\sigma_{\beta\delta(\tau)}^2 + \sigma_e^2$
$SS(\beta\delta(\tau))$	$a(b-1)(c-1)$	$n\sigma_{\beta\delta(\tau)}^2 + \sigma_e^2$
$SS(e)$	$abc(n-1)$	σ_e^2

$i = 1, 2, \ldots, a$; $j = 1, 2, \ldots, b$; $l = 1, 2, \ldots, c$, and $s = 1, 2, \ldots, n$, where the quantities in (9.1.1) are as in (5.5.1). It is assumed that $\tau_i \sim N(0, \sigma_\tau^2)$, $\beta_{i(j)} \sim N(0, \sigma_{\beta(\tau)}^2)$, $\delta_{i(l)} \sim N(0, \sigma_{\delta(\tau)}^2)$, $(\beta\delta)_{i(jl)} \sim N(0, \sigma_{\beta\delta(\tau)}^2)$, $e_{ijl(s)} \sim N(0, \sigma_e^2)$ and all the random variables are independent. Denoting the sums of squares by $SS(\tau)$, $SS(\beta(\tau))$, and so on, the expected mean squares appear in Table 9.1.

It is clear that exact F-tests exist for testing the significance of all the variance components except σ_τ^2. In Section 9.3, we show that the generalized P-value approach, outlined in Section 9.2, can be used to test a hypothesis for σ_τ^2. Quite generally, hypotheses concerning variance components in balanced mixed models, when exact F-tests do not exist, can be tested using this new approach. Furthermore, our test reduces to the usual F-test, in case an exact F-test exists.

Another problem for which the generalized P-value is applicable is that of comparing the random effects variance components in two independent balanced mixed models. A canonical form of this problem is as follows. Consider the independent random variables

$$S_1 \sim (\sigma_*^2 + \lambda_1 \sigma_1^2)\chi_{e_1}^2, \qquad S_2 \sim (\sigma_{**}^2 + \lambda_2 \sigma_2^2)\chi_{e_2}^2,$$
$$S_* \sim \sigma_*^2 \chi_{e_*}^2 \text{ and } S_{**} \sim \sigma_{**}^2 \chi_{e_{**}}^2, \tag{9.1.2}$$

where λ_1 and λ_2 are known positive real numbers and $\sigma_*^2, \sigma_{**}^2, \sigma_1^2$ and σ_2^2 are unknown variance components. The problem is to test $H_0 : \sigma_1^2 \leq \sigma_2^2$ vs $H_1 : \sigma_1^2 > \sigma_2^2$. Such a problem occurs, for example, when we have two independent balanced one-way random models with σ_*^2 and σ_{**}^2 denoting the error variances and σ_1^2 and σ_2^2 denoting the variances of the random treatment effects in the two models. If $\sigma_*^2 = \sigma_{**}^2$ and $\lambda_1 = \lambda_2$, then clearly, an F-test based on S_1 and S_2 can be used to test $H_0 : \sigma_1^2 \leq \sigma_2^2$ vs $H_1 : \sigma_1^2 > \sigma_2^2$. If λ_1 and λ_2 are different, or if σ_*^2 and σ_{**}^2 are not equal, it appears that such an exact F-test is not possible. However, the generalized P-value approach again leads to an exact test for this problem.

Solutions to the above problems are described in Section 9.3. In Section 9.4, some extensions to the unbalanced case are discussed. Two examples are presented in Section 9.5. The first example deals with testing the significance of a main effect variance component in model (9.1.1) using the unbalanced data in Table 5.1. Table 9.1 corresponds to a general balanced case of this example. The second example deals with the comparison of random effects variance components in two independent two-fold nested models with mixed effects. The data for the latter example are based on an experiment to compare two types of tubes used on military tanks for firing ammunition. The problem is to test if tube-to-tube dispersion is less in a new tube compared to a control tube. In both examples, exact F-tests do not exist and we have used the generalized P-value to carry out the test and draw conclusions.

Referring to Table 9.1, the standard practice for testing H_0: $\sigma_\tau^2 = 0$ is to use an approximate F-test based on Satterthwaite's approximation (see Chapter 2, Section 2.6). However, it is known that Satterthwaite's approximation is not always satisfactory, especially when certain sums of squares appear with a negative coefficient in the test statistic. This has been pointed out by Gaylor and Hopper (1969) through a simulation. Even when all the sums of squares appear with a positive coefficient, Satterthwaite's approximation may still be unsatisfactory. A measure to evaluate the accuracy of the approximation is derived in Khuri (1995). Section 9.6 deals with a comparison of the tests based on Satterthwaite's approximation and the generalized P-value for testing the significance of a variance component in a simple setup (see (9.6.1)). Simulated Type I error probabilities reveal that the fixed level test based on Satterthwaite's approximation can have Type I error probabilities much larger than the nominal significance level, whereas the fixed level test based on the generalized P-value, most of the time, has Type I error probabilities much less than the nominal significance level. A similar observation on the Type I error probabilities of the test based on the generalized P-value has been noted for the Behrens–Fisher problem by Thursby (1992) and Griffiths and Judge (1992). It thus appears that if the primary concern of the experimenter is to control the Type I error probability, tests based on the generalized P-value should be preferred.

9.2. THE GENERALIZED P-VALUE

The concept of a *generalized P-value* has been introduced by Tsui and Weerahandi (1989) to deal with some testing problems where nuisance parameters are present and it is difficult or impossible to obtain a nontrivial test with a fixed level of significance. The setup is as follows. Suppose θ is a scalar parameter and we are interested in testing $H_0 : \theta \leq \theta_0$ versus $H_1 : \theta > \theta_0$. Let δ be a nuisance parameter (scalar or vector), and x be the observed data based on a random variable X. Suppose it is difficult or impossible to identify a test statistic $T(X)$ whose distribution at θ_0 is independent of the nuisance parameter δ. We then consider a random variable $T(X; x, \theta, \delta)$, which also depends on the observed value and the parameters, and satisfies the following conditions:

(a) The distribution of $T(X;x,\theta,\delta)$ is free of the nuisance parameter δ,

(b) The observed value of $T(X;x,\theta,\delta)$ (i.e., $T(x;x,\theta,\delta)$, is free of δ), and

(c) $T(X;x,\theta,\delta)$ is stochastically increasing in θ (i.e., $P[T(X;x,\theta,\delta) \geq t]$ is nondecreasing in θ), for fixed x and δ. (9.2.1)

Under conditions (9.2.1)*(a)–(c)*, we can compute the generalized P-value defined by

$$p = P[T(X;x,\theta_0,\delta) \geq t], \tag{9.2.2}$$

where $t = T(x;x,\theta_0,\delta)$. This P-value, namely p, can be used to test H_0 versus H_1. We shall refer to $T(X;x,\theta,\delta)$ as a test variable. Note that x is to be treated as fixed in (9.2.2).

We now briefly explain the significance of conditions *(a)–(c)* given above. The quantity $T(X;x,\theta,\delta)$ takes the role of a test statistic, except that it depends on the observed value x and the parameters θ and δ. We may assume without loss of generality that a 'large' observed value of $T(X;x,\theta_0,\delta)$ suggests evidence against H_0. Following the usual definition of a P-value, we may thus use the quantity

$$p = P[T(X;x,\theta_0,\delta) \geq t] = \sup_{\theta \leq \theta_0} P[T(X;x,\theta,\delta) \geq t] \tag{9.2.3}$$

to test H_0 versus H_1: We reject H_0 if p in (9.2.3) is "small." In (9.2.3), $t = T(x;x,\theta_0,\delta)$, which is free of δ, in view of condition (b). Condition (c) guarantees that the two expressions in (9.2.3) are equal. Furthermore, $P[T(X;x,\theta_0,\delta) \geq t]$ in (9.2.3) is free of any unknown parameters, in view of conditions (a) and (b), and this permits us to compute the generalized P-value. Condition (c) also implies that $\mathrm{P}[T(X;x,\theta,\delta) \geq t]$ becomes large as $(\theta - \theta_0)$ increases and this is clearly a natural requirement.

In an earlier paper, Weerahandi (1987), the generalized P-value approach is used to test the equality of regression coefficients in regression models with unequal error variances. In another article, Weerahandi (1991), the generalized P-value is used to test $H_0 : \sigma_\tau^2 \leq \epsilon$ versus $H_1 : \sigma_\tau^2 > \epsilon$ for a specified $\epsilon > 0$, where σ_τ^2 is the variance of the random treatment effects in a one-way balanced random effects model. Furthermore, in Weerahandi (1995a), ANOVA under unequal error variances is carried out using the generalized P-value. Several other applications are also given in Tsui and Weerahandi (1989), Zhou and Mathew (1994), and Weerahandi (1995b, Chapter 9). This chapter is based almost entirely on the article by Zhou and Mathew (1994). It is also possible to obtain *generalized confidence intervals* following the idea of the generalized P-value, a topic which is not pursued here; see Weerahandi (1993) for details.

9.3. TESTS USING GENERALIZED *P*-VALUES IN THE BALANCED CASE

The first problem that we shall consider deals with hypothesis testing for variance components in a general mixed model with balanced data, where exact F-tests do not exist. For a discussion of this along with examples, we refer to Montgomery (1991, Chapter 8; see also Chapter 6, Section 2.6). We shall first describe the typical setup for this problem, based on the sums of squares in the ANOVA decomposition. We shall use some of the properties of balanced mixed models given in Section 2.3 of Chapter 2.

Consider a general balanced mixed model that has n random effects and consequently n variance components denoted by σ_i^2 $(i = 1, 2, \ldots, n)$. Suppose we are interested in testing hypotheses concerning the variance component σ_1^2. The testing problem we shall consider is

$$H_0 : \sigma_1^2 \le \sigma_{10}^2 \quad \text{versus} \quad H_1 : \sigma_1^2 > \sigma_{10}^2, \tag{9.3.1}$$

for a specified σ_{10}^2. In applications, we usually have $\sigma_{10}^2 = 0$ in (9.3.1). Note that for a balanced mixed model involving n random effects, the n sums of squares corresponding to the random effects have expected values which are linear combinations of the variance components. Furthermore, every variance component has a unique unbiased estimator based on these sums of squares. The latter property follows from the fact that in a balanced mixed model, the sums of squares due to the random effects are complete and sufficient for the variance components (see Chapter 2, Section 2.5). For a general balanced mixed model involving n variance components σ_i^2 $(i = 1, 2, \ldots, n)$, suppose the unbiased estimator of σ_1^2 involves p sums of squares $(p \le n)$ denoted by S_i $(i = 1, 2, \ldots, p)$. The S_i's are independent random variables and using the properties of balanced mixed models, the distribution of the S_i's can be specified as follows.

$$S_1 \sim \left(a_1\sigma_1^2 + \sum_{j=2}^{n} a_j\sigma_j^2 \right) \chi^2_{e_1}, \quad S_i \sim \left(\sum_{j=2}^{n} b_{ij}\sigma_j^2 \right) \chi^2_{e_i}, \quad i = 2, \cdots, p, \tag{9.3.2}$$

where a_j and b_{ij} are known nonnegative scalars $(i = 1, 2, \ldots, n;\ j = 2, 3, \ldots, p)$. For example, for the model (9.1.1), if $\sigma_1^2 = \sigma_\tau^2$, then referring to Table 9.1 we see that even though there are five sums of squares and five variance components corresponding to the five random effects, only four sums of squares, namely $SS(\tau)$, $SS(\beta(\tau))$, $SS(\delta(\tau))$ and $SS(\beta\delta(\tau))$, are necessary to obtain the unbiased estimator of σ_τ^2. Thus $n = 5$ and $p = 4$. In fact,

$$\hat{\sigma}_\tau^2 = \frac{1}{bcn} \left[\frac{SS(\tau)}{a-1} + \frac{SS(\beta\delta(\tau))}{a(b-1)(c-1)} - \frac{SS(\beta(\tau))}{a(b-1)} - \frac{SS(\delta(\tau))}{a(c-1)} \right] \tag{9.3.3}$$

is the unbiased estimator of σ_τ^2. Also, for this example, taking $S_1 = SS(\tau)$, $S_2 = SS(\beta\delta(\tau))$, $S_3 = SS(\beta(\tau))$, $S_4 = SS(\delta(\tau))$, $\sigma_1^2 = \sigma_\tau^2$, $\sigma_2^2 = \sigma_{\beta(\tau)}^2$, $\sigma_3^2 =$

$\sigma^2_{\delta(\tau)}$, $\sigma^2_4 = \sigma^2_{\beta\delta(\tau)}$, and $\sigma^2_5=\sigma^2_e$, we can easily write down the a_j's and b_{ij}'s in (9.3.2). For example, $a_1 = bcn$, $a_2 = cn$, $a_3 = bn$, $a_4 = n$, $a_5 = 1$; $b_{22} = b_{23} = 0$, $b_{24} = n$, $b_{25} = 1$, and so on. Note that if $b_{ij} \neq 0$ for some i and j, then $b_{ij} = a_j$. Also, $b_{ij} = b_{i'j}$ whenever these quantities are both nonzero. These can be easily verified for the above example (in the context of the model (9.1.1)) and they follow in general from the properties of balanced mixed models. The distributions in (9.3.2) imply that

$$\mathrm{E}(S_1/e_1) = a_1\sigma_1^2 + \sum_{j=2}^{n} a_j\sigma_j^2, \quad \mathrm{E}(S_i/e_i) = \sum_{j=2}^{n} b_{ij}\sigma_j^2, \quad i = 2, \cdots, p. \tag{9.3.4}$$

Similar to $\hat{\sigma}^2_\tau$ in (9.3.3), the unbiased estimator of σ^2_1 will be a linear combination of S_i/e_i $(i = 1, 2, \ldots, p)$ and the coefficients in the linear combination will be 1 or -1. Hence, without loss of generality, the unbiased estimator of σ^2_1, say $\hat{\sigma}^2_1$, can be expressed as

$$\hat{\sigma}_1^2 = \frac{1}{a_1}\left[\frac{S_1}{e_1} + \sum_{i=2}^{r}\frac{S_i}{e_i} - \sum_{i=r+1}^{p}\frac{S_i}{e_i}\right], \tag{9.3.5}$$

for some r satisfying $2 \leq r \leq p$. For a particular balanced mixed model, it is easy to write down such an unbiased estimator by inspecting the expected values of the various mean squares. This is how $\hat{\sigma}^2_\tau$ in (9.3.3) was obtained. Since $\mathrm{E}(\hat{\sigma}^2_1) = \sigma^2_1$, we get the following relation:

$$\sum_{i=2}^{r}\sum_{j=2}^{n} b_{ij}\sigma_j^2 + \sum_{j=2}^{n} a_j\sigma_j^2 = \sum_{i=r+1}^{p}\sum_{j=2}^{n} b_{ij}\sigma_j^2, \tag{9.3.6}$$

for all $\sigma^2_j \geq 0$ $(j = 1, 2, \ldots, n)$. Actually, (9.3.6) is a consequence of the property $a_j = b_{ij} = b_{i'j}$ whenever b_{ij} and $b_{i'j}$ are both nonzero. In the context of the model (9.1.1), the reader can easily verify (9.3.6) for $\hat{\sigma}^2_\tau$ given in (9.3.3).

We shall now describe the generalized P-value approach for the testing problem given in (9.3.1). Note that an exact F-test for $H_0 : \sigma^2_1 = 0$ versus $H_1 : \sigma^2_1 > 0$ exists if in (9.3.2) $a_j = b_{ij}$ $(j = 2, \ldots, n)$ for some i $(2 \leq i \leq p)$, so that $\sum_{j=2}^n a_j\sigma^2_j = \sum_{j=2}^n b_{ij}\sigma^2_j$. It is readily verified that for the example presented in Table 9.1, this condition holds for all the variance components except σ^2_τ. If this condition is violated, such an exact F-test does not exist. For the more general testing problem (9.3.1), consider the test variable

$$\begin{aligned} T &= T(S_1, \cdots, S_p; s_1, \cdots, s_p, \sigma_1^2, \cdots, \sigma_n^2) \\ &= \frac{a_1\sigma_1^2 + \sum_{i=r+1}^{p}(\sum_{j=2}^{n} b_{ij}\sigma_j^2)s_i/S_i}{(a_1\sigma_1^2 + \sum_{j=2}^{n} a_j\sigma_j^2)s_1/S_1 + \sum_{i=2}^{r}(\sum_{j=2}^{n} b_{ij}\sigma_j^2)s_i/S_i}, \end{aligned} \tag{9.3.7}$$

where s_i denotes the observed value of the random variable S_i $(i = 1, 2, \ldots, p)$. Note that, in view of (9.3.6), the observed value of T is

$$t = T(s_1, \cdots, s_p; s_1, \cdots, s_p, \sigma_1^2, \cdots, \sigma_n^2) = 1. \tag{9.3.8}$$

Thus T satisfies condition (9.2.1)(b). From (9.3.2), it is clear that the distributions of $S_1/(a_1\sigma_1^2 + \sum_{j=2}^{n} a_j\sigma_j^2)$ and $S_i/(\sum_{j=2}^{n} b_{ij}\sigma_j^2)$ $(i = 2, 3, \ldots, p)$ are free of any unknown parameter. Hence, the distribution of $T(S_1, \ldots, S_p; s_1, \ldots, s_p, \sigma_1^2, \sigma_2^2, \ldots, \sigma_n^2)$ is free of the nuisance parameters $\sigma_2^2, \ldots, \sigma_n^2$. Thus T in (9.3.7) satisfies condition (9.2.1)(a). Since σ_1^2 appears with a positive coefficient in the numerator of T in (9.3.7), it is clear that T also satisfies the condition (9.2.1)(c). Following (9.2.2), the generalized P-value for the testing problem (9.3.1) can be defined as

$$p = P\{T(S_1, \cdots, S_p; s_1, \cdots, s_p, \sigma_{10}^2, \sigma_2^2, \cdots, \sigma_n^2) \geq 1\}. \tag{9.3.9}$$

In general, the P-value in (9.3.9) cannot be evaluated analytically; see the discussion following (9.3.14) regarding its computation. It is easy to see that if $\sigma_{10}^2 = 0$ and if an exact F-test exists for testing $H_0 : \sigma_1^2 = 0$ vs $H_1 : \sigma_1^2 > 0$, then the test using the P-value in (9.3.9) reduces to the F-test. We shall illustrate this for testing H_0: $\sigma_{\beta\delta(\tau)}^2 = 0$ in Table 9.1. In this case, the unbiased estimator of $\sigma_{\beta\delta(\tau)}^2$ is based on $S_1 = SS_{\beta\delta(\tau)}$ and $S_2 = SS(e)$ and an exact F-test for testing H_0: $\sigma_{\beta\delta(\tau)}^2 = 0$ can obviously be constructed using these two sums of squares. Recall from Table 9.1 that

$$\begin{aligned} S_1 &= SS(\beta\delta(\tau)) \sim (n\sigma_{\beta\delta(\tau)}^2 + \sigma_e^2)\chi_{(a(b-1)(c-1))}^2 \\ \text{and} \quad S_2 &= SS(e) \sim \sigma_e^2\chi_{(abc(n-1))}^2. \end{aligned}$$

Hence, T in (9.3.7) becomes

$$T = T[SS(\beta\delta(\tau)), SS(e); ss(\beta\delta(\tau)), ss(e), \sigma_{\beta\delta(\tau)}^2, \sigma_e^2] = \frac{n\sigma_{\beta\delta(\tau)}^2 + \sigma_e^2\frac{ss(e)}{SS(e)}}{(n\sigma_{\beta\delta(\tau)}^2 + \sigma_e^2)\frac{ss(\beta\delta(\tau))}{SS(\beta\delta(\tau))}}.$$

With $\sigma_1^2 = \sigma_{\beta\delta(\tau)}^2$ and $\sigma_{10}^2 = 0$ in (9.3.9), the generalized P-value for testing H_0: $\sigma_{\beta\delta(\tau)}^2 = 0$, is now given by

$$\begin{aligned} p &= P\{\frac{\sigma_e^2 ss(e)/SS(e)}{\sigma_e^2 ss(\beta\delta(\tau))/SS(\beta\delta(\tau))} \geq 1\} \\ &= P\{SS(\beta\delta(\tau))/SS(e) \geq ss(\beta\delta(\tau))/ss(e)\} \\ &= P\{\frac{SS(\beta\delta(\tau))/[a(b-1)(c-1)]}{SS(e)/[abc(n-1)]} \geq \frac{ss(\beta\delta(\tau))/[a(b-1)(c-1)]}{ss(e)/[abc(n-1)]}\}, \end{aligned}$$

which is the usual P-value based on the F-statistic $SS(\beta\delta(\tau))/[a(b-1)(c-1)]/SS(e)/[abc(n-1)]$.

The second problem that we shall address deals with the comparison of the variance components σ_1^2 and σ_2^2 in model (9.1.2). The testing problem we shall consider is

$$H_0 : \sigma_1^2 \leq \sigma_2^2 \quad \text{versus} \quad H_1 : \sigma_1^2 > \sigma_2^2, \tag{9.3.10}$$

or, equivalently,

$$H_0 : \frac{\sigma_1^2}{\sigma_2^2} \leq 1 \quad \text{versus} \quad H_1 : \frac{\sigma_1^2}{\sigma_2^2} > 1. \tag{9.3.11}$$

Consider the test variable T given by

$$\begin{aligned} T &= T(S_1, S_2, S_*, S_{**}; s_1, s_2, s_*, s_{**}, \sigma_1^2, \sigma_2^2, \sigma_*^2, \sigma_{**}^2) \\ &= \frac{\lambda_1(\sigma_{**}^2 + \lambda_2\sigma_1^2)\frac{s_2}{S_2} + \lambda_2\sigma_*^2\frac{s_*}{S_*}}{\lambda_2(\sigma_*^2 + \lambda_1\sigma_1^2)\frac{s_1}{S_1} + \lambda_1\sigma_{**}^2\frac{s_{**}}{S_{**}}} \\ &= \frac{\left(\frac{\sigma_{**}^2/\sigma_2^2 + \lambda_2\sigma_1^2/\sigma_2^2}{\sigma_{**}^2/\sigma_2^2 + \lambda_2}\right)\lambda_1(\sigma_{**}^2 + \lambda_2\sigma_2^2)\frac{s_2}{S_2} + \lambda_2\sigma_*^2\frac{s_*}{S_*}}{\lambda_2(\sigma_*^2 + \lambda_1\sigma_1^2)\frac{s_1}{S_1} + \lambda_1\sigma_{**}^2\frac{s_{**}}{S_{**}}} \\ &= \frac{\theta\lambda_1(\sigma_{**}^2 + \lambda_2\sigma_2^2)\frac{s_2}{S_2} + \lambda_2\sigma_*^2\frac{s_*}{S_*}}{\lambda_2(\sigma_*^2 + \lambda_1\sigma_1^2)\frac{s_1}{S_1} + \lambda_1\sigma_{**}^2\frac{s_{**}}{S_{**}}}, \end{aligned} \tag{9.3.12}$$

where $\theta = [\sigma_{**}^2/\sigma_2^2 + \lambda_2\sigma_1^2/\sigma_2^2]/[\sigma_{**}^2/\sigma_2^2 + \lambda_2]$, and as before, s_1, s_2, s_*, and s_{**} denote the observed values of S_1, S_2, S_*, and S_{**}, respectively. Note that (9.3.11) is equivalent to H_0: $\theta \leq 1$ versus H_1: $\theta > 1$. We shall first verify that T in (9.3.12) satisfies the conditions (a)–(c) in (9.2.1). Note that the observed value of T is one and hence T satisfies (9.2.1)(b). Since the distributions of $S_1/(\sigma_*^2 + \lambda_1\sigma_1^2)$, $S_2/(\sigma_{**}^2 + \lambda_2\sigma_2^2)$, S_*/σ_*^2, and S_{**}/σ_{**}^2 are free of any unknown parameters, it follows that the distribution of T in (9.3.12) is free of any nuisance parameters for any specified value of θ. Thus T satisfies (9.2.1)(a). From the expression in (9.3.12), it is clear that T increases as θ increases and consequently $P(T \geq t)$ is an increasing function of θ. Thus T satisfies (9.2.1)(c). Note that $\theta = 1$ is equivalent to $\sigma_1^2 = \sigma_2^2$. The generalized P-value for the testing problem (9.3.11) is thus defined as

$$p = P[T(S_1, S_2, S_*, S_{**}; s_1, s_2, s_*, s_{**}, \sigma_1^2, \sigma_1^2, \sigma_*^2, \sigma_{**}^2) \geq 1]. \tag{9.3.13}$$

If $\sigma_*^2 = \sigma_{**}^2 = \sigma_0^2$, define $S_0 = S_* + S_{**}$ and the test variable T we shall use to define the generalized P-value for (9.3.11) is

$$T = T(S_1, S_2, S_0; s_1, s_2, s_0, \sigma_1^2, \sigma_2^2, \sigma_0^2) = \frac{\lambda_1(\sigma_0^2 + \lambda_2\sigma_1^2)\frac{s_2}{S_2} + \lambda_2\sigma_0^2\frac{s_0}{S_0}}{\lambda_2(\sigma_0^2 + \lambda_1^2\sigma_1^2)\frac{s_1}{S_1} + \lambda_1\sigma_0^2\frac{s_0}{S_0}}. \tag{9.3.14}$$

In Weerahandi (1995b, Section 9.6), the more general problem of testing H_0: $\sigma_1^2/\sigma_2^2 \le \theta_0$ versus H_1: $\sigma_1^2/\sigma_2^2 \ge \theta_0$ is addressed for the setup (9.1.2).

The generalized P-values in (9.3.9) and (9.3.13) will have to be computed numerically. This computation can be accomplished using the statistical software package, XPro (1994); see Remark 9.3.1 below. It is also possible to compute them by simulation. We shall now give the details of this simulation for computing (9.3.9); those for (9.3.13) are quite similar. We shall first rewrite the random quantity (9.3.7), used in defining (9.3.9), in terms of chi-squared random variables as follows. Let

$$R_1 = S_1 / \left(a_1\sigma_1^2 + \sum_{j=2}^{n} a_j\sigma_j^2 \right), \quad R_i = S_i / \left(\sum_{j=2}^{n} b_{ij}\sigma_j^2 \right), \quad i = 2, \cdots, p, \tag{9.3.15}$$

where the S_i's $(i = 1, 2, \ldots, p)$ are as in (9.3.2). From (9.3.2), it is clear that $R_i \sim \chi^2_{e_i}$ $(i = 1, 2, \ldots, p)$ and T in (9.3.7) can be written as

$$T = \frac{a_1\sigma_1^2 + \sum_{i=r+1}^{p} s_i/R_i}{s_1/R_1 + \sum_{i=2}^{r} s_i/R_i}. \tag{9.3.16}$$

Clearly, the generalized P-value p in (9.3.9) is simply $\mathrm{P}(T \ge 1 \big| \sigma_1^2 = \sigma_{10}^2)$, where T is given by (9.3.16) and σ_{10}^2 is specified by the hypotheses in (9.3.1). Note that once the data set is available to us, the values of s_1, s_2, ..., s_p, occurring in (9.3.16), are known. In order to compute p in (9.3.9), we may proceed as follows. Generate a chi-squared random variable based on each of the p chi-squared distributions in (9.3.15). This will give us an observed value of each of the chi-squared random variables R_1, R_2, ..., R_p in (9.3.15). One can then compute T in (9.3.16) with $\sigma_1^2 = \sigma_{10}^2$, and verify if the computed value is more than or equal to one. This can be repeated a large number of times and an approximate value of p in (9.3.9) is simply the proportion of times T is greater than or equal to one. The computation of p in (9.3.13) is quite similar and is based on generating observations from the appropriate chi-squared distributions. We will follow the above steps to compute the generalized P-values in Sections 9.5 and 9.6. To generate values of a chi-squared random variable with ν degrees of freedom, we used the SAS (1989) function RANGAM.

REMARK 9.3.1. Even though the procedure described above and the computations that we have carried out in Sections 9.5 and 9.6 are based on simulation, the same computations can be easily accomplished using the XPro (1994) software package. XPro specializes in exact parametric inference, including the computation of P-values and confidence intervals in various linear models, and provides an easy to use interface to carry out the analysis. In particular, the generalized P-values defined in this section, as well as some of the generalized confidence intervals defined in Weerahandi (1993), can all be easily computed using XPro.

REMARK 9.3.2. In practical applications where the generalized P-value can be used to test a hypothesis, one can decide to reject the null hypothesis when the generalized P-value is small, say less than 0.05. However, this does not mean that the test has an exact Type I error probability of 0.05. This is due to the fact that in repeated sampling, the generalized P-value does not have a uniform distribution under the null hypothesis, unlike the classical P-value. For example, p in (9.3.9) depends on the observed values s_1, s_2, ..., s_p and hence its repeated sampling property depends on the distribution of the random variables S_1, S_2, ..., S_p, which involves the unknown variance components σ_1^2, σ_2^2, ..., σ_p^2. In other words, p in (9.3.9) does not have a uniform distribution. Consequently, in a particular problem where the generalized P-value is applicable, we may not get a test with an exact Type I error probability by repeating the same experiment. However, this latter property will hold if we consider several independent testing problems. In other words, after a large number of independent situations of testing hypotheses at a fixed level α, the proportion of times the experimenter would have incorrectly rejected the null hypothesis is less than or equal to α. This fact is formally proved in Weerahandi (1993) and is clearly a property of practical interest. Note that in spite of the above, it is useful to have some idea about the Type I error probability in a given testing situation. This facilitates comparison among competing tests.

9.4. EXTENSIONS TO THE UNBALANCED CASE

The generalized P-value defined in (9.3.9) for the balanced case can be extended to general random ANOVA models with unbalanced data provided the unbalancedness results from unequal cell frequencies in the last stage. This is so because by a suitable transformation, applied to the unbalanced model, the testing problem can be reduced to the same in a balanced model, as shown in Chapter 5. Such a reduction will produce exact F-tests for many hypotheses (for details, we refer to Chapter 5). In case an exact F-test does not exist, we can use the generalized P-value. An example of this type is given in the next section.

An unbalanced setup of (9.1.2) is as follows. Consider the independent random variables $U_*, U_1, \ldots, U_r, V_*, V_1, \ldots, V_s$ distributed as

$$\begin{aligned} U_* &\sim \sigma_*^2\chi_{e_*}^2, \quad U_i \sim (\sigma_*^2 + \lambda_i\sigma_1^2)\chi_{e_i}^2, \quad (i = 1, \cdots, r) \\ V_* &\sim \sigma_{**}^2\chi_{f_*}^2, \quad V_j \sim (\sigma_{**}^2 + \eta_j\sigma_2^2)\chi_{f_j}^2, \quad (j = 1, \cdots, s), \end{aligned} \tag{9.4.1}$$

where λ_i's and η_j's are known positive real numbers and $\sigma_*^2, \sigma_{**}^2, \sigma_1^2$ and σ_2^2 are the unknown variance components. The testing problem is once again

(9.3.10) (or, equivalently, (9.3.11)) for model (9.4.1). For inference on the variance components σ_*^2 and σ_1^2, U_* and the U_i's in (9.4.1) represent the canonical form in an unbalanced mixed model involving the two variance components σ_*^2 and σ_1^2. To see this, consider an $N \times 1$ normally distributed vector of observations $\mathbf{y}$ following such a model given by

$$\mathrm{E}(\mathbf{y}) = \boldsymbol{X\beta}, \quad \mathrm{Var}(\mathbf{y}) = \sigma_*^2 \boldsymbol{I} + \sigma_1^2 \boldsymbol{V}, \tag{9.4.2}$$

where $\boldsymbol{X}$ is a known $N \times m$ matrix of rank m, $\boldsymbol{\beta}$ is a vector of unknown parameters, and $\boldsymbol{V}$ is a known nonnegative definite matrix. Such a model is also given in Section 6.2.2; see (6.2.11). Let $\boldsymbol{Z}$ be an $N \times (N - m)$ matrix satisfying $\boldsymbol{Z}'\boldsymbol{X} = 0$ and $\boldsymbol{Z}'\boldsymbol{Z} = \boldsymbol{I}_{N-r}$. If $\mathbf{u} = \boldsymbol{Z}'\mathbf{y}$, then

$$\mathrm{E}(\mathbf{u}) = \mathbf{0}, \quad \mathrm{Var}(\mathbf{u}) = \sigma_*^2 \boldsymbol{I}_{N-m} + \sigma_1^2 \boldsymbol{V}_1, \tag{9.4.3}$$

where $\boldsymbol{V}_1 = \boldsymbol{Z}'\boldsymbol{V}\boldsymbol{Z}$. For inference on the variance components σ_*^2 and σ_1^2, it is natural to consider procedures based only on $\mathbf{u}$, since $\mathbf{u}$ is invariant with respect to the group of transformations $\mathbf{y} \to \mathbf{y} + \boldsymbol{X\alpha}$, $\boldsymbol{\alpha}$ being an arbitrary $m \times 1$ vector. Let λ_i $(i = 1, 2, \ldots, r)$ denote the distinct nonzero eigenvalues of $\boldsymbol{V}_1$ and let e_i denote the multiplicity of λ_i. Furthermore, let $\boldsymbol{F} = [\boldsymbol{F}_1 : \boldsymbol{F}_2 : \ldots : \boldsymbol{F}_r]$ be an $(N - m) \times (N - m)$ orthogonal matrix, where $\boldsymbol{F}_i$ is an $(N - m) \times e_i$ matrix satisfying $\boldsymbol{V}_1\boldsymbol{F}_i = \lambda_i\boldsymbol{F}_i$ (i.e., the columns of $\boldsymbol{F}_i$ are the e_i orthonormal eigenvectors corresponding to the eigenvalue λ_i). We then have the spectral decomposition $\boldsymbol{V}_1 = \sum_{i=1}^r \lambda_i \boldsymbol{F}_i\boldsymbol{F}_i' = \sum_{i=1}^r \lambda_i \boldsymbol{E}_i$, where $\boldsymbol{E}_i = \boldsymbol{F}_i\boldsymbol{F}_i'$. Note that $\mathrm{rank}(\boldsymbol{E}_i) = \mathrm{rank}(\boldsymbol{F}_i) = e_i$. It is readily verified that $U_i = \mathbf{u}'\boldsymbol{E}_i\mathbf{u} \sim (\sigma_*^2 + \lambda_i\sigma_1^2)\chi_{e_i}^2$, and are independent for $i = 1, 2, \ldots, r$. Furthermore, if $\boldsymbol{E}_* = \boldsymbol{I} - \sum_{i=1}^r \boldsymbol{E}_i$, then $U_* = \mathbf{u}'\boldsymbol{E}_*\mathbf{u} \sim \sigma_*^2\chi_{e_*}^2$ and is independent of the U_i's, where e_*=rank$(\boldsymbol{E}_*)$. It is also known that $U_*, U_1, \ldots, U_r$ form a set of minimal sufficient statistics for the normal family of distributions of $\mathbf{u}$ in (9.4.3) (see Olsen, Seely, and Birkes, 1976, p. 880). Thus (9.4.1) is the canonical form for comparing the random effects variance components σ_1^2 and σ_2^2 in two independent unbalanced mixed models of the type (9.4.2). Similar to (9.3.12), define

$$\begin{aligned} T &= T(U_*, U_1, \cdots, U_r, V_*, V_1, \cdots, V_s; \\ &\qquad u_*, u_1, \cdots, u_r, v_*, v_1, \cdots, v_s, \sigma_*^2, \sigma_{**}^2, \sigma_1^2, \sigma_2^2) \\ &= \frac{(\sum_{i=1}^r \lambda_i) \sum_{j=1}^s (\sigma_{**}^2 + \eta_j\sigma_1^2)\frac{v_j}{V_j} + r(\sum_{j=1}^s \eta_j)\sigma_*^2 \frac{u_*}{U_*}}{\sum_{j=1}^s \eta_j \sum_{i=1}^r (\sigma_*^2 + \lambda_i\sigma_1^2)\frac{u_i}{U_i} + s(\sum_{i=1}^r \lambda_i)\sigma_{**}^2 \frac{v_*}{V_*}}, \end{aligned} \tag{9.4.4}$$

where u_*, v_*, u_i's, and v_j's denote the observed values of the corresponding random variables U_*, V_*, U_i's, and V_j's. The verification of conditions (9.2.1)(a)–(c) for T in (9.4.4) is similar to that for T in (9.3.12). Note that the observed value of T is one. Thus the generalized P-value given by $P(T \geq 1)$ (computed under σ_1^2=σ_2^2) can be used to test (9.3.10) under model (9.4.1). However, the test is not unique. For example,

$$T = \frac{(\sum_{i=1}^{r} c_i\lambda_i)\sum_{j=1}^{s} d_j(\sigma_{**}^2 + \eta_j\sigma_1^2)\frac{v_j}{V_j} + (\sum_{i=1}^{r} c_i)(\sum_{j=1}^{s} d_j\eta_j)\sigma_*^2\frac{u_*}{U_*}}{(\sum_{j=1}^{s} d_j\eta_j)\sum_{i=1}^{r} c_i(\sigma_*^2 + \lambda_i\sigma_1^2)\frac{u_i}{U_i} + (\sum_{j=1}^{s} d_j)(\sum_{i=1}^{r} c_i\lambda_i)\sigma_{**}^2\frac{v_*}{V_*}} \tag{9.4.5}$$

also satisfies conditions (9.2.1)(a)–(c) and its observed value is also one, where the c_i's and d_j's are arbitrary nonnegative real numbers.

The same nonuniqueness also occurs in the unbalanced case of the application considered in Weerahandi (1991). The problem is to test $H_0 : \sigma_1^2 \leq \sigma_{10}^2$ versus $H_1 : \sigma_1^2 > \sigma_{10}^2$ using U_* and the U_i's in (9.4.1). For this problem,

$$T = \sum_{i=1}^{r} c_i \frac{U_i(\sigma_*^2/U_* + \lambda_i\sigma_1^2/u_*)}{\sigma_*^2 + \lambda_i\sigma_1^2} \tag{9.4.6}$$

satisfies conditions (9.2.1)(a)–(c), where the c_i's are arbitrary nonnegative real numbers. The observed value of T in (9.4.6) is $\sum_{i=1}^{r} c_i\frac{u_i}{u_*}$. At present, we do not have well-defined guidelines to prefer a particular test in such situations.

9.5. TWO EXAMPLES

Our first example is taken from Milliken and Johnson (1984, pp. 264–266) and deals with a study of the efficiency of workers in assembly lines in several plants. A balanced version of this example is given in the introduction. Here we shall analyze the unbalanced data given in Milliken and Johnson (1984). The same example is also discussed in Section 5.5 and the data are given in Table 5.1. The unbalanced data are based on three randomly selected plants, and within each plant, four assembly sites and three workers were randomly selected. The data, given in Table 5.1, consist of the efficiency scores. If y_{ijls} denotes the sth efficiency score for the lth worker at the jth site in the ith factory, the model used to analyze the data is given in (9.1.1), where $i = 1, 2, 3; j = 1, 2, 3, 4; l = 1, 2, 3$, and the suffix s takes different values depending on i, j, and l, indicating that the data is unbalanced. We make the same assumptions on the random effects as we did in the introduction. Using the ANOVA table given in Table 21.3 in Milliken and Johnson (1984), one can construct an exact F-test only for testing the significance of the interaction variance component $\sigma_{\beta\delta(\tau)}^2$. However, using the transformation given in Chapter 5, exact F-tests can be constructed for testing the significance of all the variance components except σ_τ^2. This transformation is given in Section 5.5; in particular, see Table 5.2. The following table gives the sum of squares and the expected values of the mean sum of squares so obtained for the unbalanced setup mentioned above (see also Table 5.3).

Table 9.2. Sums of Squares and Expected Mean Squares for Model (9.1.1) for Data in Table 5.1

Sum of Squares	d.f.	Expected Mean Squares
$ss(\tau) = 1265.96$	2	$12\sigma_\tau^2 + 3\sigma_{\beta(\tau)}^2 + 4\sigma_{\delta(\tau)}^2 + \sigma_{\beta\delta(\tau)}^2 + \sigma_e^2$
$ss(\beta(\tau)) = 332.313$	9	$3\sigma_{\beta(\tau)}^2 + \sigma_{\beta\delta(\tau)}^2 + \sigma_e^2$
$ss(\delta(\tau)) = 733.949$	6	$4\sigma_{\delta(\tau)}^2 + \sigma_{\beta\delta(\tau)}^2 + \sigma_e^2$
$ss(\beta\delta(\tau)) = 668.634$	18	$\sigma_{\beta\delta(\tau)}^2 + \sigma_e^2$
$ss(e) = 246.245$	47	σ_e^2

It should be noted that the various sums of squares (including the error sum of squares) in Table 9.2 are different from the corresponding quantities in Milliken and Johnson (1984, Table 21.3). This is so since the sums of squares in Milliken and Johnson (1984, Table 21.3) are obtained from the usual ANOVA decomposition based on the model (9.1.1), whereas those in Table 9.2 are obtained using the procedure described in Chapter 5. From Table 9.2, it is clear that an exact F-test exists for the various hypotheses except $H_0 : \sigma_\tau^2 = 0$. For testing $H_0 : \sigma_\tau^2 = 0$ versus $H_1 : \sigma_\tau^2 > 0$, we shall use the generalized P-value obtained by using the test variable T given by (see (9.3.7))

$$T = \frac{12\sigma_\tau^2 + (3\sigma_{\beta(\tau)}^2 + \sigma_{\beta\delta(\tau)}^2 + \sigma_e^2)\frac{ss(\beta(\tau))}{SS(\beta(\tau))} + (4\sigma_{\delta(\tau)}^2 + \sigma_{\beta\delta(\tau)}^2 + \sigma_e^2)\frac{ss(\delta(\tau))}{SS(\delta(\tau))}}{(12\sigma_\tau^2 + 3\sigma_{\beta(\tau)}^2 + 4\sigma_{\delta(\tau)}^2 + \sigma_{\beta\delta(\tau)}^2 + \sigma_e^2)\frac{ss(\tau)}{SS(\tau)} + (\sigma_{\beta\delta(\tau)}^2 + \sigma_e^2)\frac{ss(\beta\delta(\tau))}{SS(\beta\delta(\tau))}}, \tag{9.5.1}$$

where $ss(\tau), ss(\beta(\tau)$, and so on denote the observed values given in Table 9.2. Writing $S_1 = SS(\tau)$, $S_2 = SS(\beta\delta(\tau))$, $S_3 = SS(\beta(\tau))$, and $S_4 = SS(\delta(\tau))$, it is readily verified that (9.5.1) is of the form (9.3.7) with $p = 4$, $r = 2$, and $n = 5$. The generalized P-value for testing $H_0 : \sigma_\tau^2 = 0$ is then given by $P(T \geq 1)$, computed under H_0. Using the fact that $SS(\tau)/(3\sigma_{\beta(\tau)}^2 + 4\sigma_{\delta(\tau)}^2 + \sigma_{\beta\delta(\tau)}^2 + \sigma_e^2) \sim \chi_2^2$ (under $H_0 : \sigma_\tau^2 = 0$), $SS(\beta(\tau))/(3\sigma_{\beta(\tau)}^2 + \sigma_{\beta\delta(\tau)}^2 + \sigma_e^2) \sim \chi_9^2$, and so on, we computed $P(T \geq 1)$, under $H_0 : \sigma_\tau^2 = 0$, using 50,000 simulations. The simulated generalized P-value turned out to be 0.0518. Thus the data do not provide strong evidence to conclude that $\sigma_\tau^2 > 0$. Based on Satterthwaite's approximate F-test, the same conclusion was earlier arrived at in Section 5.5.

Our second example deals with an experiment to compare a new tube (NT) with a control tube (CT) to be used for firing ammunition from tanks. The problem is to test if tube-to-tube variability is less for the new tube compared to the control tube. The experiment was carried out at the U.S. Army Ballistic Research Laboratory, Aberdeen Proving Ground, Maryland. Twenty new tubes and twenty control tubes were randomly selected for the experiment with 4 tanks each for mounting the new tubes and the control tubes. Five new tubes were mounted on each of 4 tanks and 5 control tubes were mounted on each of the other 4 tanks. Three rounds were fired from each tube and the observations consisted of a miss distance (the unit used was 6400 mils per 365 degrees).

Let CT_{ij} and NT_{ij} respectively denote the jth control tube and the jth new tube mounted on the ith tank ($j = 1,2,3,4,5; i = 1,2,3,4$). The three measurements (the miss distances) corresponding to each CT_{ij} and NT_{ij} are given in Table 9.3. Let y_{ijk} and z_{ijk} respectively denote the kth observation corresponding to CT_{ij} and NT_{ij}, α_i denote the effect due to the ith tank on which a control tube was mounted, γ_i denote the effect due to the ith tank on which a new tube was mounted, $\tau_{j(i)}$ denote the effect due to CT_{ij}, and $\delta_{j(i)}$ denote the effect due to NT_{ij}. The linear models to be used for analyzing the data in Table 9.3 are

$$y_{ijk} = \mu_1 + \alpha_i + \tau_{i(j)} + e_{ijk} \tag{9.5.2}$$

and

$$z_{ijk} = \mu_2 + \gamma_i + \delta_{i(j)} + f_{ijk}, \tag{9.5.3}$$

where μ_1 and μ_2 are overall means and e_{ijk} and f_{ijk} denote random error terms. The tank effects α_i and γ_i ($i = 1,2,3,4$) are fixed unknown parameters. We also assume that $\tau_{i(j)} \sim N(0, \sigma^2_{\tau(\alpha)})$, $e_{ijk} \sim N(0, \sigma^2_e)$, $\delta_{i(j)} \sim N(0, \sigma^2_{\delta(\gamma)})$, $f_{ijk} \sim N(0, \sigma^2_f)$, and all the random variables are independent. Note that models (9.5.2) and (9.5.3) are two-fold nested models with mixed effects. To assess whether tube-to-tube dispersion is less among the new tubes compared to the control tubes, we have to test the hypothesis

$$H_0 : \sigma^2_{\tau(\alpha)} \leq \sigma^2_{\delta(\gamma)} \quad \text{versus} \quad H_1 : \sigma^2_{\tau(\alpha)} > \sigma^2_{\delta(\gamma)}. \tag{9.5.4}$$

The ANOVA tables based on (9.5.2) and (9.5.3) using the data in Table 9.3 are given in Tables 9.4 and 9.5.

Since

$$\begin{aligned} SS(CT) &\sim (\sigma^2_e + 3\sigma^2_{\tau(\alpha)})\chi^2_{16},\ SS(e) \sim \sigma^2_e \chi^2_{40}, \\ SS(NT) &\sim (\sigma^2_f + 3\sigma^2_{\delta(\gamma)})\chi^2_{16},\ SS(f) \sim \sigma^2_f \chi^2_{40}, \end{aligned} \tag{9.5.5}$$

for testing (9.5.4), we are in the setup (9.1.2). The generalized P-value can be simulated using (9.3.13) and the observed values of the various sums of squares given in Tables 9.4 and 9.5. The generalized P-value, obtained using 50,000 simulations, is 0.36. Thus our decision is not to reject H_0 in (9.5.4) . Consequently, we cannot conclude that the new tubes have a smaller tube-to-tube variability compared to the control tubes.

Note that if $\sigma^2_e = \sigma^2_f$, then for testing (9.5.4), we have an exact F-test based on $SS(CT)$ and $SS(NT)$. An F-test for testing the equality of σ^2_e and σ^2_f using the error sum of squares in Table 9.4 and Table 9.5 yielded a P-value of 0.033. This suggests that σ^2_e and σ^2_f are perhaps different and it is

Table 9.3. Data for Example 2

Tank i	CT_{i1}	CT_{i2}	CT_{i3}	CT_{i4}	CT_{i5}	NT_{i1}	NT_{i2}	NT_{i3}	NT_{i4}	NT_{i5}
	2.76	1.83	1.60	1.53	2.20	1.92	1.98	2.28	1.52	1.28
$i = 1$	2.10	1.65	1.56	2.29	2.59	1.77	1.56	1.90	1.82	1.61
	1.61	1.76	1.73	2.06	1.91	1.37	1.83	2.10	1.79	1.48
	1.35	1.15	1.68	1.70	1.34	1.70	1.61	1.78	1.60	1.69
$i = 2$	1.64	1.83	1.71	1.26	1.26	1.82	1.71	2.31	1.65	1.72
	1.56	1.92	1.63	1.64	1.69	1.65	1.28	1.73	1.26	1.76
	1.33	1.65	1.94	1.72	1.81	1.79	1.64	1.84	1.80	1.73
$i = 3$	1.28	1.76	1.86	1.56	2.13	1.39	1.88	1.67	1.49	1.83
	1.40	1.81	2.00	1.91	1.86	1.52	1.60	1.64	1.92	1.79
	1.64	1.77	1.01	1.04	1.27	1.49	1.88	1.77	1.46	2.10
$i = 4$	1.80	1.63	1.63	1.78	1.38	1.60	1.60	1.56	1.29	1.46
	1.89	1.51	1.46	1.86	1.55	1.63	1.61	1.62	1.72	1.60

Source: L. Zhou and T. Mathew (1994). Reproduced with permission of the American Statistical Association.

Table 9.4. ANOVA for the CT Data

	Sum of Squares	d.f.
Tanks	1.5719	3
CT (within tanks)	$ss(CT) = 1.9706$	16
Error	$ss(e) = 2.7271$	40

Source: L. Zhou and T. Mathew (1994). Reproduced with permission of the American Statistical Association.

Table 9.5. ANOVA for the NT Data

	Sum of Squares	d.f.
Tanks	0.1133	3
NT (within tanks)	$ss(NT) = 1.1913$	16
Error	$ss(f) = 1.5109$	40

Source: L. Zhou and T. Mathew (1994). Reproduced with permission of the American Statistical Association.

more appropriate to use the generalized P-value approach for testing (9.5.4), instead of the exact F-test based on $SS(CT)$ and $SS(NT)$.

For further examples, we refer to Weerahandi (1995b, Chapter 9).

9.6. COMPARISON WITH SATTERTHWAITE'S APPROXIMATION

Satterthwaite's approximation (Satterthwaite, 1941, 1946) is a widely used procedure for testing the significance of variance components when exact F-

tests do not exist; for example, it can be used for testing H_0: $\sigma_\tau^2 = 0$ in Table 9.1 or Table 9.2. The approximation is described in Section 2.6 of Chapter 2. An attractive feature of Satterthwaite's approximation is that the approximate tests are quite easy to carry out. Thus, a natural question is whether the use of the generalized P-value has any advantage over Satterthwaite's approximation. In order to investigate this, we simulated the Type I error probabilities of the two tests in the following simple setup. Let S_1, S_2, and S_3 be independent random variables having the distributions

$$S_1 \sim (\sigma_1^2 + \sigma_2^2 + \sigma_3^2)\chi_{e_1}^2, \; S_2 \sim \sigma_2^2\chi_{e_2}^2 \text{ and } S_3 \sim \sigma_3^2\chi_{e_3}^2, \tag{9.6.1}$$

where σ_1^2, σ_2^2, and σ_3^2 are unknown variance components. The testing problem is

$$H_0 : \sigma_1^2 = 0 \text{ versus } H_1 : \sigma_1^2 > 0. \tag{9.6.2}$$

For this problem, the test based on Satterthwaite's approximation rejects H_0 for large values of the statistic

$$F = \frac{S_1/e_1}{S_2/e_2 + S_3/e_3}. \tag{9.6.3}$$

Under H_0: $\sigma_1^2 = 0$, F in (9.6.3) has an approximate F-distribution with ν_1 and ν_2 degrees of freedom, where

$$\nu_1 = e_1 \text{ and } \nu_2 = \frac{(S_2/e_2 + S_3/e_3)^2}{S_2^2/e_2^3 + S_3^2/e_3^3}. \tag{9.6.4}$$

In order to define the generalized P-value for the testing problem (9.6.2), following (9.3.7), let

$$T = \frac{\sigma_1^2 + \sigma_2^2 s_2/S_2 + \sigma_3^2 s_3/S_3}{(\sigma_1^2 + \sigma_2^2 + \sigma_3^2)s_1/S_1}, \tag{9.6.5}$$

where the lowercase letters in (9.6.5) denote the observed values of the corresponding random variables. The generalized P-value, say $p(s_1, s_2, s_3)$, for testing (9.6.2) is then defined as

$$p(s_1, s_2, s_3) = P(T \geq 1 | \sigma_1^2 = 0). \tag{9.6.6}$$

Let f denote the observed value of F in (9.6.3) and define

$$q(f) = P(F \geq f), \tag{9.6.7}$$

where the probability in (9.6.7) is computed using the approximate F-distribution of F in (9.6.3). If α is the chosen significance level, then an approximate size α test based on the statistic F in (9.6.3) will reject H_0: $\sigma_1^2 = 0$ when

$$q(f) < \alpha. \tag{9.6.8}$$

The fixed level test based on the generalized P-value will reject H_0: $\sigma_1^2 = 0$ when

$$p(s_1, s_2, s_3) < \alpha. \tag{9.6.9}$$

We simulated the Type I error probabilities of the tests based on the rejection regions (9.6.8) and (9.6.9) for two different sets of values of (e_1, e_2, e_3) in (9.6.1), namely $(e_1, e_2, e_3) = (30, 4, 4)$ and $(e_1, e_2, e_3) = (2, 9, 6)$. Also, σ_2^2 was assumed to be one and σ_3^2 was allowed to take the values 0.01, 1, 4, 9, 16, 25, and 100. Note that the Type I error probabilities will depend on σ_2^2 and σ_3^2 since the distributions of $q(F)$ (for F in (9.6.3)) and $p(S_1, S_2, S_3)$ depend on these variance components. The simulations were carried out using the above parameter values for $\alpha = 0.01$ and 0.05. The chi-squared random variables in (9.6.1) were generated using the SAS (1989) function RANGAM and for each observed value f of F in (9.6.3), $q(f)$ in (9.6.7) was computed using the approximate F-distribution of F mentioned earlier. Based on 10,000 simulations, the simulated Type I error probability was then the proportion of times $q(f)$ was below α, according to the rejection rule (9.6.8). In order to simulate the Type I error probability of the generalized P-value test, we first generated one set of values of S_1, S_2, S_3 (given in (9.6.1)), say s_1, s_2, s_3, and keeping s_1, s_2, s_3 fixed, simulated $p(s_1, s_2, s_3)$ in (9.6.6) based on 10,000 simulations. Ten thousand values of $p(s_1, s_2, s_3)$ were further generated and the proportion of such values below α gave the Type I error probability, according to the rejection rule (9.6.9). The simulated Type I error probabilities are given in Table 9.6.

It is quite clear that the Type I error probabilities of the test based on Satterthwaite's approximation can be much larger than the assumed value of

Table 9.6. Simulated Type I Error Probabilities of the Approximate F-Test and the Generalized P-Value Test for testing H_0: $\sigma_1^2 = 0$ in (9.6.1) for the Parameter Values $\sigma_2^2 = 1$, $\sigma_3^2 = 0.01, 1, 4, 9, 16, 25$, and 100, and $\alpha = 0.01$ and 0.05

Table 9.6a. $(e_1, e_2, e_3) = (30, 4, 4)$

	$\alpha = 0.01$		$\alpha = 0.05$	
σ_3^2	Approximate F-Test	Generalized P-Value Test	Approximate F-Test	Generalized P-Value Test
0.01	0.0161	0.0099	0.0579	0.0505
1	0.0068	0.0060	0.0392	0.0112
4	0.0120	0.0014	0.0522	0.0160
9	0.0204	0.0030	0.0654	0.0272
16	0.0235	0.0062	0.0667	0.0369
25	0.0231	0.0084	0.0642	0.0423
100	0.0174	0.0111	0.0560	0.0487

Source: L. Zhou and T. Mathew (1994). Reproduced with permission of the American Statistical Association.

Table 9.6b. (e_1, e_2, e_3) = (2, 9, 6)

	$\alpha = 0.01$		$\alpha = 0.05$	
σ_3^2	Approximate F-Test	Generalized P-Value Test	Approximate F-Test	Generalized P-Value Test
0.01	0.0123	0.0114	0.0536	0.0515
1	0.0100	0.0040	0.0483	0.0326
4	0.0122	0.0062	0.0509	0.0368
9	0.0128	0.0078	0.0518	0.0423
16	0.0126	0.0083	0.0529	0.0460
25	0.0124	0.0082	0.0529	0.0483
100	0.0102	0.0092	0.0526	0.0513

Source: L. Zhou and T. Mathew (1994). Reproduced with permission of the American Statistical Association.

α. This is especially true in the case of $(e_1, e_2, e_3) = (30, 4, 4)$ and $\alpha = 0.01$. In most cases, the Type I error probability of the generalized P-value test is much below the assumed value of α. We did not simulate the powers of these tests, since their Type I error probabilities can be substantially different. The simulated Type I error probabilities suggest that the generalized P-value test should be preferred if the major concern is to control the Type I error probability. Thursby (1992) and Griffiths and Judge (1992) arrive at a similar conclusion for the Behrens–Fisher problem.

EXERCISES

9.1. Consider the balanced one-way random model $y_{ij} = \mu + \tau_i + e_{ij}$ $(i = 1, 2, \ldots, v;\ j = 1, 2, \ldots, n)$, where the τ_i's and e_{ij}'s are all independent having the distributions $\tau_i \sim N(0, \sigma_\tau^2)$ and $e_{ij} \sim N(0, \sigma_e^2)$. Explain how you will test H_0: $\sigma_\tau^2 \le \delta\sigma_e^2$ versus H_1: $\sigma_\tau^2 > \delta\sigma_e^2$ $(\delta > 0$, known).

9.2. For testing the hypothesis in Exercise 9.1, derive a test in the unbalanced one-way random model.

9.3. Consider the balanced two-way random models

$$\begin{aligned} y_{i_1 j_1 k_1} &= \mu_1 + \tau_{i_1} + \beta_{j_1} + (\tau\beta)_{i_1 j_1} + e_{i_1 j_1 k_1} \\ &\quad i_1 = 1, 2, \cdots, v_1;\ j_1 = 1, 2, \cdots, b_1;\ k_1 = 1, 2, \cdots, n_1 \\ z_{i_2 j_2 k_2} &= \mu_2 + \gamma_{i_2} + \delta_{j_2} + (\gamma\delta)_{i_2 j_2} + f_{i_2 j_2 k_2} \\ &\quad i_2 = 1, 2, \cdots, v_2;\ j_2 = 1, 2, \cdots, b_2;\ k_2 = 1, 2, \cdots, n_2, \end{aligned}$$

where $\tau_{i_1} \sim N(0, \sigma_\tau^2)$, $\beta_{j_1} \sim N(0, \sigma_\beta^2)$, $(\tau\beta)_{i_1 j_1} \sim N(0, \sigma_{\tau\beta}^2)$, $e_{i_1 j_1 k_1} \sim N(0, \sigma_e^2)$, $\gamma_{i_2} \sim N(0, \sigma_\gamma^2)$, $\delta_{j_2} \sim N(0, \sigma_\delta^2)$, $(\gamma\delta)_{i_2 j_2} \sim N(0, \sigma_{\gamma\delta}^2)$ and $f_{i_2 j_2 k_2} \sim$

$N(0, \sigma_f^2)$, where all the random variables are assumed to be independent. Derive tests based on generalized P-values for testing (i) H_0: $\sigma_{\tau\beta}^2 \leq \sigma_{\gamma\delta}^2$ versus H_1: $\sigma_{\tau\beta}^2 > \sigma_{\gamma\delta}^2$, (ii) H_0: $\sigma_\tau^2 \leq \sigma_\gamma^2$ versus H_1: $\sigma_\tau^2 > \sigma_\gamma^2$.

9.4. Derive a test based on the generalized P-value for testing H_0: $\sigma_1^2/\sigma_2^2 \leq \delta$ versus H_1: $\sigma_1^2/\sigma_2^2 > \delta$ ($\delta > 0$, known), in the setup (9.1.2).

9.5. Using the sums of squares given in Table 9.2, compute the generalized P-values for testing (i) H_0: $\sigma_{\beta\delta(\tau)}^2 \leq 30$ versus H_1: $\sigma_{\beta\delta(\tau)}^2 > 30$, (ii) H_0: $\sigma_{\delta(\tau)}^2 \leq 20$ versus H_1: $\sigma_{\delta(\tau)}^2 > 20$.

BIBLIOGRAPHY

Gaylor, D. W. and Hopper, F. N. (1969). "Estimating the degrees of freedom for linear combinations of mean squares by Satterthwaite's formula." *Technometrics*, 11, 691–706.

Griffiths, W. and Judge, G. (1992). "Testing and estimating location vectors when the error covariance matrix is unknown."*Journal of Econometrics*, 54, 121–138.

Khuri, A. I. (1995). "A measure to evaluate the closeness of Satterthwaite's approximation." *Biometrical Journal*, 37, 547–563.

Milliken, G. A. and Johnson, D. E. (1984). *Analysis of Messy Data*. Lifetime Learning Publications, Belmont, California.

Montgomery, D. C. (1991). *Design and Analysis of Experiments*, Third Edition. Wiley, New York.

Olsen, A., Seely, J., and Birkes, D. (1976). "Invariant quadratic unbiased estimation for two variance components." *The Annals of Statistics*, 5, 878–890.

SAS User's Guide: Statistics, 1989 edition, SAS Institute, Inc., Cary, North Carolina.

Satterthwaite, F. E. (1941). "Synthesis of variance." *Psychometrika*, 6, 309–316.

Satterthwaite, F. E. (1946). "An approximate distribution of estimates of variance components." *Biometrics Bulletin*, 2, 110–114.

Thursby, J. G. (1992). "A comparison of several exact and approximate tests for structural shift under heteroscedasticity." *Journal of Econometrics*, 53, 363–386.

Tsui, K. W. and Weerahandi, S. (1989). "Generalized P-values in siginificance testing of hypotheses in the presence of nuisance parameters." *Journal of the American Statistical Association*, 84, 602–607.

Weerahandi, S. (1987). "Testing regression equality with unequal variances." *Econometrica*, 55, 1211–1215.

Weerahandi, S. (1991). "Testing variance components in mixed models with generalized P-values." *Journal of the American Statistical Association*, 86, 151–153.

Weerahandi, S. (1993). "Generalized confidence intervals." *Journal of the American Statistical Association*, 88, 899–905.

Weerahandi, S. (1995a). "ANOVA under unequal error variances." *Biometrics*, 51, 589–599.

Weerahandi, S. (1995b). *Exact Statistical Methods for Data Analysis*. Springer-Verlag, New York.

XPro Software Package, 1994, X-Techniques, Inc. Millington, New Jersey.

Zhou, L. and Mathew, T. (1994). "Some tests for variance components using generalized P-values." *Technometrics*, 36, 394–402.

CHAPTER 10

Multivariate Mixed and Random Models

10.1. INTRODUCTION

It is fairly straightforward to formulate multivariate generalizations of the univariate mixed and random models that we have considered in the previous chapters. The observations, the various effects, and the experimental error terms will all now be written as vectors, and their variances will be expressed as matrices. The general theory for multivariate balanced mixed and random models is developed in Sections 10.2 and 10.3, and is an extension of the corresponding univariate results in Chapter 2. In Section 10.3, we also describe a multivariate analogue of Satterthwaite's approximation. In some applications, it is more appropriate to have a multivariate formulation of some familiar univariate models. This formulation and the relevant analysis are covered in Section 10.4. The derivation of exact and optimum tests for some unbalanced models is covered in Sections 10.5 and 10.6. It is shown that all the exact F-tests in the univariate case have straightforward multivariate extensions, as will be seen in Section 10.5. However, this is not the case when it comes to the optimality properties of tests. In Section 10.6, we discuss optimum tests in the special case of a multivariate model involving only one random effect.

10.2. THE GENERAL BALANCED MODEL

The general theory for univariate balanced random and mixed models was developed in Chapter 2. In this section, we present an extension of this theory to multivariate balanced models with mixed effects.

Consider the multivariate model

$$\mathbf{Y} = \sum_{i=0}^{\nu} \mathbf{H}_i \mathbf{B}_i + \mathbf{E}, \tag{10.2.1}$$

where $\mathbf{Y}$ is a data matrix of order $N \times r$ whose rows represent a random sample of N observations on r responses of interest denoted by $y_1, y_2, \ldots, y_r$, $\mathbf{H}_i$ is a known matrix of order $N \times c_i$ and rank c_i, $\mathbf{B}_i$ is the ith effect matrix of order $c_i \times r$ $(i = 0, 1, \ldots, \nu)$, and $\mathbf{E}$ is an $N \times r$ random error matrix whose rows are independent r-variate normal vectors with mean zero and a variance–covariance matrix $\boldsymbol{\Sigma}_e$. The vector of observations on the jth response is given by the jth column of $\mathbf{Y}$, which we denote by $\mathbf{y}_j$, and is represented by the univariate model

$$\mathbf{y}_j = \sum_{i=0}^{\nu} \mathbf{H}_i \boldsymbol{\beta}_{ij} + \mathbf{e}_j, \quad j = 1, 2, \cdots, r, \tag{10.2.2}$$

where $\boldsymbol{\beta}_{ij}$ and $\mathbf{e}_j$ are the jth columns of $\mathbf{B}_i$ and $\mathbf{E}$, respectively $(i = 0, 1, \ldots, \nu,$ $j = 1, 2, \ldots, r)$. This model is of the same form as model (2.2.3) in Chapter 2.

By definition, model (10.2.1) is balanced if the univariate model in (10.2.2) is balanced for all j $(= 1, 2, \ldots, r)$. In this case, the $\mathbf{H}_i$ matrices in (10.2.1) are expressible as direct products of vectors of ones and/or identity matrices (see formulas (2.2.4) and (2.2.5) in Chapter 2).

Example 10.2.1. Consider the balanced multivariate one-way model

$$\tilde{\mathbf{y}}_{k\ell} = \boldsymbol{\mu} + \boldsymbol{\tau}_k + \boldsymbol{\epsilon}_{k\ell}, \quad k = 1, 2, \cdots, v; \ \ell = 1, 2, \cdots, n, \tag{10.2.3}$$

where $\tilde{\mathbf{y}}_{k\ell} = (y_{1k\ell}, y_{2k\ell}, \ldots, y_{rk\ell})'$ is the ℓth sample observation vector on r response variables obtained under the kth treatment $(k = 1, 2, \ldots, v;\ \ell = 1, 2, \ldots, n)$, $\boldsymbol{\mu} = (\mu_1, \mu_2, \ldots, \mu_r)'$ is a vector of unknown constants, $\boldsymbol{\tau}_k = (\tau_{1k}, \tau_{2k}, \ldots, \tau_{rk})'$ is a vector representing the kth treatment effect, and $\boldsymbol{\epsilon}_{k\ell}$ is a vector of random errors. Thus for the ℓth observation on the jth response from the kth treatment, we have the univariate one-way model,

$$y_{jk\ell} = \mu_j + \tau_{jk} + e_{jk\ell}, \tag{10.2.4}$$

where $e_{jk\ell}$ is the jth element of $\boldsymbol{\epsilon}_{k\ell}$ $(j = 1, 2, \ldots, r;\ k = 1, 2, \ldots, v;\ \ell = 1, 2, \ldots, n)$.

Now, let $\mathbf{y}_{jk} = (y_{jk1}, y_{jk2}, \ldots, y_{jkn})'$ be the vector of observations on the jth response from the kth treatment $(j = 1, 2, \ldots, r;\ k = 1, 2, \ldots, v)$. Then, from (10.2.4) we have the model

$$\mathbf{y}_{jk} = (\mu_j + \tau_{jk})\mathbf{1}_n + \mathbf{e}_{jk},$$

where $\mathbf{e}_{jk} = (e_{jk1}, e_{jk2}, \ldots, e_{jkn})'$. By combining the observation vectors on the jth response from all v treatments, we obtain the vector $\mathbf{y}_j = [\mathbf{y}'_{j1} : \mathbf{y}'_{j2} :$

$\ldots : \mathbf{y}'_{jv}]'$ for which we have the model

$$\mathbf{y}_j = \mu_j(\mathbf{1}_v \otimes \mathbf{1}_n) + (\mathbf{I}_v \otimes \mathbf{1}_n)\boldsymbol{\theta}_j + \mathbf{e}_j, \quad j = 1, 2, \cdots, r, \tag{10.2.5}$$

where $\boldsymbol{\theta}_j = (\tau_{j1}, \tau_{j2}, \ldots, \tau_{jv})'$, $\otimes$ is the symbol of direct product of matrices, and $\mathbf{e}_j$ is defined in the same way as $\mathbf{y}_j$. This model can be expressed as

$$\mathbf{y}_j = \mathbf{H}_0\boldsymbol{\beta}_{0j} + \mathbf{H}_1\boldsymbol{\beta}_{1j} + \mathbf{e}_j, \quad j = 1, 2, \cdots, r, \tag{10.2.6}$$

where $\mathbf{H}_0 = \mathbf{1}_v \otimes \mathbf{1}_n$, $\boldsymbol{\beta}_{0j} = \mu_j$, $\mathbf{H}_1 = \mathbf{I}_v \otimes \mathbf{1}_n$, $\boldsymbol{\beta}_{1j} = \boldsymbol{\theta}_j$. From (10.2.6) we obtain the single multivariate model

$$\mathbf{Y} = \mathbf{H}_0\mathbf{B}_0 + \mathbf{H}_1\mathbf{B}_1 + \mathbf{E},$$

where $\mathbf{B}_0 = [\boldsymbol{\beta}_{01} : \boldsymbol{\beta}_{02} : \ldots : \boldsymbol{\beta}_{0r}]$, $\mathbf{B}_1 = [\boldsymbol{\beta}_{11} : \boldsymbol{\beta}_{12} : \ldots : \boldsymbol{\beta}_{1r}]$, which is of the same form as model (10.2.1).

Example 10.2.2. Suppose that n measurements were made on each of r response variables in a two-factor experiment involving factors A and B having v and b levels, respectively. Let $\tilde{\mathbf{y}}_{k\ell m} = (y_{1k\ell m}, y_{2k\ell m}, \ldots, y_{rk\ell m})'$ be the mth vector of observations on the r response variables obtained under the combination of the kth level of A and the ℓth level of B ($k = 1, 2, \ldots, v$; $\ell = 1, 2, \ldots, b$; $m = 1, 2, \ldots, n$). Correspondingly, we have the multivariate two-way model

$$\begin{aligned} \tilde{\mathbf{y}}_{k\ell m} = \boldsymbol{\mu} + \boldsymbol{\tau}_k + \boldsymbol{\beta}_\ell + (\boldsymbol{\tau\beta})_{k\ell} + \mathbf{e}_{k\ell m}, \quad & k = 1, 2, \cdots, v; \\ & \ell = 1, 2, \cdots, b; \; m = 1, 2, \cdots, n, \end{aligned} \tag{10.2.7}$$

where $\boldsymbol{\mu}$ is a vector of unknown constants, $\boldsymbol{\tau}_k$ and $\boldsymbol{\beta}_\ell$ are the main effects vectors representing levels k and ℓ of factors A and B, respectively, $(\boldsymbol{\tau\beta})_{k\ell}$ is the interaction effect vector, and $\mathbf{e}_{k\ell m}$ is a random error vector. For the jth element of $\tilde{\mathbf{y}}_{k\ell m}$ we have then the univariate two-way model

$$y_{jk\ell m} = \mu_j + \tau_{jk} + \beta_{j\ell} + (\tau\beta)_{jk\ell} + e_{jk\ell m}, \quad j = 1, 2, \cdots, r,$$

where the terms on the right-hand side are the jth elements of the corresponding vectors in model (10.2.7).

Let us now define $\mathbf{y}_{jk\ell}$ as the vector $(y_{jk\ell 1}, y_{jk\ell 2}, \ldots, y_{jk\ell n})'$. If, for a fixed j, we place the $\mathbf{y}_{jk\ell}$'s one after another for $k = 1, 2, \ldots, v$; $\ell = 1, 2, \ldots, b$, we obtain the vector $\mathbf{y}_j$ for which we have the model

$$\mathbf{y}_j = \mathbf{H}_0\boldsymbol{\beta}_{0j} + \mathbf{H}_1\boldsymbol{\beta}_{1j} + \mathbf{H}_2\boldsymbol{\beta}_{2j} + \mathbf{H}_3\boldsymbol{\beta}_{3j} + \mathbf{e}_j, \quad j = 1, 2, \cdots, r,$$

where

$$\begin{aligned}
\mathbf{H}_0 &= \mathbf{1}_v \otimes \mathbf{1}_b \otimes \mathbf{1}_n, \quad \boldsymbol{\beta}_{0j} = \mu_j, \\
\mathbf{H}_1 &= \mathbf{I}_v \otimes \mathbf{1}_b \otimes \mathbf{1}_n, \quad \boldsymbol{\beta}_{1j} = (\tau_{j1}, \tau_{j2}, \cdots, \tau_{jv})', \\
\mathbf{H}_2 &= \mathbf{1}_v \otimes \mathbf{I}_b \otimes \mathbf{1}_n, \quad \boldsymbol{\beta}_{2j} = (\beta_{j1}, \beta_{j2}, \cdots, \beta_{jb})', \\
\mathbf{H}_3 &= \mathbf{I}_v \otimes \mathbf{I}_b \otimes \mathbf{1}_n, \quad \boldsymbol{\beta}_{3j} = [(\tau\beta)_{j11}, (\tau\beta)_{j12}, \cdots, (\tau\beta)_{jvb}]'.
\end{aligned}$$

The vector of random errors $\mathbf{e}_j$ is defined in the same way as $\mathbf{y}_j$. The models for $\mathbf{y}_j$ can be combined resulting in the single multivariate model

$$\mathbf{Y} = \mathbf{H}_0\mathbf{B}_0 + \mathbf{H}_1\mathbf{B}_1 + \mathbf{H}_2\mathbf{B}_2 + \mathbf{H}_3\mathbf{B}_3 + \mathbf{E},$$

where $\mathbf{B}_i$ is the matrix having $\boldsymbol{\beta}_{ij}$ as its jth column ($i = 0, 1, 2, 3;\; j = 1, 2, \ldots, r$). This has the same form as model (10.2.1).

10.3. PROPERTIES OF BALANCED MULTIVARIATE MIXED MODELS

Suppose that model (10.2.1) contains fixed as well as random effects. It can therefore be written as

$$\mathbf{Y} = \mathbf{XG} + \mathbf{ZU}, \tag{10.3.1}$$

where

$$\begin{aligned}
\mathbf{XG} &= \sum_{i=0}^{\nu-p} \mathbf{H}_i\mathbf{B}_i, \\
\mathbf{ZU} &= \sum_{i=\omega}^{\nu+1} \mathbf{H}_i\mathbf{B}_i
\end{aligned}$$

are, respectively, the fixed and random portions of the model. Here, $\omega = \nu - p + 1$, where p denotes the number of random effects excluding the error term, $1 \leq p \leq \nu$. Note that for $i = \nu + 1, \mathbf{H}_i = \mathbf{I}_N$ and $\mathbf{B}_i = \mathbf{E}$. For $i = 0, 1, \ldots, \nu - p$, the elements of $\mathbf{B}_i$ are fixed unknown parameters. For $i = \omega, \omega + 1, \ldots, \nu$, $\mathbf{B}_i$ is a random matrix whose rows are assumed to be distributed independently as $N(\mathbf{0}, \boldsymbol{\Sigma}_i)$. Moreover, it is assumed that $\mathbf{B}_\omega, \mathbf{B}_{\omega+1}, \ldots, \mathbf{B}_\nu$ are independent of one another and of the error term $\mathbf{E}$ whose rows, if we recall, are distributed as $N(\mathbf{0}, \boldsymbol{\Sigma}_{\nu+1})$, where $\boldsymbol{\Sigma}_{\nu+1} = \boldsymbol{\Sigma}_e$.

Let $\boldsymbol{\lambda}$ be a nonzero vector of order $r \times 1$. By postmultiplying the terms in model (10.3.1) by $\boldsymbol{\lambda}$ we obtain the univariate model

$$\mathbf{y}_\lambda = \mathbf{X}\mathbf{g}_\lambda + \mathbf{Z}\mathbf{u}_\lambda, \tag{10.3.2}$$

where $\mathbf{y}_\lambda = \mathbf{Y}\boldsymbol{\lambda}$, $\mathbf{g}_\lambda = \mathbf{G}\boldsymbol{\lambda}$, and $\mathbf{u}_\lambda = \mathbf{U}\boldsymbol{\lambda}$. If we denote $\mathbf{B}_i\boldsymbol{\lambda}$ by $\boldsymbol{\beta}_{i\lambda}$, then

$$\mathbf{y}_\lambda = \sum_{i=0}^{\nu-p} \mathbf{H}_i \boldsymbol{\beta}_{i\lambda} + \sum_{i=\omega}^{\nu+1} \mathbf{H}_i \boldsymbol{\beta}_{i\lambda}.$$

Note that for $i = \omega, \omega+1, \ldots, \nu+1$, $\boldsymbol{\beta}_{i\lambda}$ has the normal distribution $N(\mathbf{0}, \sigma_{i\lambda}^2 \mathbf{I}_{c_i})$, where $\sigma_{i\lambda}^2 = \boldsymbol{\lambda}'\boldsymbol{\Sigma}_i\boldsymbol{\lambda}$ and c_i is the number of rows of $\mathbf{B}_i$ (or columns of $\mathbf{H}_i$). If model (10.2.1) is balanced, then so is model (10.3.2), and vice versa. Therefore, by Lemma 2.3.4 we have

$$\mathbf{A}_j \mathbf{P}_i = \kappa_{ij} \mathbf{P}_i, \quad i, j = 0, 1, \cdots, \nu+1, \tag{10.3.3}$$

where κ_{ij} is given by formula (2.3.9), $\mathbf{A}_j = \mathbf{H}_j\mathbf{H}_j'$, and $\mathbf{P}_i$ is an idempotent matrix of rank m_i such that $\mathbf{y}_\lambda' \mathbf{P}_i \mathbf{y}_\lambda$ is the sum of squares for the ith effect $(i = 0, 1, \ldots, \nu+1)$ in the univariate model (10.3.2). Furthermore, $\mathbf{P}_i\mathbf{P}_j = \mathbf{0}$ for $i \neq j$, and $\sum_{i=0}^{\nu+1} \mathbf{P}_i = \mathbf{I}_N$ (see Section 2.3).

The mean square for the ith effect in model (10.3.2) is denoted by $s_{i\lambda}$, that is,

$$\begin{aligned} s_{i\lambda} &= \frac{1}{m_i} \mathbf{y}_\lambda' \mathbf{P}_i \mathbf{y}_\lambda, \\ &= \frac{1}{m_i} \boldsymbol{\lambda}' \mathbf{Y}' \mathbf{P}_i \mathbf{Y} \boldsymbol{\lambda}, \quad i = 0, 1, \cdots, \nu+1. \end{aligned}$$

The matrix

$$\mathbf{S}_i = \frac{1}{m_i} \mathbf{Y}' \mathbf{P}_i \mathbf{Y}, \quad i = 0, 1, \cdots, \nu+1 \tag{10.3.4}$$

is called the mean square matrix for the ith effect in the multivariate model (10.3.1). Since the $\mathbf{P}_i$'s commute, there exists an orthogonal matrix $\mathbf{Q}$ such that $\mathbf{Q}'\mathbf{P}_i\mathbf{Q} = \boldsymbol{\Lambda}_i$ is a diagonal matrix with m_i diagonal elements equal to one and the remaining elements equal to zero. Thus $\mathbf{P}_i = \mathbf{Q}_i\mathbf{Q}_i'$, where $\mathbf{Q}_i$ is of order $N \times m_i$, and consists of the columns of $\mathbf{Q}$ corresponding to the nonzero diagonal elements of $\boldsymbol{\Lambda}_i$ $(i = 0, 1, \ldots, \nu+1)$. This representation of $\mathbf{P}_i$ will be utilized to derive the distribution of $\mathbf{Y}'\mathbf{P}_i\mathbf{Y}$, as will be seen in the next section.

10.3.1. Distribution of $\mathbf{Y}'\mathbf{P}_i\mathbf{Y}$

Let us consider model (10.3.1), which we assume to be balanced and satisfy the same assumptions made earlier concerning the distribution of the random effects.

Lemma 10.3.1. Let $\mathbf{P}_i = \mathbf{Q}_i\mathbf{Q}_i'$ be the idempotent matrix associated with the mean square for the ith effect in model 10.3.1, where $\mathbf{Q}_i$ is of order $N \times m_i$

such that $\mathbf{Q}_i'\mathbf{Q}_i = \mathbf{I}_{m_i}$ $(i = 0, 1, \ldots, \nu+1)$. Furthermore, let σ_{ijk} denote the $(j,k)^{th}$ element of $\mathbf{\Sigma}_i$, the variance–covariance matrix associated with the ith random effect $(i = \omega, \omega+1, \ldots, \nu+1;\ j,k = 1,2,\ldots,r)$. Then, for any two columns, $\mathbf{y}_j$ and $\mathbf{y}_k$, of $\mathbf{Y}$ we have

$$\text{Cov}(\mathbf{Q}_i'\mathbf{y}_j, \mathbf{Q}_i'\mathbf{y}_k) = \left(\sum_{\ell=\omega}^{\nu+1} \kappa_{i\ell}\sigma_{\ell jk}\right)\mathbf{I}_{m_i}, \quad i = 0, 1, \cdots, \nu+1; \quad j,k = 1,2,\cdots,r, \tag{10.3.5}$$

where $\kappa_{i\ell}$ is the constant used in formula (10.3.3).

Proof. From model (10.2.2) we have

$$\begin{aligned}\text{Cov}(\mathbf{Q}_i'\mathbf{y}_j, \mathbf{Q}_i'\mathbf{y}_k) = \mathbf{Q}_i' \,\text{Cov}(\mathbf{y}_j, \mathbf{y}_k)\mathbf{Q}_i &= \mathbf{Q}_i' \,\text{Cov}\left(\sum_{\ell=\omega}^{\nu+1}\mathbf{H}_\ell\boldsymbol{\beta}_{\ell j}, \sum_{m=\omega}^{\nu+1}\mathbf{H}_m\boldsymbol{\beta}_{mk}\right)\mathbf{Q}_i,\\ i &= 0, 1, \cdots, \nu+1;\ j,k = 1,2,\cdots,r,\end{aligned}$$

where $\mathbf{H}_{\nu+1} = \mathbf{I}_N$ and $\boldsymbol{\beta}_{\nu+1j} = \mathbf{e}_j$, $\boldsymbol{\beta}_{\nu+1k} = \mathbf{e}_k$. But,

$$\text{Cov}(\boldsymbol{\beta}_{\ell j}, \boldsymbol{\beta}_{mk}) = \begin{cases} \mathbf{0}, & \text{if } \ell \neq m \\ \sigma_{\ell jk}\mathbf{I}_{c_\ell}, & \text{if } \ell = m \end{cases},$$

where c_ℓ is the number of columns of $\mathbf{H}_\ell$ $(\ell = \omega, \omega+1, \ldots, \nu+1)$. Hence,

$$\begin{aligned}\text{Cov}(\mathbf{Q}_i'\mathbf{y}_j, \mathbf{Q}_i'\mathbf{y}_k) = \mathbf{Q}_i'\left(\sum_{\ell=\omega}^{\nu+1}\sigma_{\ell jk}\mathbf{A}_\ell\right)\mathbf{Q}_i,\quad i &= 0, 1, \cdots, \nu+1;\\ j,k &= 1,2,\cdots,r,\end{aligned} \tag{10.3.6}$$

where $\mathbf{A}_\ell = \mathbf{H}_\ell\mathbf{H}_\ell'$. Using formula (10.3.3) and the fact that $\mathbf{P}_i = \mathbf{Q}_i\mathbf{Q}_i'$, we obtain

$$\mathbf{A}_\ell\mathbf{Q}_i = \kappa_{i\ell}\mathbf{Q}_i, \quad i,\ell = 0, 1, \cdots, \nu+1, \tag{10.3.7}$$

since $\mathbf{Q}_i'\mathbf{Q}_i = \mathbf{I}_{m_i}$. By making the substitution in formula (10.3.6), we get formula (10.3.5). This completes the proof of Lemma 10.3.1. □

Corollary 10.3.1. The rows of $\mathbf{Q}_i'\mathbf{Y}$ are independent and have the normal distribution with a common variance–covariance matrix given by

$$\mathbf{\Delta}_i = \sum_{\ell=\omega}^{\nu+1}\kappa_{i\ell}\mathbf{\Sigma}_\ell, \quad i = 0, 1, \cdots, \nu+1. \tag{10.3.8}$$

Proof. It is clear from formula (10.3.5) that the rows of

$$\mathbf{Q}_i'\mathbf{Y} = [\mathbf{Q}_i'\mathbf{y}_1 : \mathbf{Q}_i'\mathbf{y}_2 : \cdots : \mathbf{Q}_i'\mathbf{y}_r]$$

are independent. Furthermore, for any row of $\mathbf{Q}_i'\mathbf{Y}$, the $(j,k)^{th}$ element of the corresponding variance–covariance matrix is given by $\sum_{\ell=\omega}^{\nu+1} \kappa_{i\ell}\sigma_{\ell jk}$. It follows that the rows of $\mathbf{Q}_i'\mathbf{Y}$ have a common variance–covariance matrix $\mathbf{\Delta}_i$ as shown in formula (10.3.8). □

Theorem 10.3.1.

(a) If the ith effect is fixed, then $\mathbf{Y}'\mathbf{P}_i\mathbf{Y}$ has the noncentral Wishart distribution $W_r(m_i, \mathbf{\Delta}_i, \mathbf{\Gamma}_i)$, where $\mathbf{\Delta}_i$ is given by formula (10.3.8) and $\mathbf{\Gamma}_i$ is the noncentrality parameter matrix

$$\mathbf{\Gamma}_i = \mathbf{\Delta}_i^{-1/2}\mathbf{G}'\mathbf{X}'\mathbf{P}_i\mathbf{X}\mathbf{G}\mathbf{\Delta}_i^{-1/2}, \quad i = 0, 1, \cdots, \nu - p. \qquad (10.3.9)$$

(b) If the ith effect is random, then $\mathbf{Y}'\mathbf{P}_i\mathbf{Y}$ has the central Wishart distribution $W_r(m_i, \mathbf{\Delta}_i)$, $i = \omega, \omega + 1, \ldots, \nu + 1$.

(c) $\mathbf{Y}'\mathbf{P}_i\mathbf{Y}$ and $\mathbf{Y}'\mathbf{P}_j\mathbf{Y}$ are independent for $i \neq j$ $(i, j = 0, 1, \ldots, \nu + 1)$.

(d) $E(\mathbf{Y}'\mathbf{P}_i\mathbf{Y}) = m_i\mathbf{\Delta}_i + \mathbf{G}'\mathbf{X}'\mathbf{P}_i\mathbf{X}\mathbf{G}$, $i = 0, 1, \ldots, \nu + 1$.

Proof.

(a) By Corollary 10.3.1, the rows of $\mathbf{Q}_i'\mathbf{Y}$ are independent and have the normal distribution with a common variance–covariance matrix $\mathbf{\Delta}_i$ $(i = 0, 1, \ldots, \nu + 1)$. Furthermore, from model (10.3.1),

$$E(\mathbf{Q}_i'\mathbf{Y}) = \mathbf{Q}_i'\mathbf{X}\mathbf{G}. \qquad (10.3.10)$$

It follows that $\mathbf{Y}'\mathbf{P}_i\mathbf{Y} = \mathbf{Y}'\mathbf{Q}_i\mathbf{Q}_i'\mathbf{Y}$ has the noncentral Wishart distribution $W_r(m_i, \mathbf{\Delta}_i, \mathbf{\Gamma}_i)$ with the noncentrality parameter matrix

$$\begin{aligned}\mathbf{\Gamma}_i &= \mathbf{\Delta}_i^{-1/2}\mathbf{G}'\mathbf{X}'\mathbf{Q}_i\mathbf{Q}_i'\mathbf{X}\mathbf{G}\mathbf{\Delta}_i^{-1/2} \\ &= \mathbf{\Delta}_i^{-1/2}\mathbf{G}'\mathbf{X}'\mathbf{P}_i\mathbf{X}\mathbf{G}\mathbf{\Delta}_i^{-1/2}\end{aligned}$$

(see, for example, Seber, 1984, Section 2.3.3).

(b) If the ith effect is random, then we have the same results as in (a), except that, in this case, the noncentrality parameter matrix is equal to zero. This latter assertion results from applying Theorem 2.4.1(b) to the univariate model (10.3.2). In doing so we find that $\mathbf{g}_\lambda'\mathbf{X}'\mathbf{P}_i\mathbf{X}\mathbf{g}_\lambda = 0$ for $i = \omega, \omega + 1, \ldots, \nu + 1$, and for all $\boldsymbol{\lambda} \neq \mathbf{0}$. Since $\mathbf{g}_\lambda = \mathbf{G}\boldsymbol{\lambda}$, it follows that for the ith random effect,

$$\mathbf{G}'\mathbf{X}'\mathbf{P}_i\mathbf{X}\mathbf{G} = \mathbf{0}, \quad i = \omega, \omega + 1, \cdots, \nu + 1.$$

Hence, $\mathbf{\Gamma}_i = \mathbf{0}$ and, consequently, $\mathbf{Y}'\mathbf{P}_i\mathbf{Y}$ has the central Wishart distribution $W_r(m_i, \mathbf{\Delta}_i)$ for $i = \omega, \omega+1, \ldots, \nu+1$.

(c) Consider $\mathbf{Y}'\mathbf{P}_i\mathbf{Y}$ and $\mathbf{Y}'\mathbf{P}_j\mathbf{Y}$ for $i \neq j$ $(i, j = 0, 1, \ldots, \nu+1)$. Let us write $\mathbf{P}_i = \mathbf{Q}_i\mathbf{Q}_i'$ and $\mathbf{P}_j = \mathbf{Q}_j\mathbf{Q}_j'$, where the columns of $\mathbf{Q}_i$ and those of $\mathbf{Q}_j$ are orthonormal. Furthermore, since $\mathbf{P}_i\mathbf{P}_j = 0$ for $i \neq j$, $\mathbf{Q}_i'\mathbf{Q}_j = \mathbf{0}$ for $i \neq j$. Now, let $\mathbf{u}_{i\ell}'$ and $\mathbf{u}_{j\ell}'$ denote the ℓth rows of $\mathbf{Q}_i'\mathbf{Y}$ and $\mathbf{Q}_j'\mathbf{Y}$, respectively. Then,

$$\begin{aligned} \mathbf{Y}'\mathbf{P}_i\mathbf{Y} &= \mathbf{Y}'\mathbf{Q}_i\mathbf{Q}_i'\mathbf{Y} \\ &= \sum_{\ell=1}^{m_i} \mathbf{u}_{i\ell}\mathbf{u}_{i\ell}', \end{aligned} \tag{10.3.11}$$

$$\begin{aligned} \mathbf{Y}'\mathbf{P}_j\mathbf{Y} &= \mathbf{Y}'\mathbf{Q}_j\mathbf{Q}_j'\mathbf{Y} \\ &= \sum_{\ell=1}^{m_j} \mathbf{u}_{j\ell}\mathbf{u}_{j\ell}'. \end{aligned} \tag{10.3.12}$$

As in the proof of formula (10.3.6), it can be shown that for any two columns, $\mathbf{y}_q$ and $\mathbf{y}_s$, of $\mathbf{Y}$,

$$\begin{aligned} \mathrm{Cov}(\mathbf{Q}_i'\mathbf{y}_q, \mathbf{Q}_j'\mathbf{y}_s) &= \mathbf{Q}_i'\left(\sum_{\ell=\omega}^{\nu+1} \sigma_{\ell qs}\mathbf{A}_\ell\right)\mathbf{Q}_j \\ &= \mathbf{0}, \end{aligned}$$

since, by formula (10.3.7),

$$\mathbf{Q}_i'\mathbf{A}_\ell\mathbf{Q}_j = \kappa_{j\ell}\mathbf{Q}_i'\mathbf{Q}_j = \mathbf{0}.$$

Hence, the set $\{\mathbf{u}_{i1}, \mathbf{u}_{i2}, \ldots, \mathbf{u}_{im_i}; \mathbf{u}_{j1}, \mathbf{u}_{j2}, \ldots, \mathbf{u}_{jm_j}\}$ consists of mutually independent normal vectors. It follows from formulas (10.3.11) and (10.3.12) that $\mathbf{Y}'\mathbf{P}_i\mathbf{Y}$ and $\mathbf{Y}'\mathbf{P}_j\mathbf{Y}$ are independent for $i \neq j$ $(i, j = 0, 1, \ldots, \nu+1)$.

(d) Since $\mathbf{P}_i = \mathbf{Q}_i\mathbf{Q}_i'$, we get

$$\begin{aligned} E(\mathbf{Y}'\mathbf{P}_i\mathbf{Y}) &= E(\mathbf{Y}'\mathbf{Q}_i\mathbf{Q}_i'\mathbf{Y}) \\ &= m_i\mathbf{\Delta}_i + \mathbf{G}'\mathbf{X}'\mathbf{Q}_i\mathbf{Q}_i'\mathbf{X}\mathbf{G} \\ &= m_i\mathbf{\Delta}_i + \mathbf{G}'\mathbf{X}'\mathbf{P}_i\mathbf{X}\mathbf{G}, \quad i = 0, 1, \cdots, \nu+1. \end{aligned} \tag{10.3.13}$$

Note that (10.3.13) was derived by applying the following result (see, for example, the corollary to Lemma 1.1 in Seber, 1984, p. 7): If the rows of the random matrix $\mathbf{T}$ are independent and have a common variance–covariance matrix $\mathbf{\Delta}_t$, and if $\mathbf{A}$ is a constant symmetric matrix, then

$$E(\mathbf{T}'\mathbf{A}\mathbf{T}) = tr(\mathbf{A})\mathbf{\Delta}_t + E(\mathbf{T})'\mathbf{A}\,E(\mathbf{T}).$$

In our case, $\mathbf{T} = \mathbf{Q}_i'\mathbf{Y}$, $\mathbf{A} = \mathbf{I}_{m_i}$, $\boldsymbol{\Delta}_t = \boldsymbol{\Delta}_i$, and $E(\mathbf{T}) = \mathbf{Q}_i'E(\mathbf{Y}) = \mathbf{Q}_i'\mathbf{XG}$. We may recall from (b) that if $\mathbf{Y}'\mathbf{P}_i\mathbf{Y}$ is associated with a random effect, then $\mathbf{G}'\mathbf{X}'\mathbf{P}_i\mathbf{XG} = \mathbf{0}$. Hence,

$$E(\mathbf{Y}'\mathbf{P}_i\mathbf{Y}) = m_i\boldsymbol{\Delta}_i, \quad i = \omega, \omega + 1, \cdots, \nu + 1. \tag{10.3.14}$$

This completes the proof of Theorem 10.3.1. □

10.3.2. Hypothesis Testing

Estimates of the $\boldsymbol{\Sigma}_\ell$'s in formula (10.3.8) can be easily obtained by equating the terms $\mathbf{Y}'\mathbf{P}_i\mathbf{Y}$ in formula (10.3.14) to their expected values and then solving the resulting equations $(i, \ell = \omega, \omega + 1, \ldots, \nu + 1)$. We thus have from (10.3.8) and (10.3.14),

$$\mathbf{Y}'\mathbf{P}_i\mathbf{Y} = m_i\left(\sum_{\ell=\omega}^{\nu+1} \kappa_{i\ell}\hat{\boldsymbol{\Sigma}}_\ell\right), \quad i = \omega, \omega + 1, \cdots, \nu + 1.$$

The estimates $\hat{\boldsymbol{\Sigma}}_\ell$ $(\ell = \omega, \omega + 1, \ldots, \nu + 1)$ so obtained are unbiased and translation invariant (see Mathew, 1989). They may not, however, be nonnegative definite and may have to be modified in order to satisfy the nonnegativity requirement. Anderson (1985) presented such a modification in the case of the balanced multivariate random one-way model (see Example 10.2.1). For this model, formula (10.3.14) gives

$$\begin{aligned} E(\mathbf{Y}'\mathbf{P}_1\mathbf{Y}) &= (v - 1)(n\boldsymbol{\Sigma}_1 + \boldsymbol{\Sigma}_2) \\ E(\mathbf{Y}'\mathbf{P}_2\mathbf{Y}) &= v(n - 1)\boldsymbol{\Sigma}_2, \end{aligned}$$

where $\boldsymbol{\Sigma}_1$ and $\boldsymbol{\Sigma}_2$ are the variance–covariance matrices for the random effect $\boldsymbol{\tau}_k$ $(k = 1, 2, \ldots, v)$ and the error term, respectively, in model (10.2.3). Hence,

$$\begin{aligned} \hat{\boldsymbol{\Sigma}}_1 &= \frac{1}{n}\left[\frac{\mathbf{Y}'\mathbf{P}_1\mathbf{Y}}{v - 1} - \frac{\mathbf{Y}'\mathbf{P}_2\mathbf{Y}}{v(n - 1)}\right] \\ \hat{\boldsymbol{\Sigma}}_2 &= \frac{\mathbf{Y}'\mathbf{P}_2\mathbf{Y}}{v(n - 1)}. \end{aligned}$$

Note that $\hat{\boldsymbol{\Sigma}}_1$ is not necessarily positive semidefinite. Anderson (1985) proposed the following modification: If $v(n - 1) \geq r$, then $\mathbf{Y}'\mathbf{P}_2\mathbf{Y}$ is positive definite with probability one (see, for example, Seber, 1984, p. 411). In this case, there exists a nonsingular matrix $\mathbf{V}$ and a diagonal matrix $\mathbf{D}$, both of order $r \times r$, such that

$$\begin{aligned} \frac{1}{v - 1}\mathbf{Y}'\mathbf{P}_1\mathbf{Y} &= \mathbf{VDV}', \\ \frac{1}{v(n - 1)}\mathbf{Y}'\mathbf{P}_2\mathbf{Y} &= \mathbf{VV}'. \end{aligned}$$

The diagonal elements of $\mathbf{D}$ are the eigenvalues of $[\mathbf{Y}'\mathbf{P}_2\mathbf{Y}/v(n-1)]^{-1}$ $[\mathbf{Y}'\mathbf{P}_1\mathbf{Y}/(v-1)]$. We then have

$$\hat{\boldsymbol{\Sigma}}_1 = \frac{1}{n}\mathbf{V}(\mathbf{D}-\mathbf{I}_r)\mathbf{V}'.$$

In order for $\hat{\boldsymbol{\Sigma}}_1$ to be positive semidefinite, the diagonal elements of $\mathbf{D}$ must be greater than or equal to one. In general, this may not be the case. Let r_0 be the number of the diagonal elements of $\mathbf{D}$ that are greater than one (the probability that a diagonal element is exactly one is zero). Without loss of generality, we consider that these elements are written first in $\mathbf{D}$ and thus form a diagonal matrix, $\mathbf{D}^*$, of order $r_0 \times r_0$. Let $\mathbf{V}^*$ be a matrix of order $r \times r_0$ consisting of the first r_0 columns of $\mathbf{V}$. Then a positive semidefinite estimator of $\boldsymbol{\Sigma}_1$ is given by

$$\hat{\boldsymbol{\Sigma}}_1^* = \frac{1}{n}\mathbf{V}^*(\mathbf{D}^*-\mathbf{I}_{r_0})\mathbf{V}^{*\prime}.$$

In the event $r_0 = 0$, $\hat{\boldsymbol{\Sigma}}_1^*$ is set equal to zero. Some other procedures for obtaining a positive semidefinite estimator of $\boldsymbol{\Sigma}_1$ are given in Amemiya (1985) and Mathew et al. (1994).

Let us now consider tests concerning the fixed and random effects for model (10.3.1). For the ith fixed effect we have from Theorem 10.3.1(d),

$$E(\mathbf{Y}'\mathbf{P}_i\mathbf{Y}) = m_i\boldsymbol{\Delta}_i + \mathbf{G}'\mathbf{X}'\mathbf{P}_i\mathbf{X}\mathbf{G}, \quad i = 0, 1, \cdots, \nu - p.$$

Suppose that there exists a random effect with the associated term $\mathbf{Y}'\mathbf{P}_{i_0}\mathbf{Y}$ $(\omega \leq i_0 \leq \nu + 1)$ such that

$$E(\mathbf{Y}'\mathbf{P}_{i_0}\mathbf{Y}) = m_{i_0}\boldsymbol{\Delta}_i.$$

We assume that $m_{i_0} \geq r$ so that $\mathbf{Y}'\mathbf{P}_{i_0}\mathbf{Y}$ is positive definite with probability 1. In this case, an exact test concerning the hypothesis

$$H_0 : \mathbf{P}_i\mathbf{X}\mathbf{G} = \mathbf{0} \tag{10.3.15}$$

can be obtained by using any standard multivariate test statistic. For example, Roy's largest root, namely, $\lambda_{\max}[(\mathbf{Y}'\mathbf{P}_i\mathbf{Y})(\mathbf{Y}'\mathbf{P}_{i_0}\mathbf{Y})^{-1}]$ is such a statistic; see, for example, Seber (1984, Section 8.6.2). Here, $\lambda_{\max}(\cdot)$ denotes the largest eigenvalue. This test can be easily derived by applying the union-intersection principle to the univariate model (10.3.2) for which an exact F statistic exists for testing the hypothesis

$$H_{0\lambda} : \boldsymbol{\lambda}'\mathbf{G}'\mathbf{X}'\mathbf{P}_i\mathbf{X}\mathbf{G}\boldsymbol{\lambda} = 0,$$

namely,

$$F_\lambda = \frac{\boldsymbol{\lambda}'\mathbf{Y}'\mathbf{P}_i\mathbf{Y}\boldsymbol{\lambda}/m_i}{\boldsymbol{\lambda}'\mathbf{Y}'\mathbf{P}_{i_0}\mathbf{Y}\boldsymbol{\lambda}/m_{i_0}},$$

where $\boldsymbol{\lambda}$ is any nonzero vector. Large values of Roy's largest root test statistic lead to the rejection of H_0. Other multivariate tests include Wilks' likelihood ratio: $|\mathbf{Y}'\mathbf{P}_{i_0}\mathbf{Y}|/|\mathbf{Y}'\mathbf{P}_i\mathbf{Y}+\mathbf{Y}'\mathbf{P}_{i_0}\mathbf{Y}|$, Hotelling-Lawley's trace: $\text{tr}[(\mathbf{Y}'\mathbf{P}_i\mathbf{Y})(\mathbf{Y}'\mathbf{P}_{i_0}\mathbf{Y})^{-1}]$, and Pillai's trace: $\text{tr}[(\mathbf{Y}'\mathbf{P}_i\mathbf{Y})(\mathbf{Y}'\mathbf{P}_i\mathbf{Y}+\mathbf{Y}'\mathbf{P}_{i_0}\mathbf{Y})^{-1}]$. See, for example, Seber (1984, Section 8.6.2).

Similar tests can be easily derived for the random effects. For the ith of such effects, we have from formulas (10.3.8) and (10.3.14),

$$\begin{aligned} E(\mathbf{Y}'\mathbf{P}_i\mathbf{Y}) &= m_i\boldsymbol{\Delta}_i \\ &= m_i(\kappa_{ii}\boldsymbol{\Sigma}_i + \boldsymbol{\Delta}_i^*), \quad i = \omega, \omega+1, \cdots, \nu, \end{aligned}$$

where $\boldsymbol{\Delta}_i^*$ is a linear combination of $\boldsymbol{\Sigma}_\ell$'s, $\ell \neq i$. Suppose that there exists another random effect with an associated term $\mathbf{Y}'\mathbf{P}_{i_*}\mathbf{Y}$ such that

$$E(\mathbf{Y}'\mathbf{P}_{i_*}\mathbf{Y}) = m_{i_*}\boldsymbol{\Delta}_i^*.$$

Furthermore, we assume that $m_{i_*} \geq r$ so that $\mathbf{Y}'\mathbf{P}_{i*}\mathbf{Y}$ will be positive definite with probability 1. In this case, Roy's largest root, $\lambda_{\max}[(\mathbf{Y}'\mathbf{P}_i\mathbf{Y})(\mathbf{Y}'\mathbf{P}_{i_*}\mathbf{Y})^{-1}]$, can be used to test the hypothesis

$$H_0 : \boldsymbol{\Sigma}_i = \mathbf{0}. \qquad (10.3.16)$$

As before, this test is derived by an application of the union-intersection principle to the univariate model (10.3.2). The other multivariate tests mentioned earlier can also be used.

Example 10.3.1. Consider again the multivariate two-way model (10.2.7) given in Example 10.2.2. Suppose that $\boldsymbol{\tau}_k$ is fixed and $\boldsymbol{\beta}_\ell$ is random. In this case, $\mathbf{XG} = \mathbf{H}_0\mathbf{B}_0 + \mathbf{H}_1\mathbf{B}_1$. The hypothesis (10.3.15) concerning $\boldsymbol{\tau}_k$ can then be written as

$$H_0 : \mathbf{P}_1(\mathbf{H}_0\mathbf{B}_0 + \mathbf{H}_1\mathbf{B}_1) = \mathbf{0}.$$

But, $\mathbf{P}_1\mathbf{H}_0 = \mathbf{0}$ by formula (10.3.3) and the fact that $\mathbf{H}_0\mathbf{H}_0'\mathbf{P}_1 = \mathbf{A}_0\mathbf{P}_1 = \mathbf{0}$. Hence,

$$H_0 : \mathbf{P}_1\mathbf{H}_1\mathbf{B}_1 = \mathbf{0}.$$

Furthermore, by applying Lemma 2.3.1 we obtain

$$\mathbf{P}_1 = \frac{\mathbf{H}_1\mathbf{H}_1'}{bn} - \frac{\mathbf{H}_0\mathbf{H}_0'}{bvn}.$$

Therefore,

$$\begin{aligned}
\mathbf{P}_1\mathbf{H}_1 &= \left[\frac{\mathbf{H}_1\mathbf{H}_1'}{bn} - \frac{\mathbf{H}_0\mathbf{H}_0'}{bvn}\right]\mathbf{H}_1 \\
&= \mathbf{H}_1 - \frac{\mathbf{H}_0\mathbf{H}_0'\mathbf{H}_1}{bvn} \\
&= \mathbf{I}_v \otimes \mathbf{1}_b \otimes \mathbf{1}_n - \frac{\mathbf{J}_v \otimes \mathbf{1}_b \otimes \mathbf{1}_n}{v} \\
&= \left(\mathbf{I}_v - \frac{\mathbf{J}_v}{v}\right) \otimes \mathbf{1}_{bn}.
\end{aligned}$$

Since $\mathbf{B}_1 = [\boldsymbol{\beta}_{11} : \boldsymbol{\beta}_{12} : \ldots : \boldsymbol{\beta}_{1r}]$ with $\boldsymbol{\beta}_{1j} = (\tau_{j1}, \tau_{j2}, \ldots, \tau_{jv})'$, the hypothesis H_0 is true if and only if H_{0j} is true for all j, where

$$H_{0j} : \left(\mathbf{I}_v - \frac{\mathbf{J}_v}{v}\right)\boldsymbol{\beta}_{1j} = \mathbf{0}, \quad j = 1, 2, \cdots, r.$$

In turn, H_{0j} is true if and only if the elements of $\boldsymbol{\beta}_{1j}$ are equal, that is, $\tau_{j1} = \tau_{j2} = \ldots = \tau_{jv}$ for $j = 1, 2, \ldots, r$. In other words, H_0 is equivalent to the hypothesis that the means of all v levels of factor A are equal.

10.3.3. Satterthwaite's Approximation

We recall that Satterthwaite's approximation was discussed earlier in Section 2.6.1 of Chapter 2 in connection with univariate models. The derivation of any of the multivariate tests described in Section 10.3.2 is based on the presumption that there exists a random effect whose mean square matrix has the same expected value as that of the effect under consideration, when the null hypothesis is true. Such a random effect, however, may not always exist. In this case, a linear combination of mean square matrices of random effects can be used instead.

Let us consider the balanced multivariate mixed model (10.3.1). Let $\hat{\boldsymbol{\Phi}}$ denote a nonnegative linear combination of mean square matrices of random effects. Thus

$$\hat{\boldsymbol{\Phi}} = \sum_{i=\omega}^{\nu+1} a_i \mathbf{S}_i, \tag{10.3.17}$$

where $\mathbf{S}_i$ is the ith random effect's mean square matrix (see formula 10.3.4), and the a_i's are nonnegative constants. Formula (10.3.17) can be written as

$$\hat{\boldsymbol{\Phi}} = \sum_{i=\omega}^{\nu+1} \frac{a_i}{m_i} \mathbf{Y}'\mathbf{P}_i\mathbf{Y}. \tag{10.3.18}$$

We recall from Theorem 10.3.1 that for $i = \omega, \omega+1, \ldots, \nu+1$, $\mathbf{Y}'\mathbf{P}_i\mathbf{Y}$ have independent central Wishart distributions $W_r(m_i, \boldsymbol{\Delta}_i)$, where $\boldsymbol{\Delta}_i$ is defined by formula (10.3.8). Tan and Gupta (1983) approximated the distribution of $\hat{\boldsymbol{\Phi}}$ with a central Wishart distribution $W_r(m, \boldsymbol{\Delta})$, where m and $\boldsymbol{\Delta}$ are determined by equating the expected value and generalized variance of $\hat{\boldsymbol{\Phi}}$ to those of $W_r(m, \boldsymbol{\Delta})$. Such a procedure results in the so-called *multivariate Satterthwaite's approximation*. In order to facilitate the understanding of the derivation of the formulas for m and $\boldsymbol{\Delta}$, the following lemma is needed.

Lemma 10.3.2. Let $\hat{\Phi}_{jk}$ and $d_{jk}^{(i)}$ denote the $(j,k)^{th}$ elements of $\hat{\boldsymbol{\Phi}}$ and $\boldsymbol{\Delta}_i$, respectively. Then,

(a) $E(\hat{\Phi}_{jk}) = \sum_{i=\omega}^{\nu+1} a_i d_{jk}^{(i)}, \quad 1 \le j \le k \le r.$

(b) $\text{Cov}(\hat{\Phi}_{jk}, \hat{\Phi}_{st}) = \sum_{i=\omega}^{\nu+1} (a_i^2/m_i)[d_{js}^{(i)} d_{kt}^{(i)} + d_{jt}^{(i)} d_{ks}^{(i)}], \quad 1 \le j \le k \le r, \quad 1 \le s \le t \le r.$

Proof.

(a) This follows directly from formula (10.3.18) and the fact that $E(\mathbf{Y}'\mathbf{P}_i\mathbf{Y}) = m_i \boldsymbol{\Delta}_i$ for $i = \omega, \omega+1, \ldots, \nu+1$.

(b) We recall from Section 10.3.1 that

$$\mathbf{Y}'\mathbf{P}_i\mathbf{Y} = \sum_{\ell=1}^{m_i} \mathbf{u}_{i\ell}\mathbf{u}_{i\ell}', \quad i = \omega, \omega+1, \cdots, \nu+1,$$

where the $\mathbf{u}_{i\ell}$'s are mutually independent and have the normal distribution with a zero mean and a variance–covariance matrix $\boldsymbol{\Delta}_i$ (see formula 10.3.11 and Corollary 10.3.1). Let $y_{jk}^{(i)}$ denote the $(j,k)^{th}$ element of $\mathbf{Y}'\mathbf{P}_i\mathbf{Y}$, then

$$y_{jk}^{(i)} = \sum_{\ell=1}^{m_i} u_{ji\ell} u_{ki\ell}, \quad 1 \le j \le k \le r,$$

where $u_{ji\ell}$ and $u_{ki\ell}$ are the jth and k^{th} elements of $\mathbf{u}_{i\ell}$, respectively. The product $u_{ji\ell}\, u_{ki\ell}$ can be written as

$$u_{ji\ell}\, u_{ki\ell} = \mathbf{u}_{i\ell}' \mathbf{L}_{jk} \mathbf{u}_{i\ell}, \quad 1 \le j \le k \le r,$$

where $\mathbf{L}_{jk}$ is a symmetric matrix of order $r \times r$ whose elements are equal to zero except for the $(j,k)^{th}$ and $(k,j)^{th}$ elements, which are equal to $1/2$. We then have

$$\text{Cov}(y_{jk}^{(i)}, y_{st}^{(i)}) = \sum_{\ell=1}^{m_i} \text{Cov}(\mathbf{u}_{i\ell}'\mathbf{L}_{jk}\mathbf{u}_{i\ell}, \mathbf{u}_{i\ell}'\mathbf{L}_{st}\mathbf{u}_{i\ell}),$$

since the $\mathbf{u}_{i\ell}$'s are independent for $\ell = 1, 2, \ldots, m_i$. But, by formula (58) in Searle (1971, p. 66),

$$\text{Cov}(\mathbf{u}'_{i\ell}\mathbf{L}_{jk}\mathbf{u}_{i\ell}, \mathbf{u}'_{i\ell}\mathbf{L}_{st}\mathbf{u}_{i\ell}) = 2tr(\mathbf{L}_{jk}\mathbf{\Delta}_i\mathbf{L}_{st}\mathbf{\Delta}_i), \quad \ell = 1, 2, \cdots, m_i.$$

It can be verified that

$$tr(\mathbf{L}_{jk}\mathbf{\Delta}_i\mathbf{L}_{st}\mathbf{\Delta}_i) = \frac{1}{2}\left[d_{js}^{(i)}d_{kt}^{(i)} + d_{jt}^{(i)}d_{ks}^{(i)}\right].$$

Hence,

$$\text{Cov}(y_{jk}^{(i)}, y_{st}^{(i)}) = m_i\left[d_{js}^{(i)}d_{kt}^{(i)} + d_{jt}^{(i)}d_{ks}^{(i)}\right].$$

Now, from formula (10.3.18) we have

$$\hat{\Phi}_{jk} = \sum_{i=\omega}^{\nu+1}\frac{a_i}{m_i}y_{jk}^{(i)}, \quad 1 \le j \le k \le r,$$

$$\hat{\Phi}_{st} = \sum_{\ell=\omega}^{\nu+1}\frac{a_\ell}{m_\ell}y_{st}^{(\ell)}, \quad 1 \le s \le t \le r,$$

It follows that

$$\begin{aligned}\text{Cov}(\hat{\Phi}_{jk}, \hat{\Phi}_{st}) &= \sum_{i=\omega}^{\nu+1}\frac{a_i^2}{m_i^2}\,\text{Cov}(y_{jk}^{(i)}, y_{st}^{(i)}) \\ &= \sum_{i=\omega}^{\nu+1}\frac{a_i^2}{m_i}\left[d_{js}^{(i)}d_{kt}^{(i)} + d_{jt}^{(i)}d_{ks}^{(i)}\right],\end{aligned}$$

since for $i \neq \ell, \mathbf{Y}'\mathbf{P}_i\mathbf{Y}$ and $\mathbf{Y}'\mathbf{P}_\ell\mathbf{Y}$ are independent resulting in $y_{jk}^{(i)}$ and $y_{st}^{(\ell)}$ being uncorrelated. This completes the proof of Lemma 10.3.2. □

If $\hat{\mathbf{\Phi}}$ in formula (10.3.18) is approximated with a central Wishart distribution $W_r(m, \mathbf{\Delta})$, then $E(\hat{\Phi}_{jk})$ in Lemma 10.3.2(a) is approximately equal to $m\, d_{jk}$, where d_{jk} is the $(j, k)^{th}$ element of $\mathbf{\Delta}$. Furthermore, $\text{Cov}(\hat{\Phi}_{jk}, \hat{\Phi}_{st})$ is approximately equal to $m(d_{js}d_{kt} + d_{jt}d_{ks})$. The latter expression can be derived in a manner similar to the derivation of $\text{Cov}(y_{jk}^{(i)}, y_{st}^{(i)})$ in the proof of Lemma 10.3.2(b).

The generalized variance of $\hat{\mathbf{\Phi}}$ is defined as the determinant of the $r^* \times r^*$ matrix $\mathbf{V}_\phi$, where $r^* = \frac{1}{2}r(r+1)$ (see Tan and Gupta, 1983). The elements of $\mathbf{V}_\phi$ are values of $\text{Cov}(\hat{\Phi}_{jk}, \hat{\Phi}_{st})$ arranged in lexicographic order, that is, $\text{Cov}(\hat{\Phi}_{j_1k_1}, \hat{\Phi}_{s_1t_1})$ appears before $\text{Cov}(\hat{\Phi}_{j_2k_2}, \hat{\Phi}_{s_2t_2})$ in a row if $k_1 < k_2$, or $k_1 =$

k_2 and $j_1 < j_2$. Similarly, the former covariance value appears before the latter in a column if $t_1 < t_2$, or $t_1 = t_2$ and $s_1 < s_2$. Using Lemma 10.3.2(b), the matrix $\mathbf{V}_\phi$ can be expressed as

$$\mathbf{V}_\phi = \mathbf{K}_r^- \left[\sum_{i=\omega}^{\nu+1} \frac{a_i^2}{m_i} (\mathbf{\Delta}_i \otimes \mathbf{\Delta}_i) \right] \mathbf{K}_r,$$

where $\mathbf{K}_r^-$ is the Moore-Penrose inverse of $\mathbf{K}_r$. The latter matrix is of order $r^2 \times r^*$ with a typical element

$$(\mathbf{K}_r)_{ij,gh} = \frac{1}{2}(\delta_{ig}\delta_{jh} + \delta_{ih}\delta_{jg}) \quad i \le r, \quad j \le r, \quad g \le h \le r,$$

and δ_{ij} is Kronecker's delta, that is,

$$\begin{aligned} \delta_{ij} &= 1, \quad \text{if } i = j \\ &= 0, \quad \text{if } i \neq j. \end{aligned}$$

The matrices $\mathbf{K}_r$ and $\mathbf{K}_r^-$ have the property that

$$|\mathbf{K}_r^- (\mathbf{B} \otimes \mathbf{B}) \mathbf{K}_r| = |\mathbf{B}|^{r+1}$$

for any symmetric matrix $\mathbf{B}$ of order $r \times r$. For more details concerning the matrix $\mathbf{K}_r$, see Browne (1974, p. 3), Nel (1980, pp. 158–159, 178), and Nel and Van der Merwe (1986). It follows that the generalized variance of $\hat{\mathbf{\Phi}}$ is given by

$$|\mathbf{V}_\phi| = \left| \mathbf{K}_r^- \left[\sum_{i=\omega}^{\nu+1} \frac{a_i^2}{m_i} (\mathbf{\Delta}_i \otimes \mathbf{\Delta}_i) \right] \mathbf{K}_r \right|.$$

On the other hand, since $\text{Cov}(\hat{\Phi}_{jk}, \hat{\Phi}_{st})$ is approximated with $m(d_{js}d_{kt} + d_{jt}d_{ks})$, the generalized variance of the approximating central Wishart distribution $W_r(m, \mathbf{\Delta})$ is $|\mathbf{V}_w|$, where

$$|\mathbf{V}_w| = |\mathbf{K}_r^- [m(\mathbf{\Delta} \otimes \mathbf{\Delta})] \mathbf{K}_r| = m^{r^*} |\mathbf{\Delta}|^{r+1}. \tag{10.3.19}$$

Now, by equating the expected values and generalized variances of $\hat{\mathbf{\Phi}}$ to those of $W_r(m, \mathbf{\Delta})$, we obtain

$$\sum_{i=\omega}^{\nu+1} a_i \mathbf{\Delta}_i = m\mathbf{\Delta}$$

$$|\mathbf{V}_\phi| = m^{r^*} |\mathbf{\Delta}|^{r+1}.$$

Hence,

$$
\begin{aligned}
\boldsymbol{\Delta} &= \frac{1}{m} \sum_{i=w}^{\nu+1} a_i \boldsymbol{\Delta}_i \\
|\mathbf{V}_\phi| &= m^{r^*} \left| \frac{1}{m} \sum_{i=\omega}^{\nu+1} a_i \boldsymbol{\Delta}_i \right|^{r+1} \\
&= \frac{m^{r^*}}{m^{r(r+1)}} \left| \sum_{i=\omega}^{\nu+1} a_i \boldsymbol{\Delta}_i \right|^{r+1} \qquad (10.3.20) \\
&= \frac{1}{m^{r^*}} \left| \sum_{i=\omega}^{\nu+1} a_i \boldsymbol{\Delta}_i \right|^{r+1} \\
m &= \left[\frac{|\sum_{i=\omega}^{\nu+1} a_i \boldsymbol{\Delta}_i|^{r+1}}{|\mathbf{V}_\phi|} \right]^{1/r^*}.
\end{aligned}
$$

As in the univariate case, the $\boldsymbol{\Delta}_i$'s, which are unknown, can be replaced by their estimates, namely the $\mathbf{S}_i$'s, in order to get estimates of m and $\boldsymbol{\Delta}$.

10.3.4. Closeness of Satterthwaite's Approximation

The closeness of the approximation of the distribution of $\hat{\boldsymbol{\Phi}}$ in formula (10.3.17) with the central Wishart distribution was investigated by Khuri et al. (1994). In particular, the following theorem was proved:

Theorem 10.3.2. Let t be the number of nonzero a_i's in formula 10.3.17, which are denoted by $a_{(1)}, a_{(2)}, \ldots, a_{(t)}$. The corresponding mean square matrices are denoted by $\mathbf{S}_{(1)}, \mathbf{S}_{(2)}, \ldots, \mathbf{S}_{(t)}$. Then,

(a) A necessary and sufficient condition for $\hat{\boldsymbol{\Phi}}$ to have the central Wishart distribution is

$$\frac{a_{(1)}}{m_{(1)}} \boldsymbol{\Delta}_{(1)} = \frac{a_{(2)}}{m_{(2)}} \boldsymbol{\Delta}_{(2)} = \cdots = \frac{a_{(t)}}{m_{(t)}} \boldsymbol{\Delta}_{(t)}, \qquad (10.3.21)$$

where $m_{(i)}$ is the number of degrees of freedom for $\mathbf{S}_{(i)}$, and $\boldsymbol{\Delta}_{(i)}$ is the corresponding expected value ($i = 1, 2, \ldots, t$).

(b) If condition (10.3.21) is true, then the approximate number of degrees of freedom, m, in formula (10.3.20) reduces to $m = \sum_{i=1}^{t} m_{(i)}$. □

In practice, condition (10.3.21) cannot be verified since the $\boldsymbol{\Delta}_i$'s are unknown matrices. We can, however, use the data matrix $\mathbf{Y}$ to determine if there is a significant departure from this condition. Let us therefore consider testing the hypothesis

$$H_0 : \frac{a_{(1)}}{m_{(1)}}\mathbf{\Delta}_{(1)} = \frac{a_{(2)}}{m_{(2)}}\mathbf{\Delta}_{(2)} = \cdots = \frac{a_{(t)}}{m_{(t)}}\mathbf{\Delta}_{(t)}.$$

We assume that $m_{(i)} \geq r$ so that $\mathbf{Y}'\mathbf{P}_{(i)}\mathbf{Y}$ will be positive definite with probability 1, where $\mathbf{P}_{(i)}$ is such that $\mathbf{Y}'\mathbf{P}_{(i)}\mathbf{Y} = m_{(i)}\mathbf{S}_{(i)}$ $(i = 1, 2, \ldots, t)$. From formula (10.3.11) we have

$$\mathbf{Y}'\mathbf{P}_{(i)}\mathbf{Y} = \sum_{\ell=1}^{m_{(i)}} \mathbf{u}_{(i)\ell}\mathbf{u}'_{(i)\ell}, \quad i = 1, 2, \cdots, t,$$

where, by Corollary 10.3.1, $\mathbf{u}_{(i)\ell}$, for $\ell = 1, 2, \ldots, m_{(i)}$, are independent with each having the normal distribution $N(\mathbf{0}, \mathbf{\Delta}_{(i)})$. Furthermore, as in the proof of Theorem 10.3.1(c), $\mathbf{u}_{(i)\ell_1}$ and $\mathbf{u}_{(j)\ell_2}$ are independent for $i \neq j$ and any $\ell_1, \ell_2,\ 1 \leq \ell_1 \leq m_{(i)},\ 1 \leq \ell_2 \leq m_{(j)}$. Consequently, if $\mathbf{v}_{(i)\ell}$ is defined as

$$\mathbf{v}_{(i)\ell} = \left[\frac{a_{(i)}}{m_{(i)}}\right]^{1/2} \mathbf{u}_{(i)\ell}, \quad i = 1, 2, \cdots, t, \quad \ell = 1, 2, \cdots, m_{(i)},$$

then $\mathbf{v}_{(i)\ell}$ is distributed as $N(\mathbf{0}, a_{(i)}/m_{(i)}\mathbf{\Delta}_{(i)})$, and all such random vectors are independent. Testing H_0 is therefore equivalent to testing the equality of the variance–covariance matrices of t multivariate normal distributions with $m_{(i)}$ sample observations from the ith distribution $(i = 1, 2, \ldots, t)$. A test statistic for such a hypothesis is given by the likelihood ratio (see Seber, 1984, p. 448)

$$L = \frac{f^{fr/2}}{\prod_{i=1}^{t} m_{(i)}^{rm_{(i)}/2}} \frac{\prod_{i=1}^{t} |\mathbf{R}_{(i)}|^{m_{(i)}/2}}{|\mathbf{R}|^{f/2}},$$

where $f = \sum_{i=1}^{t} m_{(i)},\ \mathbf{R}_{(i)} = a_{(i)}/m_{(i)}\mathbf{Y}'\mathbf{P}_{(i)}\mathbf{Y}, i = 1, 2, \ldots, t$, and $\mathbf{R} = \sum_{i=1}^{t}\mathbf{R}_{(i)}$. When H_0 is true and f is large, $-2 \log L$ is distributed approximately as a chi-squared variate with $\frac{1}{2}r(r+1)(t-1)$ degrees of freedom. Large values of $-2 \log L$ are significant. We can therefore regard the value of $-2 \log L$ as a measure of closeness of Wishart approximation of the distribution of $\hat{\mathbf{\Phi}}$.

Example 10.3.2. Consider the multivariate two-way model (10.2.7), where $\tilde{\mathbf{y}}_{k\ell m}$ is a vector of $r = 2$ observations and $v = 4,\ b = 4,\ n = 3$. We assume that $\boldsymbol{\tau}_k,\ \boldsymbol{\beta}_\ell,\ (\boldsymbol{\tau\beta})_{k\ell}$, and $\mathbf{e}_{k\ell m}$ are independent and normally distributed with mean vectors $\mathbf{0}$ and variance–covariance matrices given by $\mathbf{\Sigma}_\tau, \mathbf{\Sigma}_\beta,\ \mathbf{\Sigma}_{\tau\beta}$, and $\mathbf{\Sigma}_e$, respectively. The expected values of the corresponding mean square matrices are given in Table 10.1. Let $\mathbf{\Phi} = \mathbf{\Sigma}_\tau + \mathbf{\Sigma}_\beta + \mathbf{\Sigma}_{\tau\beta} + \mathbf{\Sigma}_e$ denote the total variance–covariance matrix. This can be expressed in terms of $\mathbf{\Delta}_\tau, \mathbf{\Delta}_\beta, \mathbf{\Delta}_{\tau\beta}$, and $\mathbf{\Delta}_e$ given in Table 10.1 as

$$\mathbf{\Phi} = \frac{1}{12}\mathbf{\Delta}_\tau + \frac{1}{12}\mathbf{\Delta}_\beta + \frac{1}{6}\mathbf{\Delta}_{\tau\beta} + \frac{2}{3}\mathbf{\Delta}_e.$$

Table 10.1. Expected Values of Mean Square Matrices

Mean Square	d.f.	Expected Value of Mean Square
$\mathbf{S}_\tau$	3	$\boldsymbol{\Delta}_\tau = 12\boldsymbol{\Sigma}_\tau + 3\boldsymbol{\Sigma}_{\tau\beta} + \boldsymbol{\Sigma}_e$
$\mathbf{S}_\beta$	3	$\boldsymbol{\Delta}_\beta = 12\boldsymbol{\Sigma}_\beta + 3\boldsymbol{\Sigma}_{\tau\beta} + \boldsymbol{\Sigma}_e$
$\mathbf{S}_{\tau\beta}$	9	$\boldsymbol{\Delta}_{\tau\beta} = 3\boldsymbol{\Sigma}_{\tau\beta} + \boldsymbol{\Sigma}_e$
$\mathbf{S}_e$	32	$\boldsymbol{\Delta}_e = \boldsymbol{\Sigma}_e$

An unbiased estimator of $\boldsymbol{\Phi}$ is given by

$$\hat{\boldsymbol{\Phi}} = \frac{1}{12}\mathbf{S}_\tau + \frac{1}{12}\mathbf{S}_\beta + \frac{1}{6}\mathbf{S}_{\tau\beta} + \frac{2}{3}\mathbf{S}_e.$$

Then, $\hat{\boldsymbol{\Phi}}$ is approximately distributed as $W_2(m, \boldsymbol{\Delta})$, where

$$\boldsymbol{\Delta} = \frac{1}{m}\left(\frac{1}{12}\boldsymbol{\Delta}_\tau + \frac{1}{12}\boldsymbol{\Delta}_\beta + \frac{1}{6}\boldsymbol{\Delta}_{\tau\beta} + \frac{2}{3}\boldsymbol{\Delta}_e\right),$$

and by formula (10.3.20),

$$m = \left[\frac{|\frac{1}{12}\boldsymbol{\Delta}_\tau + \frac{1}{12}\boldsymbol{\Delta}_\beta + \frac{1}{6}\boldsymbol{\Delta}_{\tau\beta} + \frac{2}{3}\boldsymbol{\Delta}_e|^3}{|\mathbf{V}_\phi|}\right]^{\frac{1}{3}},$$

$$\mathbf{V}_\phi = \mathbf{K}_2^- \left[\frac{(\frac{1}{12})^2}{3}(\boldsymbol{\Delta}_\tau \otimes \boldsymbol{\Delta}_\tau) + \frac{(\frac{1}{12})^2}{3}(\boldsymbol{\Delta}_\beta \otimes \boldsymbol{\Delta}_\beta) + \frac{(\frac{1}{6})^2}{9}(\boldsymbol{\Delta}_{\tau\beta} \otimes \boldsymbol{\Delta}_{\tau\beta}) + \frac{(\frac{2}{3})^2}{32}(\boldsymbol{\Delta}_e \otimes \boldsymbol{\Delta}_e)\right]\mathbf{K}_2,$$

where

$$\mathbf{K}_2 = \begin{bmatrix} 1 & 0 & 0 \\ 0 & \frac{1}{2} & 0 \\ 0 & \frac{1}{2} & 0 \\ 0 & 0 & 1 \end{bmatrix},$$

$$\mathbf{K}_2^- = \begin{bmatrix} 1 & 0 & 0 & 0 \\ 0 & 1 & 1 & 0 \\ 0 & 0 & 0 & 1 \end{bmatrix}$$

(see Nel and Van der Merwe, 1986, p. 3730).

In this example, the null hypothesis stated in Theorem 10.3.2 is of the form

$$H_0 : \frac{1}{36}\boldsymbol{\Delta}_\tau = \frac{1}{36}\boldsymbol{\Delta}_\beta = \frac{1}{54}\boldsymbol{\Delta}_{\tau\beta} = \frac{1}{48}\boldsymbol{\Delta}_e.$$

In order to demonstrate the testing of this hypothesis, we generate a data set of normal variates from model (10.2.7). As an illustration, we take $\boldsymbol{\mu} = \mathbf{0}$,

Table 10.2. A Generated Data Set for Model (10.2.7)

	ℓ							
	1		2		3		4	
k	y_1	y_2	y_1	y_2	y_1	y_2	y_1	y_2
1	0.554	2.200	3.101	−0.545	−0.379	0.433	−1.828	0.499
	0.790	3.271	3.117	−2.103	−1.476	−0.872	−0.285	1.061
	−1.239	1.131	3.633	−0.694	−1.325	−0.636	0.940	2.570
2	−3.987	1.230	−1.440	0.980	−1.925	−1.113	−3.297	−1.146
	−2.800	0.330	−1.503	1.380	−1.803	−0.230	−4.727	−2.777
	−3.088	0.553	−0.904	0.098	−2.465	−0.766	−4.486	−4.513
3	−3.940	−0.893	−3.033	−4.044	−6.301	−2.383	−5.945	−2.037
	−4.488	−2.313	−3.305	−2.353	−3.966	−1.168	−4.358	−0.310
	−6.739	−2.245	−3.578	−1.401	−6.558	−4.712	−4.207	−1.532
4	−2.822	−1.658	0.307	−1.839	−1.939	−2.856	−1.632	−2.031
	−3.627	−0.917	0.165	−0.127	−1.225	−2.748	−1.172	−0.998
	−2.573	−1.041	1.811	−1.131	−4.297	−4.580	−0.780	−1.443

$\boldsymbol{\Sigma}_\tau = \boldsymbol{\Sigma}_\beta = \begin{pmatrix} 2 & 1 \\ 1 & 2 \end{pmatrix}, \boldsymbol{\Sigma}_{\tau\beta} = \begin{pmatrix} 1 & 0.6 \\ 0.6 & 1 \end{pmatrix}$, and $\boldsymbol{\Sigma}_e = \begin{pmatrix} 1 & 0.4 \\ 0.4 & 1 \end{pmatrix}$. The generated data are shown in Table 10.2.

The corresponding value of the approximate chi-squared test statistic is $-2 \log L = 102.755$ with $\frac{1}{2}r(r+1)(t-1) = 9$ degrees of freedom. This is a highly significant result and H_0 is therefore rejected.

Using the given values of $\boldsymbol{\Sigma}_\tau, \boldsymbol{\Sigma}_\beta, \boldsymbol{\Sigma}_{\tau\beta}$, and $\boldsymbol{\Sigma}_e$, we find that $m = 9.844$ and $\boldsymbol{\Delta} = \begin{pmatrix} 0.609 & 0.305 \\ 0.305 & 0.609 \end{pmatrix}$. Upon replacing $\boldsymbol{\Delta}_\tau, \boldsymbol{\Delta}_\beta, \boldsymbol{\Delta}_{\tau\beta}$, and $\boldsymbol{\Delta}_e$, in the formulas for m and $\boldsymbol{\Delta}$, by their corresponding estimates, namely, $\mathbf{S}_\tau, \mathbf{S}_\beta, \mathbf{S}_{\tau\beta}$, and $\mathbf{S}_e$, respectively, from Table 10.1, we get the following estimated values for m and $\boldsymbol{\Delta} : \hat{m} = 11.151, \hat{\boldsymbol{\Delta}} = \begin{pmatrix} 0.662 & 0.219 \\ 0.219 & 0.310 \end{pmatrix}$.

10.4. A MULTIVARIATE APPROACH TO UNIVARIATE BALANCED MIXED MODELS

The analysis described in Chapter 2 of the balanced mixed model is based on certain assumptions concerning the model's random effects (see Section 2.4). In particular, the levels of each factor with a random effect are assumed to be independently distributed and have equal variances. Such assumptions impose rigid restrictions on the variances and covariances associated with the response values under consideration. Consequently, they may be difficult to justify in practice.

Different sets of assumptions can be found in the statistical literature concerning the random effects in a mixed model (see, for example, Hocking, 1973). The least stringent assumptions are those considered by Scheffé (1956; 1959, Chapter 8) in connection with the balanced mixed two-way model. More specifically, the random effects are assumed to have a general variance–covariance structure. In this case, tests concerning the fixed effects in the model are no longer based on the usual ANOVA-based F-statistics. Instead, Scheffé proposed an exact multivariate test using Hotelling's T^2 statistic. The use of this multivariate approach was more recently adopted by Alalouf (1980) who presented multivariate tests concerning the random effects of a two-way model.

Let us now begin by reviewing the multivariate testing procedure concerning the fixed effects in a balanced mixed two-way model under Scheffé's less stringent assumptions.

10.4.1. Testing the Fixed Effects in a Balanced Mixed Two-Way Model

Consider the balanced mixed two-way model,

$$y_{ijk} = \mu + \tau_i + \beta_j + (\tau\beta)_{ij} + e_{ijk} \quad i = 1, 2, \cdots, v; \quad j = 1, 2, \cdots, b; \quad k = 1, 2, \cdots, n, \tag{10.4.1}$$

where τ_i is a fixed unknown parameter and β_j is a random variable. The usual assumptions concerning the random effects in this model are that the $\beta_j, (\tau\beta)_{ij}$, and the e_{ijk} are independently distributed as normal variates with zero means and variances, $\sigma^2_\beta, \sigma^2_{\tau\beta}$, and σ^2_e, respectively. We refer to model (10.4.1) under these assumptions as Model A. Scheffé (1956; 1959, Chapter 8) considered less restrictive assumptions concerning the same random effects. More specifically, let $\boldsymbol{\zeta}_j$ be a random vector defined as

$$\boldsymbol{\zeta}_j = \beta_j \mathbf{1}_v + (\boldsymbol{\tau\beta})_j, \quad j = 1, 2, \cdots, b, \tag{10.4.2}$$

where $(\boldsymbol{\tau\beta})_j = [(\tau\beta)_{1j}, (\tau\beta)_{2j}, \ldots, (\tau\beta)_{vj}]'$. The $\boldsymbol{\zeta}_j$ are assumed to be normally and independently distributed with zero means and a common variance–covariance matrix $\boldsymbol{\Sigma}$. They are also assumed to be independent of the e_{ijk}, which are independently distributed as $N(0, \sigma^2_e)$. We refer to model (10.4.1) under these more general assumptions as Model B. Note that if Model A's assumptions are valid, then

$$\boldsymbol{\Sigma} = \sigma^2_\beta \mathbf{J}_v + \sigma^2_{\tau\beta} \mathbf{I}_v. \tag{10.4.3}$$

Let us now consider testing the null hypothesis

$$H_\tau : \tau_1 = \tau_2 = \cdots = \tau_v$$

concerning the fixed effects given Model B's assumptions. For this purpose, let $\bar{y}_{ij.}$ denote the $(i,j)^{th}$ cell mean. Then,

$$\bar{y}_{ij.} = \mu + \tau_i + \beta_j + (\tau\beta)_{ij} + \bar{e}_{ij.}, \quad i = 1,2,\cdots,v; \quad j = 1,2,\cdots,b, \tag{10.4.4}$$

where $\bar{e}_{ij.} = \frac{1}{n}\sum_{k=1}^{n} e_{ijk}$. This can be written as

$$\bar{\mathbf{y}}_j = \boldsymbol{\mu}_\tau + \boldsymbol{\zeta}_j + \bar{\mathbf{e}}_j, \quad j = 1,2,\cdots,b, \tag{10.4.5}$$

where $\boldsymbol{\mu}_\tau = (\mu + \tau_1, \mu + \tau_2, \ldots, \mu + \tau_v)'$; $\bar{\mathbf{y}}_j$ and $\bar{\mathbf{e}}_j$ are vectors whose i^{th} elements are $\bar{y}_{ij.}$ and $\bar{e}_{ij.}$, respectively $(i = 1,2,\ldots,v)$. The mean and variance–covariance matrix of $\bar{\mathbf{y}}_j$ are

$$\begin{aligned} E(\bar{\mathbf{y}}_j) &= \boldsymbol{\mu}_\tau, \quad j = 1,2,\cdots,b & (10.4.6)\\ \text{Var}(\bar{\mathbf{y}}_j) &= \boldsymbol{\Sigma} + \frac{\sigma_e^2}{n}\mathbf{I}_v, \quad j = 1,2,\cdots,b. & (10.4.7)\end{aligned}$$

Let $\mathbf{d}_j$ be a vector defined as follows:

$$\mathbf{d}_j = \mathbf{C}\,\bar{\mathbf{y}}_j, \quad j = 1,2,\cdots,b,$$

where $\mathbf{C} = [\mathbf{I}_{v-1} : -\mathbf{1}_{v-1}]$. The $\mathbf{d}_j$'s form a sample of independently and normally distributed random vectors with a common mean and a common variance–covariance matrix given by

$$\begin{aligned} E(\mathbf{d}_j) &= \mathbf{C}\boldsymbol{\mu}_\tau, \quad j = 1,2,\cdots,b\\ \text{Var}(\mathbf{d}_j) &= \mathbf{C}\left(\boldsymbol{\Sigma} + \frac{\sigma_e^2}{n}\mathbf{I}_v\right)\mathbf{C}', \quad j = 1,2,\cdots,b.\end{aligned}$$

Under H_τ, $\mathbf{C}\boldsymbol{\mu}_\tau = \mathbf{0}$. In this case, the statistic

$$T^2 = b\,\bar{\mathbf{d}}'\mathbf{S}^{-1}\bar{\mathbf{d}} \tag{10.4.8}$$

has Hotelling's T^2 distribution with $b-1$ degrees of freedom, where $\bar{\mathbf{d}} = \frac{1}{b}\sum_{j=1}^{b}\mathbf{d}_j$ and $\mathbf{S}$ is the sample variance–covariance matrix,

$$\mathbf{S} = \frac{1}{b-1}\sum_{j=1}^{b}(\mathbf{d}_j - \bar{\mathbf{d}})(\mathbf{d}_j - \bar{\mathbf{d}})'. \tag{10.4.9}$$

This test requires that $b \geq v$ so that $\mathbf{S}$ will be positive definite with probability 1. An F-statistic can then be obtained using T^2, namely,

$$F_d = \frac{b-v+1}{(b-1)(v-1)}T^2,$$

which, under H_τ, has the F-distribution with $v - 1$ and $b - v + 1$ degrees of freedom (see, for example, Seber, 1984, p. 63). A large value of F_d results in the rejection of H_τ.

It should be noted that the rows of the matrix $\mathbf{C}$ form a basis for the $(v - 1)$-dimensional orthogonal complement (in the v-dimensional Euclidean space) of the one-dimensional space spanned by $\mathbf{1}_v$. Hence, the elements of $\mathbf{C}\boldsymbol{\mu}_\tau$ form a basis for the space of all contrasts among $\mu + \tau_1, \mu + \tau_2, \ldots, \mu + \tau_v$. It is easy to show that T^2 in formula (10.4.8) is invariant to the choice of $\mathbf{C}$ provided that the rows of $\mathbf{C}$ form such a basis.

The main disadvantage of the previously mentioned multivariate procedure for testing H_τ is that F_d has little power when $b - v + 1$ is small. We may recall from Chapter 2 that the traditional ANOVA approach for testing H_τ uses the statistic $F = MS(\tau)/MS(\tau\beta)$, where $MS(\tau)$ and $MS(\tau\beta)$ are the usual mean squares corresponding to τ_i and $(\tau\beta)_{ij}$, respectively, in model (10.4.1). Under Model A's assumptions, this statistic has the F-distribution with $v - 1$ and $(b - 1)(v - 1)$ degrees of freedom when H_τ is true. Under Model B's assumptions, however, F no longer has the F-distribution, even when H_τ is true. This is attributed to the fact that both $MS(\tau)$ and $MS(\tau\beta)$ are expressible as linear combinations of independent chi-squared variates (see Imhof, 1962, Section 3). Imhof (1962) considered an approximation of the distribution of F by applying Satterthwaite's approximation (see Section 2.6.1) to both $MS(\tau)$ and $MS(\tau\beta)$.

Scheffé's (1959, Chapter 8) multivariate approach in the case of the balanced mixed two-way model can be easily extended to higher-order crossed-classification models provided that there is only one factor having random effects in the model. In general, however, the analysis becomes considerably more complicated in the presence of more than one of such factors. Imhof (1960) provided an extension of Scheffé's approach to a balanced mixed three-way model with two random-effect factors.

More recently, Khatri and Patel (1992) used the likelihood ratio principle to obtain an approximate test for testing the fixed effects in a mixed two-way model under Model B's assumptions. Their model is actually more general than the one considered by Scheffé (1959, Chapter 8) in the sense that it deals with unbalanced data and accommodates one or more covariates.

10.4.2. Testing the Random Effects Under Model B's Assumptions

Alalouf (1980) and Schott (1985) considered the following hypothesis concerning $\boldsymbol{\Sigma}$, the variance–covariance matrix of $\boldsymbol{\zeta}_j$ in formula (10.4.2):

$$H_0 : \mathbf{L}_1'\boldsymbol{\Sigma}\mathbf{L}_1 = \mathbf{0},$$

where $\mathbf{L}_1$ is a $v \times v_1 (v_1 \leq v)$ constant matrix such that $\mathbf{L}_1'\mathbf{L}_1 = \mathbf{I}_{v_1}$. This particular hypothesis represents a generalization of two hypotheses defined by

Scheffé (1959, p. 264) concerning the quantities σ_B^2 and σ_{AB}^2, where

$$\sigma_B^2 = \frac{1}{v^2}\sum_{j=1}^{v}\sum_{k=1}^{v}\sigma_{jk}$$

$$\sigma_{AB}^2 = \frac{1}{v-1}\sum_{j=1}^{v}(\sigma_{jj}-\sigma_B^2),$$

and σ_{jk} is the $(j,k)^{th}$ element of $\mathbf{\Sigma}$.

Let us now proceed to develop a test statistic for H_0. For this purpose, we premultiply the two sides of model (10.4.5) by $\mathbf{L}_1'$. As a result, we obtain

$$\mathbf{L}_1'\bar{\mathbf{y}}_j = \mathbf{L}_1'\boldsymbol{\mu}_\tau + \mathbf{L}_1'\boldsymbol{\zeta}_j + \mathbf{L}_1'\bar{\mathbf{e}}_j, \quad j=1,2,\cdots,b.$$

The mean and variance–covariance matrix of $\mathbf{L}_1'\bar{\mathbf{y}}_j$ can then be expressed as

$$E(\mathbf{L}_1'\bar{\mathbf{y}}_j) = \mathbf{L}_1'\boldsymbol{\mu}_\tau, \quad j=1,2,\cdots,b$$

$$\text{Var}(\mathbf{L}_1'\bar{\mathbf{y}}_j) = \mathbf{L}_1'\mathbf{\Sigma}\mathbf{L}_1 + \frac{\sigma_e^2}{n}\mathbf{I}_{v_1}, \quad j=1,2,\cdots,b,$$

since $\mathbf{L}_1'\mathbf{L}_1 = \mathbf{I}_{v_1}$. Thus, the hypothesis H_0 is equivalent to

$$H_0': \ \text{Var}(\mathbf{L}_1'\bar{\mathbf{y}}_j) = \frac{\sigma_e^2}{n}\mathbf{I}_{v_1}, \quad j=1,2,\cdots,b.$$

Hence, under H_0', the vectors $\mathbf{L}_1'\bar{\mathbf{y}}_j$, $j=1,2,\ldots,b$, which are independent and normally distributed, have a common diagonal variance–covariance matrix with diagonal elements equal to σ_e^2/n. Testing H_0' is therefore equivalent to testing the so-called sphericity hypothesis (see, for example, Muirhead, 1982, Section 8.3; Seber, 1984, p. 93) with regard to the variance–covariance matrix of $\mathbf{L}_1'\bar{\mathbf{y}}_j$, $j=1,2,\ldots,b$. The corresponding test statistic is obtained on the basis of the likelihood ratio principle and is of the form

$$\Lambda = \lambda_\ell^{2/b} = \frac{|\mathbf{H}|}{[tr(\mathbf{H})/v_1]^{v_1}},$$

where λ_ℓ is the likelihood ratio statistic, and $\mathbf{H}$ is given by

$$\mathbf{H} = \sum_{j=1}^{b}(\mathbf{L}_1'\bar{\mathbf{y}}_j - \mathbf{L}_1'\bar{\mathbf{y}})(\mathbf{L}_1'\bar{\mathbf{y}}_j - \mathbf{L}_1'\bar{\mathbf{y}})',$$

where $\bar{\mathbf{y}} = \frac{1}{b}\sum_{j=1}^{b}\bar{\mathbf{y}}_j$. This test requires that $b-1 \geq v_1$ so that $\mathbf{H}$ will be positive definite with probability 1 (see Muirhead, 1982, Theorem 8.3.2).

Small values of Λ are significant. The exact distribution of Λ and a table of its percentage points are given by Nagarsenker and Pillai (1973). Under H_0' (or equivalently, H_0) and for large b, $-n_1 \log \Lambda$ has approximately the chi-squared distribution with $\frac{1}{2}(v_1+2)(v_1-1)$ degrees of freedom, where $n_1 = b - 1 - \frac{1}{6v_1}(2v_1^2 + v_1 + 2)$ (see Muirhead, 1982, Theorem 8.3.7; Seber, 1984, p. 93). In this case, H_0' is rejected for large values of $-n_1 \log \Lambda$.

Note that when $v_1 = 1$, that is, when $\mathbf{L}_1$ is a vector, $\Lambda = 1$ no matter what the data set is. This occurs, for example, when testing a hypothesis concerning σ_B^2, which can be written as $\sigma_B^2 = \frac{1}{v^2}\mathbf{1}_v'\boldsymbol{\Sigma}\mathbf{1}_v$. The test statistic Λ should therefore not be used in such a case. Instead, we can use the following test statistic whose derivation is given below.

We have that $\mathbf{L}_1'\bar{\mathbf{y}}_j (j = 1, 2, \ldots, b)$ are independent and normally distributed with a common mean and a common variance σ_e^2/n, if H_0' is true. Under the alternative hypothesis, the variance exceeds σ_e^2/n. Hence, under H_0', $n\mathbf{H}/\sigma_e^2 \sim \chi_{b-1}^2$. Furthermore, the residual sum of squares, $SS(e) = \sum_{i=1}^{v} \sum_{j=1}^{b} \sum_{k=1}^{n} (y_{ijk} - \bar{y}_{ij\cdot})^2$, is independent of $\bar{\mathbf{y}}_j$ for $j = 1, 2, \ldots, b$, and hence of $\mathbf{H}$, with $SS(e)/\sigma_e^2$ being distributed as $\chi_{bv(n-1)}^2$ (see the Appendix in Khuri, 1989). It follows that the statistic $n\mathbf{H}/[(b-1)MS(e)]$, where $MS(e) = SS(e)/[bv(n-1)]$, has under H_0' the F-distribution with $b-1$ and $bv(n-1)$ degrees of freedom. Large values of this statistic are significant.

10.4.2.1. Testing a Particular Variance–Covariance Structure. Another hypothesis concerning $\boldsymbol{\Sigma}$ was considered by Khuri (1989), namely the hypothesis

$$H_\sigma : \boldsymbol{\Sigma} = a\mathbf{J}_v + c\mathbf{I}_v,$$

where a and c are nonnegative constants. Testing this hypothesis provides a check on the validity of Model A's assumptions under which $\boldsymbol{\Sigma}$ has the form given by formula (10.4.3). We now explain how this can be done.

Consider again the random vector $\bar{\mathbf{y}}_j$ given by formula (10.4.5). The mean and variance–covariance matrix of $\bar{\mathbf{y}}_j$ are described in formulas (10.4.6) and (10.4.7), respectively ($j = 1, 2, \ldots, b$). It can be verified that H_σ is equivalent to the hypothesis

$$H_\sigma' : \boldsymbol{\Gamma} = \sigma^2\rho\mathbf{J}_v + \sigma^2(1-\rho)\mathbf{I}_v,$$

where $\boldsymbol{\Gamma} = \boldsymbol{\Sigma} + (\sigma_e^2/n)\,\mathbf{I}_v$, $\sigma^2 = a + c + \sigma_e^2/n$, and $\rho = a/\sigma^2$. We note that H_σ' is a special case of the more general hypothesis

$$H_\sigma'' : \boldsymbol{\Gamma} = \sigma^2\rho^*\mathbf{J}_v + \sigma^2(1-\rho^*)\mathbf{I}_v,$$

where ρ^* is a correlation coefficient which is not necessarily nonnegative as ρ is.

A test concerning the hypothesis H''_σ can be obtained by applying the likelihood ratio principle based on the sample, $\bar{\mathbf{y}}_1, \bar{\mathbf{y}}_2, \ldots, \bar{\mathbf{y}}_b$ of random vectors. The corresponding likelihood ratio test statistic, denoted by γ_ℓ, is such that (see, for example, Seber, 1984, p. 95)

$$M = \gamma_\ell^{2/b} = \frac{v^v (v-1)^{v-1} |\hat{\mathbf{\Gamma}}|}{(\mathbf{1}'_v \hat{\mathbf{\Gamma}} \mathbf{1}_v)[v \text{ tr } (\hat{\mathbf{\Gamma}}) - \mathbf{1}'_v \hat{\mathbf{\Gamma}} \mathbf{1}_v]^{v-1}},$$

where

$$\hat{\mathbf{\Gamma}} = \frac{1}{b-1} \sum_{j=1}^{b} (\bar{\mathbf{y}}_j - \bar{\mathbf{y}})(\bar{\mathbf{y}}_j - \bar{\mathbf{y}})'$$

is the usual sample variance–covariance matrix which estimates $\mathbf{\Gamma}$ unbiasedly, and $\bar{\mathbf{y}} = \frac{1}{b} \sum_{j=1}^{b} \bar{\mathbf{y}}_j$. Small values of γ_ℓ (or M) are significant. The exact null distribution of the statistic M was determined by Wilks (1946) for $v = 2, 3$. He showed that the probability integral of M for $v = 2$ and that of $\sqrt{M}$ for $v = 3$ are incomplete beta functions with parameters $\frac{1}{2}(b-2)$ and $\frac{1}{2}$, for $v = 2$, and $b - 3$ and 2, for $v = 3$. Nagarsenker (1975) derived the exact distribution of M and provided a tabulation of its 5% and 1% points for several values of b and v. For large values of b and under the null hypothesis H''_σ, the statistic $-\{b - 1 - [v(v+1)^2(2v-3)]/[6(v-1)(v^2+v-4)]\} \log M$ is asymptotically distributed as a chi-squared variate with $\frac{1}{2}[v(v+1) - 4]$ degrees of freedom (see Box, 1950, p. 375; Seber, 1984, p. 95).

Rejection of the hypothesis H''_σ supports the contention that $\mathbf{\Gamma}$ cannot have the form given under H'_σ. This implies invalidity of Model A's assumptions regarding $\mathbf{\Sigma}$. Failure to reject H''_σ, however, does not necessarily mean that Model A's assumptions are valid. This is due to the fact that H'_σ is only a special case of H''_σ.

The problem of assessing whether $\mathbf{\Sigma}$ deviates from the structure specified in formula (10.4.3) was more recently considered by Oman (1995). His approach is based on using residual plots associated with the random main and interaction effects in the model. These residuals are derived in terms of the vectors $\bar{\mathbf{y}}_j$ $(j = 1, 2, \ldots, b)$. In some cases, the residual plots can provide additional information concerning systematic departures from Model A's assumptions regarding $\mathbf{\Sigma}$.

Example 10.4.1. We consider an example given by Hald (1952, p.471) and reproduced in Khuri (1989, Section 3). In this example, measurements were obtained on the waterproof quality of sheets of material manufactured by $v = 3$ different machines over a period of $b = 9$ days. Three sheets were selected for each machine-day combination. The response values, namely, the logarithms of the permeability in seconds for the sheets, are given in Table 10.3. The effects associated with machines and days are considered fixed and random, respectively.

Table 10.3. The Logarithms of the Permeabilities in Seconds

	Machine		
Day	1	2	3
	1.404	1.306	1.932
1	1.346	1.628	1.674
	1.618	1.140	1.399
	1.447	1.241	1.426
2	1.569	1.185	1.768
	1.820	1.516	1.859
	1.914	1.506	1.382
3	1.477	1.575	1.690
	1.894	1.649	1.361
	1.887	1.673	1.721
4	1.485	1.372	1.528
	1.392	1.114	1.371
	1.772	1.227	1.320
5	1.728	1.397	1.489
	1.545	1.531	1.336
	1.665	1.404	1.633
6	1.539	1.452	1.612
	1.680	1.627	1.359
	1.918	1.229	1.328
7	1.931	1.508	1.802
	2.129	1.436	1.385
	1.845	1.583	1.689
8	1.790	1.627	2.248
	2.042	1.282	1.795
	1.540	1.636	1.703
9	1.428	1.067	1.370
	1.704	1.384	1.839

Source: A. Hald (1952, Table 16.32, p. 472). Reproduced with permission of John Wiley & Sons, Inc.

Let us first test the fixed effects hypothesis concerning the three machines, that is,

$$H_\tau : \tau_1 = \tau_2 = \tau_3.$$

In this case, the value of Hotelling's T^2 statistic in (10.4.8) is $T^2 = 25.8096$, and the F-statistic value of F_d in Sectin 10.4.1 is

$$F_d = \frac{b - v + 1}{(b-1)(v-1)} T^2 = 11.2917 \text{ with 2 and 7 degrees of freedom.}$$

The corresponding P-value is 0.0064. There is therefore a significant difference among the means of the three machines.

We now consider testing the two hypotheses concerning the random effects, namely, σ_B^2 and σ_{AB}^2 in Section 10.4.2. These hypotheses can be expressed as

$$H_0 : \mathbf{L}_1' \boldsymbol{\Sigma} \mathbf{L}_1 = \mathbf{0}$$

with $\mathbf{L}_1 = (1/\sqrt{3})\mathbf{1}_3$ for σ_B^2, and

$$\mathbf{L}_1 = \begin{bmatrix} \frac{1}{\sqrt{2}} & -\frac{1}{\sqrt{6}} \\ -\frac{1}{\sqrt{2}} & -\frac{1}{\sqrt{6}} \\ 0 & \frac{2}{\sqrt{6}} \end{bmatrix}$$

for σ_{AB}^2 (see Alalouf, 1980, Section 7). For the latter quantity, the test statistic described in Section 10.4.2 is given by

$$\Lambda = \frac{|\mathbf{H}|}{[tr(\mathbf{H})/2]^2},$$

and has the value $\Lambda = 0.8508$. Using the large-sample approximation, $-n_1 \log \Lambda$ has approximately the chi-squared distribution with $\frac{1}{2}(v_1 + 2)(v_1 - 1) = 2$ (since $v_1 = 2$) degrees of freedom, where

$$n_1 = b - 1 - \frac{1}{6v_1}(2v_1^2 + v_1 + 2) = 7.$$

Hence, $-n_1 \log \Lambda = 1.1313$. This is not a significant result. As for the test concerning σ_B^2, the appropriate test statistic is $F = n\mathbf{H}/[(b-1)MS(e)]$, as was seen in Section 10.4.2, and has the value $F = 1.8539$ with 8 and 54 degrees of freedom. The corresponding P-value is 0.0869. The hypothesis that $\sigma_B^2 = 0$ can therefore be rejected at the 0.09 level of significance.

Next, let us consider testing the hypothesis H_σ'' in Section 10.4.2.1 concerning the particular structure of the variance–covariance matrix $\boldsymbol{\Sigma}$. The value of the corresponding test statistic is $M = 0.5737$, which exceeds the 10% critical value 0.2997. The latter value was obtained from Beyer's (1968, p. 256) tables of percentage points of the beta distribution (recall that for $v = 3$, $\sqrt{M}$ has the beta distribution with parameters $b - 3 = 6$ *and* 2). Thus, H_σ'' cannot be rejected at the 10% level of significance. Consequently, we cannot conclude invalidity of Model A's assumptions regarding $\boldsymbol{\Sigma}$. Of course, this does not necessarily mean that Model A's assumptions are valid in this example.

10.5. THE DERIVATION OF EXACT TESTS

Whenever an exact F-test exists for testing the significance of a fixed effect or a variance component in a univariate mixed or random model, balanced or unbalanced, exact tests also exist for testing a similar hypothesis in the corresponding multivariate model. We shall illustrate this using the unbalanced multivariate one-way random model. This model is the multivariate analogue of model (4.2.1). Let $\mathbf{y}_{ij}$ $(i = 1, 2, \ldots, v;\ j = 1, 2, \ldots, n_i)$ be $r \times 1$ random vectors following the model

$$\mathbf{y}_{ij} = \boldsymbol{\mu} + \boldsymbol{\tau}_i + \mathbf{e}_{ij}, \tag{10.5.1}$$

where $\boldsymbol{\mu}$ is a fixed unknown $r \times 1$ parameter vector and the $\boldsymbol{\tau}_i$'s and $\mathbf{e}_{ij}$'s are $r \times 1$ independent random vectors with $\boldsymbol{\tau}_i \sim N(\mathbf{0}, \boldsymbol{\Sigma}_\tau)$ and $\mathbf{e}_{ij} \sim N(\mathbf{0}, \boldsymbol{\Sigma}_e)$ $(i = 1, 2, \ldots, v;\ j = 1, 2, \ldots, n_i)$. The $\boldsymbol{\tau}_i$'s are the random treatment effect vectors and the $\mathbf{e}_{ij}$'s are the experimental error terms. Suppose we are interested in the testing problem

$$H_\tau : \boldsymbol{\Sigma}_\tau = \mathbf{0} \text{ versus } H_1 : \boldsymbol{\Sigma}_\tau \neq \mathbf{0}. \tag{10.5.2}$$

Write $\mathbf{Y} = [\mathbf{y}_{11} : \mathbf{y}_{12} : \ldots : \mathbf{y}_{1n_1} : \ldots : \mathbf{y}_{v1} : \mathbf{y}_{v2} : \ldots : \mathbf{y}_{vn_v}]'$, $\boldsymbol{\tau} = [\boldsymbol{\tau}_1 : \boldsymbol{\tau}_2 : \ldots : \boldsymbol{\tau}_v]'$, $\mathbf{E} = [\mathbf{e}_{11} : \mathbf{e}_{12} : \ldots : \mathbf{e}_{1n_1} : \ldots : \mathbf{e}_{v1} : \mathbf{e}_{v2} : \ldots : \mathbf{e}_{vn_v}]'$ and $n_. = \sum_{i=1}^{v} n_i$. Then $\mathbf{Y}$ and $\mathbf{E}$ are $n_. \times r$ matrices, $\boldsymbol{\tau}$ is a $v \times r$ matrix and the model (10.5.1) can be written as

$$\mathbf{Y} = \mathbf{1}_{n.}\boldsymbol{\mu}' + \mathbf{diag}(\mathbf{1}_{n_1}, \mathbf{1}_{n_2}, \cdots, \mathbf{1}_{n_v})\boldsymbol{\tau} + \mathbf{E}. \tag{10.5.3}$$

For an $m \times n$ matrix $\mathbf{A}$, let vec($\mathbf{A}$) denote the $mn \times 1$ vector obtained by writing the columns of $\mathbf{A}$ one below the other. Using this notation, the distributional assumption on $\boldsymbol{\tau}$ and $\mathbf{E}$ can be written as

$$\text{vec}(\boldsymbol{\tau}') \sim N(\mathbf{0}, \mathbf{I}_v \otimes \boldsymbol{\Sigma}_\tau) \text{ and } \text{vec}(\mathbf{E}') \sim N(\mathbf{0}, \mathbf{I}_{n.} \otimes \boldsymbol{\Sigma}_e). \tag{10.5.4}$$

Model (10.5.3) now gives

$$E(\mathbf{Y}) = \mathbf{1}_{n.}\boldsymbol{\mu}' \text{ and } \text{Var}(\text{vec}(\mathbf{Y}')) = \mathbf{diag}(\mathbf{J}_{n_1}, \mathbf{J}_{n_2}, \cdots, \mathbf{J}_{n_v}) \otimes \boldsymbol{\Sigma}_\tau + \mathbf{I}_{n.} \otimes \boldsymbol{\Sigma}_e. \tag{10.5.5}$$

Write $\mathbf{Y}_i = [\mathbf{y}_{i1} : \mathbf{y}_{i2} : \ldots : \mathbf{y}_{in_i}]'$, $\mathbf{y}_{i.} = \sum_{j=1}^{n_i} \mathbf{y}_{ij}$ $(i = 1, 2, \ldots, v)$ and $\mathbf{y}_{..} = \sum_{i=1}^{v} \sum_{j=1}^{n_i} \mathbf{y}_{ij}$. Also, let $\mathbf{SS}(\boldsymbol{\tau})$ and $\mathbf{SS}(\mathbf{e})$, respectively, denote the sum of squares and sum of products matrices corresponding to treatments (i.e., the $\boldsymbol{\tau}_i$'s) and error based on model (10.5.1). Table 10.4 gives the multivariate analysis of variance (MANOVA) for model (10.5.1), along with the expressions for $\mathbf{SS}(\boldsymbol{\tau})$ and $\mathbf{SS}(\mathbf{e})$. The table is the multivariate generalization of Table 3.1.

Table 10.4. MANOVA for Model (10.5.1)

Source	Sum of Squares and Products	d.f.	Expected Mean Squares
Treatments	$\mathbf{SS}(\boldsymbol{\tau})=\sum_{i=1}^{v}\mathbf{y}_{i\cdot}\mathbf{y}'_{i\cdot}/n_i-\mathbf{y}_{\cdot\cdot}\mathbf{y}'_{\cdot\cdot}/n.$	$v-1$	$\frac{1}{v-1}\left(n.-\frac{1}{n.}\sum_{i=1}^{v}n_i^2\right)\boldsymbol{\Sigma}_\tau+\boldsymbol{\Sigma}_e$
Error	$\mathbf{SS}(\mathbf{e})=\sum_{i=1}^{v}\sum_{j=1}^{n_i}\mathbf{y}_{ij}\mathbf{y}'_{ij}-\sum_{i=1}^{v}\mathbf{y}_{i\cdot}\mathbf{y}'_{i\cdot}/n_i$	$n.-v$	$\boldsymbol{\Sigma}_e$

The univariate distribution results for the sums of squares in Table 3.1 have immediate multivariate generalizations. Thus, $\mathbf{SS}(\mathbf{e}) \sim W_r(n. - v, \boldsymbol{\Sigma}_e)$ and under H_τ: $\boldsymbol{\Sigma}_\tau = 0$, $\mathbf{SS}(\boldsymbol{\tau}) \sim W_r(v-1, \boldsymbol{\Sigma}_e)$. Furthermore, $\mathbf{SS}(\mathbf{e})$ and $\mathbf{SS}(\boldsymbol{\tau})$ are independently distributed. (The proofs of the above are omitted since they are similar to the proofs of the corresponding univariate results.) Here, we assume that $n. - v \geq r$, so that $\mathbf{SS}(\mathbf{e})$ is positive definite with probability one. The expected values in Table 10.4 suggest that suitable scalar-valued functions of $[\mathbf{SS}(\mathbf{e})]^{-1}\mathbf{SS}(\boldsymbol{\tau})$ can be used to test the hypothesis H_τ in (10.5.2). Some standard choices are given below:

$$
\begin{aligned}
T_1 &= \frac{1}{|\mathbf{I}_r + [\mathbf{SS}(\mathbf{e})]^{-1}\mathbf{SS}(\boldsymbol{\tau})|} = \frac{|\mathbf{SS}(\mathbf{e})|}{|\mathbf{SS}(\mathbf{e}) + \mathbf{SS}(\boldsymbol{\tau})|} \quad \begin{pmatrix}\text{Wilks' likelihood}\\ \text{ratio criterion}\end{pmatrix}\\
T_2 &= \operatorname{tr}\left([\mathbf{SS}(\mathbf{e})]^{-1}\mathbf{SS}(\boldsymbol{\tau})\right) \quad \text{(Hotelling-Lawley's trace criterion)}\\
T_3 &= \operatorname{tr}\left(\mathbf{SS}(\boldsymbol{\tau})[\mathbf{SS}(\mathbf{e}) + \mathbf{SS}(\boldsymbol{\tau})]^{-1}\right) \qquad (10.5.6)\\
&= \operatorname{tr}\left([\mathbf{SS}(\mathbf{e})]^{-1}\mathbf{SS}(\boldsymbol{\tau})\{\mathbf{I}_r + [\mathbf{SS}(\mathbf{e})]^{-1}\mathbf{SS}(\boldsymbol{\tau})\}\right) \quad \text{(Pillai's trace criterion)}\\
T_4 &= \lambda_{\max}\left([\mathbf{SS}(\mathbf{e})]^{-1}\mathbf{SS}(\boldsymbol{\tau})\right) \quad \text{(Roy's largest root criterion).}
\end{aligned}
$$

In the definition of T_4, $\lambda_{\max}(.)$ denotes the largest eigenvalue of the matrix argument. If one is using T_1 in (10.5.6) for testing (10.5.2), the decision is to reject H_τ for small values of T_1. For the other criteria, we reject H_τ if T_i is large ($i = 2, 3, 4$). Tables of the percentage points of the null distributions of the T_i's are available. For these tables and further details, we refer to Anderson (1984).

Note that if the $\boldsymbol{\tau}_i$'s in (10.5.1) are fixed effects, then the statistics T_1–T_4 in (10.5.6) are appropriate for testing H_τ: $\boldsymbol{\tau}_1 = \boldsymbol{\tau}_2 = \ldots = \boldsymbol{\tau}_v$. In fact, the test based on T_1 is the likelihood ratio test (LRT) for testing H_τ: $\boldsymbol{\tau}_1 = \boldsymbol{\tau}_2 = \ldots = \boldsymbol{\tau}_v$. However, when the $\boldsymbol{\tau}_i$'s are random, the test based on T_1 is not the LRT for testing (10.5.2). This is due to the fact that in order to derive the LRT for testing (10.5.2), it is necessary to maximize the likelihood function subject to the constraint that $\boldsymbol{\Sigma}_\tau$ and $\boldsymbol{\Sigma}_e$ are nonnegative definite matrices and the resulting test turns out to be different from the one based on T_1. The LRT for testing (10.5.2) is derived in Zhou and Mathew (1993) and is not given here.

All the univariate exact F-tests concerning the fixed effects or variance components, described in the previous chapters, have corresponding multivariate generalizations resulting in test statistics similar to T_1–T_4 in (10.5.6). The generalization is very straightforward and should be clear from the computations given previously for testing H_τ in (10.5.2). Essentially, the sums of squares in the univariate case should be replaced by the corresponding sum of squares and sum of products matrices in the multivariate case and then statistics similar to T_1–T_4 in (10.5.6) can be immediately constructed.

REMARK 10.5.1. It should be pointed out that there are some problems in the multivariate case that do not arise in the univariate case. For example, regarding model (10.5.1), one may be interested in testing the hypothesis that the rank of $\mathbf{\Sigma}_\tau$ is a specified number. The likelihood ratio test for this problem is derived in Schott and Saw (1984) and Anderson et al. (1986).

REMARK 10.5.2. Multivariate tests that recover inter-block information can be derived in a manner similar to that in Chapter 7. For details, see Zhou and Mathew (1994).

REMARK 10.5.3. Procedures based on the generalized P-value, described in Chapter 9, are currently not available for the corresponding multivariate models.

10.6. OPTIMUM TESTS

Optimality results are in general unavailable in multivariate mixed and random models, except in some very special situations. Recall that even in the univariate case, optimum tests are known in general mixed and random models only in the balanced case; see Chapter 2. In unbalanced univariate models, optimum tests have been derived only for some specific models; see Chapters 4 and 6. The multivariate optimality results given below apply only to models involving exactly one multivariate random effect so that there are only two multivariate components of variance—one corresponding to the random effect and a second one corresponding to the error term. The results given below are based on Zhou and Mathew (1993).

The model that we shall consider is a multivariate generalization of model (6.2.11). Let $\mathbf{Y}$ be an $N \times r$ matrix following the model

$$\mathbf{Y} = \mathbf{X}_1\mathbf{B} + \mathbf{X}_2\boldsymbol{\tau} + \mathbf{E}, \tag{10.6.1}$$

where $\mathbf{X}_1$: $N \times m$ and $\mathbf{X}_2$: $N \times p$ are known design matrices, $\mathbf{B}$ is an $m \times r$ matrix of fixed effects, $\boldsymbol{\tau}$ is a $p \times r$ matrix of random effects, and $\mathbf{E}$ is the

$N \times r$ matrix of experimental error terms. We shall assume without loss of generality that $\text{rank}(\mathbf{X}_1) = m$. It is also assumed that $\boldsymbol{\tau}$ and $\mathbf{E}$ are independently distributed, the rows of $\boldsymbol{\tau}$ are independent random vectors distributed as $N(\mathbf{0}, \boldsymbol{\Sigma}_\tau)$, and the rows of $\mathbf{E}$ are independent random vectors distributed as $N(\mathbf{0}, \boldsymbol{\Sigma}_e)$. Thus

$$\text{E}(\mathbf{Y}) = \mathbf{X}_1\mathbf{B} \text{ and } \text{Var}(\text{vec}((\mathbf{Y}')) = \mathbf{X}_2\mathbf{X}_2' \otimes \boldsymbol{\Sigma}_\tau + \mathbf{I}_N \otimes \boldsymbol{\Sigma}_e. \tag{10.6.2}$$

The multivariate one-way random model discussed in the previous section is clearly a special case of model (10.6.2). The problem that we shall address, and the problem for which optimality results are available, is stated as

$$\mathbf{H}_\tau : \boldsymbol{\Sigma}_\tau = \mathbf{0} \text{ versus } \mathbf{H}_1 : \boldsymbol{\Sigma}_\tau \neq \mathbf{0}. \tag{10.6.3}$$

In a balanced version of model (10.6.2), Das and Sinha (1987, 1988) addressed this problem under the assumption that $\boldsymbol{\Sigma}_\tau = \sigma_\tau^2 \boldsymbol{\Sigma}_e$, so that the problem is to test H_τ: $\sigma_\tau^2 = 0$. We shall first derive the optimum test for the testing problem (10.6.3) for model (10.6.2) without any further assumptions, as was done in Zhou and Mathew (1993), and then give the result obtained by Das and Sinha (1987, 1988).

We first note that the testing problem (10.6.3) in model (10.6.2) is invariant under the action of the group of transformations $\{(\mathbf{A}, \mathbf{L})$: $\mathbf{A}$ is an $r \times r$ nonsingular matrix and $\mathbf{L}$ is an $m \times r$ matrix$\}$ acting on $\mathbf{Y}$ as

$$\mathbf{Y} \longrightarrow \mathbf{Y}\mathbf{A}' + \mathbf{X}_1\mathbf{L}. \tag{10.6.4}$$

An LBI test exists under the action of this group and is given by the following theorem.

Theorem 10.6.1. Consider the testing problem (10.6.3) for model (10.6.2) and let $\mathbf{P}_{\mathbf{X}_1} = \mathbf{X}_1(\mathbf{X}_1'\mathbf{X}_1)^{-1}\mathbf{X}_1'$. Under the group action (10.6.4), the LBI test rejects H_τ for large values of the statistic $\text{tr}\big[\mathbf{Y}'(\mathbf{I} - \mathbf{P}_{\mathbf{X}_1})\mathbf{X}_2\mathbf{X}_2'(\mathbf{I} - \mathbf{P}_{\mathbf{X}_1})\mathbf{Y}\{\mathbf{Y}'(\mathbf{I} - \mathbf{P}_{\mathbf{X}_1})\mathbf{Y}\}^{-1}\big]$. □

The derivation of the LBI test, which is quite technical, is given in Zhou and Mathew (1993) and is omitted here. Note that since rank $(\mathbf{I} - \mathbf{P}_{\mathbf{X}_1}) = N -$ rank $(\mathbf{X}_1)$, we need the assumption, $N -$ rank $(\mathbf{X}_1) \geq r$, so that the $r \times r$ matrix $\mathbf{Y}'(\mathbf{I} - \mathbf{P}_{\mathbf{X}_1})\mathbf{Y}$ (whose inverse appears in Theorem 10.6.1) is positive definite with probability one.

In order to see the structure of the test in Theorem 10.6.1, we shall specialize to the unbalanced one-way random model (10.5.3) and the testing problem (10.5.2). Comparing with (10.6.2), we see that for model (10.5.3),

$$N = n_., \; \mathbf{X}_1 = \mathbf{1}_{n_.}, \text{ and } \mathbf{X}_2 = \mathbf{diag}(\mathbf{1}_{n_1}, \mathbf{1}_{n_2}, \cdots, \mathbf{1}_{n_v}). \tag{10.6.5}$$

Hence, $\mathbf{P}_{\mathbf{X}_1} = (1/n.)\,\mathbf{1}_{n.}\mathbf{1}'_{n.}$. Define $\bar{\mathbf{y}}_{i.} = (1/n_i)\sum_{j=1}^{n_i}\mathbf{y}_{ij}$ and $\bar{\mathbf{y}}_{..} = (1/n.)\sum_{i=1}^{v}\sum_{j=1}^{n_i}\mathbf{y}_{ij}$. It is readily verified that

$$\begin{aligned}
\mathbf{Y}'(\mathbf{I}-\mathbf{P}_{\mathbf{X}_1})\mathbf{X}_2\mathbf{X}_2'(\mathbf{I}-\mathbf{P}_{\mathbf{X}_1})\mathbf{Y} &= \sum_{i=1}^{v} n_i^2(\bar{\mathbf{y}}_{i.}-\bar{\mathbf{y}}_{..})(\bar{\mathbf{y}}_{i.}-\bar{\mathbf{y}}_{..})', \\
\mathbf{Y}'(\mathbf{I}-\mathbf{P}_{\mathbf{X}_1})\mathbf{Y} &= \sum_{i=1}^{v}\sum_{j=1}^{n_i}(\mathbf{y}_{ij}-\bar{\mathbf{y}}_{..})(\mathbf{y}_{ij}-\bar{\mathbf{y}}_{..})' \\
&= \sum_{i=1}^{v}\sum_{j=1}^{n_i}(\mathbf{y}_{ij}-\bar{\mathbf{y}}_{i.})(\mathbf{y}_{ij}-\bar{\mathbf{y}}_{i.})' \\
&\quad + \sum_{i=1}^{v} n_i(\bar{\mathbf{y}}_{i.}-\bar{\mathbf{y}}_{..})(\bar{\mathbf{y}}_{i.}-\bar{\mathbf{y}}_{..})' \\
&= \mathbf{SS}(\mathbf{e}) + \mathbf{SS}(\boldsymbol{\tau}),
\end{aligned} \tag{10.6.6}$$

where $\mathbf{SS}(\mathbf{e})$ and $\mathbf{SS}(\boldsymbol{\tau})$ are the sum of squares and sum of products matrices given in Table 10.4. Using (10.6.6), it follows from Theorem 10.6.1 that the LBI test rejects H_τ in (10.5.2) for large values of the statistic

$$\mathrm{tr}\left[\left\{\sum_{i=1}^{v} n_i^2(\bar{\mathbf{y}}_{i.}-\bar{\mathbf{y}}_{..})(\bar{\mathbf{y}}_{i.}-\bar{\mathbf{y}}_{..})'\right\}\{\mathbf{SS}(\mathbf{e})+\mathbf{SS}(\boldsymbol{\tau})\}^{-1}\right]. \tag{10.6.7}$$

We note that the above test statistic has a structure similar to the corresponding univariate test statistic (4.2.8). In the balanced case (i.e., $n_i = n$ for $i = 1, 2, \ldots, v$), the test statistic in (10.6.7) can be written as

$$n\ \mathrm{tr}\left[\mathbf{SS}(\boldsymbol{\tau})\{\mathbf{SS}(\mathbf{e})+\mathbf{SS}(\boldsymbol{\tau})\}^{-1}\right], \tag{10.6.8}$$

which is equivalent to the statistic T_3 corresponding to Pillai's trace test given in the previous section. In other words, Pillai's trace test is the LBI test for testing the hypotheses in (10.5.2) only in the balanced case. In the unbalanced case, the LBI test is based on the statistic (10.6.7) and is different from Pillai's trace test. It should be noted that if the $\boldsymbol{\tau}_i$'s in (10.5.1) are fixed effects, then Pillai's trace test is the LBI test for testing the significance of the $\boldsymbol{\tau}_i$'s in the unbalanced case also. Such a result is known to be valid for testing a linear hypothesis in a multivariate linear model with fixed effects; see Kariya and Sinha (1989, Corollary 5.3).

We shall now discuss the construction of some exact tests for testing (10.6.3) for model (10.6.2). Our purpose is to derive Pillai's trace test for this problem and derive a condition under which the LBI test in Theorem 10.6.1 will coincide with Pillai's trace test. The balancedness condition turned out to be sufficient for this in the one-way random model. Our goal is to derive a similar balancedness condition in the more general model (10.6.2). The derivation that follows will result in sum of squares and sum of products matrices similar to $\mathbf{SS}(\boldsymbol{\tau})$ and $\mathbf{SS}(\mathbf{e})$ in Table 10.4. Test statistics similar to T_1–T_4 in the previous section can then be constructed. The following computations are similar to those in Section 4.3.1 leading to the F-ratio in (4.3.10).

Since $\mathbf{X}_1$ in (10.6.1) is an $N \times m$ matrix of rank m, let $\mathbf{Z}_1$ be an $N \times (N - m)$ matrix satisfying $\mathbf{Z}_1'\mathbf{X}_1 = \mathbf{0}$ and $\mathbf{Z}_1'\mathbf{Z}_1 = \mathbf{I}_{N-m}$. Clearly, $\mathbf{Z}_1\mathbf{Z}_1' = (\mathbf{I}_N - \mathbf{P}_{\mathbf{X}_1})$. Then the model for $\mathbf{U}_1 = \mathbf{Z}_1'\mathbf{Y}$ is

$$\mathbf{U}_1 = \mathbf{Z}_1'\mathbf{X}_2\boldsymbol{\tau} + \mathbf{Z}_1'\mathbf{E}. \tag{10.6.9}$$

Hence,

$$\mathrm{E}(\mathbf{U}_1) = \mathbf{0} \text{ and } \mathrm{Var}(\mathrm{vec}(\mathbf{U}_1')) = \mathbf{Z}_1'\mathbf{X}_2\mathbf{X}_2'\mathbf{Z}_1 \otimes \boldsymbol{\Sigma}_\tau + \mathbf{I}_{N-m} \otimes \boldsymbol{\Sigma}_e. \tag{10.6.10}$$

Thus $\mathbf{SS}(\boldsymbol{\tau}|\mathbf{B})$, the sum of squares and sum of products matrix for $\boldsymbol{\tau}$, adjusted for $\mathbf{B}$ and based on model (10.6.1), is the same as the sum of squares and sum of products matrix for $\boldsymbol{\tau}$ computed using (10.6.9). Furthermore, $\mathbf{SS}(\mathbf{e})$, the sum of squares and sum of products matrix due to error is the same, whether it is computed using model (10.6.1) or model (10.6.9). These quantities are in fact given by

$$\begin{aligned} \mathbf{SS}(\boldsymbol{\tau}|\mathbf{B}) &= \mathbf{U}_1'\mathbf{Z}_1'\mathbf{X}_2(\mathbf{X}_2'\mathbf{Z}_1\mathbf{Z}_1'\mathbf{X}_2)^{-}\mathbf{X}_2'\mathbf{Z}_1\mathbf{U}_1 \\ &= \mathbf{Y}'(\mathbf{I} - \mathbf{P}_{\mathbf{X}_1})\mathbf{X}_2(\mathbf{X}_2'(\mathbf{I} - \mathbf{P}_{\mathbf{X}_1})\mathbf{X}_2)^{-}\mathbf{X}_2'(\mathbf{I} - \mathbf{P}_{\mathbf{X}_1})\mathbf{Y}, \\ \mathbf{SS}(\mathbf{e}) &= \mathbf{U}_1'\mathbf{U}_1 - \mathbf{SS}(\boldsymbol{\tau}|\mathbf{B}). \end{aligned} \tag{10.6.11}$$

Let

$$\begin{aligned} m_\tau &= \mathrm{rank}(\mathbf{X}_2'\mathbf{Z}_1) = \mathrm{rank}[\mathbf{X}_2'(\mathbf{I} - \mathbf{P}_{\mathbf{X}_1})\mathbf{X}_2], \\ m_e &= N - m - m_\tau. \end{aligned} \tag{10.6.12}$$

Using (10.6.10), we then get

$$\begin{aligned} \mathrm{E}(\mathbf{SS}(\boldsymbol{\tau}|\mathbf{B})) &= m_\tau\boldsymbol{\Sigma}_e + \mathrm{tr}[\mathbf{X}_2'(\mathbf{I} - \mathbf{P}_{\mathbf{X}_1})\mathbf{X}_2]\boldsymbol{\Sigma}_\tau, \\ \mathrm{E}(\mathbf{SS}(\mathbf{e})) &= m_e\boldsymbol{\Sigma}_e. \end{aligned} \tag{10.6.13}$$

Also, $\mathbf{SS}(\boldsymbol{\tau}|\mathbf{B})$ and $\mathbf{SS}(\mathbf{e})$ are independently distributed as $\mathbf{SS}(\boldsymbol{\tau}|\mathbf{B}) \sim W_r(m_\tau, \boldsymbol{\Sigma}_e)$ (under H_τ: $\boldsymbol{\Sigma}_\tau = \mathbf{0}$) and $\mathbf{SS}(\mathbf{e}) \sim W_r(m_e, \boldsymbol{\Sigma}_e)$. Various functions of $[\mathbf{SS}(\mathbf{e})]^{-1}\mathbf{SS}(\boldsymbol{\tau}|\mathbf{B})$ can be used to test H_τ: $\boldsymbol{\Sigma}_\tau = \mathbf{0}$ and statistics similar to T_1–T_4 can be constructed. In particular, the statistic corresponding to Pillai's trace is given by $T_3 = \mathrm{tr}[\mathbf{SS}(\boldsymbol{\tau}|\mathbf{B})\{\mathbf{SS}(\mathbf{e}) + \mathbf{SS}(\boldsymbol{\tau}|\mathbf{B})\}^{-1}]$. From (10.6.11), it follows that $\mathbf{SS}(\mathbf{e}) + \mathbf{SS}(\boldsymbol{\tau}|\mathbf{B}) = \mathbf{U}_1'\mathbf{U}_1 = \mathbf{Y}'\mathbf{Z}_1\mathbf{Z}_1'\mathbf{Y} = \mathbf{Y}'(\mathbf{I} - \mathbf{P}_{\mathbf{X}_1})\mathbf{Y}$. Hence,

$$T_3 = \mathrm{tr}[\mathbf{SS}(\boldsymbol{\tau}|\mathbf{B})\{\mathbf{Y}'(\mathbf{I} - \mathbf{P}_{\mathbf{X}_1})\mathbf{Y}\}^{-1}]. \tag{10.6.14}$$

As already noted (using the unbalanced one-way model), the test based on (10.6.14) is in general not the same as the LBI test in Theorem 10.6.1. Comparing T_3 with the expression for the LBI test statistic in Theorem 10.6.1, it follows that the two tests coincide if $\mathbf{SS}(\boldsymbol{\tau}|\mathbf{B})$ is a scalar multiple of $\mathbf{Y}'(\mathbf{I}-\mathbf{P}_{\mathbf{X}_1})\mathbf{X}_2\mathbf{X}_2'(\mathbf{I}-\mathbf{P}_{\mathbf{X}_1})\mathbf{Y}$, or equivalently,

$$(\mathbf{I}-\mathbf{P}_{\mathbf{X}_1})\mathbf{X}_2\mathbf{X}_2'(\mathbf{I}-\mathbf{P}_{\mathbf{X}_1}) = c(\mathbf{I}-\mathbf{P}_{\mathbf{X}_1})\mathbf{X}_2[\mathbf{X}_2'(\mathbf{I}-\mathbf{P}_{\mathbf{X}_1})\mathbf{X}_2]^{-}\mathbf{X}_2'(\mathbf{I}-\mathbf{P}_{\mathbf{X}_1}), \tag{10.6.15}$$

for some $c > 0$. The condition (10.6.15) follows by using the expression for $\mathbf{SS}(\boldsymbol{\tau}|\mathbf{B})$ given in (10.6.11). Since $(\mathbf{I}-\mathbf{P}_{\mathbf{X}_1})\mathbf{X}_2[\mathbf{X}_2'(\mathbf{I}-\mathbf{P}_{\mathbf{X}_1})\mathbf{X}_2]^{-}\mathbf{X}_2'(\mathbf{I}-\mathbf{P}_{\mathbf{X}_1})$ is the orthogonal projection matrix onto the vector space spanned by the columns of $(\mathbf{I}-\mathbf{P}_{\mathbf{X}_1})\mathbf{X}_2$, (10.6.15) can hold if and only if the nonzero eigenvalues of $(\mathbf{I}-\mathbf{P}_{\mathbf{X}_1})\mathbf{X}_2\mathbf{X}_2'(\mathbf{I}-\mathbf{P}_{\mathbf{X}_1})$ are all equal. Note that the nonzero eigenvalues of $(\mathbf{I}-\mathbf{P}_{\mathbf{X}_1})\mathbf{X}_2\mathbf{X}_2'(\mathbf{I}-\mathbf{P}_{\mathbf{X}_1})$ are the same as those of $\mathbf{X}_2'(\mathbf{I}-\mathbf{P}_{\mathbf{X}_1})\mathbf{X}_2$. We thus have the following result.

Corollary 10.6.1. Consider the setup of Theorem 10.6.1 and suppose the nonzero eigenvalues of $\mathbf{X}_2'(\mathbf{I}-\mathbf{P}_{\mathbf{X}_1})\mathbf{X}_2$ are all equal. Then, for the testing problem (10.6.3), under the group action in (10.6.4), the LBI test coincides with Pillai's trace test and is based on the statistic T_3 in (10.6.14). □

The situation where the nonzero eigenvalues of $\mathbf{X}_2'(\mathbf{I}-\mathbf{P}_{\mathbf{X}_1})\mathbf{X}_2$ are all equal can be considered as a "balanced situation." Indeed, this condition holds in the balanced multivariate one-way random model and Pillai's trace test is the LBI test for testing H_τ: $\boldsymbol{\Sigma}_\tau = \mathbf{0}$, as was already noted. The above balancedness condition also appears in Theorem 6.2.2.

In order to apply the LBI test in practice, we need the percentiles of the null distribution of the statistic in Theorem 10.6.1. When the balancedness condition holds (i.e., when the nonzero eigenvalues of $\mathbf{X}_2'(\mathbf{I}-\mathbf{P}_{\mathbf{X}_1})\mathbf{X}_2$ are all equal), this does not present any difficulty, since the test coincides with Pillai's trace test and tables for the percentiles of the null distribution of the trace statistic are available; see Table 3 in Anderson (1984, p. 630). When the balancedness condition does not hold, the percentiles of the null distribution of the LBI test statistic will have to be obtained numerically. This was done in Zhou and Mathew (1993) for some unbalanced multivariate one-way random models. The simulation results in Zhou and Mathew (1993) show that for local alternatives, the LBI test does have an advantage (in terms of power) over the tests based on Wilks' likelihood ratio, Pillai's trace, and Roy's largest root. Of course, this is to be expected.

A balanced setup of model (10.6.1) is considered in Das and Sinha (1987), with the additional assumption that $\boldsymbol{\Sigma}_\tau = \sigma_\tau^2\boldsymbol{\Sigma}_e$. The problem now is to test H_τ: $\sigma_\tau^2 = 0$. Das and Sinha (1987) show that in the balanced case, the LBI test for this problem coincides with Pillai's trace test based on the statistic T_3 in (10.6.14).

EXERCISES

10.1. Let $\mathbf{T}$ be a random matrix whose rows are independent and have a common variance–covariance matrix $\boldsymbol{\Delta}_t$. Show that if $\mathbf{A}$ is a constant symmetric matrix, then

$$E(\mathbf{T}'\mathbf{A}\mathbf{T}) = tr(\mathbf{A})\boldsymbol{\Delta}_t + E(\mathbf{T})'\mathbf{A}E(\mathbf{T}).$$

10.2. Consider Theorem 10.3.2. Show that if condition (10.3.21) is valid, then $m = \sum_{i=1}^{t} m_{(i)}$, where m is the approximate number of degrees of freedom given by formula (10.3.20).

10.3. Consider the statistic T^2 given by formula (10.4.8), where $\bar{\mathbf{d}} = (1/b) \sum_{j=1}^{b} \mathbf{d}_j$ and $\mathbf{d}_j = \mathbf{C}\bar{\mathbf{y}}_j$, $j = 1, 2, \cdots, b$. Show that T^2 is invariant to the choice of $\mathbf{C}$ provided that the rows of $\mathbf{C}$ form a basis for the $(v-1)$-dimensional orthogonal complement of the one-dimensional space spanned by $\mathbf{1}_v$.

10.4. Consider the hypothesis H_0 described in formula (10.3.15), and the associated ratio

$$F_\lambda = \frac{\boldsymbol{\lambda}'\mathbf{Y}'\mathbf{P}_i\mathbf{Y}\boldsymbol{\lambda}/m_i}{\boldsymbol{\lambda}'\mathbf{Y}'\mathbf{P}_{i_0}\mathbf{Y}\boldsymbol{\lambda}/m_{i_0}},$$

where $\boldsymbol{\lambda}$ is a nonzero vector.

(a) Show that $\sup_{\boldsymbol{\lambda}\neq\mathbf{0}} F_\lambda = (m_{i_0}/m_i)\lambda_{\max}[(\mathbf{Y}'\mathbf{P}_i\mathbf{Y})(\mathbf{Y}'\mathbf{P}_{i_0}\mathbf{Y})^{-1}]$, where $\lambda_{\max}$ denotes the largest eigenvalue of the matrix inside brackets.

(b) Deduce from part (a) that H_0 can be rejected for large values of

$$\lambda_{\max}[(\mathbf{Y}'\mathbf{P}_i\mathbf{Y})(\mathbf{Y}'\mathbf{P}_{i_0}\mathbf{Y})^{-1}].$$

10.5. Consider the multivariate one-way model (10.2.3). Let $\mathbf{P}_0$, $\mathbf{P}_1$, and $\mathbf{P}_2$ be idempotent matrices of ranks m_0, m_1, and m_2, respectively, such that $\mathbf{S}_0 = (1/m_0)\mathbf{Y}'\mathbf{P}_0\mathbf{Y}$, $\mathbf{S}_1 = (1/m_1)\mathbf{Y}'\mathbf{P}_1\mathbf{Y}$, and $\mathbf{S}_2 = (1/m_2)\mathbf{Y}'\mathbf{P}_2\mathbf{Y}$ are the mean square matrices associated with $\boldsymbol{\mu}$, $\boldsymbol{\tau}_k$, and the error term, respectively.

(a) Show that

$$\begin{aligned}
\mathbf{P}_0 &= \frac{1}{vn}\mathbf{J}_v \otimes \mathbf{J}_n, \\
\mathbf{P}_1 &= \frac{1}{n}\mathbf{I}_v \otimes \mathbf{J}_n - \frac{1}{vn}\mathbf{J}_v \otimes \mathbf{J}_n \\
\mathbf{P}_2 &= \mathbf{I}_v \otimes \mathbf{I}_n - \frac{1}{n}\mathbf{I}_v \otimes \mathbf{J}_n.
\end{aligned}$$

[Hint: Use Lemma 2.3.1 in Chapter 2.]

(b) Verify that $m_0 = 1$, $m_1 = v - 1$, $m_2 = v(n-1)$.

(c) Show that

$$\begin{aligned}
\mathbf{Y}'\mathbf{P}_1\mathbf{Y} &= n\sum_{k=1}^{v}\left(\bar{\tilde{\mathbf{y}}}_{k\cdot} - \bar{\tilde{\mathbf{y}}}_{\cdot\cdot}\right)\left(\bar{\tilde{\mathbf{y}}}_{k\cdot} - \bar{\tilde{\mathbf{y}}}_{\cdot\cdot}\right)' \\
\mathbf{Y}'\mathbf{P}_2\mathbf{Y} &= \sum_{k=1}^{v}\sum_{\ell=1}^{n}\left(\tilde{\mathbf{y}}_{k\ell} - \bar{\tilde{\mathbf{y}}}_{k\cdot}\right)\left(\tilde{\mathbf{y}}_{k\ell} - \bar{\tilde{\mathbf{y}}}_{k\cdot}\right)',
\end{aligned}$$

where $\bar{\tilde{\mathbf{y}}}_{k\cdot} = \frac{1}{n}\sum_{\ell=1}^{n}\tilde{\mathbf{y}}_{k\ell}$, $\bar{\tilde{\mathbf{y}}}_{\cdot\cdot} = \frac{1}{vn}\sum_{k=1}^{v}\sum_{\ell=1}^{n}\tilde{\mathbf{y}}_{k\ell}$.

10.6. Consider the multivariate two-way model (10.2.7). Let $\mathbf{P}_1$, $\mathbf{P}_2$, $\mathbf{P}_3$, and $\mathbf{P}_4$ be idempotent matrices associated with the mean square matrices for factors A, B, their interaction $A * B$, and the error term, respectively.

(a) Show that

$$\begin{aligned}
\mathbf{P}_1 &= \frac{1}{bn}\mathbf{I}_v \otimes \mathbf{J}_b \otimes \mathbf{J}_n - \frac{1}{bvn}\mathbf{J}_v \otimes \mathbf{J}_b \otimes \mathbf{J}_n \\
\mathbf{P}_2 &= \frac{1}{vn}\mathbf{J}_v \otimes \mathbf{I}_b \otimes \mathbf{J}_n - \frac{1}{bvn}\mathbf{J}_v \otimes \mathbf{J}_b \otimes \mathbf{J}_n \\
\mathbf{P}_3 &= \frac{1}{n}\mathbf{I}_v \otimes \mathbf{I}_b \otimes \mathbf{J}_n - \frac{1}{bn}\mathbf{I}_v \otimes \mathbf{J}_b \otimes \mathbf{J}_n - \frac{1}{vn}\mathbf{J}_v \otimes \mathbf{I}_b \otimes \mathbf{J}_n \\
&\quad + \frac{1}{bvn}\mathbf{J}_v \otimes \mathbf{J}_b \otimes \mathbf{J}_n \\
\mathbf{P}_4 &= \mathbf{I}_v \otimes \mathbf{I}_b \otimes \mathbf{I}_n - \frac{1}{n}\mathbf{I}_v \otimes \mathbf{I}_b \otimes \mathbf{J}_n.
\end{aligned}$$

(b) Show that

$$\begin{aligned}
\mathbf{Y}'\mathbf{P}_1\mathbf{Y} &= bn\sum_{k=1}^{v}\left(\bar{\tilde{\mathbf{y}}}_{k\cdot\cdot} - \bar{\tilde{\mathbf{y}}}_{\cdot\cdot\cdot}\right)\left(\bar{\tilde{\mathbf{y}}}_{k\cdot\cdot} - \bar{\tilde{\mathbf{y}}}_{\cdot\cdot\cdot}\right)' \\
\mathbf{Y}'\mathbf{P}_2\mathbf{Y} &= vn\sum_{\ell=1}^{b}\left(\bar{\tilde{\mathbf{y}}}_{\cdot\ell\cdot} - \bar{\tilde{\mathbf{y}}}_{\cdot\cdot\cdot}\right)\left(\bar{\tilde{\mathbf{y}}}_{\cdot\ell\cdot} - \bar{\tilde{\mathbf{y}}}_{\cdot\cdot\cdot}\right)' \\
\mathbf{Y}'\mathbf{P}_3\mathbf{Y} &= n\sum_{k=1}^{v}\sum_{\ell=1}^{b}\left(\bar{\tilde{\mathbf{y}}}_{k\ell\cdot} - \bar{\tilde{\mathbf{y}}}_{k\cdot\cdot} - \bar{\tilde{\mathbf{y}}}_{\cdot\ell\cdot} + \bar{\tilde{\mathbf{y}}}_{\cdot\cdot\cdot}\right)\left(\bar{\tilde{\mathbf{y}}}_{k\ell\cdot} - \bar{\tilde{\mathbf{y}}}_{k\cdot\cdot} - \bar{\tilde{\mathbf{y}}}_{\cdot\ell\cdot} + \bar{\tilde{\mathbf{y}}}_{\cdot\cdot\cdot}\right)' \\
\mathbf{Y}'\mathbf{P}_4\mathbf{Y} &= \sum_{k=1}^{v}\sum_{\ell=1}^{b}\sum_{m=1}^{n}\left(\tilde{\mathbf{y}}_{k\ell m} - \bar{\tilde{\mathbf{y}}}_{k\ell\cdot}\right)\left(\tilde{\mathbf{y}}_{k\ell m} - \bar{\tilde{\mathbf{y}}}_{k\ell\cdot}\right)',
\end{aligned}$$

where:

$$\bar{\tilde{\mathbf{y}}}_{k\ell\cdot} = \frac{1}{n}\sum_{m=1}^{n}\tilde{\mathbf{y}}_{k\ell m},$$

$$\bar{\tilde{\mathbf{y}}}_{k..} = \frac{1}{bn}\sum_{\ell=1}^{b}\sum_{m=1}^{n}\tilde{\mathbf{y}}_{k\ell m},$$

$$\bar{\tilde{\mathbf{y}}}_{.\ell.} = \frac{1}{vn}\sum_{k=1}^{v}\sum_{m=1}^{n}\tilde{\mathbf{y}}_{k\ell m},$$

$$\bar{\tilde{\mathbf{y}}}_{...} = \frac{1}{bvn}\sum_{k=1}^{v}\sum_{\ell=1}^{b}\sum_{m=1}^{n}\tilde{\mathbf{y}}_{k\ell m}.$$

(c) Assuming that the effects of A and B are fixed and random, respectively,
   (i) obtain the expected values of $\mathbf{Y}'\mathbf{P}_i\mathbf{Y}(i = 1,2,3,4)$.
   (ii) give test statistics for testing the significance of A, B, and $A * B$.

10.7. Records for milk and fat yields for 574 daughters of 12 sires in three herds were reported in Meyer (1985, p. 159). The following table gives the totals of MY (milk yield) and FY (fat yield) for the sire $\times$ herd subclasses:

	Herd 1		Herd 2		Herd 3	
Sire	MY	FY	MY	FY	MY	FY
1	62312	2434	83649	3084	61283	2267
2	51583	1995	60905	2169	56574	2202
3	46245	1814	69433	2519	63068	2405
4	98333	3755	98657	3871	94499	3644
5	58665	2131	72200	2630	82572	2874
6	129134	4605	94571	3372	45611	1613
7	60732	2202	71859	2624	102405	3558
8	58989	2102	57079	2129	95845	3647
9	82909	3200	81687	3043	73211	2808
10	73757	2931	86288	3234	47915	1931
11	62817	2375	56081	2438	91181	3827
12	75513	2854	62786	2573	105979	3911

Source: K. Meyer (1985, Table 1, p. 159). Reproduced with permission of the International Biometric Society.

Assuming that the sire effect is random and the herd effect is fixed, test the significance of both effects.

10.8. Timm (1980, p. 72) used the following data set, which was obtained in a study concerning the relative effectiveness of two orthopedic adjustments of the mandible. Nine subjects were assigned to each of two orthopedic treatments, called activator treatments. On each of three time points, three dependent variables, y_1, y_2, y_3, were observed. These variables reflected the vertical position and size of the mandible. The same data set was analyzed by Boik (1988):

		Time								
		1			2			3		
Treatment	Subject	y_1	y_2	y_3	y_1	y_2	y_3	y_1	y_2	y_3
1	1	117.0	59.0	10.5	117.5	59.0	16.5	118.5	60.0	16.5
	2	109.0	60.0	30.5	110.5	61.5	30.5	111.0	61.5	30.5
	3	117.0	60.0	23.5	120.0	61.5	23.5	120.5	62.0	23.5
	4	122.0	67.5	33.0	126.0	70.5	32.0	127.0	71.5	32.5
	5	116.0	61.5	24.5	118.5	62.5	24.5	119.5	63.5	24.5
	6	123.0	65.5	22.0	126.0	61.5	22.0	127.0	67.5	22.0
	7	130.5	68.5	33.0	132.0	69.5	32.5	134.5	71.0	32.0
	8	126.5	69.0	20.0	128.5	71.0	20.0	130.5	73.0	20.0
	9	113.0	58.0	25.0	116.5	59.0	25.0	118.0	60.5	24.5
2	1	128.0	67.0	24.0	129.0	67.5	24.0	131.5	69.0	24.0
	2	116.5	63.5	28.5	120.0	65.0	29.5	121.5	66.0	29.5
	3	121.5	64.5	26.5	125.5	67.5	27.0	127.0	69.0	27.0
	4	109.5	54.0	18.0	112.0	55.5	18.5	114.0	57.0	19.0
	5	133.0	72.0	34.5	136.0	73.5	34.5	137.5	75.5	34.5
	6	120.0	62.5	26.0	124.5	65.0	26.0	126.0	66.0	26.0
	7	129.5	65.0	18.5	133.5	68.0	18.5	134.5	69.0	18.5
	8	122.0	64.5	18.5	124.0	65.5	18.5	125.5	66.0	18.5
	9	125.0	65.5	21.5	127.0	66.5	21.5	128.0	67.0	21.6

Source: N.H. Timm (1980, Table 7.2, p. 72). Reproduced with permission of Elsevier Science B.V.

(a) Write a model for this experiment.
(b) Test the significance of the treatment by time interaction.
(c) Test the significance of the treatment and time effects.

10.9. Consider the data set in Exercise 10.7. Let $\mathbf{\Sigma}_\tau$ and $\mathbf{\Sigma}_e$ denote the variance–covariance matrices associated with the sire effect and the random error, respectively. Let $\mathbf{S}_\tau$ and $\mathbf{S}_e$ be the corresponding mean square matrices.

(a) Obtain the expected values of $\mathbf{S}_\tau$ and $\mathbf{S}_e$.
(b) Let $\mathbf{\Phi} = \mathbf{\Sigma}_\tau + \mathbf{\Sigma}_e$. Obtain an unbiased estimator $\hat{\mathbf{\Phi}}$ of $\mathbf{\Phi}$.
(c) Find the values of m and $\mathbf{\Delta}$ such that $\hat{\mathbf{\Phi}}$ is approximately distributed as $W_2(m, \mathbf{\Delta})$.
(d) On the basis of the given data, do we have sufficient evidence to conclude that Satterthwaite's approximation in part (c) is inadequate? Use $\alpha = 0.10$.

10.10. Consider model (10.4.1), where τ_i is fixed and β_j is random. Write this model in vector form as

$$\mathbf{y} = \mu \mathbf{1}_{vbn} + (\mathbf{I}_v \otimes \mathbf{1}_b \otimes \mathbf{1}_n)\boldsymbol{\tau} + (\mathbf{1}_v \otimes \mathbf{I}_b \otimes \mathbf{1}_n)\boldsymbol{\beta} + (\mathbf{I}_v \otimes \mathbf{I}_b \otimes \mathbf{1}_n)(\boldsymbol{\tau\beta}) + \mathbf{e},$$

where:

$$\begin{aligned} \boldsymbol{\tau} &= (\tau_1, \tau_2, \cdots, \tau_v)', \\ \boldsymbol{\beta} &= (\beta_1, \beta_2, \cdots, \beta_b)', \\ (\boldsymbol{\tau\beta}) &= [(\tau\beta)_{11}, \cdots, (\tau\beta)_{1b}, \\ &\quad (\tau\beta)_{21}, \cdots, (\tau\beta)_{2b}, \cdots, (\tau\beta)_{vb}]'; \end{aligned}$$

$\mathbf{y}$ and $\mathbf{e}$ are the vectors of y_{ijk}'s and e_{ijk}'s, respectively.

Let $MS(\tau)$, $MS(\beta)$, $MS(\tau\beta)$, and $MS(e)$ denote the mean squares corresponding to $\tau_i, \beta_j, (\tau\beta)_{ij}$, and the error term, respectively, from the usual ANOVA table.

(a) Show that under Model B's assumptions,

$$\begin{aligned} E(\mathbf{y}) &= (\mu \mathbf{1}_v + \boldsymbol{\tau}) \otimes \mathbf{1}_{bn} \\ \text{Var}(\mathbf{y}) &= \boldsymbol{\Sigma} \otimes \mathbf{I}_b \otimes \mathbf{J}_n + \sigma_e^2 \mathbf{I}_{vbn}, \end{aligned}$$

where $\boldsymbol{\Sigma}$ is the common variance–covariance matrix of $\boldsymbol{\zeta}_j$ in formula (10.4.2), $j = 1, 2, \cdots, b$.

(b) Obtain the expected values of $MS(\tau)$, $MS(\beta)$, $MS(\tau\beta)$, and $MS(e)$ under Model B's assumptions.

(c) Verify that $MS(\tau)$, $MS(\beta)$, and $MS(\tau\beta)$ in part (b) are expressible as linear combinations of independent chi-squared variates.

BIBLIOGRAPHY

Alalouf, I. S. (1980). "A multivariate approach to a mixed linear model." *Journal of the American Statistical Association*, 75, 194–200.

Amemiya, Y. (1985). "What should be done when an estimated between-group covariance matrix is not nonnegative definite?" *The American Statistician*, 39, 112–117.

Anderson, B. M., Anderson, T. W., and Olkin, I. (1986). "Maximum likelihood estimators and likelihood ratio criteria in multivariate components of variance." *The Annals of Statistics*, 14, 405–417.

Anderson, T. W. (1984). *An Introduction to Multivariate Statistical Analysis*, Second Edition. Wiley, New York.

Anderson, T. W. (1985). "Components of variance in MANOVA." In: *Multivariate Analysis-VI* (P.R. Krishnaiah, Ed.), North-Holland, Amsterdam, 1–8.

Beyer, W. H. (1968). *Handbook of Tables for Probability and Statistics*, Second Edition. The Chemical Rubber Company, Cleveland, Ohio.

Boik, R. J. (1988). "The mixed model for multivariate repeated measures: Validity conditions and an approximate test." *Psychometrika*, 53, 469–486.

Box, G. E. P. (1950). "Problems in the analysis of growth and wear curves." *Biometrics*, 6, 362–389.

Browne, M. W. (1974). "Generalized least squares estimators in the analysis of covariance structures." *South African Statistical Journal*, 8, 1–24.

Das, R. and Sinha, B. K. (1987). "Robust optimum invariant unbiased tests for variance components." In: *Proceedings of the Second International Tampere Conference in Statistics* (T. Pukkila, S. Puntanen, Eds.), University of Tampere, Finland, 317–342.

Das, R. and Sinha, B. K. (1988). "Optimum invariant tests in random MANOVA models." *Canadian Journal of Statistics*, 16, 193–200.

Hald, A. (1952). *Statistical Theory with Engineering Applications*. Wiley, New York.

Hocking, R. R. (1973). "A discussion of the two-way mixed model." *The American Statistician*, 27, 148–152.

Imhof, J. P. (1960). "A mixed model for the complete three-way layout with two random-effects factors." *Annals of Mathematical Statistics*, 31, 906–928.

Imhof, J. P. (1962). "Testing the hypothesis of no fixed main-effects in Scheffé's mixed model." *Annals of Mathematical Statistics*, 33, 1085–1095.

Kariya, T. and Sinha, B. K. (1989). *Robustness of Statistical Tests*. Academic Press, Boston.

Khatri, C. G. and Patel, H. I. (1992). "Analysis of a multicenter trial using a multivariate approach to a mixed linear model." *Communications in Statistics-Theory and Methods*, 21, 21–39.

Khuri, A. I. (1989). "Testing a covariance matrix structure in a mixed model with no empty cells." *Journal of Statistical Planning and Inference*, 22, 117–125.

Khuri, A. I., Mathew, T., and Nel, D. G. (1994). "A test to determine closeness of multivariate Satterthwaite's approximation." *Journal of Multivariate Analysis*, 51, 201–209.

Mathew, T. (1989). "MANOVA in the multivariate components of variance model." *Journal of Multivariate Analysis*, 29, 30–38.

Mathew, T., Niyogi, A., and Sinha, B. K. (1994). "Improved nonnegative estimation of variance components in balanced multivariate mixed models." *Journal of Multiviariate Analysis*, 51, 83–101.

Meyer, K. (1985). "Maximum likelihood estimation of variance components for a multivariate mixed model with equal design matrices." *Biometrics*, 41, 153–165.

Muirhead, R. J. (1982). *Aspects of Multivariate Statistical Theory*. Wiley, New York.

Nagarsenker, B. N. (1975). "Percentage points of Wilks' L_{vc} criterion." *Communications in Statistics-Theory and Methods*, 4, 629–641.

Nagarsenker, B. N. and Pillai, K. C. S. (1973). "The distribution of the sphericity test criterion." *Journal of Multivariate Analysis*, 3, 226–235.

Nel, D. G. (1980). "On matrix differentiation in statistics." *South African Statistical Journal*, 14, 137–193.

Nel, D. G. and Van Der Merwe, C. A. (1986). "A solution to the multivariate Behrens–Fisher problem." *Communications in Statistics—Theory and Methods*, 15, 3719–3735.

Oman, S. D. (1995). "Checking the assumptions in mixed-model analysis of variance: A residual analysis approach." *Computational Statistics and Data Analysis*, 20, 309–330.

Scheffé, H. (1956). "A mixed model for the analysis of variance." *Annals of Mathematical Statistics*, 27, 23–36.

Scheffé, H. (1959). *The Analysis of Variance.* Wiley, New York.

Schott, J. R. (1985). "Multivariate maximum likelihood estimators for the mixed linear model." *Sankhyā, Series B*, 47, 179–185.

Schott, J. R. and Saw, J. G. (1984). "A multivariate one-way classification model with random effects." *Journal of Multivariate Analysis*, 15, 1–12.

Searle, S. R. (1971). *Linear Models.* Wiley, New York.

Seber, G. A. F. (1984). *Multivariate Observations.* Wiley, New York.

Tan, W. Y. and Gupta, R. P. (1983). "On approximating a linear combination of central Wishart matrices with positive coefficients." *Communications in Statistics—Theory and Methods*, 12, 2589–2600.

Timm, N. H. (1980). "Multivariate analysis of variance of repeated measurements." In: *Handbook of Statistics*, Volume 1 (P. R. Krishnaiah, Ed.), Elsevier Science B.V., Amsterdam, 41–87.

Wilks, S. S. (1946). "Sample criteria for testing equality of means, equality of variances, and equality of covariances in a normal multivariate distribution." *Annals of Mathematical Statistics*, 17, 257–281.

Zhou, L. and Mathew, T. (1993). "Hypotheses tests for variance components in some multivariate mixed models." *Journal of Statistical Planning and Inference*, 37, 215–227.

Zhou, L. and Mathew, T. (1994). "Combining independent tests in multivariate linear models." *Journal of Multivariate Analysis*, 51, 265–275.

APPENDIX

Solutions to Selected Exercises

CHAPTER 2

2.1. We have that

$$\frac{(\sigma_e^2 + 4\sigma_\beta^2)MS(\tau)}{(\sigma_e^2 + 6\sigma_\tau^2)MS(\beta)} \sim F_{3,5}$$

Hence,

$$\begin{aligned} E\left[\frac{MS(\tau)}{MS(\beta)}\right] &= \frac{1+6\gamma_1}{1+4\gamma_2}E(F_{3,5}) \\ &= \frac{5(1+6\gamma_1)}{3(1+4\gamma_2)}. \end{aligned}$$

Thus, $\frac{3}{5}MS(\tau)/MS(\beta)$ is unbiased for λ. Using the fact that this estimator is a function of complete and sufficient statistics, it must be the UMVUE of λ by the Lehmann–Scheffé theorem.

2.2. **(a)** The population structure is $[(i)(j:k)]:\ell$.

(b) The ANOVA table is

Source	d.f.	MS	$E(MS)$
α_i	2	$MS(\alpha)$	$30\sum_{i=1}^{3}\alpha_i^2 + \sigma_e^2$
β_j	4	$MS(\beta)$	$9\sum_{j=1}^{5}\beta_j^2 + 9\sigma_{\delta(\beta)}^2 + \sigma_e^2$
$\delta_{j(k)}$	15	$MS(\delta(\beta))$	$9\sigma_{\delta(\beta)}^2 + \sigma_e^2$
Residual	158	$MS(e)$	σ_e^2

The test statistic for the age effect is $F = MS(\alpha)/MS(e)$, which, under the null hypothesis, has the F-distribution with 2 and 158 degrees of freedom. Under the alternative hypothesis, it has the noncentral F-distribution with 2 and 158 degrees of freedom and a noncentrality parameter $\zeta = (60/\sigma_e^2)\sum_{i=1}^3 \alpha_i^2$. Its power function is given by

$$\text{Power} = P[F'_{2,158}(\zeta) > F_{\alpha,2,158}].$$

(c) $\bar{y}_{\cdot j\cdot\cdot} - \bar{y}_{\cdot j'\cdot\cdot} \mp \left[\frac{2}{36}MS(\delta(\beta))\right]^{1/2} t_{\frac{\alpha}{2},15}$.

(d) An unbiased estimator of λ is given by

$$\hat{\lambda} = \frac{1}{9}MS(\delta(\beta)) + \frac{8}{9}MS(e).$$

Using Satterthwaite's approximation, an approximate $100(1-\alpha)\%$ confidence interval for λ is of the form

$$\left(\frac{f\hat{\lambda}}{\chi^2_{\alpha/2,f}}, \frac{f\hat{\lambda}}{\chi^2_{1-\alpha/2,f}}\right),$$

where

$$f = \frac{[MS(\delta(\beta)) + 8MS(e)]^2}{[MS(\delta(\beta))]^2/15 + [8MS(e)]^2/158}.$$

It is also possible to obtain an exact, but conservative, confidence interval for λ.

2.3. (a) The population structure is $[(i:j)(k)]:\ell$, where i, j, k, and ℓ are the subscripts associated with plant, bobbin, distance, and replications, respectively.

(b)

$$y_{ijk\ell} = \mu + \tau_i + \beta_{i(j)} + \delta_k + (\tau\delta)_{ik} + (\beta\delta)_{i(jk)} + e_{ijk(\ell)},$$

where τ_i is the effect of the ith plant $(i = 1,2)$, $\beta_{i(j)}$ is the effect of the jth bobbin within the ith plant $(j = 1,2,\ldots,6)$, and δ_k is the effect of the kth distance $(k = 1,2,\ldots,5)$.

2.4. (a) The population structure is $[(i:j)(k:\ell)]:m$, where subscripts i and k are associated with type A and type B processes, respectively; subscripts j and ℓ are associated with the sublevels nested within type A and type B processes, respectively; and m is a replication subscript. The corresponding model is therefore of the form

$$y_{ijk\ell m} = \mu + \tau_i + \beta_{i(j)} + \delta_k + \gamma_{k(\ell)} + (\tau\delta)_{ik} + (\beta\delta)_{i(jk)} + (\tau\gamma)_{k(i\ell)} + (\beta\gamma)_{ik(j\ell)} + e_{ijk\ell(m)},$$

where τ_i and δ_k are the effects associated with type A and type B processes, and $\beta_{i(j)}$ and $\gamma_{k(\ell)}$ are the corresponding nested effects.

(b)

$$
\begin{aligned}
E[MS(\tau)] &= 120\sum_{i=1}^{4}\tau_i^2 + 72\sigma^2_{\beta(\tau)} + 18\sigma^2_{\beta(\tau)\delta} + 15\sigma^2_{\tau\gamma(\delta)} \\
&\quad + 3\sigma^2_{\beta(\tau)\gamma(\delta)} + \sigma^2_e \\
E[MS(\beta(\tau))] &= 72\sigma^2_{\beta(\tau)} + 18\sigma^2_{\beta(\tau)\delta} + 3\sigma^2_{\beta(\tau)\gamma(\delta)} + \sigma^2_e \\
E[MS(\delta)] &= 120\sum_{k=1}^{4}\delta_k^2 + 60\sigma^2_{\gamma(\delta)} + 18\sigma^2_{\beta(\tau)\delta} + 15\sigma^2_{\tau\gamma(\delta)} \\
&\quad + 3\sigma^2_{\beta(\tau)\gamma(\delta)} + \sigma^2_e \\
E[MS(\gamma(\delta))] &= 60\sigma^2_{\gamma(\delta)} + 15\sigma^2_{\tau\gamma(\delta)} + 3\sigma^2_{\beta(\tau)\gamma(\delta)} + \sigma^2_e \\
E[MS(\tau\delta)] &= 10\sum_{i=1}^{4}\sum_{k=1}^{4}(\tau\delta)^2_{ik} + 18\sigma^2_{\beta(\tau)\delta} + 15\sigma^2_{\tau\gamma(\delta)} \\
&\quad + 3\sigma^2_{\beta(\tau)\gamma(\delta)} + \sigma^2_e \\
E[MS(\beta(\tau)\delta)] &= 18\sigma^2_{\beta(\tau)\delta} + 3\sigma^2_{\beta(\tau)\gamma(\delta)} + \sigma^2_e \\
E[MS(\tau\gamma(\delta))] &= 15\sigma^2_{\tau\gamma(\delta)} + 3\sigma^2_{\beta(\tau)\gamma(\delta)} + \sigma^2_e \\
E[MS(\beta(\tau)\gamma(\delta))] &= 3\sigma^2_{\beta(\tau)\gamma(\delta)} + \sigma^2_e \\
E[MS(e)] &= \sigma^2_e.
\end{aligned}
$$

2.5. (a) The population structure is $[(i:j)(k)]:\ell$, where subscripts i, j, k, and ℓ are associated with method, coil, location, and replications, respectively. The corresponding model is

$$
y_{ijk\ell} = \mu + \tau_i + \beta_{i(j)} + \delta_k + (\tau\delta)_{ik} + (\beta\delta)_{i(jk)} + e_{ijk(\ell)},
$$

where τ_i and δ_k are associated with method and location, respectively, and $\beta_{i(j)}$ denotes the nested coil effect.

(b) The effects τ_i and δ_k are fixed and $\beta_{i(j)}$ is random. Hence, the complete ANOVA table is

Source	d.f.	MS	$E(MS)$	F	P-value
τ_i	1	2646.00	$12\sum_{i=1}^{2}\tau_i^2 + 4\sigma^2_{\beta(\tau)} + 2\sigma^2_{\beta(\tau)\delta} + \sigma^2_e$	1.09	0.3552
$\beta_{i(j)}$	4	2425.33	$4\sigma^2_{\beta(\tau)} + 2\sigma^2_{\beta(\tau)\delta} + \sigma^2_e$	45.83	0.0013
δ_k	1	1872.67	$12\sum_{k=1}^{2}\delta_k^2 + 2\sigma^2_{\beta(\tau)\delta} + \sigma^2_e$	35.39	0.0040
$(\tau\delta)_{ik}$	1	16.67	$6\sum_{i=1}^{2}\sum_{k=1}^{2}(\tau\delta)^2_{ik} + 2\sigma^2_{\beta(\tau)\delta} + \sigma^2_e$	0.31	0.6046
$(\beta\delta)_{i(jk)}$	4	52.92	$2\sigma^2_{\beta(\tau)\delta} + \sigma^2_e$	0.50	0.7362
Error	12	105.75	σ^2_e		

(c)

$$
\frac{SS(\tau)}{4\sigma^2_{\beta(\tau)} + 2\sigma^2_{\beta(\tau)\delta} + \sigma^2_e} \sim \chi_1'^2(\zeta_1),
$$

where

$$\zeta_1 = \frac{12\sum_{i=1}^{2}\tau_i^2}{4\sigma^2_{\beta(\tau)} + 2\sigma^2_{\beta(\tau)\delta} + \sigma^2_e}.$$

$$\frac{SS(\beta(\tau))}{4\sigma^2_{\beta(\tau)} + 2\sigma^2_{\beta(\tau)\delta} + \sigma^2_e} \sim \chi^2_4.$$

$$\frac{SS(\delta)}{2\sigma^2_{\beta(\tau)\delta} + \sigma^2_e} \sim \chi'^2_1(\zeta_3),$$

where

$$\zeta_3 = \frac{12\sum_{k=1}^{2}\delta_k^2}{2\sigma^2_{\beta(\tau)\delta} + \sigma^2_e}.$$

$$\frac{SS(\tau\delta)}{2\sigma^2_{\beta(\tau)\delta} + \sigma^2_e} \sim \chi'^2_1(\zeta_4),$$

where

$$\zeta_4 = \frac{6\sum_{i=1}^{2}\sum_{k=1}^{2}(\tau\delta)^2_{ik}}{2\sigma^2_{\beta(\tau)\delta} + \sigma^2_e}.$$

$$\frac{SS(\beta(\tau)\delta)}{2\sigma^2_{\beta(\tau)\delta} + \sigma^2_e} \sim \chi^2_4.$$

$$\frac{SS(e)}{\sigma^2_e} \sim \chi^2_{12}.$$

(d) See the ANOVA table in part (b).

2.6. (a) $i : [(j)(k : \ell)]$.

(b)

Source	d.f.	MS	$E(MS)$
α_i	2	$MS(\alpha)$	$18\sum_{i=1}^{3}\alpha_i^2 + 12\sigma^2_\beta + 9\sigma^2_\gamma + 3\sigma^2_\delta + 3\sigma^2_{\beta\gamma} + \sigma^2_e$
$\beta_{i(j)}$	6	$MS(\beta(\alpha))$	$12\sigma^2_\beta + 3\sigma^2_{\beta\gamma} + \sigma^2_e$
$\gamma_{i(k)}$	9	$MS(\gamma(\alpha))$	$9\sigma^2_\gamma + 3\sigma^2_\delta + 3\sigma^2_{\beta\gamma} + \sigma^2_e$
$\delta_{ik(\ell)}$	24	$MS(\delta(\beta))$	$3\sigma^2_\delta + \sigma^2_e$
$(\beta\gamma)_{i(jk)}$	18	$MS(\beta(\alpha)\gamma(\alpha))$	$3\sigma^2_{\beta\gamma} + \sigma^2_e$
Error	48	$MS(e)$	σ^2_e

(c) **(i)** $F = [MS(\alpha) + MS(\beta(\alpha)\gamma(\alpha))]/[MS(\beta(\alpha)) + MS(\gamma(\alpha))]$.

(ii) $F = MS(\delta(\beta))/7MS(e)$.
This is a two-sided test.

(d) Let $\alpha^* = 1 - (1-\alpha)^{1/2}$. Consider the following confidence intervals for θ_1 and θ_2, with a confidence coefficient $1 - \alpha^*$ each,

$$u_1 < \theta_1 < u_2$$
$$v_1 < \theta_2 < v_2,$$

where

$$u_1 = \frac{6MS(\beta(\alpha))}{\chi^2_{\alpha^*/2,6}}$$
$$u_2 = \frac{6MS(\beta(\alpha))}{\chi^2_{1-\alpha^*/2,6}}$$
$$v_1 = \frac{18MS(\beta(\alpha)\gamma(\alpha))}{\chi^2_{\alpha^*/2,18}}$$
$$v_2 = \frac{18MS(\beta(\alpha)\gamma(\alpha))}{\chi^2_{1-\alpha^*/2,18}}.$$

Thus, the rectangular region, $u_1 < \theta_1 < u_2, v_1 < \theta_2 < v_2$ defines a confidence region for $\boldsymbol{\theta} = (\theta_1, \theta_2)'$ with a confidence coefficient $(1 - \alpha)$.

(e)

$$P(\hat{\sigma}^2_\beta < 0) = P[MS(\beta(\alpha)) < MS(\beta(\alpha)\gamma(\alpha))]$$
$$= P\left[F_{6,18} < \frac{1}{12\Delta + 1}\right].$$

2.8. (a)

Source	d.f.	SS	MS	$E(MS)$	F
α_i	$a-1$	$SS(\alpha)$	$MS(\alpha)$	$\frac{bcn}{a-1}\sum_{i=1}^a \alpha_i^2 + cn\sigma^2_\beta + n\sigma^2_\gamma + \sigma^2_e$	$\frac{MS(\alpha)}{MS(\beta(\alpha))}$
$\beta_{i(j)}$	$a(b-1)$	$SS(\beta(\alpha))$	$MS(\beta(\alpha))$	$cn\sigma^2_\beta + n\sigma^2_\gamma + \sigma^2_e$	$\frac{MS(\beta(\alpha))}{MS(\gamma(\beta))}$
$\gamma_{ij(k)}$	$ab(c-1)$	$SS(\gamma(\beta))$	$MS(\gamma(\beta))$	$n\sigma^2_\gamma + \sigma^2_e$	$\frac{MS(\gamma(\beta))}{MS(e)}$
Error	$abc(n-1)$	$SS(e)$	$MS(e)$	σ^2_e	

(b)

$$P(\hat{\sigma}^2_\beta < 0) = P[MS(\beta(\alpha)) < MS(\gamma(\beta))]$$
$$= P\left[F_{a(b-1),ab(c-1)} < \frac{n\sigma^2_\gamma + \sigma^2_e}{cn\sigma^2_\beta + n\sigma^2_\gamma + \sigma^2_e}\right].$$

(c)

$$\hat{\sigma}^2_\beta = \frac{MS(\beta(\alpha)) - MS(\gamma(\beta))}{cn}$$
$$Var(\hat{\sigma}^2_\beta) = \frac{1}{c^2n^2}\{Var[MS(\beta(\alpha))] + Var[MS(\gamma(\beta))]\}$$
$$= \frac{1}{c^2n^2}\left\{\left[\frac{cn\sigma^2_\beta + n\sigma^2_\gamma + \sigma^2_e}{a(b-1)}\right]^2 Var(\chi^2_{a(b-1)})\right.$$
$$\left. + \left[\frac{n\sigma^2_\gamma + \sigma^2_e}{ab(c-1)}\right]^2 Var(\chi^2_{ab(c-1)})\right\}$$

$$= \frac{2}{c^2n^2}\left\{\frac{(cn\sigma_\beta^2 + n\sigma_\gamma^2 + \sigma_e^2)^2}{a(b-1)} + \frac{(n\sigma_\gamma^2 + \sigma_e^2)^2}{ab(c-1)}\right\}.$$

An estimate of this variance can be obtained by replacing the variance components by their ANOVA estimates.

(d) $\bar{y}_{1\ldots} - \bar{y}_{2\ldots} \mp \left[\frac{2}{bcn}MS(\beta(\alpha))\right]^{1/2} t_{\frac{\alpha}{2},a(b-1)}$.

2.9. From (2.3.3), we get for any fixed j

$$\begin{aligned}\sum_{\psi_i \subset \psi_j} \mathbf{P}_i &= \sum_{\psi_i \subset \psi_j} \sum_{l=0}^{\nu+1} \frac{\mathbf{A}_l}{b_l} \lambda_{il} \\ &= \sum_{l=0}^{\nu+1} \frac{\mathbf{A}_l}{b_l} \sum_{\psi_i \subset \psi_j} \lambda_{il} \\ &= \frac{\mathbf{A}_j}{b_j},\end{aligned}$$

since $\sum_{\psi_i \subset \psi_j} \lambda_{il} = 1$ when $l = j$, and 0 otherwise.

2.10. Rewriting (2.3.7) as $\mathbf{A}_j = b_j(\sum_{\psi_l \subset \psi_j} \mathbf{P}_l)$ and using the orthogonal properties of the idempotent matrices $\mathbf{P}_i$'s, we get $\mathbf{A}_j\mathbf{P}_i = b_j(\sum_{\psi_l \subset \psi_j} \mathbf{P}_l\mathbf{P}_i) = \kappa_{ij}\mathbf{P}_i$, where $\kappa_{ij} = 0$ if $\psi_i \not\subset \psi_j$, and b_j if $\psi_i \subset \psi_j$.

2.11. Using the fact that $\mathbf{VP}_i = \delta_i\mathbf{P}_i$, and that $\text{rank}(\mathbf{P}_i) = m_i$, it follows that δ_i is an eigenvalue of $\mathbf{V}$ with multiplicity m_i. Since this holds for all i, the result follows.

Incidentally, for $\nu = 0$, another proof can be given as follows. Since $\mathbf{P}_0 + \mathbf{P}_1 = \mathbf{I}_N$, we can write $\mathbf{V} = \delta_1\mathbf{I}_N + (\delta_0 - \delta_1)\mathbf{P}_0$. Using the fact that there always exists an orthogonal matrix $\mathbf{Q}$ satisfying $\mathbf{QP}_0\mathbf{Q}' = \mathbf{D}$, where $\mathbf{D}$ is a diagonal matrix with m_0 diagonal elements as 1 and the rest as 0, it clearly follows that $\mathbf{QVQ}'$ is a diagonal matrix with m_0 elements equal to δ_0 and the rest equal to δ_1. Hence, $|\mathbf{V}| = \delta_0^{m_0}\delta_1^{m_1}$.

2.13. Clearly, $\mathbf{Ay}$ is a linear combination of $y_{..}$ and $(y_{11} - y_{12} - y_{21} - y_{22})$, and likewise, $\mathbf{By}$ is linear in the contrasts of $y_{i.}$'s and $y_{.j}$'s where $y_{i.} = \sum_j y_{ij}$ and $y_{.j} = \sum_i y_{ij}$, and y_{ij}'s follow the two-way crossed-classification without interaction random effects model given in (2.2.11). Noting that $\mathbf{y}_{i.} = (y_{i1}, \ldots, y_{ib})'$ is distributed as $N_b[\mu\mathbf{1}_b, (\sigma_\beta^2 + \sigma_e^2)\mathbf{I}_b + \sigma_\tau^2\mathbf{J}]$, it immediately follows that $\text{cov}[\boldsymbol{\delta}'\mathbf{y}_{i.}, \mathbf{1}'\mathbf{y}_{i.}] = 0$ whenever $\boldsymbol{\delta}'\mathbf{1} = 0$. Similarly, it also follows that $\text{cov}[\boldsymbol{\eta}'\mathbf{y}_{.j}, \mathbf{1}'\mathbf{y}_{.j}] = 0$ whenever $\boldsymbol{\eta}'\mathbf{1} = 0$ where $\mathbf{y}_{.j} = (y_{1j}, \ldots, y_{vj})'$ is distributed as $N_v[\mu\mathbf{1}_v, (\sigma_\tau^2 + \sigma_e^2)\mathbf{I}_v + \sigma_\beta^2\mathbf{J}]$. By a similar argument, any contrast among $y_{i.}$'s and also among $y_{.j}$'s is independent of $y_{..}$. Hence, the independence between $\mathbf{Ay}$ and $\mathbf{By}$ follows.

CHAPTER 3

3.1. (a) We have that $SS(\tau) = \mathbf{y}'\mathbf{Q}\mathbf{y}$, where $\mathbf{Q}$ is given by formula (3.2.2), $\mathbf{y} \sim N(\mu\mathbf{1}_{n.}, \mathbf{\Sigma})$, and

$$\mathbf{\Sigma} = \sigma_\tau^2(\oplus_{i=1}^{v}\mathbf{J}_{n_i}) + \sigma_e^2\mathbf{I}_{n.}.$$

By applying formula 40 in Searle (1971, Chapter 2), we obtain

$$\begin{aligned} E[SS(\tau)] &= \mu^2\mathbf{1}'_{n.}\mathbf{Q}\mathbf{1}_{n.} + tr(\mathbf{Q}\mathbf{\Sigma}) \\ &= tr\{\mathbf{Q}[\sigma_\tau^2(\oplus_{i=1}^{v}\mathbf{J}_{n_i}) + \sigma_e^2\mathbf{I}_{n.}]\}, \end{aligned}$$

since $\mathbf{Q}\mathbf{1}_{n.} = \mathbf{0}$. Hence,

$$\begin{aligned} E[SS(\tau)] &= \sigma_\tau^2 tr[\oplus_{i=1}^{v}\mathbf{J}_{n_i} - \frac{1}{n.}\mathbf{J}_{n.}(\oplus_{i=1}^{v}\mathbf{J}_{n_i})] + \sigma_e^2(v-1) \\ &= (n. - \frac{1}{n.}\sum_{i=1}^{v} n_i^2)\sigma_\tau^2 + (v-1)\sigma_e^2. \end{aligned}$$

(b) $SS(e) = \mathbf{y}'\left[\mathbf{I}_{n.} - \oplus_{i=1}^{v}\left(\frac{\mathbf{J}_{n_i}}{n_i}\right)\right]\mathbf{y}$. By theorem 4 in Searle (1971, Chapter 2), it is sufficient to show that

$$\mathbf{Q}\mathbf{\Sigma}\left[\mathbf{I}_{n.} - \oplus_{i=1}^{v}\left(\frac{\mathbf{J}_{n_i}}{n_i}\right)\right] = \mathbf{0}. \tag{A.3.1}$$

Now,

$$\mathbf{Q}\mathbf{\Sigma} = \sigma_\tau^2\mathbf{Q}(\oplus_{i=1}^{v}\mathbf{J}_{n_i}) + \sigma_e^2\mathbf{Q}, \tag{A.3.2}$$

and

$$(\oplus_{i=1}^{v}\mathbf{J}_{n_i})\left[\mathbf{I}_{n.} - \oplus_{i=1}^{v}\left(\frac{\mathbf{J}_{n_i}}{n_i}\right)\right] = \mathbf{0} \tag{A.3.3}$$

$$\begin{aligned} \mathbf{Q}\left[\mathbf{I}_{n.} - \oplus_{i=1}^{v}\left(\frac{\mathbf{J}_{n_i}}{n_i}\right)\right] &= \oplus_{i=1}^{v}\left(\frac{\mathbf{J}_{n_i}}{n_i}\right) \\ &\quad -\frac{\mathbf{J}_{n.}}{n.} - \oplus_{i=1}^{v}\left(\frac{\mathbf{J}_{n_i}}{n_i}\right) + \frac{\mathbf{J}_{n.}}{n.} = \mathbf{0}. \end{aligned} \tag{A.3.4}$$

From (A.3.2), (A.3.3), and (A.3.4.), we conclude (A.3.1).

(c) Using formula (3.2.5), we find that

$$\begin{aligned} Cov(\hat{\sigma}_\tau^2, \hat{\sigma}_e^2) &= \frac{1}{d}\, Cov[MS(\tau) - MS(e), MS(e)] \\ &= -\frac{1}{d}Var[MS(e)], \end{aligned}$$

since $MS(\tau)$ and $MS(e)$ are independent by part (b). But, $\frac{1}{\sigma_e^2}SS(e) \sim \chi^2_{n.-v}$, and

$$Var[MS(e)] = \frac{\sigma_e^4}{(n.-v)^2} Var(\chi^2_{n.-v}) = \frac{2\sigma_e^4}{n.-v}.$$

Thus,

$$Cov(\hat{\sigma}_\tau^2, \hat{\sigma}_e^2) = -\frac{2\sigma_e^4}{d(n.-v)}.$$

(d) Using formula (3.2.5), the moment generating function of $\hat{\sigma}_\tau^2$ is given by

$$\begin{aligned} E[\exp(t\hat{\sigma}_\tau^2)] &= E\left[\exp\left\{\frac{t}{d}[MS(\tau) - MS(e)]\right\}\right] \\ &= E\left[\exp\left\{\frac{t}{d}MS(\tau)\right\}\right] E\left[\exp\left\{-\frac{t}{d}MS(e)\right\}\right]. \end{aligned} \quad \text{(A.3.5)}$$

But, by formula (3.2.4),

$$\begin{aligned} E\left[\exp\left\{\frac{t}{d}MS(\tau)\right\}\right] &= E[\exp\{\frac{t}{d(v-1)}\sum_{i=1}^{s}\lambda_i\chi^2_{m_i}\}] \\ &= \Pi_{i=1}^{s} E[\exp\{\frac{t\lambda_i}{d(v-1)}\chi^2_{m_i}\}] \\ &= \Pi_{i=1}^{s}\left[1 - \frac{2t\lambda_i}{n. - \frac{1}{n.}\sum_{i=1}^{v} n_i^2}\right]^{-m_i/2}. \end{aligned} \quad \text{(A.3.6)}$$

Furthermore,

$$\begin{aligned} E\left[\exp\{-\frac{t}{d}MS(e)\}\right] &= E\left[\exp\{-\frac{t}{d}\,\frac{\sigma_e^2}{(n.-v)}\chi^2_{n.-v}\}\right] \\ &= \left[1 + \frac{2t\sigma_e^2}{d(n.-v)}\right]^{-(n.-v)/2} \\ &= \left[1 + \frac{2t(v-1)\sigma_e^2}{(n.-v)(n. - \frac{1}{n.}\sum_{i=1}^{v} n_i^2)}\right]^{-(n.-v)/2}. \end{aligned} \quad \text{(A.3.7)}$$

From (A.3.5), (A.3.6), and (A.3.7), we conclude that

$$E[\exp(t\hat{\sigma}_\tau^2)] =$$

$$\left[1+\frac{2t(v-1)\sigma_e^2}{(n.-v)(n.-\frac{1}{n.}\sum_{i=1}^v n_i^2)}\right]^{-(n.-v)/2} \Pi_{i=1}^s \left[1-\frac{2t\lambda_i}{n.-\frac{1}{n.}\sum_{i=1}^v n_i^2}\right]^{-m_i/2}.$$

3.2. **(a)** $E\left[\frac{SS(\tau)}{SS(e)}\right] = E\left[\frac{1}{SS(e)}\right]E\left[SS(\tau)\right]$

$$E\left[SS(\tau)\right] = (n. - \frac{1}{n.}\sum_{i=1}^{v} n_i^2)\sigma_\tau^2 + (v-1)\sigma_e^2. \qquad \text{(A.3.8)}$$

$$\begin{aligned} E\left[\frac{1}{SS(e)}\right] &= \frac{1}{\sigma_e^2}E\left[\frac{1}{\chi^2_{n.-v}}\right] \\ &= \frac{1}{\sigma_e^2}\int_0^\infty \frac{1}{2^{\frac{n.-v}{2}}\Gamma\left(\frac{n.-v}{2}\right)} e^{-\frac{x}{2}} x^{\frac{n.-v}{2}-2} dx \\ &= \frac{1}{\sigma_e^2(n.-v-2)}, \quad \text{if } n.-v-2>0. \qquad \text{(A.3.9)} \end{aligned}$$

Using (A.3.8) and (A.3.9), we get

$$E\left[\frac{SS(\tau)}{SS(e)}\right] = \frac{(n.-\frac{1}{n.}\sum_{i=1}^v n_i^2)\sigma_\tau^2 + (v-1)\sigma_e^2}{(n.-v-2)\sigma_e^2}.$$

(b) $\frac{1}{d}E\left[\frac{n.-v-2}{n.-v}\ \frac{MS(\tau)}{MS(e)} - 1\right] = \frac{1}{d}\left[\frac{(n.-v-2)(n.-v)}{(n.-v)(v-1)}E\left[\frac{SS(\tau)}{SS(e)}\right] - 1\right] = \frac{\sigma_\tau^2}{\sigma_e^2}.$
This follows from using part (a) and the fact that $d = \frac{1}{v-1}(n. - \frac{1}{n.}\sum_{i=1}^v n_i^2)$.

3.3. **(a)** In this case, the variance–covariance matrix of $\mathbf{y}$ is of the form

$$\boldsymbol{\Sigma} = \sigma_\tau^2(\oplus_{i=1}^v \mathbf{J}_{n_i}) + \oplus_{i=1}^v(\sigma_i^2 \mathbf{I}_{n_i}).$$

It follows that

$$SS(\tau) = \sum_{i=1}^{s}\lambda_i\chi^2_{m_i}, \qquad \text{(A.3.10)}$$

where $\lambda_1, \lambda_2, \ldots, \lambda_s$ are the distinct nonzero eigenvalues of $\mathbf{Q\Sigma}$ with multiplicities $m_1, m_2, \ldots, m_s$, respectively. The matrix $\mathbf{Q}$ is given by formula (3.2.2) and the χ^2's in (A.3.10) are independent. Note that

$$\mathbf{Q\Sigma} = \left[\oplus_{i=1}^v\left(\frac{\mathbf{J}_{n_i}}{n_i}\right) - \frac{\mathbf{J}_{n.}}{n.}\right]\left[\oplus_{i=1}^v(\sigma_\tau^2\mathbf{J}_{n_i} + \sigma_i^2\mathbf{I}_{n_i})\right].$$

Similarly,

$$\begin{aligned} SS(e) &= \mathbf{y}'\left[\mathbf{I}_{n.} - \oplus_{i=1}^{v}\left(\frac{\mathbf{J}_{n_i}}{n_i}\right)\right]\mathbf{y} \\ &= \sum_{i=1}^{v}\sigma_i^2\chi_{n_i-1}^2, \end{aligned}$$

since

$$\begin{aligned} &\left[\mathbf{I}_{n.} - \oplus_{i=1}^{v}\left(\frac{\mathbf{J}_{n_i}}{n_i}\right)\right]\left[\oplus_{i=1}^{v}\left(\sigma_\tau^2\mathbf{J}_{n_i} + \sigma_i^2\mathbf{I}_{n_i}\right)\right] \\ &\qquad = \oplus_{i=1}^{v}\sigma_i^2(\mathbf{I}_{n_i} - \frac{1}{n_i}\mathbf{J}_{n_i}), \end{aligned}$$

and the nonzero eigenvalues of the matrix on the right-hand side of this equation are $\sigma_1^2, \sigma_2^2, \ldots, \sigma_v^2$ with multiplicities $n_1 - 1, n_2 - 1, \ldots, n_v - 1$, respectively.

(b)

$$\begin{aligned} &\left[\mathbf{I}_{n.} - \oplus_{i=1}^{v}\left(\frac{\mathbf{J}_{n_i}}{n_i}\right)\right]\boldsymbol{\Sigma}\left[\oplus_{i=1}^{v}\left(\frac{\mathbf{J}_{n_i}}{n_i}\right) - \frac{\mathbf{J}_{n.}}{n.}\right] \\ &= \left[\oplus_{i=1}^{v}\sigma_i^2\left(\mathbf{I}_{n_i} - \frac{1}{n_i}\mathbf{J}_{n_i}\right)\right]\left[\oplus_{i=1}^{v}\left(\frac{\mathbf{J}_{n_i}}{n_i}\right) - \frac{\mathbf{J}_{n.}}{n.}\right] = \mathbf{0}. \end{aligned}$$

Hence, $SS(\tau)$ and $SS(e)$ are still independent under assumptions (i)–(iii).

(c)

$$\begin{aligned} E\left(\hat{\sigma}_\tau^2\right) &= \frac{1}{d}\left[E\left[MS(\tau)\right] - E\left[MS(e)\right]\right] \\ &= \frac{1}{d}\left[\frac{1}{v-1}\,tr\left\{\left[\oplus_{i=1}^{v}\left(\frac{\mathbf{J}_{n_i}}{n_i}\right) - \frac{\mathbf{J}_{n.}}{n.}\right]\left[\oplus_{i=1}^{v}\left(\sigma_\tau^2\mathbf{J}_{n_i} + \sigma_i^2\mathbf{I}_{n_i}\right)\right]\right\}\right. \\ &\quad \left. - \frac{1}{n.-v}tr\left\{\left[\mathbf{I}_{n.} - \oplus_{i=1}^{v}\left(\frac{\mathbf{J}_{n_i}}{n_i}\right)\right]\left[\oplus_{i=1}^{v}(\sigma_\tau^2\mathbf{J}_{n_i} + \sigma_i^2\mathbf{I}_{n_i})\right]\right\}\right] \\ &= \frac{1}{d}\left[\frac{1}{v-1}\left(\sigma_\tau^2 n. + \sum_{i=1}^{v}\sigma_i^2 - \frac{\sigma_\tau^2}{n.}\sum_{i=1}^{v}n_i^2\right.\right. \\ &\quad \left.\left. - \frac{1}{n.}\sum_{i=1}^{v}n_i\sigma_i^2\right) - \frac{1}{n.-v}\left(\sum_{i=1}^{v}n_i\sigma_i^2 - \sum_{i=1}^{v}\sigma_i^2\right)\right] \\ &= \sigma_\tau^2 + \frac{1}{d}\left[\frac{1}{v-1}\sum_{i=1}^{v}\left(1 - \frac{n_i}{n.}\right)\sigma_i^2 - \frac{1}{n.-v}\sum_{i=1}^{v}(n_i - 1)\sigma_i^2\right]. \end{aligned}$$

Thus, $\hat{\sigma}_\tau^2$ is a biased estimator of σ_τ^2.

3.4. (a) $U_i = c_1(\mu + \tau_i + \bar{e}_{i.}) + c_{2i}\sum_{j=1}^{n_i}\ell_{ij}e_{ij}$. Hence, $E(U_i) = c_1\mu, \quad i = 1, 2, 3, 4.$

(b)
$$
\begin{aligned}
Var(U_i) &= c_1^2\sigma_\tau^2 + c_1^2\frac{\sigma_e^2}{n_i} + c_{2i}^2\sum_{j=1}^{n_i}\ell_{ij}^2\sigma_e^2 \\
&= c_1^2\sigma_\tau^2 + \left[\frac{c_1^2}{n_i} + c_{2i}^2\sum_{j=1}^{n_i}\ell_{ij}^2\right]\sigma_e^2, \quad i = 1,2,3,4.
\end{aligned}
$$

(c) It is clear that U_i is normally distributed. Furthermore, for $i \neq k$,

$$Cov(U_i, U_k) =$$

$$Cov\left[c_1(\tau_i + \bar{e}_{i\cdot}) + c_{2i}\sum_{j=1}^{n_i}\ell_{ij}e_{ij}, c_1(\tau_k + \bar{e}_{k\cdot}) + c_{2k}\sum_{j=1}^{n_k}\ell_{kj}e_{kj}\right] = 0.$$

The U_i's are independent since they are uncorrelated and normally distributed.

(d) By comparing Var(U_i) in part (b) to $\sigma_\tau^2 + 2\sigma_e^2$, we get

$$c_1^2 = 1 \text{ and } \frac{c_1^2}{n_i} + c_{2i}^2\sum_{j=1}^{n_i}\ell_{ij}^2 = 2.$$

For $i = 1$, choose $\ell_{11} = \ell_{12} = 1, \ell_{13} = 0, \ell_{14} = \ell_{15} = -1 \Rightarrow c_{21}^2 = \frac{9}{20}$.
For $i = 2$, choose $\ell_{21} = \ell_{22} = 1, \ell_{23} = \ell_{24} = -1 \Rightarrow c_{22}^2 = \frac{7}{16}$.
For $i = 3$, choose $\ell_{31} = \ell_{32} = 1, \ell_{33} = \ell_{34} = -1 \Rightarrow c_{23}^2 = \frac{7}{16}$.
Finally, for $i = 4$, choose $\ell_{41} = \ell_{42} = \ell_{43} = 1$, $\ell_{44} = \ell_{45} = \ell_{46} = -1 \Rightarrow c_{24}^2 = \frac{11}{36}$.

(e) On the basis of parts (a)–(d), the U_i's are i.i.d $N(c_1\mu, \sigma_\tau^2 + 2\sigma_e^2)$. Hence, an unbiased estimator of $\sigma_\tau^2 + 2\sigma_e^2$ is given by $\frac{1}{3}\sum_{i=1}^4(U_i - \overline{U})^2$, where $\overline{U} = \frac{1}{4}\sum_{i=1}^4 U_i$.

(f)
$$\frac{1}{\sigma_\tau^2 + 2\sigma_e^2}\sum_{i=1}^4(U_i - \overline{U})^2 \sim \chi_3^2.$$

Therefore, an exact $(1-\alpha)100\%$ confidence interval for $\sigma_\tau^2 + 2\sigma_e^2$ is given by

$$\left[\frac{\sum_{i=1}^4(U_i - \overline{U})^2}{\chi^2_{\frac{\alpha}{2},3}}, \frac{\sum_{i=1}^4(U_i - \overline{U})^2}{\chi^2_{1-\frac{\alpha}{2},3}}\right].$$

3.6. **(a)** Δ can also be written as

$$
\begin{aligned}
\Delta &= \frac{\text{rank}(\mathbf{Q\Sigma})tr[(\mathbf{Q\Sigma})^2]}{[tr(\mathbf{Q\Sigma})]^2} \\
&= \frac{(v-1)tr[(\mathbf{Q\Sigma})^2]}{[tr(\mathbf{Q\Sigma})]^2}. \qquad \text{(A.3.11)}
\end{aligned}
$$

The matrix $\mathbf{Q\Sigma}$ is given by formula (3.2.3).

(i) For $\frac{\sigma_\tau^2}{\sigma_e^2} = 3.0, v = 6$, and $n_1 = 5$, $n_2 = 3$, $n_3 = 6$, $n_4 = 5$, $n_5 = 3$, $n_6 = 8$, and by substitution in (A.3.11), we find that $\Delta - 1 = 0.0845$.

(ii) Similarly, for $\frac{\sigma_\tau^2}{\sigma_e^2} = 3.0, v = 6$, and $n_1 = 2$, $n_2 = 3$, $n_3 = 2$, $n_4 = 3$, $n_5 = 3$, and $n_6 = 17, \Delta - 1 = 0.3779$.

We note that the degree of imbalance for the design in (ii) is higher than that in (i). This is reflected in the larger value of $\Delta - 1$ in (ii).

(b) $SS^*(\tau) = \mathbf{y}'\mathbf{Q}^*\mathbf{y}$, where

$$\mathbf{Q}^* = \tilde{n}\ \mathbf{diag}\left(\frac{\mathbf{1}_{n_1}}{n_1}, \frac{\mathbf{1}_{n_2}}{n_2}, \ldots, \frac{\mathbf{1}_{n_v}}{n_v}\right)\left[\mathbf{I}_v - \frac{1}{v}\mathbf{J}_v\right] \times \mathbf{diag}\left(\frac{\mathbf{1}'_{n_1}}{n_1}, \frac{\mathbf{1}'_{n_2}}{n_2}, \ldots, \frac{\mathbf{1}'_{n_v}}{n_v}\right).$$

Redoing (i) and (ii) in part (a), using $\mathbf{Q}^*$ in place of $\mathbf{Q}$, we obtain

(i) $\Delta - 1 = 0.00049$.

(ii) $\Delta - 1 = 0.00155$.

We note that these values are considerably smaller than the corresponding values in part (a). This indicates that the accuracy of Satterthwaite's approximation in the case of $SS^*(\tau)$ is higher than in the case of $SS(\tau)$. This agrees with the conclusion in Thomas and Hultquist (1978) that the chi-squared distribution is an excellent approximation to the true distribution of $SS^*(\tau)$. See also Burdick and Graybill (1992, p. 70).

3.7. (a) (i) $n_1 = 19, n_2 = 8, n_3 = 11, n_4 = 8, n_5 = 4$. The exact ϕ value for this design is 0.7987.

(ii) $n_1 = 15, n_2 = 5, n_3 = 14, n_4 = 3, n_5 = 13$. The exact ϕ value is 0.8013.

(b) $\mathcal{F} = 0.15958$, for the design in (i), and $\mathcal{F} = 0.166963$, for the design in (ii). The design in (i) is slightly better since it has the smaller $\mathcal{F}$ value.

CHAPTER 4

4.1. We shall first prove Lemma 4.3.1 when all the n_{ij}'s are nonzero. The lemma is actually a simple consequence of the expressions for $\boldsymbol{X}_1$, $\boldsymbol{X}_2$, and $\boldsymbol{X}_3$ given in (4.3.3).

(a) When $n_{ij} \neq 0$ for all i and j, we need to show that $\text{rank}(\boldsymbol{X}_3) = vb$. Note that $\boldsymbol{X}_3$ is a block diagonal matrix with $\mathbf{1}_{n_{ij}}$'s as the diagonal blocks. Hence the columns of $\boldsymbol{X}_3$ are linearly independent. Consequently, $\text{rank}(\boldsymbol{X}_3) = bv$.

(b) Part (b) follows from part (c).

(c) Suppose the n_{ij}'s are all nonzero. It is easy to verify that $\boldsymbol{X}_1 = \boldsymbol{X}_3(\boldsymbol{I}_v \otimes \mathbf{1}_b)$ and $\boldsymbol{X}_2 = \boldsymbol{X}_3(\mathbf{1}_v \otimes \boldsymbol{I}_b)$. Hence $\mathcal{C}(\boldsymbol{X}_i) \in \mathcal{C}(\boldsymbol{X}_3)$, $i = 1, 2$.

(d) If we add all the columns in $\boldsymbol{X}_1$, $\boldsymbol{X}_2$ or $\boldsymbol{X}_3$, we get the vector $\mathbf{1}_{n_{..}}$. Hence $\mathbf{1}_{n_{..}} \in \mathcal{C}(\boldsymbol{X}_i)$, $i = 1, 2, 3$.

When some n_{ij}'s are zeros, the expression for $\boldsymbol{X}_3$ will need slight modification; some columns of $\boldsymbol{X}_3$ will be all zeros. However, the above proof will go through with obvious changes.

4.2. The model is given by

$$y_{ijkl} = \mu + \tau_i + \beta_j + \gamma_k + e_{ijkl}, \qquad \text{(A.4.1)}$$

$i = 1, 2, \ldots, v$; $j = 1, 2, \ldots, b$; $k = 1, 2, \ldots, c$; $l = 1, 2, \ldots, n_{ijk}$. Write $\mathbf{y}_{ijk} = (y_{ijk1}, y_{ijk2}, y_{ijkn_{ijk}})'$ and $\mathbf{y} = (\mathbf{y}'_{111}, \mathbf{y}'_{112}, \ldots, \mathbf{y}'_{11c}, \mathbf{y}'_{121}, \mathbf{y}'_{122}, \mathbf{y}'_{12c}, \ldots, \mathbf{y}'_{vb1}, \mathbf{y}'_{vb2}, \ldots, \mathbf{y}'_{vbc})'$. The model (A.4.1) can be written as

$$\mathbf{y} = \mu \mathbf{1}_{n_{...}} + \boldsymbol{X}_1 \boldsymbol{\tau} + \boldsymbol{X}_2 \boldsymbol{\beta} + \boldsymbol{X}_3 \boldsymbol{\gamma} + \mathbf{e}, \qquad \text{(A.4.2)}$$

where $\mathbf{e}$ is defined similar to $\mathbf{y}$, $n_{...} = \sum_{i=1}^{v} \sum_{j=1}^{b} \sum_{k=1}^{c} n_{ijk}$, $\boldsymbol{\tau}$ is the vector of τ_i's etc., and $\boldsymbol{X}_1$, $\boldsymbol{X}_2$, and $\boldsymbol{X}_3$ are appropriate design matrices. We assume $\boldsymbol{\tau} \sim N(0, \sigma_\tau^2 \boldsymbol{I}_v)$, $\boldsymbol{\beta} \sim N(0, \sigma_\beta^2 \boldsymbol{I}_b)$, $\boldsymbol{\gamma} \sim N(0, \sigma_\gamma^2 \boldsymbol{I}_c)$, and $\mathbf{e} \sim N(0, \sigma_e^2 \boldsymbol{I}_{n_{...}})$, and all the random variables are independent. Consider the problem of testing H_τ: $\sigma_\tau^2 = 0$. Let $\text{rank}(\boldsymbol{X}_2 : \boldsymbol{X}_3) = r_{23}$ and let $\boldsymbol{Z}_{23}$ be an $n_{...} \times (n_{...} - r_{23})$ matrix satisfying $\boldsymbol{Z}'_{23}(\boldsymbol{X}_2 : \boldsymbol{X}_3) = 0$ and $\boldsymbol{Z}'_{23}\boldsymbol{Z}_{23} = \boldsymbol{I}_{n_{...} - r_{23}}$. Define $\mathbf{u}_{23} = \boldsymbol{Z}'_{23}\mathbf{y}$. We then have the model

$$\mathbf{u}_{23} = \boldsymbol{Z}'_{23}\boldsymbol{X}_1\boldsymbol{\tau} + \boldsymbol{Z}'_{23}\mathbf{e}. \qquad \text{(A.4.3)}$$

In order to arrive at (A.4.3), we have also used the fact that the vector $\mathbf{1}_{n_{...}}$ belongs to the vector space generated by the columns of $\boldsymbol{X}_2$ (or $\boldsymbol{X}_3$) and hence $\boldsymbol{Z}'_{23}\mathbf{1}_{n_{...}} = 0$. (The proof of this is similar to the proof of Lemma 4.3.1(d), given above.) Note that (A.4.3) is similar to the model (4.3.6). Using the arguments following (4.3.6), we conclude that

$$SS(\tau|\beta, \gamma) = \mathbf{u}_{23'}\boldsymbol{Z}'_{23}\boldsymbol{X}_1(\boldsymbol{X}'_1\boldsymbol{Z}_{23}\boldsymbol{Z}'_{23}\boldsymbol{X}_1)^{-}\boldsymbol{X}'_1\boldsymbol{Z}_{23}\mathbf{u}_{23}. \qquad \text{(A.4.4)}$$

Let m_v denote the degrees of freedom associated with $SS(\tau|\beta, \gamma)$ and m_e denote the degrees of freedom associated with $SS(e)$. Then

$$\begin{aligned} m_v &= \text{rank}(\boldsymbol{X}'_1\boldsymbol{Z}_{23}) = \text{rank}[\boldsymbol{X}'_1(\boldsymbol{I} - \boldsymbol{P}_{(X_2, X_3)}\boldsymbol{X}_1], \\ m_e &= n_{...} - \text{rank}(\boldsymbol{X}_1 : \boldsymbol{X}_2 : \boldsymbol{X}_3). \end{aligned}$$

Following the arguments that lead to (4.3.10), H_τ: $\sigma_\tau^2 = 0$ can be tested using the F-statistic

$$F_\tau = \frac{SS(\tau|\beta,\gamma)/m_v}{SS(e)/m_e}. \tag{A.4.5}$$

We note that the above F-ratio coincides with the F-ratio that can be obtained based on the ANOVA decomposition in the fixed effects case. Following the arguments in Section 1.2 of Chapter 1, we conclude that the test based on F_τ is also the Wald's variance component test. Similar tests can be obtained for testing H_β: $\sigma_\beta^2 = 0$ and H_γ: $\sigma_\gamma^2 = 0$.

4.3. Referring to the model (4.3.5) and using the definition of $\boldsymbol{X}_1$ and $\boldsymbol{X}_2$ in (4.3.3), we conclude that

$$\begin{aligned} \boldsymbol{X}_1'\boldsymbol{X}_1 &= \mathbf{diag}(n_{1.}, n_{2.}, \ldots, n_{v.}) \\ \boldsymbol{X}_2'\boldsymbol{X}_2 &= \mathbf{diag}(n_{.1}, n_{.2}, \ldots, n_{.b}). \end{aligned} \tag{A.4.6}$$

Also define

$$\mathbf{T} = \boldsymbol{X}_1'\mathbf{y}, \; \mathbf{B} = \boldsymbol{X}_2'\mathbf{y}, \; \text{ and } \; \boldsymbol{N} = \boldsymbol{X}_1'\boldsymbol{X}_2. \tag{A.4.7}$$

Note that the ith component of $\mathbf{T}$ is $\sum_{j=1}^{b}\sum_{k=1}^{n_{ij}} y_{ijk}$ and the jth component of $\mathbf{B}$ is $\sum_{i=1}^{v}\sum_{k=1}^{n_{ij}} y_{ijk}$. From the definition of $\boldsymbol{Z}_2$ (used to arrive at model (4.3.6)), it follows that $\boldsymbol{Z}_2\boldsymbol{Z}_2' = \boldsymbol{I} - \boldsymbol{X}_2(\boldsymbol{X}_2'\boldsymbol{X}_2)^{-1}\boldsymbol{X}_2'$. In order to simplify the expression for $SS(\tau|\beta)$ in (4.3.7), note that

$$\begin{aligned} \boldsymbol{X}_1'\boldsymbol{Z}_2\mathbf{u}_2 &= \boldsymbol{X}_1'\boldsymbol{Z}_2\boldsymbol{Z}_2'\mathbf{y} = \boldsymbol{X}_1'\left[\boldsymbol{I} - \boldsymbol{X}_2(\boldsymbol{X}_2'\boldsymbol{X}_2)^{-1}\boldsymbol{X}_2'\right]\mathbf{y} \\ &= \mathbf{T} - \boldsymbol{N}\mathbf{diag}(1/n_{.1}, 1/n_{.2}, \ldots, 1/n_{.b})\mathbf{B} = \mathbf{q} \text{ (say)} \\ \boldsymbol{X}_1'\boldsymbol{Z}_2\boldsymbol{Z}_2'\boldsymbol{X}_1 &= \boldsymbol{X}_1'\left[\boldsymbol{I} - \boldsymbol{X}_2(\boldsymbol{X}_2'\boldsymbol{X}_2)^{-1}\boldsymbol{X}_2'\right]\boldsymbol{X}_1 \\ &= \mathbf{diag}(n_{1.}, n_{2.}, \ldots, n_{v.}) \\ &\quad - \boldsymbol{N}\mathbf{diag}(1/n_{.1}, 1/n_{.2}, \ldots, 1/n_{.b})\boldsymbol{N}' = \boldsymbol{C} \text{ (say)}. \end{aligned} \tag{A.4.8}$$

Hence, from (4.3.7) and (A.4.8),

$$SS(\tau|\beta) = \mathbf{u}_2'\boldsymbol{Z}_2'\boldsymbol{X}_1(\boldsymbol{X}_1'\boldsymbol{Z}_2\boldsymbol{Z}_2'\boldsymbol{X}_1)^{-}\boldsymbol{X}_1'\boldsymbol{Z}_2\mathbf{u}_2 = \mathbf{q}'\boldsymbol{C}^{-}\mathbf{q},$$

which is the sum of squares due to $\boldsymbol{\tau}$, adjusted for $\boldsymbol{\beta}$, computed using the model (4.3.4). Some details of this also appear in Section 7.2 of Chapter 7, for a special case of model (4.3.4); see (7.2.4)–(7.2.9). This proves (a). **(b)** is proved similarly.

4.4. Consider the balanced case (i.e., the n_{ij}'s are all equal). Let n denote the common value of the n_{ij}'s. As already noted in Remark 4.3.4, in the balanced case, $\boldsymbol{D}_0 = \frac{1}{n}\boldsymbol{I}_{bv-1}$, and hence, $\mathbf{w}_\tau = \mathbf{z}_\tau$, $\mathbf{w}_\beta = \mathbf{z}_\beta$, and $\mathbf{w}_{\tau\beta} = \mathbf{z}_{\tau\beta}$.

Thus we need to show that $n\mathbf{z}_\tau'\mathbf{z}_\tau$, $n\mathbf{z}_\beta'\mathbf{z}_\beta$, and $n\mathbf{z}_{\tau\beta}'\mathbf{z}_{\tau\beta}$ are the balanced ANOVA sum of squares for $\boldsymbol{\tau}$, $\boldsymbol{\beta}$, and $(\boldsymbol{\tau\beta})$, respectively. We shall prove this for $n\mathbf{z}_\tau'\mathbf{z}_\tau$; the rest of them will follow similarly. Note that $\bar{y}_{ij.} = \frac{1}{n}y_{ij.}$, where $y_{ij.} = \sum_{k=1}^{n} y_{ijk}$. We shall first give an explicit representation for the matrix $\boldsymbol{P}$ satisfying (4.3.17). Toward this, let $\boldsymbol{Z}_\tau$ and $\boldsymbol{Z}_\beta$ be, respectively, $v \times (v-1)$ and $b \times (b-1)$ matrices satisfying $\boldsymbol{Z}_\tau'\mathbf{1}_v = 0$, $\boldsymbol{Z}_\tau'\boldsymbol{Z}_\tau = \boldsymbol{I}_{v-1}$, $\boldsymbol{Z}_\beta'\mathbf{1}_b = 0$, and $\boldsymbol{Z}_\beta'\boldsymbol{Z}_\beta = \boldsymbol{I}_{b-1}$. Note that $\boldsymbol{Z}_\tau\boldsymbol{Z}_\tau' = \boldsymbol{I}_v - \frac{1}{v}\mathbf{1}_v\mathbf{1}_v'$ and $\boldsymbol{Z}_\beta\boldsymbol{Z}_\beta' = \boldsymbol{I}_b - \frac{1}{b}\mathbf{1}_b\mathbf{1}_b'$. Then

$$\boldsymbol{P}' = \left(\frac{1}{\sqrt{bv}}\mathbf{1}_{bv} : \boldsymbol{Z}_\tau \otimes \frac{1}{\sqrt{b}}\mathbf{1}_b : \frac{1}{\sqrt{v}}\mathbf{1}_v \otimes \boldsymbol{Z}_\beta : \boldsymbol{Z}_\tau \otimes \boldsymbol{Z}_\beta\right). \qquad \text{(A.4.9)}$$

It is readily verified that $\boldsymbol{P}'$ is a $bv \times bv$ orthogonal matrix and it satisfies (4.3.17). In particular, the vector $\mathbf{z}_\tau$ defined following (4.3.17) is given by

$$\mathbf{z}_\tau = \left(\boldsymbol{Z}_\tau' \otimes \frac{1}{\sqrt{b}}\mathbf{1}_b'\right)\bar{\mathbf{y}}.$$

Hence

$$\begin{aligned}
\mathbf{z}_\tau'\mathbf{z}_\tau &= \bar{\mathbf{y}}'(\boldsymbol{Z}_\tau\boldsymbol{Z}_\tau' \otimes \frac{1}{b}\mathbf{1}_b\mathbf{1}_b')\bar{\mathbf{y}} = \bar{\mathbf{y}}'\left[(\boldsymbol{I}_v - \frac{1}{v}\mathbf{1}_v\mathbf{1}_v') \otimes \frac{1}{b}\mathbf{1}_b\mathbf{1}_b'\right]\bar{\mathbf{y}} \\
&= \bar{\mathbf{y}}'(\boldsymbol{I}_v \otimes \frac{1}{b}\mathbf{1}_b\mathbf{1}_b')\bar{\mathbf{y}} - \frac{1}{bv}\bar{\mathbf{y}}'\mathbf{1}_{bv}\mathbf{1}_{bv}'\bar{\mathbf{y}} \\
&= \frac{1}{b}\sum_{i=1}^{v}(\sum_{j=1}^{b}\bar{y}_{ij.})^2 - \frac{1}{bv}(\sum_{i=1}^{v}\sum_{j=1}^{b}\bar{y}_{ij.})^2 \\
&= \frac{1}{n^2 b}\sum_{i=1}^{v} y_{i..}^2 - \frac{1}{bvn^2}y_{...}^2,
\end{aligned}$$

where the last expression follows since $\bar{y}_{ij.} = \frac{1}{n}y_{ij.}$. From the above, it is clear that $n\mathbf{z}_\tau'\mathbf{z}_\tau$ is the usual balanced ANOVA sum of squares due to $\boldsymbol{\tau}$. Since

$$\mathbf{z}_\beta'\mathbf{z}_\beta = \bar{\mathbf{y}}'\left[\frac{1}{v}\mathbf{1}_v\mathbf{1}_v' \otimes (\boldsymbol{I}_b - \frac{1}{b}\mathbf{1}_b\mathbf{1}_b')\right]\bar{\mathbf{y}},$$

$$\text{and } \mathbf{z}_{\tau\beta}'\mathbf{z}_{\tau\beta} = \bar{\mathbf{y}}'\left[(\boldsymbol{I}_v - \frac{1}{v}\mathbf{1}_v\mathbf{1}_v') \otimes (\boldsymbol{I}_b - \frac{1}{b}\mathbf{1}_b\mathbf{1}_b')\right]\bar{\mathbf{y}},$$

it can similarly be proved that $n\mathbf{z}_\beta'\mathbf{z}_\beta$ and $n\mathbf{z}_{\tau\beta}'\mathbf{z}_{\tau\beta}$ are the balanced ANOVA sum of squares due to $\boldsymbol{\beta}$ and $(\boldsymbol{\tau\beta})$, respectively.

4.6. Consider the balanced case of the model (4.4.1) and let n denote the common value of the n_{ij}'s and let b_0 denote the common value of the

b_i's. Referring to the notations in Section 4.4.2, we get the following simplified expressions in the balanced case.

$$\boldsymbol{O}_1 = \frac{1}{\sqrt{b_0}}(\boldsymbol{I}_v \otimes \mathbf{1}'_{b_0}), \text{ and } \boldsymbol{O}_2 = \boldsymbol{I}_v \otimes \boldsymbol{Z}'_0, \tag{A.4.10}$$

where $\boldsymbol{Z}_0$ is a $b_0 \times (b_0 - 1)$ matrix satisfying $\boldsymbol{Z}'_0\boldsymbol{Z}_0 = \boldsymbol{I}_{b_0-1}$ and $\boldsymbol{Z}'_0\mathbf{1}_{b_0} = 0$. It is readily verified that $\boldsymbol{O}' = (\boldsymbol{O}'_1 : \boldsymbol{O}'_2)'$ is a $vb_0 \times vb_0$ orthogonal matrix. Also, in the balanced case, the $vb_0 \times vb_0$ orthogonal matrix $\boldsymbol{Q}$ (defined following equation (4.4.11)) has the expression

$$\boldsymbol{Q} = \begin{pmatrix} \frac{1}{\sqrt{v}}\mathbf{1}'_v : 0 \\ \boldsymbol{Q}'_1 \end{pmatrix}, \tag{A.4.11}$$

where $\boldsymbol{Q}_1$, which is $vb_0 \times (vb_0 - 1)$, is further partitioned as

$$\boldsymbol{Q}'_1 = (\boldsymbol{Q}'_{11} : \boldsymbol{Q}'_{12}). \tag{A.4.12}$$

In (A.4.12), $\boldsymbol{Q}_{11}$ is $v \times (vb_0 - 1)$ and $\boldsymbol{Q}_{12}$ is $v(b_0 - 1) \times (vb_0 - 1)$. Due to the orthogonality of $\boldsymbol{Q}$ in (A.4.11), we get $\boldsymbol{Q}'_1\boldsymbol{Q}_1 = \boldsymbol{I}_{vb_0-1}$ and

$$\boldsymbol{Q}_1\boldsymbol{Q}'_1 = \boldsymbol{I}_{vb_0} - \begin{pmatrix} \frac{1}{v}\mathbf{1}_v\mathbf{1}'_v & 0 \\ 0 & 0 \end{pmatrix}. \tag{A.4.13}$$

From (A.4.12) and (A.4.13),

$$\boldsymbol{Q}_{11}\boldsymbol{Q}'_{11} = \boldsymbol{I}_v - \frac{1}{v}\mathbf{1}_v\mathbf{1}'_v, \text{ and } \boldsymbol{Q}_{11}\boldsymbol{Q}'_{12} = 0. \tag{A.4.14}$$

Furthermore,

$$\boldsymbol{Q}'_1\mathbf{diag}(b_1, b_2, \ldots, b_v, 0)\boldsymbol{Q}_1 = b_0\boldsymbol{Q}'_{11}\boldsymbol{Q}_{11}. \tag{A.4.15}$$

The matrix $\boldsymbol{S}$ is a $(vb_0 - 1) \times (vb_0 - 1)$ orthogonal matrix that diagonalizes $b_0\boldsymbol{Q}'_{11}\boldsymbol{Q}_{11}$. Since $\boldsymbol{Q}_{11}\boldsymbol{Q}'_{11} = \boldsymbol{I}_v - (1/v)\mathbf{1}_v\mathbf{1}'_v$, it follows that $\boldsymbol{Q}'_{11}\boldsymbol{Q}_{11}$ has $(v - 1)$ nonzero eigenvalues equal to one. Hence $\boldsymbol{S}$ can be chosen so that

$$\boldsymbol{Q}'_{11}\boldsymbol{Q}_{11} = \boldsymbol{S}\mathbf{diag}(\boldsymbol{I}_{v-1}, 0)\boldsymbol{S}'. \tag{A.4.16}$$

Write $\boldsymbol{S} = (\boldsymbol{S}_1 : \boldsymbol{S}_2)$, where $\boldsymbol{S}_1$ is $(vb_0 - 1) \times (v - 1)$. From (A.4.16),

$$\boldsymbol{Q}'_{11}\boldsymbol{Q}_{11} = \boldsymbol{S}_1\boldsymbol{S}'_1. \tag{A.4.17}$$

Since $\boldsymbol{E}_0 = (1/n)\boldsymbol{I}_b$, we have $\mathbf{v} = \boldsymbol{S}'\mathbf{u}$; see (4.4.14). Hence $\mathbf{v}_\tau = \boldsymbol{S}'_1\mathbf{u}$. Also recall that $\mathbf{u} = \boldsymbol{Q}'_1\mathbf{z} = \boldsymbol{Q}'_1\boldsymbol{O}\bar{\mathbf{y}}$. Hence

$$\mathbf{v}'_\tau\mathbf{v}_\tau = \mathbf{u}'\boldsymbol{S}_1\boldsymbol{S}'_1\mathbf{u} = \mathbf{u}'\boldsymbol{Q}'_{11}\boldsymbol{Q}_{11}\mathbf{u} \text{ (from (A.4.17))}$$

$$
\begin{aligned}
&= \bar{\mathbf{y}}'\boldsymbol{O}'\boldsymbol{Q}_1\boldsymbol{Q}'_{11}\boldsymbol{Q}_{11}\boldsymbol{Q}'_1\boldsymbol{O}\bar{\mathbf{y}} \\
&= \bar{\mathbf{y}}'\boldsymbol{O}'\begin{pmatrix} \boldsymbol{I}_v - \frac{1}{v}\mathbf{1}_v\mathbf{1}'_v & 0 \\ 0 & 0 \end{pmatrix}\boldsymbol{O}\bar{\mathbf{y}} \text{ (using (A.4.14))} \\
&= \bar{\mathbf{y}}'\boldsymbol{O}'_1(\boldsymbol{I}_v - \frac{1}{v}\mathbf{1}_v\mathbf{1}'_v)\boldsymbol{O}_1\bar{\mathbf{y}} \\
&= \bar{\mathbf{y}}'\left[(\boldsymbol{I}_v - \frac{1}{v}\mathbf{1}_v\mathbf{1}'_v) \otimes \frac{1}{b_0}\mathbf{1}_{b_0}\mathbf{1}'_{b_0}\right]\bar{\mathbf{y}} \text{ (using (A.4.10))} \\
&= \frac{1}{b_0}\sum_{i=1}^{v}\left(\sum_{j=1}^{b_0}\bar{y}_{ij.}\right)^2 - \frac{1}{vb_0}\left(\sum_{i=1}^{v}\sum_{j=1}^{b_0}\bar{y}_{ij.}\right)^2 \\
&= \frac{1}{b_0 n^2}\sum_{i=1}^{v} y_{i..}^2 - \frac{1}{vb_0n^2}y_{...}^2, \qquad \text{(A.4.18)}
\end{aligned}
$$

where the last expression follows by noting that $\bar{y}_{ij.} = 1/n\sum_{k=1}^{n} y_{ijk}$. Also, $y_{i..} = \sum_{j=1}^{b_0}\sum_{k=1}^{n} y_{ijk}$ and $y_{...} = \sum_{i=1}^{v}\sum_{j=1}^{b_0}\sum_{k=1}^{n} y_{ijk}$. From (A.4.18), it is clear that $n\mathbf{v}'_\tau\mathbf{v}_\tau$ is the sum of squares due to the nesting effect in the balanced case of the model (4.4.1).

Due to the orthogonality of $\boldsymbol{S}$, we also get

$$
\begin{aligned}
\mathbf{v}'_\tau\mathbf{v}_\tau + \mathbf{v}'_\beta\mathbf{v}_\beta &= \mathbf{v}'\mathbf{v} = \bar{\mathbf{y}}'\boldsymbol{O}'\boldsymbol{Q}_1\boldsymbol{Q}'_1\boldsymbol{O}\bar{\mathbf{y}} \\
&= \bar{\mathbf{y}}'\boldsymbol{O}'_1(\boldsymbol{I}_v - \frac{1}{v}\mathbf{1}_v\mathbf{1}'_v)\boldsymbol{O}_1\bar{\mathbf{y}} + \bar{\mathbf{y}}'\boldsymbol{O}'_2\boldsymbol{O}_2\bar{\mathbf{y}}. \qquad \text{(A.4.19)}
\end{aligned}
$$

We already proved in (A.4.18) that $\mathbf{v}'_\tau\mathbf{v}_\tau = \bar{\mathbf{y}}'\boldsymbol{O}'_1(\boldsymbol{I}_v - \frac{1}{v}\mathbf{1}_v\mathbf{1}'_v)\boldsymbol{O}_1\bar{\mathbf{y}}$. Hence, from (A.4.19), we get

$$
\begin{aligned}
\mathbf{v}'_\beta\mathbf{v}_\beta &= \bar{\mathbf{y}}'\boldsymbol{O}'_2\boldsymbol{O}_2\bar{\mathbf{y}} = \bar{\mathbf{y}}'\left[\boldsymbol{I}_v \otimes \left(\boldsymbol{I}_{b_0} - \frac{1}{b_0}\mathbf{1}_{b_0}\mathbf{1}'_{b_0}\right)\right]\bar{\mathbf{y}} \\
&= \sum_{i=1}^{v}\sum_{j=1}^{b_0}\bar{y}_{ij.}^2 - \frac{1}{b_0}\sum_{i=1}^{v}\left(\sum_{j=1}^{b_0}\bar{y}_{ij.}\right)^2 = \frac{1}{n^2}\sum_{i=1}^{v}\left[\sum_{j=1}^{b_0} y_{ij.}^2 - \frac{1}{b_0}y_{i..}^2\right].
\end{aligned}
$$

Hence $n\mathbf{v}'_\beta\mathbf{v}_\beta$ is the sum of squares due to the nested effect in the balanced case of the model (4.4.1).

CHAPTER 5

5.1. (a)

$$
\begin{aligned}
\mathbf{H}_0 &= \mathbf{1}_{a_1} \otimes \mathbf{1}_{a_2} \otimes \mathbf{1}_{a_3} \\
\mathbf{H}_1 &= \mathbf{I}_{a_1} \otimes \mathbf{1}_{a_2} \otimes \mathbf{1}_{a_3} \\
\mathbf{H}_2 &= \mathbf{I}_{a_1} \otimes \mathbf{I}_{a_2} \otimes \mathbf{1}_{a_3} \\
\mathbf{H}_3 &= \mathbf{1}_{a_1} \otimes \mathbf{1}_{a_2} \otimes \mathbf{I}_{a_3} \\
\mathbf{H}_4 &= \mathbf{I}_{a_1} \otimes \mathbf{1}_{a_2} \otimes \mathbf{I}_{a_3} \\
\mathbf{H}_5 &= \mathbf{I}_{a_1} \otimes \mathbf{I}_{a_2} \otimes \mathbf{I}_{a_3}.
\end{aligned}
$$

(b)

$$\mathbf{P}_1 = \frac{\mathbf{A}_1}{a_2 a_3} - \frac{\mathbf{A}_0}{a_1 a_2 a_3}$$
$$\mathbf{P}_2 = \frac{\mathbf{A}_2}{a_3} - \frac{\mathbf{A}_1}{a_2 a_3}$$
$$\mathbf{P}_3 = \frac{\mathbf{A}_3}{a_1 a_2} - \frac{\mathbf{A}_0}{a_1 a_2 a_3},$$

where $\mathbf{A}_i = \mathbf{H}_i \mathbf{H}_i' (i = 0, 1, 2, 3)$.

(c)

$$\mathbf{A}_1 \mathbf{P}_2 = \mathbf{0}, \quad \mathbf{A}_2 \mathbf{P}_1 = a_3 \mathbf{P}_1$$
$$\mathbf{A}_2 \mathbf{P}_2 = a_3 \mathbf{P}_2, \quad \mathbf{A}_3 \mathbf{P}_1 = \mathbf{0}.$$

5.2. (a) We have that

$$\frac{MS_3}{4\sigma^2_{\delta(\tau)} + \sigma^2_{\beta\delta(\tau)} + \sigma^2_e} \frac{\sigma^2_{\beta\delta(\tau)} + \sigma^2_e}{MS_4} \sim F_{6,18}.$$

The 95% confidence interval for $\frac{\sigma^2_{\delta(\tau)}}{\sigma^2_{\beta\delta(\tau)} + \sigma^2_e}$ is then given by

$$\left[\frac{1}{4}\left(\frac{MS_3}{MS_4 F_{0.025,6,18}} - 1\right), \; \frac{1}{4}\left(\frac{MS_3 F_{0.025,18,6}}{MS_4} - 1\right)\right].$$

(b) $F = MS_4/4MS_2(e)$, which, under H_0, has the F-distribution with 18 and 47 degrees of freedom. This is a two-sided test.

(c)

$$P\left(\hat{\sigma}^2_{\beta(\tau)} < 0\right) = P\left(MS_2 < MS_4\right)$$
$$= P\left(F_{9,18} < \frac{\sigma^2_{\beta\delta(\tau)} + \sigma^2_e}{3\sigma^2_{\beta(\tau)} + \sigma^2_{\beta\delta(\tau)} + \sigma^2_e}\right).$$

5.3. (a) The population structure is $[(i : j)(k)] : \ell$, where i, j, k, ℓ are subscripts associated with metropolitan area, supermarket, food item, and replications, respectively.

(b)

$$y_{ijk\ell} = \mu + \tau_i + \beta_{i(j)} + \delta_k + (\tau\delta)_{ik} + (\beta\delta)_{i(jk)} + e_{ijk(\ell)},$$

where τ_i is the effect of the ith metropolitan area $(i = 1, 2, 3, 4)$, $\beta_{i(j)}$ is the effect of the jth supermarket within the ith metropolitan area $(j = 1, 2, 3)$, and δ_k is the effect of the kth food item $(k = 1, 2, 3, 4)$. All effects in the model are random with variance components $\sigma^2_\tau, \sigma^2_{\beta(\tau)}, \sigma^2_\delta, \sigma^2_{\tau\delta}, \sigma^2_{\beta(\tau)\delta}, \sigma^2_e$. To derive the ANOVA table for this model, the following steps should be followed:

(i) Use formula (5.2.13) to obtain the matrices $\mathbf{H}_0, \mathbf{H}_1, \mathbf{H}_2, \mathbf{H}_3, \mathbf{H}_4$, and $\mathbf{H}_5$ corresponding to $\mu, \tau_i, \beta_{i(j)}, \delta_k, (\tau\delta)_{ik}$, and $(\beta\delta)_{i(jk)}$, respectively. Note that

$$\begin{aligned}
\mathbf{H}_0 &= \mathbf{1}_4 \otimes \mathbf{1}_3 \otimes \mathbf{1}_4 \\
\mathbf{H}_1 &= \mathbf{I}_4 \otimes \mathbf{1}_3 \otimes \mathbf{1}_4 \\
\mathbf{H}_2 &= \mathbf{I}_4 \otimes \mathbf{I}_3 \otimes \mathbf{1}_4 \\
\mathbf{H}_3 &= \mathbf{1}_4 \otimes \mathbf{1}_3 \otimes \mathbf{I}_4 \\
\mathbf{H}_4 &= \mathbf{I}_4 \otimes \mathbf{1}_3 \otimes \mathbf{I}_4 \\
\mathbf{H}_5 &= \mathbf{I}_4 \otimes \mathbf{I}_3 \otimes \mathbf{I}_4.
\end{aligned}$$

(ii) Use formula (5.3.3) to obtain the matrices $\mathbf{P}_0, \mathbf{P}_1, \mathbf{P}_2, \mathbf{P}_3, \mathbf{P}_4$, and $\mathbf{P}_5$:

$$\begin{aligned}
\mathbf{P}_0 &= \frac{1}{48}\mathbf{A}_0 \\
\mathbf{P}_1 &= \frac{1}{12}\mathbf{A}_1 - \frac{1}{48}\mathbf{A}_0 \\
\mathbf{P}_2 &= \frac{1}{4}\mathbf{A}_2 - \frac{1}{12}\mathbf{A}_1 \\
\mathbf{P}_3 &= \frac{1}{12}\mathbf{A}_3 - \frac{1}{48}\mathbf{A}_0 \\
\mathbf{P}_4 &= \frac{1}{3}\mathbf{A}_4 - \frac{1}{12}\mathbf{A}_1 - \frac{1}{12}\mathbf{A}_3 + \frac{1}{48}\mathbf{A}_0 \\
\mathbf{P}_5 &= \mathbf{I}_{48} - \frac{1}{4}\mathbf{A}_2 - \frac{1}{3}\mathbf{A}_4 + \frac{1}{12}\mathbf{A}_1,
\end{aligned}$$

where $\mathbf{A}_i = \mathbf{H}_i\mathbf{H}_i' (i = 0, 1, 2, 3, 4, 5)$.

(iii) Obtain the $\tilde{\mathbf{Q}}$ matrix (see formula 5.3.11). Here,

$$\tilde{\mathbf{Q}} = [\mathbf{Q}_1' : \mathbf{Q}_2' : \mathbf{Q}_3' : \mathbf{Q}_4' : \mathbf{Q}_5']'.$$

The columns of $\mathbf{Q}_i'$ are orthonormal eigenvectors of $\mathbf{P}_i$ corresponding to the eigenvalue one, since $\mathbf{P}_i = \mathbf{Q}_i'\mathbf{Q}_i (i = 1, 2, 3, 4, 5)$. The ranks of $\mathbf{P}_i$ are $m_1 = 3, m_2 = 8, m_3 = 3, m_4 = 9$, and $m_5 = 24$.

(iv) Use formula (5.3.11) to compute $\mathbf{u} = \tilde{\mathbf{Q}}\bar{\mathbf{y}}$, where $\bar{\mathbf{y}}$ is the vector of cell means of order 48×1.

(v) Use formula (5.3.13) to compute $\mathbf{G} = \tilde{\mathbf{Q}}\mathbf{K}\tilde{\mathbf{Q}}'$, where $\mathbf{K}$ is the diagonal matrix given by formula (5.2.16) whose diagonal elements are the reciprocals of the cell frequencies.

(vi) Use formula (5.3.15) to compute the matrix $\mathbf{R}$. Then, find the matrices $\mathbf{C}_1$ and $\mathbf{C}_2$ (see formula 5.3.23). The latter two matrices are of orders 118×47 and 118×23, respectively.

(vii) Compute the matrix $(\lambda_{\max}\mathbf{I}_{47} - \mathbf{G})^{1/2}$, where $\lambda_{\max}$ is the largest eigenvalue of $\mathbf{G}$. Here, $\lambda_{\max} = 0.5$. For this purpose, obtain first the matrix $\mathbf{U}_{\max}$ of orthonormal eigenvectors of $\lambda_{\max}\mathbf{I}_{47} - \mathbf{G}$. Let $\mathbf{\Lambda}_{\max}$ be the diagonal matrix with diagonal elements equal to the corresponding eigenvalues. Then, $(\lambda_{\max}\mathbf{I}_{47} - \mathbf{G})^{1/2} = \mathbf{U}_{\max}\mathbf{\Lambda}_{\max}^{1/2}\mathbf{U}'_{\max}$, where $\mathbf{\Lambda}_{\max}^{1/2}$ is a diagonal matrix whose diagonal elements are the square roots of the corresponding diagonal elements of $\mathbf{\Lambda}_{\max}$.

(viii) Use formula (5.3.24) to compute the vector $\boldsymbol{\phi}$.

(ix) Partition $\boldsymbol{\phi}$ as $[\boldsymbol{\phi}'_1 : \boldsymbol{\phi}'_2 : \boldsymbol{\phi}'_3 : \boldsymbol{\phi}'_4 : \boldsymbol{\phi}'_5]'$ in a manner that corresponds to the partitioning of $\tilde{\mathbf{Q}}$.

(x) Compute the following sums of squares $SS_i = \boldsymbol{\phi}'_i\boldsymbol{\phi}_i (i = 1, 2, 3, 4, 5)$, and the error sum of squares $SS_2(e) = \mathbf{y}'\mathbf{C}_2\mathbf{C}'_2\mathbf{y}$. Note that $SS_1, SS_2, SS_3, SS_4, SS_5$ correspond to $\tau_i, \beta_{i(j)}, \delta_k, (\tau\delta)_{ik}, (\beta\delta)_{i(jk)}$, respectively,

(xi) Steps (i)–(x) result in the following ANOVA table:

Source	d.f.	SS	MS	$E(MS)$
τ_i	3	0.4039	0.1346	$12\sigma_\tau^2 + 4\sigma_{\beta(\tau)}^2 + 3\sigma_{\tau\delta}^2 + \sigma_{\beta(\tau)\delta}^2 + 0.5\sigma_e^2$
$\beta_{i(j)}$	8	0.0178	0.0022	$4\sigma_{\beta(\tau)}^2 + \sigma_{\beta(\tau)\delta}^2 + 0.5\sigma_e^2$
δ_k	3	180.9278	60.3093	$12\sigma_\delta^2 + 3\sigma_{\tau\delta}^2 + \sigma_{\beta(\tau)\delta}^2 + 0.5\sigma_e^2$
$(\tau\delta)_{ik}$	9	0.0717	0.0080	$3\sigma_{\tau\delta}^2 + \sigma_{\beta(\tau)\delta}^2 + 0.5\sigma_e^2$
$(\beta\delta)_{i(jk)}$	24	0.0321	0.0013	$\sigma_{\beta(\tau)\delta}^2 + 0.5\sigma_e^2$
Error	23	0.0030	0.00013	σ_e^2

(c) (i)

$$F = \frac{MS(\tau)}{MS(\beta(\tau)) + MS(\tau\delta) - MS(\beta(\tau)\delta)} = 15.124,$$

with 3 and m^* degrees of freedom, where

$$m^* = \frac{[MS(\beta(\tau)) + MS(\tau\delta) - MS(\beta(\tau)\delta)]^2}{\frac{[MS(\beta(\tau))]^2}{8} + \frac{[MS(\tau\delta)]^2}{9} + \frac{[-MS(\beta(\tau)\delta)]^2}{24}} = 10.17.$$

The approximate P-value is 0.00045. There is therefore a significant variation among metropolitan areas.

(ii) $F = MS(\beta(\tau))/MS(\beta(\tau)\delta) = 1.692$, with 8 and 24 degrees of freedom. The corresponding P-value is 0.1519. There is no significant variation among supermarkets within metropolitan areas.

(d) $F = MS(\tau\delta)/MS(\beta(\tau)\delta) = 6.154$, with 9 and 24 degrees of freedom. The corresponding P-value is 0.00017. There is a significant interaction effect between metropolitan areas and food items.

CHAPTER 6

6.1. Recall that for a *BIBD*, $n_{.j} = k$ for all $j = 1, \ldots, b$. Hence the $bk \times bk$ matrix $\mathbf{J}$ becomes $\mathbf{J} = \mathbf{diag}(\mathbf{J}_k, \ldots, \mathbf{J}_k)$. Now since $\mathbf{J}_k$ is a $k \times k$ matrix with all elements unity, the action of $\boldsymbol{\Gamma}$ in terms of permuting the plots within each block will keep $\mathbf{J}_k$ and hence $\mathbf{J}$ the same. Similarly, the action of $\boldsymbol{\Gamma}$ in terms of permuting the blocks as a whole would merely reshuffle the diagonal submatrices in $\mathbf{J}$, thereby keeping it the same. Hence the conclusion follows.

6.2. Denoting the eigenvalues of the nonnegative definite matrix $\mathbf{V}$ by $\lambda_1, \ldots, \lambda_N$, it follows easily that $|\mathbf{I} + \theta\mathbf{V}| = \prod_{i=1}^{N}(1 + \theta\lambda_i) = 1 + \theta(\sum_{i=1}^{N}\lambda_i) + o(\theta)$. This completes the first part of (6.2.19). The proof of the second part is similar. To prove the last part of (6.2.19), let us denote the eigenvalues of the $(N-m) \times (N-m)$ matrix $\mathbf{Z}'\mathbf{V}\mathbf{Z}$ by $\delta_1, \ldots, \delta_{N-m}$ and assume that the $(N-m) \times (N-m)$ orthogonal matrix $\mathbf{Q}$ diagonalizes $\mathbf{Z}'\mathbf{V}\mathbf{Z}$. Then we can write $\mathbf{I} + \theta\mathbf{Z}'\mathbf{V}\mathbf{Z} = \mathbf{Q}(\mathbf{I} + \theta\,\mathbf{diag}(\delta_1, \ldots, \delta_{N-m})\mathbf{Q}'$ so that

$$\begin{aligned}
(\mathbf{I} + \theta\mathbf{Z}'\mathbf{V}\mathbf{Z})^{-1} &= \mathbf{Q}(\mathbf{I} + \theta\,\mathbf{diag}(\delta_1, \cdots, \delta_{N-m})^{-1}\mathbf{Q}' \\
&= \mathbf{Q}\mathbf{diag}(1 + \theta\,\delta_1, \cdots, 1 + \theta\,\delta_{N-m})^{-1}\mathbf{Q}' \\
&= \mathbf{Q}\mathbf{diag}(\{1 + \theta\,\delta_1\}^{-1}, \cdots, \{1 + \theta\,\delta_{N-m}\}^{-1})\mathbf{Q}' \\
&= \mathbf{Q}\mathbf{diag}(\{1 - \theta\,\delta_1 + o(\theta)\}, \cdots, \{1 - \theta\,\delta_{N-m} + o(\theta)\})\mathbf{Q}' \\
&= \mathbf{Q}(\mathbf{I} - \theta\,\mathbf{diag}(\delta_1, \cdots, \delta_{N-m}) + o(\theta))\mathbf{Q}' \\
&= \mathbf{I} - \theta\mathbf{Z}'\mathbf{V}\mathbf{Z} + o(\theta).
\end{aligned}$$

A direct multiplication now yields the desired result.

6.3. Referring to (6.2.11), the linear model for $\mathbf{Z}'\mathbf{y}$ is given by

$$E(\mathbf{Z}'\mathbf{y}) = \mathbf{0}, \quad Var(\mathbf{Z}'\mathbf{y}) = \sigma^2(\mathbf{I}_{N-m} + \theta\mathbf{Z}'\mathbf{V}\mathbf{Z})$$

since $\mathbf{Z}'\mathbf{Z} = \mathbf{I}_{N-m}$. Let, as in the text, $r = rank(\mathbf{Z}'\mathbf{V}\mathbf{Z})$, and $\mathbf{C}^* = [\mathbf{C}_1 : \mathbf{C}_2]$ be an orthogonal matrix such that $\mathbf{C}^*(\mathbf{Z}'\mathbf{V}\mathbf{Z})\mathbf{C}^{*'} = \mathbf{diag}(\lambda_1, \ldots, \lambda_r, 0, \ldots, 0)$ where $\lambda_1, \ldots, \lambda_r$ are the eigenvalues of the matrix $\mathbf{Z}'\mathbf{V}\mathbf{Z}$. Noting that, under $H_0 : \theta = 0$, $Var(\mathbf{Z}'\mathbf{y}) = \sigma^2\mathbf{I}_{N-m}$, it follows trivially that, under H_0, $\mathbf{C}_1'\mathbf{Z}'\mathbf{y}$ and $\mathbf{C}_2'\mathbf{Z}'\mathbf{y}$ are independent normal with 0 mean vectors and covariance matrices as multiples of the identity matrix. Hence, under H_0, $\mathbf{y}'\mathbf{Q}_1\mathbf{y}$ and $\mathbf{y}'\mathbf{Q}_0\mathbf{y}$, as defined in (6.2.14), are independent scaled chi-squared variables with r and $N - m - r$ degrees of freedom, respectively. Hence the result follows.

6.6. It is easy to verify that the $v(b-1) \times v(b-1)$ matrix $\mathbf{J}_v \otimes \mathbf{I}_{b-1} = \mathbf{diag}(\mathbf{J}_v, \ldots, \mathbf{J}_v)$ so that the eigenvalues of this matrix are the roots of the equation $|\mathbf{J}_v - \lambda \mathbf{I}_v| = 0$ in multiples of $(b-1)$. Trivially, $\mathbf{J}_v$ is a matrix of rank 1 and the eigenvalues are v with multiplicity 1 and 0 with multiplicity $(v-1)$. Hence the result follows.

6.7. The motivation for this result is essentially due to the variance structure of $\mathbf{u}$ given in (6.3.12), which necessitates taking care of the *unwanted* last term involving the matrix $\mathbf{L}$. Clearly, $\mathbf{u}$ is independent of $SS(e)$ defined in (6.3.4), and hence independent of $\mathbf{H}_1'\mathbf{y}$. (*a*) follows trivially from (6.3.11) and the fact that $E(\mathbf{H}_1'\mathbf{y}) = \mathbf{0}$ because of (6.3.17). (*b*) follows from (6.3.12) and the property of independence mentioned above. (*c*) follows because $Var(\mathbf{w})$ can be expressed as a diagonal matrix with the first submatrix as $[(v\sigma_\beta^2 + \sigma_{\tau\beta}^2 + \lambda_{\max}(\mathbf{L})\sigma_e^2)\mathbf{I}_{b-1}]$, and the second submatrix as $(\sigma_{\tau\beta}^2 + \lambda_{\max}(\mathbf{L})\sigma_e^2)\mathbf{I}_{(v-1)(b-1)}$.

CHAPTER 7

7.1. For a BIBD, $\boldsymbol{C}_1$ and $\boldsymbol{C}_2$ are given in (7.3.1). Let $\boldsymbol{O} = (\boldsymbol{O}_1 : 1/\sqrt{v}\mathbf{1}_v)$ be the orthogonal matrix satisfying (7.3.2). Then

$$\boldsymbol{Q}_1 = \sqrt{\frac{k}{\lambda v}}\boldsymbol{O}_1, \text{ and } \boldsymbol{Q}_2 = \frac{1}{\sqrt{r-\lambda}}\boldsymbol{O}_1 \tag{A.7.1}$$

satisfy $\boldsymbol{Q}_1'\boldsymbol{C}_1\boldsymbol{Q}_1 = \boldsymbol{I}_{v-1}$ and $\boldsymbol{Q}_2'\boldsymbol{C}_2\boldsymbol{Q}_2 = \boldsymbol{I}_{v-1}$. Using (7.3.2), we see that $\boldsymbol{A}_1$ and $\boldsymbol{A}_2$ in (7.5.2) become

$$\begin{aligned}
\boldsymbol{A}_1 &= \boldsymbol{Q}_1'\boldsymbol{C}_1\boldsymbol{O}_1 = \sqrt{\frac{k}{\lambda v}}\boldsymbol{O}_1'\boldsymbol{C}_1\boldsymbol{O}_1 = \sqrt{\frac{\lambda v}{k}}\boldsymbol{I}_{v-1} \\
\boldsymbol{A}_2 &= \boldsymbol{Q}_2'\boldsymbol{C}_2\boldsymbol{O}_1 = \frac{1}{\sqrt{r-\lambda v}}\boldsymbol{O}_1'\boldsymbol{C}_2\boldsymbol{O}_1 = \sqrt{r-\lambda}\boldsymbol{I}_{v-1}.
\end{aligned}$$

Hence

$$\boldsymbol{A}_1(\boldsymbol{A}_1'\boldsymbol{A}_1 + \boldsymbol{A}_2'\boldsymbol{A}_2)^{-1}\boldsymbol{A}_2' = \sqrt{(r-\lambda)\frac{\lambda v}{k}}\left[\frac{\lambda v}{k} + (r-\lambda)\right]^{-1}\boldsymbol{I}_{v-1}. \tag{A.7.2}$$

Also, the λ_i's in (7.5.6) are the nonzero eigenvalues of $\boldsymbol{C}_1^-\boldsymbol{C}_2$; see the discussion following (7.5.25). From (7.3.1), we see that $\boldsymbol{C}_1^- = \frac{k}{\lambda v}(\boldsymbol{I}_v - \frac{1}{v}\mathbf{1}_v\mathbf{1}_v')$ is a g-inverse of $\boldsymbol{C}_1$. Thus

$$\boldsymbol{C}_1^-\boldsymbol{C}_2 = \frac{(r-\lambda)k}{\lambda v}(\boldsymbol{I}_v - \frac{1}{v}\mathbf{1}_v\mathbf{1}_v'),$$

and consequently, $\lambda_i = \frac{(r-\lambda)k}{\lambda v}$, $i = 1, 2, \ldots, v-1$. Hence ξ in (7.5.7) becomes

$$\xi = \frac{\sqrt{(r-\lambda)}\sqrt{\frac{k}{\lambda v}}}{1 + \frac{(r-\lambda)k}{\lambda v}} = \frac{\sqrt{(r-\lambda)}\sqrt{\frac{\lambda v}{k}}}{\frac{\lambda v}{k} + (r-\lambda)}. \tag{A.7.3}$$

Substituting (A.7.2) and (A.7.3) in (7.5.7), we see that (7.5.7) simplifies to R in (7.3.7).

7.2. Using the distribution of $\mathbf{x}_1$ and $\mathbf{x}_2$ in (7.5.3) and using the fact that $\mathbf{x}_1$ and $\mathbf{x}_2$ are independent, we see that the expected value of the numerator of R in (7.5.7) is given by

$$\mathrm{E}\left[\mathbf{x}_1' \boldsymbol{A}_1(\boldsymbol{A}_1'\boldsymbol{A}_1 + \boldsymbol{A}_2'\boldsymbol{A}_2)^{-1}\boldsymbol{A}_2'\mathbf{x}_2\right] = \boldsymbol{\tau}^{*'}\boldsymbol{A}_1'\boldsymbol{A}_1(\boldsymbol{A}_1'\boldsymbol{A}_1 + \boldsymbol{A}_2'\boldsymbol{A}_2)^{-1}\boldsymbol{A}_2'\boldsymbol{A}_2\boldsymbol{\tau}^*,$$

which is nonnegative if the matrix $\boldsymbol{A}_1'\boldsymbol{A}_1(\boldsymbol{A}_1'\boldsymbol{A}_1 + \boldsymbol{A}_2'\boldsymbol{A}_2)^{-1}\boldsymbol{A}_2'\boldsymbol{A}_2$ is nonnegative definite. Using the representation (7.5.6), it is readily verified that this matrix is positive definite.

7.3. Let $\boldsymbol{Z}$ be a $k \times (k-1)$ matrix such that $\boldsymbol{Z}'\mathbf{1}_k = 0$ and $\boldsymbol{Z}'\boldsymbol{Z} = \boldsymbol{I}_{k-1}$. Hence $\boldsymbol{Z}\boldsymbol{Z}' = \boldsymbol{I}_k - \frac{1}{k}\mathbf{1}_k\mathbf{1}_k'$ and

$$\boldsymbol{O} = \left(\boldsymbol{I}_b \otimes \frac{1}{\sqrt{k}}\mathbf{1}_k : \boldsymbol{I}_b \otimes \boldsymbol{Z}\right)$$

is a $bk \times bk$ orthogonal matrix. From (7.2.2), the model for $\boldsymbol{O}'\mathbf{y}$ can be written as

$$\begin{aligned}(1/\sqrt{k})(\boldsymbol{I}_b \otimes \mathbf{1}_k')\mathbf{y} &= \sqrt{k}\mu\mathbf{1}_b + (1/\sqrt{k})\boldsymbol{N}'\boldsymbol{\tau} + \sqrt{k}\boldsymbol{\beta} + (1/\sqrt{k})(\boldsymbol{I}_b \otimes \mathbf{1}_k')\mathbf{e} \\ (\boldsymbol{I}_b \otimes \boldsymbol{Z}')\mathbf{y} &= (\boldsymbol{I}_b \otimes \boldsymbol{Z}')\boldsymbol{X}_1\boldsymbol{\tau} + (\boldsymbol{I}_b \otimes \boldsymbol{Z}')\mathbf{e}, \end{aligned} \tag{A.7.4}$$

where $\boldsymbol{N}' = (\boldsymbol{I}_b \otimes \mathbf{1}_k')\boldsymbol{X}_1$. Since $(\boldsymbol{I}_b \otimes \mathbf{1}_k')\mathbf{y} = \mathbf{B}$, the first model in (A.7.4) is equivalent to the model (7.2.11). Also, under the distributional assumption (7.2.3), the covariance between $\mathbf{B}$ and $(\boldsymbol{I}_b \otimes \boldsymbol{Z}')\mathbf{y}$ can be shown to be zero. Hence these quantities are independent. (The second model in (A.7.4) is obtained by eliminating $\boldsymbol{\beta}$ from the model (7.2.2). This is very similar to the derivation of the model (4.3.6) from (4.3.5) in Chapter 4.) By direct computations, we can verify that $\mathbf{q}_1$ in (7.2.5) is the least estimator of $\boldsymbol{C}_1\boldsymbol{\tau}$ and s_1^2 is the error sum of squares based on the model for $(\boldsymbol{I}_b \otimes \boldsymbol{Z}')\mathbf{y}$, which is a fixed effects model. (See also the solution to problem 4.3 in Chapter 4.) Hence $\mathbf{q}_1$ and s_1^2 are independently distributed. Similarly, $\mathbf{q}_2$ and s_2^2 are computed based on the fixed effects model (7.2.11). Hence $\mathbf{q}_2$ and s_2^2 are also independent. Furthermore, since $\mathbf{B}$ and $(\boldsymbol{I}_b \otimes \boldsymbol{Z}')\mathbf{y}$ are independent, it follows that $(\mathbf{q}_1, s_1^2)$ and $(\mathbf{q}_2, s_2^2)$ are independent.

7.4. For the Youden square design, let R_i, C_j and T_s denote the sum of the observations from the ith row, j^{th} column and s^{th} treatment, respectively ($i = 1, 2, \ldots, v$; $j = 1, 2, \ldots, k$; $s = 1, 2, \ldots, v$). Let $q_{1s} = T_s - \frac{1}{k} n_{si} R_i$, $i = 1, 2, \ldots, v$, where $n_{si} = 1$ if the sth treatment occurs in the ith row and $n_{si} = 0$, otherwise. Let $\mathbf{q}_1$ be the $v \times 1$ vector consisting of the q_{1s}'s. Then

$$\mathbf{q}_1 \sim N(\boldsymbol{C}_1 \boldsymbol{\tau}, \sigma_e^2 \boldsymbol{C}_1). \tag{A.7.5}$$

Here $\boldsymbol{C}_1 = \lambda v/k(\boldsymbol{I}_v - 1/v \mathbf{1}_v \mathbf{1}_v')$, where λ is the number of rows where any pair of treatments occur together. (The fact that the rows and treatments form a BIBD is used in order to obtain the expression for $\boldsymbol{C}_1$.) Furthermore, if s_1^2 denotes the usual error sum of squares, then

$$\frac{s_1^2}{\sigma_e^2} \sim \chi^2_{(v-1)(k-2)}. \tag{A.7.6}$$

Thus, for testing H_τ: $\tau_1 = \tau_2 = \ldots = \tau_v$, we have the usual F-statistic

$$F_1 = \frac{\mathbf{q}_1' \boldsymbol{C}_1^- \mathbf{q}_1/(v-1)}{s_1^2/(v-1)(k-2)} = \frac{\frac{k}{\lambda v(v-1)} \mathbf{q}_1' \mathbf{q}_1}{s_1^2/(v-1)(k-2)}, \tag{A.7.7}$$

where the last expression in (A.7.7) follows by noting that $\boldsymbol{C}_1^- = (k/\lambda v)(\boldsymbol{I}_v - (1/v)\mathbf{1}_v \mathbf{1}_v')$ is a g-inverse of $\boldsymbol{C}_1$. The F-test based on F_1 is valid irrespective of whether the row effects are fixed or random. Furthermore, F_1 in (A.7.7) is essentially the test statistic given in (7.3.5), for a BIBD, except that the degrees of freedom for s_1^2 is different. It is easy to derive a representation for F_1, similar to the second expression for F_1 in (7.3.5).

(a) The model for R_i, the sum of the observations from the ith row, is

$$R_i = \mu_1 + \sum_{s=1}^{v} n_{si} \tau_s + g_i,$$

where $\mu_1 = k\mu + \sum_{j=1}^{k} \beta_j$ and $g_i = k\alpha_i + \sum_{j=1}^{k} e_{ij}$. If $\mathbf{R}$ denotes the vector consisting of the R_i's, we have the model

$$\mathbf{R} = \mu_1 \mathbf{1}_v + \boldsymbol{N}' \boldsymbol{\tau} + \mathbf{g}, \tag{A.7.8}$$

where $\boldsymbol{N}$ is the $v \times v$ matrix having n_{si}'s as the entries, and $\mathbf{g}$ is the vector consisting of the g_i's. Note that

$$\mathbf{g} \sim N(\mathbf{0}, k(k\sigma_\alpha^2 + \sigma_e^2)\boldsymbol{I}_v). \tag{A.7.9}$$

The model (A.7.8), along with (A.7.9), is similar to (7.2.11), along with (7.2.12). If

$$\mathbf{q}_2 = \boldsymbol{N}(\boldsymbol{I}_v - \frac{1}{v}\mathbf{1}_v \mathbf{1}_v')\mathbf{R}, \text{ and } \boldsymbol{C}_2 = (r - \lambda)(\boldsymbol{I}_v - \frac{1}{v}\mathbf{1}_v \mathbf{1}_v'),$$

then

$$\mathbf{q}_2 \sim N(\boldsymbol{C}_2\boldsymbol{\tau}, k(k\sigma_\alpha^2 + \sigma_e^2)\boldsymbol{C}_2). \qquad \text{(A.7.10)}$$

Since the number of rows and the number of treatments are the same, the error sum of squares s_2^2, similar to that in (7.2.15), is nonexistent. Since the expressions for $\boldsymbol{C}_1$ and $\boldsymbol{C}_2$ are the same as those for a BIBD, the models (A.7.5), (A.7.6), and (A.7.10) can be written in the canonical form (7.3.33) and the results in Section 7.3.2 can be applied in order to derive combined tests that recover interrow information.

(b) If α_i's are also random, we can apply the results in Section 7.4 (for the case $b = v$) in order to obtain combined tests for testing H_0: $\sigma_\alpha^2 = 0$.

7.5. Suppose the block sizes are $k_1, k_2, \ldots, k_b$, where the k_i's are not all equal. The intra-block F-statistic F_1 in (7.2.16) can still be derived, with appropriate modifications in the expressions for $\mathbf{q}_1$, $\boldsymbol{C}_1$, and s_1^2. However, in the model (7.2.11), the distribution of $\mathbf{g}$ now becomes

$$\mathbf{g} \sim N(\mathbf{0}, \mathbf{diag}(k_1(k_1\sigma_\beta^2 + \sigma_e^2), k_2(k_2\sigma_\beta^2 + \sigma_e^2), \ldots, k_b(k_b\sigma_\beta^2 + \sigma_e^2)).$$

In other words, the covariance matrix of $\mathbf{g}$ is no longer a multiple of the identity matrix. In this case, it is no longer clear how to construct a test for $\boldsymbol{\tau}$ using the model (7.2.11). Our derivations in Section 7.5 depend crucially on the fact that the covariance matrix in the inter-block model (i.e., model (7.2.11)) is a multiple of the identity matrix (i.e., the block sizes are equal).

7.6. This problem is a special case of problem 7.7.

7.7. We shall express the models in a suitable canonical form, which will be similar to the model (1.1) in Zhou and Mathew (1993). The details of the combined test will be omitted and can be found in Zhou and Mathew (1993).

Without loss of generality, we assume that the matrix $\boldsymbol{K}$ is an $s \times m$ matrix of rank s. Let $\boldsymbol{K}\hat{\boldsymbol{\beta}}_i$ denote the least squares estimator of $\boldsymbol{K\beta}$ based on the model $\mathbf{y}_i \sim N(\boldsymbol{X}_i\boldsymbol{\beta}, \sigma_i^2\boldsymbol{I}_{n_i})$, $i = 1, 2, \ldots, a$. Then

$$\boldsymbol{K}\hat{\boldsymbol{\beta}}_i \sim N(\boldsymbol{K\beta}, \sigma_i^2\boldsymbol{K}(\boldsymbol{X}_i'\boldsymbol{X}_i)^-\boldsymbol{K}').$$

Let

$$\boldsymbol{A}_i = \left[\boldsymbol{K}(\boldsymbol{X}_i'\boldsymbol{X}_i)^-\boldsymbol{K}'\right]^{-1/2}, \ \mathbf{y}_{1i} = \boldsymbol{A}_i\boldsymbol{K}\hat{\boldsymbol{\beta}}_i, \ \theta = \boldsymbol{K\beta}. \qquad \text{(A.7.11)}$$

Furthermore, let s_i^2 denote the error sum of squares based on the model $\mathbf{y}_i \sim N(\boldsymbol{X}_i\boldsymbol{\beta}, \sigma_i^2\boldsymbol{I}_{n_i})$. Then the quantities $\mathbf{y}_{1i}$'s, which are $s \times 1$, and s_i^2's are all independently distributed as

$$\mathbf{y}_{1i} \sim N(\boldsymbol{A}_i\theta, \sigma_i^2\boldsymbol{I}_s), \ s_i^2/\sigma_i^2 \sim \chi_{e_i}^2, \qquad \text{(A.7.12)}$$

$i = 1, 2, \ldots, a$, where the e_i's denote the degrees of freedom associated with s_i^2. The models in (A.7.12) are the same as the models in (1.1) of Zhou and Mathew (1993). The combined test derived in Zhou and Mathew (1993) can now be applied for testing H_0: $\theta = 0$ (i.e., H_0: $\boldsymbol{K\beta} = 0$).

CHAPTER 8

8.1. Here we need to show that for testing the hypothesis H_β under the random δ_l's, model 8.2(*iv*), part (*iii*), of Lemma A.8.1 is applicable. Toward this end, let us write $\mathbf{y}$ as the $bv \times 1$ vector $\mathbf{y} = (\bar{y}_{11.}, \ldots, \bar{y}_{1v.}, \ldots, \bar{y}_{b1.}, \ldots, \bar{y}_{bv.})'$. Then, following the structure given in this section, the dispersion matrix $\mathbf{V}$ of $\mathbf{y}$ can be easily written as $a\mathbf{I} + \mathbf{V}_1$ where the matrix $\mathbf{V}_1$ has a very special structure, namely, it has *null* diagonal submatrices of order $v \times v$ and every off-diagonal element is another submatrix of order $v \times v$ which is a multiple of the identity matrix $\mathbf{I}_v$. Since, under H_β, the range of the new *design* matrix $\mathbf{X}_0$ is generated by vectors of the form $(1, 0, \ldots, 0, 1, 0, \ldots, 0, \ldots, 1, 0, \ldots, 0)', \ldots, (0, 0, \ldots, 1, 0, 0, \ldots, 1, \ldots, 0, 0, \ldots, 1)'$, it is verified directly that range$(\mathbf{V}_1) \subset$ range$(\mathbf{X}_0)$. Hence the result follows.

8.2. In order to verify (8.2.3), we note from (8.2.2) and the definition of $(\mathbf{M} \otimes \mathbf{P}_1)\mathbf{y}$ that

$$\begin{aligned} Var[(\mathbf{M} \otimes \mathbf{P}_1)\mathbf{y}] &= (\mathbf{M} \otimes \mathbf{I}_{s-1})\big[\sigma_\delta^2(\mathbf{J}_n \otimes \mathbf{I}_{s-1}) \\ &\quad + \sigma_\gamma^2(\{(\mathbf{1}_b \otimes \mathbf{I}_v)(\mathbf{1}_b' \otimes \mathbf{I}_v)\} \otimes \mathbf{I}_{s-1}) \\ &\quad + \sigma_{**}^2 \mathbf{I}_{n(s-1)}\big](\mathbf{M} \otimes \mathbf{I}_{s-1})' \\ &= \sigma_\delta^2(\mathbf{M} \otimes \mathbf{I}_{s-1})(\mathbf{J}_n \otimes \mathbf{I}_{s-1})(\mathbf{M}' \otimes \mathbf{I}_{s-1}) \\ &\quad + \sigma_\gamma^2(\mathbf{M} \otimes \mathbf{I}_{s-1})[\{(\mathbf{1}_b \otimes \mathbf{I}_v)(\mathbf{1}_b' \otimes \mathbf{I}_v)\} \otimes \mathbf{I}_{s-1}](\mathbf{M}' \otimes \mathbf{I}_{s-1}) \\ &\quad + \sigma_{**}^2(\mathbf{M} \otimes \mathbf{I}_{s-1})\mathbf{I}_{n(s-1)}(\mathbf{M}' \otimes \mathbf{I}_{s-1}). \end{aligned}$$

Using the fact that $\mathbf{M1}_n = (b^{\frac{1}{2}}\mathbf{1}_v', \mathbf{0})'$, we get

$$(\mathbf{M} \otimes \mathbf{I}_{s-1})(\mathbf{J}_n \otimes \mathbf{I}_{s-1})(\mathbf{M}' \otimes \mathbf{I}_{s-1}) = \mathbf{diag}(b\mathbf{1}_v\mathbf{1}_v', 0) \otimes \mathbf{I}_{s-1}.$$

Similarly, it follows that

$$\begin{aligned} &(\mathbf{M} \otimes \mathbf{I}_{s-1})[\{(\mathbf{1}_b \otimes \mathbf{I}_v)(\mathbf{1}_b' \otimes \mathbf{I}_v)\} \otimes \mathbf{I}_{s-1}](\mathbf{M}' \otimes \mathbf{I}_{s-1}) \\ &= \mathbf{M}[\{(\mathbf{1}_b \otimes \mathbf{I}_v)(\mathbf{1}_b' \otimes \mathbf{I}_v)\}]\mathbf{M}' \otimes \mathbf{I}_{s-1} = \mathbf{diag}(b\mathbf{I}_v, 0) \otimes \mathbf{I}_{s-1}; \end{aligned}$$

see the expressions before (8.2.3). Finally, since the matrix $\mathbf{M}$ is orthogonal, we immediately get

$$(\mathbf{M} \otimes \mathbf{I}_{s-1})\mathbf{I}_{n(s-1)}(\mathbf{M}' \otimes \mathbf{I}_{s-1}) = \mathbf{I}_{n(s-1)}.$$

Putting together all the terms, we get (8.2.3).

8.4. Recall the definition of f_{ij}^u which is 1 if the uth level of the factor A occurs in the jth whole plot of the ith block and 0 otherwise, and the fact that there are b blocks with the ith block having k_i whole plots and there are v levels of A. It then easily follows that $\mathbf{f}'_{ij}\mathbf{1}_v = 1$ for all i and j since there is only one level of A occurring in any whole plot of any block. Hence $\mathbf{F}_i\mathbf{1}_v = \mathbf{1}_{k_i}$.

Analogously, for any fixed i, summing over all the k_i whole plots, $\mathbf{F}'_i\mathbf{1}_{k_i}$ stands for the vector of replications of the v whole plot treatments (i.e., the v levels of A), namely, $(n_{1i}, \ldots, n_{vi})'$.

Again, noting that any vector $\mathbf{f}_{ij}$ has only one element as 1 and the rest are 0's so that $\mathbf{f}_{ij}\mathbf{f}'_{ij}$ is a diagonal matrix of order $v \times v$ with exactly one diagonal element as 1 and the rest are as 0's, it follows that $\mathbf{F}'_i\mathbf{F}_i = \sum_{j=1}^{j=k_i} \mathbf{f}_{ij}\mathbf{f}'_{ij} = \mathbf{diag}(n_{1i}, \ldots, n_{vi})$. Finally, by the definition of the matrix $\mathbf{F}$, we get $\mathbf{F}'\mathbf{F} = \sum_{i=1}^{b} \mathbf{F}'_i\mathbf{F}_i = \mathbf{diag}(\sum_{i=1}^{b} n_{1i}, \ldots, \sum_{i=1}^{b} n_{vi}) = \mathbf{R}$.

8.6. The verification of (8.3.16) is very similar to that of (8.2.3), which is explained in exercise **8.2** above. Note from (8.3.15) and the definition of $(\mathbf{M} \otimes \mathbf{P}_1)\mathbf{y}$ that

$$
\begin{aligned}
& Var[(\mathbf{M} \otimes \mathbf{P}_1)\mathbf{y}] \\
&= (\mathbf{M} \otimes \mathbf{I}_{s-1})[\sigma_\delta^2(\mathbf{J}_n \otimes \mathbf{I}_{s-1}) + \sigma_\gamma^2(\mathbf{FF}' \otimes \mathbf{I}_{s-1}) + \sigma_{**}^2\mathbf{I}_{n(s-1)}](\mathbf{M} \otimes \mathbf{I}_{s-1})' \\
&= \sigma_\delta^2(\mathbf{M} \otimes \mathbf{I}_{s-1})(\mathbf{J}_n \otimes \mathbf{I}_{s-1})(\mathbf{M}' \otimes \mathbf{I}_{s-1}) \\
&+ \sigma_\gamma^2(\mathbf{M} \otimes \mathbf{I}_{s-1})[\mathbf{FF}' \otimes \mathbf{I}_{s-1}](\mathbf{M}' \otimes \mathbf{I}_{s-1}) \\
&+ \sigma_{**}^2(\mathbf{M} \otimes \mathbf{I}_{s-1})\mathbf{I}_{n(s-1)}(\mathbf{M}' \otimes \mathbf{I}_{s-1}).
\end{aligned}
$$

Using the fact that $\mathbf{M1}_n = [\mathbf{r}'_0 : \mathbf{0}]'$, we get

$$(\mathbf{M} \otimes \mathbf{I}_{s-1})(\mathbf{J}_n \otimes \mathbf{I}_{s-1})(\mathbf{M}' \otimes \mathbf{I}_{s-1}) = \mathbf{diag}(\mathbf{r}_0\mathbf{r}'_0, 0) \otimes \mathbf{I}_{s-1}.$$

Similarly, since $\mathbf{MFF}'\mathbf{M}' = \mathbf{diag}(\mathbf{R}, \mathbf{0})$, it follows that

$$
\begin{aligned}
(\mathbf{M} \otimes \mathbf{I}_{s-1})[\mathbf{FF}' \otimes \mathbf{I}_{s-1}](\mathbf{M}' \otimes \mathbf{I}_{s-1}) &= [\mathbf{MFF}'\mathbf{M}'] \otimes \mathbf{I}_{s-1} \\
&= \mathbf{diag}(\mathbf{R}, 0) \otimes \mathbf{I}_{s-1}.
\end{aligned}
$$

Finally, since the matrix $\mathbf{M}$ here is also orthogonal, we immediately get $(\mathbf{M} \otimes \mathbf{I}_{s-1})\mathbf{I}_{n(s-1)}(\mathbf{M}' \otimes \mathbf{I}_{s-1}) = \mathbf{I}_{n(s-1)}$. Putting together all the terms, we get (8.3.16).

CHAPTER 9

9.1. Let $SS(\tau)$ and $SS(e)$ denote the sum of squares due to the τ_i's and the sum of squares due to error, respectively. Then $SS(\tau)$ and $SS(e)$ are independently distributed and

$$\frac{SS(\tau)}{\sigma_e^2 + n\sigma_\tau^2} \sim \chi^2_{v-1}, \text{ and } \frac{SS(e)}{\sigma_e^2} \sim \chi^2_{v(n-1)}.$$

Let

$$T = \frac{SS(\tau)/(v-1)}{SS(e)/v(n-1)} = \left(1 + n\frac{\sigma_\tau^2}{\sigma_e^2}\right) \frac{[SS(\tau)/(\sigma_e^2 + n\sigma_\tau^2)]/(v-1)}{[SS(e)/\sigma_e^2]/v(n-1)}. \tag{A.9.1}$$

From (A.9.1) it is clear that T is stochastically increasing in σ_τ^2/σ_e^2. If $ss(\tau)$ and $ss(e)$ denote the observed values of $SS(\tau)$ and $SS(e)$, respectively, the P-value for testing H_0: $\sigma_\tau^2 \leq \delta\sigma_e^2$ versus H_1: $\sigma_\tau^2 > \delta\sigma_e^2$ is given by

$$\mathrm{P}\left(T \geq \frac{ss(\tau)/(v-1)}{ss(e)/v(n-1)} \middle| \sigma_\tau^2 = \delta\sigma_e^2\right) = \mathrm{P}\left((1+n\delta)F \geq \frac{ss(\tau)/(v-1)}{ss(e)/v(n-1)}\right), \tag{A.9.2}$$

where $F = [\sigma_e^2/(\sigma_e^2 + n\sigma_\tau^2)][SS(\tau)/(v-1)]/[SS(e)/v(n-1)]$ follows the central F-distribution with degrees of freedom $(v-1, v(n-1))$. The P-value in (A.9.2) can be easily computed. This solution is also given in Weerahandi (1995, Section 9.3).

9.2. Consider the unbalanced one-way random model involving v treatments and n_i observations for the ith treatment $(i = 1, 2, \ldots, v)$. We then have the model (4.2.1), or, equivalently, (4.2.3). Note that $SS(e)/\sigma_e^2 \sim \chi^2_{(n_.-v)}$, where $n_. = \sum_{i=1}^{v} n_i$. We shall show that the statistic

$$T = \frac{\mathbf{y}'(\boldsymbol{I}_{n_.} - \frac{1}{n_.}\mathbf{1}_{n_.}\mathbf{1}'_{n_.})\mathbf{diag}(\mathbf{1}_{n_1}\mathbf{1}'_{n_1}, \mathbf{1}_{n_2}\mathbf{1}'_{n_2}, \ldots, \mathbf{1}_{n_v}\mathbf{1}'_{n_v})(\boldsymbol{I}_{n_.} - \frac{1}{n_.}\mathbf{1}_{n_.}\mathbf{1}'_{n_.})\mathbf{y}}{SS(e)} \tag{A.9.3}$$

can be used for testing H_0: $\sigma_\tau^2/\sigma_e^2 \leq \delta$ versus H_1: $\sigma_\tau^2/\sigma_e^2 > \delta$. Toward this, let $\boldsymbol{Z}$ be an $n_. \times (n_. - 1)$ matrix satisfying $\boldsymbol{Z}'\mathbf{1}_{n_.} = 0$ and $\boldsymbol{Z}'\boldsymbol{Z} = \boldsymbol{I}_{n_.-1}$. Then $\boldsymbol{Z}\boldsymbol{Z}' = \boldsymbol{I}_{n_.} - 1/n_.\mathbf{1}_{n_.}\mathbf{1}'_{n_.}$. Let $\mathbf{u} = \boldsymbol{Z}'\mathbf{y}$. We then have the model

$$\boldsymbol{E}(\mathbf{u}) = 0 \text{ and } \mathrm{Var}(\mathbf{u}) = \sigma_e^2\boldsymbol{I}_{n_.-1} + \sigma_\tau^2\boldsymbol{Z}'\mathbf{diag}(\mathbf{1}_{n_1}\mathbf{1}'_{n_1}, \mathbf{1}_{n_2}\mathbf{1}'_{n_2}, \ldots, \mathbf{1}_{n_v}\mathbf{1}'_{n_v})\boldsymbol{Z}.$$

Consider the spectral decomposition

$$\boldsymbol{Z}'\mathbf{diag}(\mathbf{1}_{n_1}\mathbf{1}'_{n_1}, \mathbf{1}_{n_2}\mathbf{1}'_{n_2}, \ldots, \mathbf{1}_{n_v}\mathbf{1}'_{n_v})\boldsymbol{Z} = \sum_{i=1}^{r} \lambda_i \boldsymbol{E}_i, \tag{A.9.4}$$

similar to the spectral decomposition of $\boldsymbol{V}_1$ in (9.4.3); see the discussion following (9.4.3). In (A.9.4), the λ_i's are the nonzero eigenvalues of $\mathbf{Z}'\mathbf{diag}(\mathbf{1}_{n_1}\mathbf{1}'_{n_1}, \mathbf{1}_{n_2}\mathbf{1}'_{n_2}, \ldots, \mathbf{1}_{n_v}\mathbf{1}'_{n_v})\mathbf{Z}$ and the $\boldsymbol{E}_i$'s are the corresponding mutually orthogonal projections. If $U_i = \mathbf{u}'\boldsymbol{E}_i\mathbf{u}$, then the U_i's are independent and $U_i/(\sigma_e^2 + \lambda_i\sigma_\tau^2) \sim \chi^2_{e_i}$, where e_i is the rank of $\boldsymbol{E}_i$, which is also the multiplicity of the eigenvalue λ_i. Thus T in (A.9.3) can be written as

$$\begin{aligned} T &= \frac{\mathbf{u}'\mathbf{Z}'\mathbf{diag}(\mathbf{1}_{n_1}\mathbf{1}'_{n_1}, \mathbf{1}_{n_2}\mathbf{1}'_{n_2}, \ldots, \mathbf{1}_{n_v}\mathbf{1}'_{n_v})\mathbf{Z}\mathbf{u}}{SS(e)} \\ &= \frac{\sum_{i=1}^r \lambda_i \mathbf{u}'\boldsymbol{E}_i\mathbf{u}}{SS(e)} \\ &= \frac{1}{\sigma_e^2}\sum_{i=1}^r \lambda_i(\sigma_e^2 + \lambda_i\sigma_\tau^2)\frac{U_i}{(\sigma_e^2 + \lambda_i\sigma_\tau^2)} \Big/ \frac{SS(e)}{\sigma_e^2} \\ &= \sum_{i=1}^r \lambda_i\left(1 + \lambda_i\frac{\sigma_\tau^2}{\sigma_e^2}\right)U_{i0}/U_0, \end{aligned} \tag{A.9.5}$$

where $U_{i0} = U_i/(\sigma_e^2 + \lambda_i\sigma_\tau^2) \sim \chi^2_{e_i}$ $(i = 1, 2, \ldots, r)$ and $U_0 = SS(e)/\sigma_e^2 \sim \chi^2_{n_.-v}$. From (A.9.5), it is clear that T is stochastically increasing in σ_τ^2/σ_e^2. Hence for testing H_0: $\sigma_\tau^2/\sigma_e^2 \le \delta$ versus H_1: $\sigma_\tau^2/\sigma_e^2 > \delta$, the P-value is given by

$$\mathrm{P}\left(T \ge t \,\middle|\, \frac{\sigma_\tau^2}{\sigma_e^2} = \delta\right),$$

where t is the observed value of t. The above P-value can also be written as

$$\mathrm{P}\left(\sum_{i=1}^r \lambda_i(1 + \lambda_i\delta)U_{i0}/U_0 \ge t\right).$$

Note that the computation of the above P-value is similar to the computation of the P-value based on the test statistic (4.2.8). For this, one can use an algorithm due to Davies (1980) or the approximation due to Hirotsu (1979). (See Appendix 3.1. of Chapter 3 where Hirotsu's approximation is given.)

The test statistic T is a linear combination of $\mathbf{u}'\boldsymbol{E}_i\mathbf{u}/SS(e)$, with coefficients λ_i. Any linear combination of $\mathbf{u}'\boldsymbol{E}_i\mathbf{u}/SS(e)$, with positive coefficients, will also provide a valid test for testing H_0: $\frac{\sigma_\tau^2}{\sigma_e^2} \le \delta$ versus H_1: $\frac{\sigma_\tau^2}{\sigma_e^2} > \delta$. In the balanced case, all such tests coincide.

9.3. Let $SS(\tau)$, $SS(\beta)$, etc. denote the various sums of squares. Then

$$SS(\tau)/(b_1n_1\sigma_\tau^2 + n_1\sigma_{\tau\beta}^2 + \sigma_e^2) \sim \chi^2_{v_1-1},$$

$$SS(\beta)/(a_1n_1\sigma_\beta^2 + n_1\sigma_{\tau\beta}^2 + \sigma_e^2) \sim \chi^2_{b_1-1},$$

$$SS(\tau\beta)/(n_1\sigma_{\tau\beta}^2 + \sigma_e^2) \sim \chi^2_{(v_1-1)(b_1-1)}, \quad SS(e)/\sigma_e^2 \sim \chi^2_{v_1b_1(n_1-1)},$$

$$SS(\gamma)/(b_2 n_2 \sigma_\gamma^2 + n_2 \sigma_{\gamma\delta}^2 + \sigma_f^2) \sim \chi^2_{v_2-1},$$
$$SS(\delta)/(a_2 n_2 \sigma_\delta^2 + n_2 \sigma_{\gamma\delta}^2 + \sigma_f^2) \sim \chi^2_{b_2-1},$$
$$SS(\gamma\delta)/(n_2 \sigma_{\gamma\delta}^2 + \sigma_f^2) \sim \chi^2_{(v_2-1)(b_2-1)}, \; SS(f)/\sigma_f^2 \sim \chi^2_{v_2 b_2 (n_2-1)},$$

where $SS(e)$ and $SS(f)$ denote the error sums of squares for the two models. Also, let $ss(\tau)$ denote the observed value of $SS(\tau)$, etc. We shall give a solution to the second testing problem and the solution to the first one follows similarly. Thus the testing problem is H_0: $\sigma_\tau^2 \leq \sigma_\gamma^2$ versus H_1: $\sigma_\tau^2 > \sigma_\gamma^2$. Consider

$$T = \frac{\frac{b_1 n_1 (b_2 n_2 \sigma_\tau^2 + n_2 \sigma_{\gamma\delta}^2 + \sigma_f^2)}{SS(\gamma)} ss(\gamma) + \frac{b_2 n_2 (n_1 \sigma_{\tau\beta}^2 + \sigma_e^2)}{SS(\tau\beta)} ss(\tau\beta)}{\frac{b_2 n_2 (b_1 n_1 \sigma_\tau^2 + n_2 \sigma_{\tau\beta}^2 + \sigma_e^2)}{SS(\tau)} ss(\tau) + \frac{b_1 n_1 (n_2 \sigma_{\gamma\delta}^2 + \sigma_f^2)}{SS(\gamma\delta)} ss(\gamma\delta)}$$

$$= \frac{(b_2 n_2 \sigma_\tau^2 + n_2 \sigma_{\gamma\delta}^2 + \sigma_f^2)}{(b_2 n_2 \sigma_\gamma^2 + n_2 \sigma_{\gamma\delta}^2 + \sigma_f^2)} \frac{\frac{b_1 n_1 (b_2 n_2 \sigma_\gamma^2 + n_2 \sigma_{\gamma\delta}^2 + \sigma_f^2)}{SS(\gamma)} ss(\gamma) + \frac{b_2 n_2 (n_1 \sigma_{\tau\beta}^2 + \sigma_e^2)}{SS(\tau\beta)} ss(\tau\beta)}{\frac{b_2 n_2 (b_1 n_1 \sigma_\tau^2 + n_2 \sigma_{\tau\beta}^2 + \sigma_e^2)}{SS(\tau)} ss(\tau) + \frac{b_1 n_1 (n_2 \sigma_{\gamma\delta}^2 + \sigma_f^2)}{SS(\gamma\delta)} ss(\gamma\delta)}$$

$$= \frac{\theta \frac{b_1 n_1 (b_2 n_2 \sigma_\gamma^2 + n_2 \sigma_{\gamma\delta}^2 + \sigma_f^2)}{SS(\gamma)} ss(\gamma) + \frac{b_2 n_2 (n_1 \sigma_{\tau\beta}^2 + \sigma_e^2)}{SS(\tau\beta)} ss(\tau\beta)}{\frac{b_2 n_2 (b_1 n_1 \sigma_\tau^2 + n_2 \sigma_{\tau\beta}^2 + \sigma_e^2)}{SS(\tau)} ss(\tau) + \frac{b_1 n_1 (n_2 \sigma_{\gamma\delta}^2 + \sigma_f^2)}{SS(\gamma\delta)} ss(\gamma\delta)}, \tag{A.9.6}$$

where

$$\theta = \left(b_2 n_2 \frac{\sigma_\tau^2}{\sigma_\gamma^2} + \frac{(n_2 \sigma_{\gamma\delta}^2 + \sigma_f^2)}{\sigma_\gamma^2} \right) \Big/ \left((b_2 n_2 + \frac{(n_2 \sigma_{\gamma\delta}^2 + \sigma_f^2)}{\sigma_\gamma^2} \right);$$

the testing problem becomes H_0: $\theta \leq 1$ versus H_0: $\theta > 1$. It is readily verified that T in (A.9.6) satisfies all the conditions in (9.2.1) and the observed value of T is one. Hence the generalized P-value can be defined similar to (9.3.9).

9.4. The statistic T is now given by

$$T = \frac{\lambda_1 (\sigma_{**}^2 + \frac{\lambda_2}{\delta} \sigma_1^2) \frac{s_2}{S_2} + \frac{\lambda_2}{\delta} \sigma_*^2 \frac{s_*}{S_*}}{\frac{\lambda_2}{\delta} (\sigma_*^2 + \lambda_1 \sigma_1^2) \frac{s_1}{S_1} + \lambda_1 \sigma_{**}^2 \frac{s_{**}}{S_{**}}}$$

$$= \frac{\theta \lambda_1 (\sigma_{**}^2 + \lambda_2 \sigma_2^2) \frac{s_2}{S_2} + \frac{\lambda_2}{\delta} \sigma_*^2 \frac{s_*}{S_*}}{\frac{\lambda_2}{\delta} (\sigma_*^2 + \lambda_1 \sigma_1^2) \frac{s_1}{S_1} + \lambda_1 \sigma_{**}^2 \frac{s_{**}}{S_{**}}}$$

where

$$\theta = \frac{\sigma_{**}^2 + \frac{\lambda_2}{\delta} \sigma_1^2}{\sigma_{**}^2 + \lambda_2 \sigma_2^2}.$$

It can be verified that T satisfies all the conditions in (9.2.1). Note that the testing problem is equivalent to H_0: $\theta \leq 1$ versus H_0: $\theta > 1$. Also, when $\theta = 1$ (i.e., $\sigma_1^2 = \delta\sigma_2^2$), the observed value of T is one. Hence the generalized P-value is $\mathrm{P}(T \geq 1 \big| \theta = 1)$. This solution is also given in Weerahandi (1995, Section 9.6).

CHAPTER 10

10.1. Let $\mathbf{t}_i'$ denote the ith row of $\mathbf{T}$. Then,

$$
\begin{aligned}
E(\mathbf{T}'\mathbf{A}\mathbf{T}) &= E(\sum_i \sum_j a_{ij}\mathbf{t}_i\mathbf{t}_j') \\
&= E\{\sum_i \sum_j a_{ij}[(\mathbf{t}_i - \boldsymbol{\mu}_i)(\mathbf{t}_j - \boldsymbol{\mu}_j)' + \boldsymbol{\mu}_i\mathbf{t}_j' + \mathbf{t}_i\boldsymbol{\mu}_j' - \boldsymbol{\mu}_i\boldsymbol{\mu}_j']\},
\end{aligned}
$$

where a_{ij} is the $(i, j)^{th}$ element of $\mathbf{A}$ and $\boldsymbol{\mu}_i$ is the mean of $\mathbf{t}_i$. It follows that

$$
E(\mathbf{T}'\mathbf{A}\mathbf{T}) = \sum_i \sum_j a_{ij}Cov(\mathbf{t}_i, \mathbf{t}_j) + \sum_i \sum_j a_{ij}\boldsymbol{\mu}_i\boldsymbol{\mu}_j'.
$$

But, $Cov(\mathbf{t}_i, \mathbf{t}_j) = \mathbf{0}$ for $i \neq j$. We then have

$$
\begin{aligned}
E(\mathbf{T}'\mathbf{A}\mathbf{T}) &= \sum_i a_{ii}Var(\mathbf{t}_i) + E(\mathbf{T})'\mathbf{A}E(\mathbf{T}) \\
&= tr(\mathbf{A})\boldsymbol{\Delta}_t + E(\mathbf{T})'\mathbf{A}E(\mathbf{T}).
\end{aligned}
$$

10.2. Suppose that $\frac{a_{(i)}}{m_{(i)}}\boldsymbol{\Delta}_{(i)} = \boldsymbol{\Delta}_0$ for $i = 1, 2, \ldots, t$. Then,

$$
\begin{aligned}
\left|\mathbf{V}_\phi\right| &= \left|\mathbf{K}_r^- \left[\sum_{i=\omega}^{\nu+1} \frac{a_i^2}{m_i}(\boldsymbol{\Delta}_i \otimes \boldsymbol{\Delta}_i)\right] \mathbf{K}_r\right| \\
&= \left|\mathbf{K}_r^- \left[\sum_{i=1}^{t} m_{(i)}(\boldsymbol{\Delta}_0 \otimes \boldsymbol{\Delta}_0)\right] \mathbf{K}_r\right| \\
&= \left[\sum_{i=1}^{t} m_{(i)}\right]^{r(r+1)/2} \left|\boldsymbol{\Delta}_0\right|^{r+1}.
\end{aligned}
$$

Furthermore,

$$
\begin{aligned}
\left|\sum_{i=\omega}^{\nu+1} a_i\boldsymbol{\Delta}_i\right|^{r+1} &= \left|\sum_{i=1}^{t} a_{(i)}\boldsymbol{\Delta}_{(i)}\right|^{r+1} \\
&= \left[\sum_{i=1}^{t} m_{(i)}\right]^{r(r+1)} \left|\boldsymbol{\Delta}_0\right|^{r+1}.
\end{aligned}
$$

Formula (10.3.20) now becomes

$$\begin{aligned} m &= \left[\frac{[\sum_{i=1}^{t} m_{(i)}]^{r(r+1)} |\mathbf{\Delta}_0|^{r+1}}{[\sum_{i=1}^{t} m_{(i)}]^{r(r+1)/2} |\mathbf{\Delta}_0|^{r+1}} \right]^{1/r^*} \\ &= \sum_{i=1}^{t} m_{(i)}. \end{aligned}$$

10.3. Let $\tilde{\mathbf{C}}$ be another matrix whose rows form a basis for the $(v-1)$-dimensional orthogonal complement of the one-dimensional space spanned by $\mathbf{1}_v$. There exists a nonsingular matrix $\tilde{\mathbf{Q}}$ such that $\mathbf{C} = \tilde{\mathbf{Q}}\tilde{\mathbf{C}}$. Let $\tilde{\mathbf{d}}_j = \tilde{\mathbf{C}}\bar{\mathbf{y}}_j$, $j = 1, 2, \ldots, b$. Then, $\mathbf{d}_j = \tilde{\mathbf{Q}}\tilde{\mathbf{d}}_j$, $\bar{\mathbf{d}} = \tilde{\mathbf{Q}}\bar{\tilde{\mathbf{d}}}$, and by formula (10.4.9),

$$\mathbf{S} = \frac{1}{b-1}\tilde{\mathbf{Q}} \sum_{j=1}^{b} (\tilde{\mathbf{d}}_j - \bar{\tilde{\mathbf{d}}})(\tilde{\mathbf{d}}_j - \bar{\tilde{\mathbf{d}}})'\tilde{\mathbf{Q}}'.$$

Making the proper substitution in (10.4.8), we get

$$T^2 = b\bar{\tilde{\mathbf{d}}}' \left[\frac{1}{b-1} \sum_{j=1}^{b} (\tilde{\mathbf{d}}_j - \bar{\tilde{\mathbf{d}}})(\tilde{\mathbf{d}}_j - \bar{\tilde{\mathbf{d}}})' \right]^{-1} \bar{\tilde{\mathbf{d}}}.$$

Thus, T^2 is not changed when $\mathbf{C}$ is replaced by $\tilde{\mathbf{C}}$.

10.4. **(a)** Since $\mathbf{Y}'\mathbf{P}_{i_0}\mathbf{Y}$ is positive definite, there exists a nonsingular matrix $\mathbf{C}_0$ such that $\mathbf{C}_0'\mathbf{Y}'\mathbf{P}_{i_0}\mathbf{Y}\mathbf{C}_0 = \mathbf{I}$ and $\mathbf{C}_0'\mathbf{Y}'\mathbf{P}_i\mathbf{Y}\mathbf{C}_0 = \mathbf{D}_0$, where $\mathbf{D}_0$ is a diagonal matrix whose diagonal elements are the eigenvalues of $(\mathbf{Y}'\mathbf{P}_i\mathbf{Y})(\mathbf{Y}'\mathbf{P}_{i_0}\mathbf{Y})^{-1}$ (see, for example, Graybill, 1983, Theorem 12.2.13, p. 408). Hence,

$$F_\lambda = \frac{m_{i_0}}{m_i} \frac{\boldsymbol{\lambda}'\mathbf{C}_0'^{-1}\mathbf{D}_0\mathbf{C}_0^{-1}\boldsymbol{\lambda}}{\boldsymbol{\lambda}'\mathbf{C}_0'^{-1}\mathbf{C}_0^{-1}\boldsymbol{\lambda}}.$$

Let $\mathbf{C}_0^{-1}\boldsymbol{\lambda} = \boldsymbol{\tau}$. Then,

$$\begin{aligned} F_\lambda &= \frac{m_{i_0}}{m_i} \frac{\boldsymbol{\tau}'\mathbf{D}_0\boldsymbol{\tau}}{\boldsymbol{\tau}'\boldsymbol{\tau}}. \\ \sup_{\boldsymbol{\lambda}\neq\mathbf{0}} F_\lambda &= \frac{m_{i_0}}{m_i} \sup_{\boldsymbol{\tau}\neq\mathbf{0}} \frac{\boldsymbol{\tau}'\mathbf{D}_0\boldsymbol{\tau}}{\boldsymbol{\tau}'\boldsymbol{\tau}}. \end{aligned}$$

But,

$$\boldsymbol{\tau}'\mathbf{D}_0\boldsymbol{\tau} \leq \boldsymbol{\tau}'\boldsymbol{\tau}\lambda_{\max}(\mathbf{D}_0).$$

Furthermore, $\lambda_{\max}(\mathbf{D}_0) = \lambda_{\max}[(\mathbf{Y}'\mathbf{P}_i\mathbf{Y})(\mathbf{Y}'\mathbf{P}_{i_0}\mathbf{Y})^{-1}]$. Thus,

$$\sup_{\boldsymbol{\tau}\neq\mathbf{0}} \frac{\boldsymbol{\tau}'\mathbf{D}_0\boldsymbol{\tau}}{\boldsymbol{\tau}'\boldsymbol{\tau}} \leq \boldsymbol{\lambda}_{\max}[(\mathbf{Y}'\mathbf{P}_i\mathbf{Y})(\mathbf{Y}'\mathbf{P}_{i_0}\mathbf{Y})^{-1}].$$

Note that if $\boldsymbol{\tau} = \boldsymbol{\tau}_0$, where $\boldsymbol{\tau}_0$ is an eigenvector of $\mathbf{D}_0$ with an eigenvalue $\lambda_{\max}[(\mathbf{Y}'\mathbf{P}_i\mathbf{Y})(\mathbf{Y}'\mathbf{P}_{i_0}\mathbf{Y})^{-1}]$, then

$$\frac{\boldsymbol{\tau}'\mathbf{D}_0\boldsymbol{\tau}}{\boldsymbol{\tau}'\boldsymbol{\tau}} = \lambda_{\max}[(\mathbf{Y}'\mathbf{P}_i\mathbf{Y})(\mathbf{Y}'\mathbf{P}_{i_0}\mathbf{Y})^{-1}].$$

It follows that

$$\sup_{\boldsymbol{\lambda}\neq\mathbf{0}} F_\lambda = \frac{m_{i_0}}{m_i}\lambda_{\max}[(\mathbf{Y}'\mathbf{P}_i\mathbf{Y})(\mathbf{Y}'\mathbf{P}_{i_0}\mathbf{Y})^{-1}].$$

(b) This is true by the fact that if k is a constant, then $F_\lambda \leq k$ for all $\boldsymbol{\lambda} \neq \mathbf{0}$ if and only if $\sup_{\boldsymbol{\lambda}\neq\mathbf{0}} F_\lambda \leq k$. Equivalently, $F_\lambda > k$ for some $\boldsymbol{\lambda} \neq \mathbf{0}$ if and only if $\sup_{\boldsymbol{\lambda}\neq\mathbf{0}} F_\lambda > k$. Thus, a large value of $\lambda_{\max}[(\mathbf{Y}'\mathbf{P}_i\mathbf{Y})(\mathbf{Y}'\mathbf{P}_{i_0}\mathbf{Y})^{-1}]$ implies that F_λ is large for some $\boldsymbol{\lambda}$. This leads to the rejection of $H_{0\lambda}$, and hence H_0, by the union-intersection principle.

10.5. (a) By Lemma 2.3.1,

$$\begin{aligned}
\mathbf{P}_0 &= \frac{1}{vn}\mathbf{A}_0 = \frac{1}{vn}\mathbf{J}_v \otimes \mathbf{J}_n, \\
\mathbf{P}_1 &= \frac{1}{n}\mathbf{A}_1 - \frac{1}{vn}\mathbf{A}_0 = \frac{1}{n}\mathbf{I}_v \otimes \mathbf{J}_n - \frac{1}{vn}\mathbf{J}_v \otimes \mathbf{J}_n, \\
\mathbf{P}_2 &= \mathbf{A}_2 - \frac{1}{n}\mathbf{A}_1 = \mathbf{I}_v \otimes \mathbf{I}_n - \frac{1}{n}\mathbf{I}_v \otimes \mathbf{J}_n.
\end{aligned}$$

(b) $m_0 = \text{rank } (\mathbf{P}_0) = 1$, $m_1 = \text{rank } (\mathbf{P}_1) = v - 1$, $m_2 = \text{rank } (\mathbf{P}_2) = v(n-1)$.

(c) A row of $\mathbf{Y}$ is of the form $\tilde{\mathbf{y}}'_{k\ell}$ for some $k, \ell (k = 1, 2, \ldots, v;\ \ell = 1, 2, \ldots, n)$. Thus, $\tilde{\mathbf{y}}_{k\ell}$ represents a typical column of $\mathbf{Y}'$. It follows that

$$\begin{aligned}
\mathbf{Y}'\mathbf{P}_1\mathbf{Y} &= \frac{1}{n}\sum_{k=1}^{v}\left(\sum_{\ell=1}^{n}\tilde{\mathbf{y}}_{k\ell}\right)\left(\sum_{\ell=1}^{n}\tilde{\mathbf{y}}_{k\ell}\right)' \\
&\quad -\frac{1}{vn}\left(\sum_{k=1}^{v}\sum_{\ell=1}^{n}\tilde{\mathbf{y}}_{k\ell}\right)\left(\sum_{k=1}^{v}\sum_{\ell=1}^{n}\tilde{\mathbf{y}}_{k\ell}\right)' \\
&= n\sum_{k=1}^{v}(\bar{\tilde{\mathbf{y}}}_{k.} - \bar{\tilde{\mathbf{y}}}_{..})(\bar{\tilde{\mathbf{y}}}_{k.} - \bar{\tilde{\mathbf{y}}}_{..})'.
\end{aligned}$$

$$\begin{aligned}
\mathbf{Y}'\mathbf{P}_2\mathbf{Y} &= \sum_{k=1}^{v}\sum_{\ell=1}^{n} \tilde{\mathbf{y}}_{k\ell}\tilde{\mathbf{y}}'_{k\ell} - n\sum_{k=1}^{v} \bar{\tilde{\mathbf{y}}}_{k\cdot}\bar{\tilde{\mathbf{y}}}'_{k\cdot} \\
&= \sum_{k=1}^{v}\sum_{\ell=1}^{n} (\tilde{\mathbf{y}}_{k\ell} - \bar{\tilde{\mathbf{y}}}_{k\cdot})(\tilde{\mathbf{y}}_{k\ell} - \bar{\tilde{\mathbf{y}}}_{k\cdot})'.
\end{aligned}$$

10.6. **(c)** The fixed portion of the model is $\mathbf{XG} = \mathbf{H}_0\mathbf{B}_0 + \mathbf{H}_1\mathbf{B}_1$. Let $\boldsymbol{\Sigma}_2, \boldsymbol{\Sigma}_3$, and $\boldsymbol{\Sigma}_4$ be the variance–covariance matrices for $\boldsymbol{\beta}_\ell$, $(\boldsymbol{\tau\beta})_{k\ell}$, and the error term, respectively, in model (10.2.7).

(i) Using part (d) of Theorem 10.3.1, we get

$$\begin{aligned}
E(\mathbf{Y}'\mathbf{P}_1\mathbf{Y}) &= (v-1)(n\boldsymbol{\Sigma}_3 + \boldsymbol{\Sigma}_4) + \mathbf{G}'\mathbf{X}'\mathbf{P}_1\mathbf{XG}, \\
E(\mathbf{Y}'\mathbf{P}_2\mathbf{Y}) &= (b-1)(vn\boldsymbol{\Sigma}_2 + n\boldsymbol{\Sigma}_3 + \boldsymbol{\Sigma}_4), \\
E(\mathbf{Y}'\mathbf{P}_3\mathbf{Y}) &= (v-1)(b-1)(n\boldsymbol{\Sigma}_3 + \boldsymbol{\Sigma}_4), \\
E(\mathbf{Y}'\mathbf{P}_4\mathbf{Y}) &= vb(n-1)\boldsymbol{\Sigma}_4.
\end{aligned}$$

(ii) $\lambda_{\max}[(\mathbf{Y}'\mathbf{P}_1\mathbf{Y})(\mathbf{Y}'\mathbf{P}_3\mathbf{Y})^{-1}]$ for A,
$\lambda_{\max}[(\mathbf{Y}'\mathbf{P}_2\mathbf{Y})(\mathbf{Y}'\mathbf{P}_3\mathbf{Y})^{-1}]$ for B,
$\lambda_{\max}[(\mathbf{Y}'\mathbf{P}_3\mathbf{Y})(\mathbf{Y}'\mathbf{P}_4\mathbf{Y})^{-1}]$ for $A * B$.

10.7. The model is

$$\tilde{\mathbf{y}}_{k\ell} = \boldsymbol{\mu} + \boldsymbol{\beta}_\ell + \boldsymbol{\tau}_k + \mathbf{e}_{k\ell}, \quad \ell = 1,2,3; \quad k = 1,2,\cdots,12,$$

where $\boldsymbol{\beta}_\ell$ and $\boldsymbol{\tau}_k$ correspond to the ℓth herd and k^{th} sire, respectively. Let $\boldsymbol{\Sigma}_\tau$ and $\boldsymbol{\Sigma}_e$ denote the variance–covariance matrices associated with $\boldsymbol{\tau}_k$ and $\mathbf{e}_{k\ell}$, respectively. The matrices $\mathbf{Y}'\mathbf{P}_i\mathbf{Y}$, $i = 1,2,3$, for herds, sires, and the error terms are, respectively,

$$\begin{aligned}
\mathbf{Y}'\mathbf{P}_1\mathbf{Y} &= \mathbf{Y}'\left(\frac{1}{12}\mathbf{I}_3 \otimes \mathbf{J}_{12} - \frac{1}{36}\mathbf{J}_3 \otimes \mathbf{J}_{12}\right)\mathbf{Y} \\
&= \begin{bmatrix} 146990246 & 5678716.17 \\ 5678716.17 & 219457.39 \end{bmatrix}, \\
\mathbf{Y}'\mathbf{P}_2\mathbf{Y} &= \mathbf{Y}'\left(\frac{1}{3}\mathbf{J}_3 \otimes \mathbf{I}_{12} - \frac{1}{36}\mathbf{J}_3 \otimes \mathbf{J}_{12}\right)\mathbf{Y} \\
&= \begin{bmatrix} 4458437630.1 & 162870218 \\ 162870218 & 6414060.3 \end{bmatrix}, \\
\mathbf{Y}'\mathbf{P}_3\mathbf{Y} &= \mathbf{Y}'\left(\mathbf{I}_3 \otimes \mathbf{I}_{12} - \frac{1}{12}\mathbf{I}_3 \otimes \mathbf{J}_{12} - \frac{1}{3}\mathbf{J}_3 \otimes \mathbf{I}_{12} + \frac{1}{36}\mathbf{J}_3 \otimes \mathbf{J}_{12}\right)\mathbf{Y} \\
&= \begin{bmatrix} 8710092904.7 & 308693037.83 \\ 308693037.83 & 11176601.94 \end{bmatrix}.
\end{aligned}$$

(i) Herd effect: Roy's largest root test statistic value is

$$\lambda_{\max}[(\mathbf{Y}'\mathbf{P}_1\mathbf{Y})(\mathbf{Y}'\mathbf{P}_3\mathbf{Y})^{-1}] = 0.0233.$$

The corresponding P-value is 0.7762. The herd effect is therefore not significant. Note that the critical values for Roy's largest root test can be obtained, for example, from Seber (1984, Appendix D14).

(ii) The sire effect: Roy's largest root test statistic value is

$$\lambda_{\max}[(\mathbf{Y}'\mathbf{P}_2\mathbf{Y})(\mathbf{Y}'\mathbf{P}_3\mathbf{Y})^{-1}] = 1.9953.$$

The corresponding P-value is 0.0028. The sire effect is significant.

10.8. (a) The model is

$$\tilde{\mathbf{y}}_{k\ell m} = \boldsymbol{\mu} + \boldsymbol{\tau}_k + \boldsymbol{\beta}_{k(\ell)} + \boldsymbol{\gamma}_m + (\boldsymbol{\tau}\boldsymbol{\gamma})_{km} + \mathbf{e}_{k\ell m},$$

where $\boldsymbol{\tau}_k$ represents the kth treatment effect ($k = 1, 2$), $\boldsymbol{\beta}_{k(\ell)}$ represents the effect of the ℓth subject within the kth treatment ($j = 1, 2, \ldots, 9$), $\boldsymbol{\gamma}_m$ represents the mth time ($m = 1, 2, 3$), and $(\boldsymbol{\tau}\boldsymbol{\gamma})_{km}$ is the treatment by time interaction effect. Here, $\boldsymbol{\tau}_k$ and $\boldsymbol{\gamma}_m$ are fixed, and $\boldsymbol{\beta}_{k(\ell)}$ is random with a variance–covariance matrix $\boldsymbol{\Sigma}_\beta$. The matrices $\mathbf{Y}'\mathbf{P}_i\mathbf{Y}$, $i = 1, 2, 3, 4, 5$, corresponding to treatments, subjects, times, treatment by time interaction, and the error term, respectively, are

$$\begin{aligned}
\mathbf{Y}'\mathbf{P}_1\mathbf{Y} &= \mathbf{Y}'\left(\frac{1}{27}\mathbf{I}_2 \otimes \mathbf{J}_9 \otimes \mathbf{J}_3 - \frac{1}{54}\mathbf{J}_2 \otimes \mathbf{J}_9 \otimes \mathbf{J}_3\right)\mathbf{Y} \\
\mathbf{Y}'\mathbf{P}_2\mathbf{Y} &= \mathbf{Y}'\left(\frac{1}{3}\mathbf{I}_2 \otimes \mathbf{I}_9 \otimes \mathbf{J}_3 - \frac{1}{27}\mathbf{I}_2 \otimes \mathbf{J}_9 \otimes \mathbf{J}_3\right)\mathbf{Y} \\
\mathbf{Y}'\mathbf{P}_3\mathbf{Y} &= \mathbf{Y}'\left(\frac{1}{18}\mathbf{J}_2 \otimes \mathbf{J}_9 \otimes \mathbf{I}_3 - \frac{1}{54}\mathbf{J}_2 \otimes \mathbf{J}_9 \otimes \mathbf{J}_3\right)\mathbf{Y} \\
\mathbf{Y}'\mathbf{P}_4\mathbf{Y} &= \mathbf{Y}'\left(\frac{1}{9}\mathbf{I}_2 \otimes \mathbf{J}_9 \otimes \mathbf{I}_3 - \frac{1}{27}\mathbf{I}_2 \otimes \mathbf{J}_9 \otimes \mathbf{J}_3 - \frac{1}{18}\mathbf{J}_2 \otimes \mathbf{J}_9 \otimes \mathbf{I}_3\right. \\
&\quad \left.+\frac{1}{54}\mathbf{J}_2 \otimes \mathbf{J}_9 \otimes \mathbf{J}_3\right)\mathbf{Y} \\
\mathbf{Y}'\mathbf{P}_5\mathbf{Y} &= \mathbf{Y}'\left(\mathbf{I}_2 \otimes \mathbf{I}_9 \otimes \mathbf{I}_3 - \frac{1}{3}\mathbf{I}_2 \otimes \mathbf{I}_9 \otimes \mathbf{J}_3 - \frac{1}{9}\mathbf{I}_2 \otimes \mathbf{J}_9 \otimes \mathbf{I}_3\right. \\
&\quad \left.+\frac{1}{27}\mathbf{I}_2 \otimes \mathbf{J}_9 \otimes \mathbf{J}_3\right)\mathbf{Y}.
\end{aligned}$$

The corresponding expected values are

$$\begin{aligned}
E(\mathbf{Y}'\mathbf{P}_1\mathbf{Y}) &= \mathbf{G}'\mathbf{X}'\mathbf{P}_1\mathbf{X}\mathbf{G} + 3\boldsymbol{\Sigma}_\beta + \boldsymbol{\Sigma}_e \\
E(\mathbf{Y}'\mathbf{P}_2\mathbf{Y}) &= 16(3\boldsymbol{\Sigma}_\beta + \boldsymbol{\Sigma}_e) \\
E(\mathbf{Y}'\mathbf{P}_3\mathbf{Y}) &= \mathbf{G}'\mathbf{X}'\mathbf{P}_3\mathbf{X}\mathbf{G} + 2\boldsymbol{\Sigma}_e
\end{aligned}$$

$$\begin{aligned} E(\mathbf{Y}'\mathbf{P}_4\mathbf{Y}) &= \mathbf{G}'\mathbf{X}'\mathbf{P}_4\mathbf{X}\mathbf{G} + 2\boldsymbol{\Sigma}_e \\ E(\mathbf{Y}'\mathbf{P}_5\mathbf{Y}) &= 32\boldsymbol{\Sigma}_e, \end{aligned}$$

where $\mathbf{XG} = \mathbf{H}_0\mathbf{B}_0 + \mathbf{H}_1\mathbf{B}_1 + \mathbf{H}_3\mathbf{B}_3 + \mathbf{H}_4\mathbf{B}_4$. Note that rank $(\mathbf{P}_1) = 1$, rank $(\mathbf{P}_2) = 16$, rank $(\mathbf{P}_3) = 2$, rank $(\mathbf{P}_4) = 2$, and rank $(\mathbf{P}_5) = 32$.

(b) Roy's largest root statistic value for testing the treatment by time interaction effect is

$$\lambda_{\max}[(\mathbf{Y}'\mathbf{P}_4\mathbf{Y})(\mathbf{Y}'\mathbf{P}_5\mathbf{Y})^{-1}] = 0.1334.$$

The P-value is 0.2678. This interaction is not significant.

(c) For the treatment effect, Roy's test statistic value is

$$\lambda_{\max}[(\mathbf{Y}'\mathbf{P}_1\mathbf{Y})(\mathbf{Y}'\mathbf{P}_2\mathbf{Y})^{-1}] = 0.1314.$$

The P-value is 0.6176. The treatment effect is not significant. For the time effect,

$$\lambda_{\max}[(\mathbf{Y}'\mathbf{P}_3\mathbf{Y})(\mathbf{Y}'\mathbf{P}_5\mathbf{Y})^{-1}] = 13.6216.$$

The P-value is 0.0001. The time effect is highly significant.

10.10. (a) Let $\mathbf{y}_{ij}$ denote the vector of observations in the $(i, j)^{th}$ cell. Then,

$$\begin{aligned} E(\mathbf{y}_{ij}) &= (\mu + \tau_i)\mathbf{1}_n \\ Var(\mathbf{y}_{ij}) &= Var\{[\beta_j + (\tau\beta)_{ij}]\mathbf{1}_n + \mathbf{e}_{ij}\} \\ &= \sigma_{ii}\mathbf{J}_n + \sigma_e^2\mathbf{I}_n, \end{aligned}$$

where σ_{ii} is the ith diagonal element of $\boldsymbol{\Sigma} = (\sigma_{ij})$, the variance–covariance matrix of $\boldsymbol{\zeta}_j$ (see formula 10.4.2), and $\mathbf{e}_{ij}$ is the vector of random errors in the $(i, j)^{th}$ cell. Furthermore,

$$\begin{aligned} Cov(\mathbf{y}_{i_1 j}, \mathbf{y}_{i_2 j}) &= \sigma_{i_1 i_2}\mathbf{J}_n, \quad i_1 \neq i_2 \\ Cov(\mathbf{y}_{i_1 j_1}, \mathbf{y}_{i_2 j_2}) &= \mathbf{0}, \quad j_1 \neq j_2. \end{aligned}$$

Let $\mathbf{y}_i$ be the vector of observations under level i of the fixed factor, that is,

$$\mathbf{y}_i = [\mathbf{y}'_{i1} : \mathbf{y}'_{i2} : \cdots : \mathbf{y}'_{ib}]'.$$

Then,

$$\begin{aligned} E(\mathbf{y}_i) &= (\mu + \tau_i)\mathbf{1}_{bn}, \quad i = 1, 2, \cdots, v \\ Var(\mathbf{y}_i) &= \sigma_{ii}(\mathbf{I}_b \otimes \mathbf{J}_n) + \sigma_e^2\mathbf{I}_{bn}, \quad i = 1, 2, \cdots, v \\ Cov(\mathbf{y}_{i_1}, \mathbf{y}_{i_2}) &= \sigma_{i_1 i_2}(\mathbf{I}_b \otimes \mathbf{J}_n), \quad i_1 \neq i_2. \end{aligned}$$

Now, $\mathbf{y} = [\mathbf{y}_1' : \mathbf{y}_2' : \ldots : \mathbf{y}_v']'$. Its expected value and variance–covariance matrix are then given by

$$\begin{aligned} E(\mathbf{y}) &= (\mu \mathbf{1}_v + \boldsymbol{\tau}) \otimes \mathbf{1}_{bn} \\ Var(\mathbf{y}) &= \boldsymbol{\Sigma} \otimes \mathbf{I}_b \otimes \mathbf{J}_n + \sigma_e^2 \mathbf{I}_{vbn}. \end{aligned}$$

(b)

$$\begin{aligned} MS(\tau) &= \frac{1}{v-1}\mathbf{y}'\mathbf{P}_1\mathbf{y} \\ MS(\beta) &= \frac{1}{b-1}\mathbf{y}'\mathbf{P}_2\mathbf{y} \\ MS(\tau\beta) &= \frac{1}{(v-1)(b-1)}\mathbf{y}'\mathbf{P}_3\mathbf{y} \\ MS(e) &= \frac{1}{bv(n-1)}\mathbf{y}'\mathbf{P}_4\mathbf{y}, \end{aligned}$$

where

$$\begin{aligned} \mathbf{P}_1 &= \frac{1}{bn}\mathbf{I}_v \otimes \mathbf{J}_b \otimes \mathbf{J}_n - \frac{1}{vbn}\mathbf{J}_v \otimes \mathbf{J}_b \otimes \mathbf{J}_n \\ \mathbf{P}_2 &= \frac{1}{vn}\mathbf{J}_v \otimes \mathbf{I}_b \otimes \mathbf{J}_n - \frac{1}{vbn}\mathbf{J}_v \otimes \mathbf{J}_b \otimes \mathbf{J}_n \\ \mathbf{P}_3 &= \frac{1}{n}\mathbf{I}_v \otimes \mathbf{I}_b \otimes \mathbf{J}_n - \frac{1}{bn}\mathbf{I}_v \otimes \mathbf{J}_b \otimes \mathbf{J}_n - \frac{1}{vn}\mathbf{J}_v \otimes \mathbf{I}_b \otimes \mathbf{J}_n + \frac{1}{vbn}\mathbf{J}_v \otimes \mathbf{J}_b \otimes \mathbf{J}_n \\ \mathbf{P}_4 &= \mathbf{I}_v \otimes \mathbf{I}_b \otimes \mathbf{I}_n - \frac{1}{n}\mathbf{I}_v \otimes \mathbf{I}_b \otimes \mathbf{J}_n. \end{aligned}$$

Hence,

$$\begin{aligned} E[MS(\tau)] &= \frac{1}{v-1}\{\mathbf{g}'\mathbf{X}'\mathbf{P}_1\mathbf{X}\mathbf{g} + tr[\mathbf{P}_1(\boldsymbol{\Sigma} \otimes \mathbf{I}_b \otimes \mathbf{J}_n + \sigma_e^2\mathbf{I}_{vbn})]\} \\ E[MS(\beta)] &= \frac{1}{b-1}tr[\mathbf{P}_2(\boldsymbol{\Sigma} \otimes \mathbf{I}_b \otimes \mathbf{J}_n + \sigma_e^2\mathbf{I}_{vbn})] \\ E[MS(\tau\beta)] &= \frac{1}{(v-1)(b-1)}tr[\mathbf{P}_3(\boldsymbol{\Sigma} \otimes \mathbf{I}_b \otimes \mathbf{J}_n + \sigma_e^2\mathbf{I}_{vbn})] \\ E[MS(e)] &= \frac{1}{vb(n-1)}tr[\mathbf{P}_4(\boldsymbol{\Sigma} \otimes \mathbf{I}_b \otimes \mathbf{J}_n + \sigma_e^2\mathbf{I}_{vbn})], \end{aligned}$$

where $\mathbf{X}\mathbf{g} = \mu\mathbf{1}_{vbn} + (\mathbf{I}_v \otimes \mathbf{1}_b \otimes \mathbf{1}_n)\boldsymbol{\tau}$. Note that

$$\mathbf{g}'\mathbf{X}'\mathbf{P}_1\mathbf{X}\mathbf{g} = bn\sum_{i=1}^{v}(\tau_i - \bar{\tau}_.)^2.$$

Furthermore,

$$\mathbf{P}_1(\mathbf{\Sigma} \otimes \mathbf{I}_b \otimes \mathbf{J}_n + \sigma_e^2 \mathbf{I}_{vbn}) = \frac{1}{b}\left(\mathbf{\Sigma} - \frac{1}{v}\mathbf{J}_v\mathbf{\Sigma}\right) \otimes \mathbf{J}_b \otimes \mathbf{J}_n + \sigma_e^2 \mathbf{P}_1 \tag{A.10.1}$$

$$\mathbf{P}_2(\mathbf{\Sigma} \otimes \mathbf{I}_b \otimes \mathbf{J}_n + \sigma_e^2 \mathbf{I}_{vbn}) = \frac{1}{v}\mathbf{J}_v\mathbf{\Sigma} \otimes \left(\mathbf{I}_b - \frac{1}{b}\mathbf{J}_b\right) \otimes \mathbf{J}_n + \sigma_e^2 \mathbf{P}_2 \tag{A.10.2}$$

$$\mathbf{P}_3(\mathbf{\Sigma} \otimes \mathbf{I}_b \otimes \mathbf{J}_n + \sigma_e^2 \mathbf{I}_{vbn}) = \left(\mathbf{\Sigma} - \frac{1}{v}\mathbf{J}_v\mathbf{\Sigma}\right) \otimes \left(\mathbf{I}_b - \frac{1}{b}\mathbf{J}_b\right) \otimes \mathbf{J}_n + \sigma_e^2 \mathbf{P}_3 \tag{A.10.3}$$

$$\mathbf{P}_4(\mathbf{\Sigma} \otimes \mathbf{I}_b \otimes \mathbf{J}_n + \sigma_e^2 \mathbf{I}_{vbn}) = \sigma_e^2 \mathbf{P}_4.$$

Hence,

$$\begin{aligned}
E[MS(\tau)] &= \frac{bn}{v-1}\sum_{i=1}^{v}(\tau_i - \bar{\tau}_.)^2 + \frac{n}{v-1}\left[tr(\mathbf{\Sigma}) - \frac{1}{v}\mathbf{1}_v'\mathbf{\Sigma}\mathbf{1}_v\right] + \sigma_e^2 \\
E[MS(\beta)] &= \frac{n}{v}\mathbf{1}_v'\mathbf{\Sigma}\mathbf{1}_v + \sigma_e^2 \\
E[MS(\tau\beta)] &= \frac{n}{v-1}\left[tr(\mathbf{\Sigma}) - \frac{1}{v}\mathbf{1}_v'\mathbf{\Sigma}\mathbf{1}_v\right] + \sigma_e^2 \\
E[MS(e)] &= \sigma_e^2.
\end{aligned}$$

(c)

$$\begin{aligned}
MS(\tau) &= \frac{1}{v-1}\sum_i \lambda_{1i}\chi^2_{m_{1i}} \\
MS(\beta) &= \frac{1}{b-1}\sum_i \lambda_{2i}\chi^2_{m_{2i}} \\
MS(\tau\beta) &= \frac{1}{(v-1)(b-1)}\sum_i \lambda_{3i}\chi^2_{m_{3i}},
\end{aligned}$$

where the λ's are, respectively, the distinct nonzero eigenvalues of the matrices on the right-hand side of formulas (A.10.1), (A.10.2), and (A.10.3). The degrees of freedom of the independent chi-squared variates are the corresponding multiplicities of these eigenvalues.

General Bibliography

Agresti, A. (1990). *Categorical Data Analysis*. Wiley, New York.

Ahrens, H. J. and Pincus, R. (1981). "On two measures of unbalancedness in a one-way model and their relation to efficiency." *Biometrical Journal*, 23, 227–235.

Ahrens, H. J. and Sanchez, J. E. (1982). "Unbalancedness and efficiency in estimating components of variance: MINQUE and ANOVA procedure." *Biometrial Journal*, 24, 649–661.

Ahrens, H. J. and Sanchez, J. E. (1988). "Unbalancedness of designs, measures of." In: *Encyclopedia of Statistical Sciences*, Vol. 9 (S. Kotz, N. L. Johnson, Eds.), Wiley, New York, 383–386.

Ahrens, H. J. and Sanchez, J. E. (1992). "Imbalance and its influence on variance component estimation." *Biometrical Journal*, 34, 539–555.

Alalouf, I. S. (1980). "A multivariate approach to a mixed linear model." *Journal of the American Statistical Association*, 75, 194–200.

Amemiya, Y. (1985). "What should be done when an estimated between-group covariance matrix is not nonnegative definite?" *The American Statistician*, 39, 112–117.

Anderson, B. M., Anderson, T. W., and Olkin, I. (1986). "Maximum likelihood estimators and likelihood ratio criteria in multivariate components of variance." *Annals of Statistics*, 14, 405–417.

Anderson, R. L. and Crump, P. P. (1967). "Comparisons of designs and estimation procedures for estimating parameters in a two-stage nested process." *Technometrics*, 9, 499–516.

Anderson, T. W. (1984). *An Introduction to Multivariate Statistical Analysis*, Second Edition. Wiley, New York.

Anderson, T. W. (1985). "Components of variance in MANOVA." In: *Multivariate Analysis-VI* (P.R. Krishnaiah, Ed.), North-Holland, Amsterdam, 1–8.

Arnold, S. F. (1981). *The Theory of Linear Models and Multivariate Analysis*. Wiley, New York.

Bartlett, M. S. (1936). "The information available in small samples." *Proceedings of the Cambridge Philosophical Society*, 32, 560–566.

Beyer, W. H. (1968). *Handbook of Tables for Probability and Statistics*, Second Edition. The Chemical Rubber Company, Cleveland, Ohio.

Boik, R. J. (1988). "The mixed model for multivariate repeated measures: Validity conditions and an approximate test." *Psychometrika*, 53, 469–486.

Box, G. E. P. (1950). "Problems in the analysis of growth and wear curves." *Biometrics*, 6, 362–389.

Browne, M. W. (1974). "Generalized least squares estimators in the analysis of covariance structures." *South African Statistical Journal*, 8, 1–24.

Burdick, R. K. and Graybill, F. A. (1992). *Confidence Intervals on Variance Components*. Dekker, New York.

Burdick, R. K. and Sielken, R. L. (1978). "Exact confidence intervals for linear combinations of variance components in nested classifications." *Journal of the American Statistical Association*, 73, 632–635.

Capen, R. C. (1991). "Exact testing procedures for unbalanced random and mixed linear models." Unpublished Ph.D. dissertation, University of Florida, Gainesville, Florida.

Caro, R. F., Grossman, M., and Fernando, R. L. (1985). "Effects of data imbalance on estimation of heritability." *Theoretical and Applied Genetics*, 69, 523–530.

Christensen, R. (1996). "Exact tests for variance components." *Biometrics*, 52, 309–314.

Cohen, A. and Sackrowitz, H. B. (1989). "Exact tests that recover inter-block information in balanced incomplete block designs." *Journal of the American Statistical Association*, 84, 556–559.

Cummings, W. B. and Gaylor, D. W. (1974). "Variance component testing in unbalanced nested designs." *Journal of the American Statistical Association*, 69, 765–771.

Damon, R. A., Jr. and Harvey, W. R. (1987). *Experimental Design, ANOVA, and Regression*. Harper and Row, New York.

Das, R. and Sinha, B. K. (1987). "Robust optimum invariant unbiased tests for variance components." In: *Proceedings of the Second International Tampere Conference in Statistics* (T. Pukkila, S. Puntanen, Eds.), University of Tampere, Finland, 317–342.

Das, R. and Sinha, B. K. (1988). "Optimum invariant tests in random MANOVA models." *Canadian Journal of Statistics*, 16, 193–200.

Davies, R. B. (1973). "Numerical inversion of a characteristic function." *Biometrika*, 60, 415–417.

Davies, R. B. (1980). "The distribution of a linear combination of χ^2 random variables." *Applied Statistics*, 29, 323–333.

Dobrohotov, I. S. (1968). "On the theory of a general linear hypothesis with unknown weights." *Trudy Mathematical Institute Steklov*, 104, 1–19.

Donner, A. and Koval, J. J. (1987). "A procedure for generating group sizes from a one-way classification with a specified degree of imbalance." *Biometrical Journal*, 29, 181–187.

Donner, A. and Koval, J. J. (1989). "The effect of imbalance on significance-testing in one-way Model II analysis of variance." *Communications in Statistics—Theory and Methods*, 18, 1239–1250.

Eaton, M. L. (1983). *Multivariate Statistics*. Wiley, New York.

Edwards, C. H., Jr. (1973). *Advanced Calculus of Several Variables*. Academic Press, New York.

El-Bassiouni, M. Y. and Seely, J. F. (1988). "On the power of Wald's variance component test in the unbalanced random one-way model." In: *Optimal Design and Analysis of Experiments* (Dodge, Y., Federov, V. V., Wynn, H. P., Eds.), North-Holland, Amsterdam, 157–165.

Feingold, M. (1985). "A test statistic for combined intra- and inter-block estimates." *Journal of Statistical Planning and Inference*, 12, 103–114.

Feingold, M. (1988). "A more powerful test for incomplete block designs." *Communications in Statistics—Theory and Methods*, 17, 3107–3119.

Ferguson, T. S. (1967). *Mathematical Statistics*. Academic Press, New York.

Fisher, R. A. (1932). *Statistical Methods for Research Workers*. Oliver and Boyd, London.

Gallo, J. and Khuri, A. I. (1990). "Exact tests for the random and fixed effects in an unbalanced mixed two-way cross-classification model." *Biometrics*, 46, 1087–1095.

Gaylor, D. W. and Hopper, F. N. (1969). "Estimating the degrees of freedom for linear combinations of mean squares by Satterthwaite's formula." *Technometrics*, 11, 691–706.

Gnot, S. and Michalski, A. (1994). "Tests based on admissible estimators in two variance components models." *Statistics*, 25, 213–223.

Graybill, F. A. (1983). *Matrices with Applications in Statistics*, Second Edition. Wadsworth, Belmont, California.

Griffiths, W. and Judge, G. (1992). "Testing and estimating location vectors when the error covariance matrix is unknown."*Journal of Econometrics*, 54, 121–138.

Hald, A. (1952). *Statistical Theory with Engineering Applications*. Wiley, New York.

Herbach, L. H. (1959). "Properties of model-II type analysis of variance tests, A: Optimum nature of the F-test for model II in the balanced case." *Annals of Mathematical Statistics*, 30, 939–959.

Hernandez, R. P. and Burdick, R. K. (1993). "Confidence intervals on the total variance in an unbalanced two-fold nested design." *Biometrical Journal*, 35, 515–522.

Hernandez, R. P., Burdick, R. K., and Birch, N. J. (1992). "Confidence intervals and tests of hypotheses on variance components in an unbalanced two-fold nested design." *Biometrical Journal*, 34, 387–402.

Hinkelmann, K. and Kempthorne, O. (1994). *Design and Analysis of Experiments*, Volume I. Wiley, New York.

Hirotsu, C. (1979). "An F approximation and its application." *Biometrika*, 66, 577–584.

Hocking, R. R. (1973). "A discussion of the two-way mixed model." *American Statistician*, 27, 148–152.

Humak, K. M. S. (1984). *Statistische Methoden dwe Modellbildung III: Statistische Inferenz fur Kovarianzparameter*. Akademie-Verlag, Berlin.

Imhof, J. P. (1960). "A mixed model for the complete three-way layout with two random-effects factors." *Annals of Mathematical Statistics*, 31, 906–928.

Imhof, J. P. (1961). "Computing the distribution of quadratic forms in normal variables." *Biometrika*, 48, 419–426.

Imhof, J. P. (1962). "Testing the hypothesis of no fixed main-effects in Scheffé's mixed model." *Annals of Mathematical Statistics*, 33, 1085–1095.

John, P. W. M. (1971). *Statistical Design and Analysis of Experiments*. Macmillan, New York.

Johnson, N. L. and Kotz, S. (1970). *Continuous Univariate Distributions—2*. Wiley, New York.

Jordan, S. M. and Krishnamoorthy, K. (1995). "On combining independent tests in linear models." *Statistics and Probability Letters*, 23, 117–122.

Kariya, T. and Sinha, B. K. (1985). Nonnull and optimality robustness of some multivariate tests." *Annals of Statistics*, 13, 1182–1197.

Kariya, T. and Sinha, B. K. (1989). *Robustness of Statistical Tests*. Academic Press, Boston.

Khatri, C. G. and Patel, H. I. (1992). "Analysis of a multicenter trial using a multivariate approach to a mixed linear model." *Communications in Statistics—Theory and Methods*, 21, 21–39.

Khattree, R. and Naik, D. N. (1990). "Optimum tests for random effects in unbalanced nested designs." *Statistics*, 21, 163–168.

Khuri, A. I. (1982). "Direct products: A powerful tool for the analysis of balanced data." *Communications in Statistics—Theory and Methods*, 11, 2903–2920.

Khuri, A. I. (1987a). "An exact test for the nesting effect's variance component in an unbalanced random two-fold nested model." *Statistics and Probability Letters*, 5, 305–311.

Khuri, A. I. (1987b). "Measures of imbalance for unbalanced models." *Biometrical Journal*, 29, 383–396.

Khuri, A. I. (1989). "Testing a covariance matrix structure in a mixed model with no empty cells." *Journal of Statistical Planning and Inference*, 22, 117–125.

Khuri, A. I. (1990). "Exact tests for random models with unequal cell frequencies in the last stage." *Journal of Statistical Planning and Inference*, 24, 177–193.

Khuri, A. I. (1995a). "A measure to evaluate the closeness of Satterthwaite's approximation." *Biometrical Journal*, 37, 547–563.

Khuri, A. I. (1995b). "A test to detect inadequacy of Satterthwaite's approximation in balanced mixed models." *Statistics*, 27, 45–54.

Khuri, A. I. (1996). "A method for determining the effect of imbalance." *Journal of Statisitcal Planning and Inference*, 55, 115–129.

Khuri, A. I. and Cornell, J. A. (1996). *Response Surfaces*, Second Edition. Dekker, New York.

Khuri, A. I. and Ghosh, M. (1990). "Minimal sufficient statistics for the unbalanced two-fold nested model." *Statistics and Probability Letters*, 10, 351–353.

Khuri, A. I. and Littell, R. C. (1987). "Exact tests for the main effects variance components in an unbalanced random two-way model." *Biometrics*, 43, 545–560.

Khuri, A. I., Mathew, T., and Nel, D. G. (1994). "A test to determine closeness of multivariate Satterthwaite's approximation." *Journal of Multivariate Analysis*, 51, 201–209.

King, M. L. (1980). "Robust tests for spherical symmetry and their applications to least squares regression." *Annals of Statistics*, 8, 1265–1271.

Kleffe, J. and Seifert, B. (1988). "On the role of MINQUE in testing of hypotheses under mixed linear models." *Communications in Statistics—Theory and Methods*, 17, 1287–1309.

LaMotte, L. R. (1976). "Invariant quadratic estimators in the random one-way ANOVA model." *Biometrics*, 32, 793–804.

Lehmann, E. L. (1986). *Testing Statistical Hypotheses*, Second Edition. Wiley, New York.

Lentner, M. and Bishop, T. (1986). *Experimental Design and Analysis*. Valley Book Company, Blacksburg, Virginia.

Lera Marqués, L. (1994). "Measures of imbalance for higher-order designs." *Biometrical Journal*, 36, 481–490.

Lin, T. H. and Harville, D. A. (1991). "Some alternatives to Wald's confidence interval and test." *Journal of the American Statistical Association*, 86, 179–187.

Linnik, Ju. V. (1966). "Characterization of tests of the Bartlett-Scheffé type." *Trudy Mathematical Institute Steklov*, 79, 32–40.

Marshall, A. W. and Olkin, I. (1979). *Inequalities: Theory of Majorization and Its Applications*. Academic Press, New York.

Mathew, T. (1989a). "MANOVA in the multivariate components of variance model." *Journal of Multivariate Analysis*, 29, 30–38.

Mathew, T. (1989b). "Optimum invariant tests in mixed linear models with two variance components." In: *Statistical Data Analysis and Inference* (Y. Dodge, Ed.), North-Holland, Amsterdam, 381–388.

Mathew, T. and Bhimasankaram, P. (1983). "On the robustness of the LRT with respect to specification errors in a linear model." *Sankhyā, Series A*, 45, 212–225.

Mathew, T., Niyogi, A., and Sinha, B. K. (1994). "Improved nonnegative estimation of variance components in balanced multivariate mixed models." *Journal of Multiviariate Analysis*, 51, 83–101.

Mathew, T. and Sinha, B. K. (1988a). "Optimum tests for fixed effects and variance components in balanced models." *Journal of the American Statistical Association*, 83, 133–135.

Mathew, T. and Sinha, B. K. (1988b). "Optimum tests in unbalanced two-way models without interaction." *Annals of Statistics*, 16, 1727–1740.

Mathew, T. and Sinha, B. K. (1992). "Exact and optimum tests in unbalanced split-plot designs under mixed and random models." *Journal of the American Statistical Association*, 87, 192–200.

Mathew, T., Sinha, B. K., and Zhou, L. (1993). "Some statistical procedures for combining independent tests." *Journal of the American Statistical Association*, 88, 912–919.

Mazuy, K. K. and Connor, W. S. (1965). "Student's t in a two-way classification with unequal variances." *Annals of Mathematical Statistics*, 36, 1248–1255.

Meyer, K. (1985). "Maximum likelihood estimation of variance components for a multivariate mixed model with equal design matrices." *Biometrics*, 41, 153–165.

Michalski, A. and Zmyslony, R. (1996). "Testing hypotheses for variance components in mixed linear models." *Statistics*, 27, 297–310.

Milliken, G. A. and Johnson, D. E. (1984). *Analysis of Messy Data*. Lifetime Learning Publications, Blemont, California.

Montgomery, D. C. (1991). *Design and Analysis of Experiments*, Third Edition. Wiley, New York.

Muirhead, R. J. (1982). *Aspects of Multivariate Statistical Theory*. Wiley, New York.

Nagarsenker, B. N. (1975). "Percentage points of Wilks' L_{vc} criterion." *Communications in Statistics—Theory and Methods*, 4, 629–641.

Nagarsenker, B. N. and Pillai, K. C. S. (1973). "The distribution of the sphericity test criterion." *Journal of Multivariate Analysis*, 3, 226–235.

Naik, D. N. and Khattree, R. (1992). "Optimum tests for treatments when blocks are nested and random." *Statistics*, 23, 101–108.

Nel, D. G. (1980). "On matrix differentiation in statistics." *South African Statistical Journal*, 14, 137–193.

Nel, D. G. and Van Der Merwe, C. A. (1986). "A solution to the multivariate Behrens-Fisher problem." *Communications in Statistics—Theory and Methods*, 15, 3719–3735.

Newton, H. J. (1993). "New developments in statistical computing." *American Statistician*, 47, 146.

Öfversten, J. (1993). "Exact tests for variance components in unbalnced mixed linear models." *Biometrics*, 49, 45–57.

Olsen, A., Seely, J., and Birkes, D. (1976). "Invariant quadratic unbiased estimation for two variance components." *Annals of Statistics*, 5, 878–890.

Oman, S. D. (1995). "Checking the assumptions in mixed-model analysis of variance: A residual analysis approach." *Computational Statistics and Data Analysis*, 20, 309–330.

Portnoy, S. (1973). "On recovery of intra-block information." *Journal of the American Statistical Association*, 68, 384–391.

Rao, C. R. (1947). "General methods of analysis for incomplete block designs." *Journal of the American Statistical Association*, 42, 541–561.

Rao, C. R. (1971). "Estimation of variance components—MINQUE theory." *Journal of Multivariate Analysis*, 1, 257–275.

Rao, C. R. (1973). *Linear Statistical Inference and its Applications*. Wiley, New York.

Rao, C. R. and Kleffe, J. (1988). *Estimation of Variance Components and Applications*. North-Holland, Amsterdam.

Rao, C. R. and Mitra, S. K. (1971). *Generalized Inverse of Matrices and its Applications*. Wiley, New York.

SAS User's Guide: Statistics, 1989 Edition, SAS Institute, Inc., Cary, North Carolina.

Satterthwaite, F. E. (1941). "Synthesis of variance." *Psychometrika*, 6, 309–316.

Satterthwaite, F. E. (1946). "An approximate distribution of estimates of variance components." *Biometrics Bulletin*, 2, 110–114.

Scheffé, H. (1943). "On the solutions of the Behrens-Fisher problem, based on the t distribution." *Annals of Mathematical Statistics*, 14, 35–44.

Scheffé, H. (1956). "A mixed model for the analysis of variance." *Annals of Mathematical Statistics*, 27, 23–36.

Scheffé, H. (1959). *The Analysis of Variance*. Wiley, New York.

Schmidt, W. H. and Thrum, R. (1981). "Contributions to asymptotic theory in regression models with linear covariance structure." *Mathematische Operationsforschung und Statistik, Series Statistics*, 12, 243–269.

Schott, J. R. (1985). "Multivariate maximum likelihood estimators for the mixed linear model." *Sankhyā, Series B*, 47, 179–185.

Schott, J. R. and Saw, J. G. (1984). "A multivariate one-way classification model with random effects." *Journal of Multivariate Analysis*, 15, 1–12.

Searle, S. R. (1971). *Linear Models*. Wiley, New York.

Searle, S. R. (1982). *Matrix Algebra Useful for Statistics*. Wiley, New York.

Searle, S. R. (1987). *Linear Models for Unbalanced Data*. Wiley, New York.

Searle, S. R., Casella, G., and McCulloch, C. E. (1992). *Variance Components*. Wiley, New York.

Seber, G. A. F. (1984). *Multivariate Observations*. Wiley, New York.

Seely, J. F. and El-Bassiouni, Y. (1983). "Applying Wald's variance component test." *Annals of Statistics*, 11, 197–201.

Seifert, B. (1978). "A note on the UMPU character of a test for the mean in balanced randomized nested classification." *Statistics*, 9, 185–189.

Seifert, B. (1979). "Optimal testing for fixed effects in general balanced mixed classification models." *Mathematics Operations: Series Statistics*, 10, 237–255.

Seifert, B. (1981). "Explicit formulae of exact tests in mixed balanced ANOVA models." *Biometrical Journal*, 23, 535–550.

Seifert, B. (1985). "Estimation and test of variance components using the MINQUE method." *Statistics*, 16, 621–635.

Seifert, B. (1992). "Exact tests in unbalanced mixed analysis of variance." *Journal of Statistical Planning and Inference*, 30, 257–266.

Shah, K. R. (1975). "Analysis of block designs." *Gujarat Statistical Review*, 2, 1–11.

Shah, K. R. (1992). "Recovery of inter-block information: An update." *Journal of Statistical Planning and Inference*, 30, 163–172.

Shelby, C. E., Harvey, W. R., Clark, R. T., Quesenberry, J. R., and Woodward, R. R. (1963). "Estimates of phenotypic and genetic parameters in ten years of Miles City R.O.P. steer data." *Journal of Animal Science*, 22, 346–353.

Singh, B. (1989). "A comparison of variance component estimators under unbalanced situations." *Sankhyā, Series B*, 51, 323–330.

Singh, B. (1992). "On the effect of unbalancedness and heteroscedasticity on the ANOVA estimator of group variance component in one-way random model." *Biometrical Journal*, 34, 91–96.

Sinha, Bikas K. (1982). "On complete classes of experiments for certain invariant problems of linear inference." *Journal of Statistical Planning and Inference*, 7, 171–180.

Smith, D. W. and Hocking, R. R. (1978). "Maximum likelihood analysis of the mixed model: The balanced case." *Communications in Statistics—Theory and Methods*, 7, 1253–1266.

Spjøtvoll, E. (1967). "Optimum invariant tests in unbalanced variance components models." *Annals of Mathematical Statistics*, 38, 422–428.

Spjøtvoll, E. (1968). "Confidence intervals and tests for variance ratios in unbalanced variance components models." *Review of International Statistical Institute*, 36, 37–42.

Tan, W. Y. and Cheng, S. S. (1984). "On testing variance components in three-stage unbalanced nested random effects models." *Sankhyā, Series B*, 46, 188–200.

Tan, W. Y. and Gupta, R. P. (1983). "On approximating a linear combination of central Wishart matrices with positive coefficients." *Communications in Statistics—Theory and Methods*, 12, 2589–2600.

Thomas, J. D. and Hultquist, R. A. (1978). "Interval estimation for the unbalanced case of the one-way random effects model." *Annals of Statistics*, 6, 582–587.

Thompson, W. A. (1955a). "The ratio of variances in a variance components model." *Annals of Mathematical Statistics*, 26, 325–329.

Thompson, W. A. (1955b). "On the ratio of variances in the mixed incomplete block model." *Annals of Mathematical Statistics*, 26, 721–733.

Thomsen, I. (1975). "Testing hypotheses in unbalanced variance components models for two-way layouts." *Annals of Statistics*, 3, 257–265.

Thursby, J. G. (1992). "A comparison of several exact and approximate tests for structural shift under heteroscedasticity." *Journal of Econometrics*, 53, 363–386.

Timm, N. H. (1980). "Multivariate analysis of variance of repeated measurements." In: *Handbook of Statistics*, Volume 1 (P. R. Krishnaiah, Ed.), Elsevier Science B.V., Amsterdam, 41–87.

Tsui, K. W. and Weerahandi, S. (1989). "Generalized P-values in siginificance testing of hypotheses in the presence of nuisance parameters." *Journal of the American Statistical Association*, 84, 602–607.

Wald, A. (1947). "A note on regression analysis." *Annals of Mathematical Statistics*, 18, 586–589.

Weerahandi, S. (1987). "Testing regression equality with unequal variances." *Econometrica*, 55, 1211–1215.

Weerahandi, S. (1991). "Testing variance components in mixed models with generalized P-values." *Journal of the American Statistical Association*, 86, 151–153.

Weerahandi, S. (1993). "Generalized confidence intervals." *Journal of the American Statistical Association*, 88, 899–905.

Weerahandi, S. (1995a). "ANOVA under unequal error variances." *Biometrics*, 51, 589–599.

Weerahandi, S. (1995b). *Exact Statistical Methods for Data Analysis*. Springer-Verlag, New York.

Westfall, P. H. (1989). "Power comparisons for invariant variance ratio tests in mixed ANOVA models." *Annals of Statistics*, 17, 318–326.

Wijsman, R. A. (1967). "Cross-section of orbits and their applications to densities of maximal invariants." In: *Fifth Berkeley Symposium on Mathematical Statistics and Probability, I*, University of California, Berkeley, 389–400.

Wilks, S. S. (1946). "Sample criteria for testing equality of means, equality of variances, and equality of covariances in a normal multivariate distribution." *Annals of Mathematical Statistics*, 17, 257–281.

XPro Software Package (1994). X-Techniques, Inc., Millington, New Jersey.

Yates, F. (1937). The Design and Analysis of Factorial Experiments. Imperial Bureau of Soil Science, Harpenden, England.

Yates, F. (1939). "The recovery of inter-block information in varietal trials arranged in three dimensional lattice." *Annals of Eugenics*, 9, 136–156.

Yates, F. (1940). "The recovery of inter-block information in balanced incomplete block designs." *Annals of Eugenics*, 10, 317–325.

Zacks, S. (1971). *The Theory of Statistical Inference.* Wiley, New York.

Zhang, Z. (1992). "Recovery tests in BIBD's with very small degrees of freedom for inter-block errors." *Statistics and Probability Letters*, 15, 197–202.

Zhou, L. and Mathew, T. (1993a). "Combining independent tests in linear models." *Journal of the American Statistical Association*, 88, 650–655.

Zhou, L. and Mathew, T. (1993b). "Hypotheses tests for variance components in some multivariate mixed models." *Journal of Statistical Planning and Inference*, 37, 215–227.

Zhou, L. and Mathew, T. (1994a). "Some tests for variance components using generalized P-values." *Technometrics*, 36, 394–402.

Zhou, L. and Mathew, T. (1994b). "Combining independent tests in multivariate linear models." *Journal of Multivariate Analysis*, 51, 265–275.

Zyskind, G. (1962). "On structure, relation, Σ and expectation of mean squares." *Sankhyā, Series A*, 24, 115–148.

Author Index

Subject Index

WILEY SERIES IN PROBABILITY AND STATISTICS

ESTABLISHED BY WALTER A. SHEWHART AND SAMUEL S. WILKS

Probability and Statistics Section

*ANDERSON · The Statistical Analysis of Time Series
ARNOLD, BALAKRISHNAN, and NAGARAJA · A First Course in Order Statistics
BACCELLI, COHEN, OLSDER, and QUADRAT · Synchronization and Linearity: An Algebra for Discrete Event Systems
BASILEVSKY · Statistical Factor Analysis and Related Methods: Theory and Applications
BERNARDO and SMITH · Bayesian Statistical Concepts and Theory
BILLINGSLEY · Convergence of Probability Measures
BOROVKOV · Asymptotic Methods in Queuing Theory
BRANDT, FRANKEN, and LISEK · Stationary Stochastic Models
CAINES · Linear Stochastic Systems
CAIROLI and DALANG · Sequential Stochastic Optimization
CONSTANTINE · Combinatorial Theory and Statistical Design
COVER and THOMAS · Elements of Information Theory
CSÖRGŐ and HORVÁTH · Weighted Approximations in Probability Statistics
CSÖRGŐ and HORVÁTH · Limit Theorems in Change Point Analysis
DETTE and STUDDEN · The Theory of Canonical Moments with Applications in Statistics, Probability, and Analysis
*DOOB · Stochastic Processes
DRYDEN and MARDIA · Statistical Analysis of Shape
DUPUIS and ELLIS · A Weak Convergence Approach to the Theory of Large Deviations
ETHIER and KURTZ · Markov Processes: Characterization and Convergence
FELLER · An Introduction to Probability Theory and Its Applications, Volume 1, *Third Edition,* Revised; Volume II, *Second Edition*
FULLER · Introduction to Statistical Time Series, *Second Edition*
FULLER · Measurement Error Models
GELFAND and SMITH · Bayesian Computation
GHOSH, MUKHOPADHYAY, and SEN · Sequential Estimation
GIFI · Nonlinear Multivariate Analysis
GUTTORP · Statistical Inference for Branching Processes
HALL · Introduction to the Theory of Coverage Processes
HAMPEL · Robust Statistics: The Approach Based on Influence Functions
HANNAN and DEISTLER · The Statistical Theory of Linear Systems
HUBER · Robust Statistics
IMAN and CONOVER · A Modern Approach to Statistics
JUREK and MASON · Operator-Limit Distributions in Probability Theory
KASS and VOS · Geometrical Foundations of Asymptotic Inference
KAUFMAN and ROUSSEEUW · Finding Groups in Data: An Introduction to Cluster Analysis

*Now available in a lower priced paperback edition in the Wiley Classics Library.

Probability and Statistics (Continued)

KELLY · Probability, Statistics, and Optimization
LINDVALL · Lectures on the Coupling Method
McFADDEN · Management of Data in Clinical Trials
MANTON, WOODBURY, and TOLLEY · Statistical Applications Using Fuzzy Sets
MARDIA and JUPP · Statistics of Directional Data, *Second Edition*
MORGENTHALER and TUKEY · Configural Polysampling: A Route to Practical Robustness
MUIRHEAD · Aspects of Multivariate Statistical Theory
OLIVER and SMITH · Influence Diagrams, Belief Nets and Decision Analysis
*PARZEN · Modern Probability Theory and Its Applications
PRESS · Bayesian Statistics: Principles, Models, and Applications
PUKELSHEIM · Optimal Experimental Design
RAO · Asymptotic Theory of Statistical Inference
RAO · Linear Statistical Inference and Its Applications, *Second Edition*
*RAO and SHANBHAG · Choquet-Deny Type Functional Equations with Applications to Stochastic Models
ROBERTSON, WRIGHT, and DYKSTRA · Order Restricted Statistical Inference
ROGERS and WILLIAMS · Diffusions, Markov Processes, and Martingales, Volume I: Foundations, *Second Edition;* Volume II: Îto Calculus
RUBINSTEIN and SHAPIRO · Discrete Event Systems: Sensitivity Analysis and Stochastic Optimization by the Score Function Method
RUZSA and SZEKELY · Algebraic Probability Theory
SCHEFFE · The Analysis of Variance
SEBER · Linear Regression Analysis
SEBER · Multivariate Observations
SEBER and WILD · Nonlinear Regression
SERFLING · Approximation Theorems of Mathematical Statistics
SHORACK and WELLNER · Empirical Processes with Applications to Statistics
SMALL and McLEISH · Hilbert Space Methods in Probability and Statistical Inference
STAPLETON · Linear Statistical Models
STAUDTE and SHEATHER · Robust Estimation and Testing
STOYANOV · Counterexamples in Probability
TANAKA · Time Series Analysis: Nonstationary and Noninvertible Distribution Theory
THOMPSON and SEBER · Adaptive Sampling
WELSH · Aspects of Statistical Inference
WHITTAKER · Graphical Models in Applied Multivariate Statistics
YANG · The Construction Theory of Denumerable Markov Processes

Applied Probability and Statistics Section

ABRAHAM and LEDOLTER · Statistical Methods for Forecasting
AGRESTI · Analysis of Ordinal Categorical Data
AGRESTI · Categorical Data Analysis
ANDERSON, AUQUIER, HAUCK, OAKES, VANDAELE, and WEISBERG · Statistical Methods for Comparative Studies
ARMITAGE and DAVID (editors) · Advances in Biometry
*ARTHANARI and DODGE · Mathematical Programming in Statistics
ASMUSSEN · Applied Probability and Queues
*BAILEY · The Elements of Stochastic Processes with Applications to the Natural Sciences
BARNETT and LEWIS · Outliers in Statistical Data, *Third Edition*

*Now available in a lower priced paperback edition in the Wiley Classics Library.

Applied Probability and Statistics (Continued)

BARTHOLOMEW, FORBES, and McLEAN · Statistical Techniques for Manpower Planning, *Second Edition*
BATES and WATTS · Nonlinear Regression Analysis and Its Applications
BECHHOFER, SANTNER, and GOLDSMAN · Design and Analysis of Experiments for Statistical Selection, Screening, and Multiple Comparisons
BELSLEY · Conditioning Diagnostics: Collinearity and Weak Data in Regression
BELSLEY, KUH, and WELSCH · Regression Diagnostics: Identifying Influential Data and Sources of Collinearity
BHAT · Elements of Applied Stochastic Processes, *Second Edition*
BHATTACHARYA and WAYMIRE · Stochastic Processes with Applications
BIRKES and DODGE · Alternative Methods of Regression
BLOOMFIELD · Fourier Analysis of Time Series: An Introduction
BOLLEN · Structural Equations with Latent Variables
BOULEAU · Numerical Methods for Stochastic Processes
BOX · Bayesian Inference in Statistical Analysis
BOX and DRAPER · Empirical Model-Building and Response Surfaces
BOX and DRAPER · Evolutionary Operation: A Statistical Method for Process Improvement
BUCKLEW · Large Deviation Techniques in Decision, Simulation, and Estimation
BUNKE and BUNKE · Nonlinear Regression, Functional Relations and Robust Methods: Statistical Methods of Model Building
CHATTERJEE and HADI · Sensitivity Analysis in Linear Regression
CHOW and LIU · Design and Analysis of Clinical Trials
CLARKE and DISNEY · Probability and Random Processes: A First Course with Applications, *Second Edition*
*COCHRAN and COX · Experimental Designs, *Second Edition*
CONOVER · Practical Nonparametric Statistics, *Second Edition*
CORNELL · Experiments with Mixtures, Designs, Models, and the Analysis of Mixture Data, *Second Edition*
*COX · Planning of Experiments
CRESSIE · Statistics for Spatial Data, *Revised Edition*
DANIEL · Applications of Statistics to Industrial Experimentation
DANIEL · Biostatistics: A Foundation for Analysis in the Health Sciences, *Sixth Edition*
DAVID · Order Statistics, *Second Edition*
*DEGROOT, FIENBERG, and KADANE · Statistics and the Law
DODGE · Alternative Methods of Regression
DOWDY and WEARDEN · Statistics for Research, *Second Edition*
DUNN and CLARK · Applied Statistics: Analysis of Variance and Regression, *Second Edition*
ELANDT-JOHNSON and JOHNSON · Survival Models and Data Analysis
EVANS, PEACOCK, and HASTINGS · Statistical Distributions, *Second Edition*
FLEISS · The Design and Analysis of Clinical Experiments
FLEISS · Statistical Methods for Rates and Proportions, *Second Edition*
FLEMING and HARRINGTON · Counting Processes and Survival Analysis
GALLANT · Nonlinear Statistical Models
GLASSERMAN and YAO · Monotone Structure in Discrete-Event Systems
GNANADESIKAN · Methods for Statistical Data Analysis of Multivariate Observations, *Second Edition*
GOLDSTEIN and LEWIS · Assessment: Problems, Development, and Statistical Issues
GREENWOOD and NIKULIN · A Guide to Chi-Squared Testing
*HAHN · Statistical Models in Engineering
HAHN and MEEKER · Statistical Intervals: A Guide for Practitioners

*Now available in a lower priced paperback edition in the Wiley Classics Library.

Applied Probability and Statistics (Continued)

HAND · Construction and Assessment of Classification Rules
HAND · Discrimination and Classification
HEIBERGER · Computation for the Analysis of Designed Experiments
HINKELMAN and KEMPTHORNE: · Design and Analysis of Experiments, Volume 1: Introduction to Experimental Design
HOAGLIN, MOSTELLER, and TUKEY · Exploratory Approach to Analysis of Variance
HOAGLIN, MOSTELLER, and TUKEY · Exploring Data Tables, Trends and Shapes
HOAGLIN, MOSTELLER, and TUKEY · Understanding Robust and Exploratory Data Analysis
HOCHBERG and TAMHANE · Multiple Comparison Procedures
HOCKING · Methods and Applications of Linear Models: Regression and the Analysis of Variables
HOGG and KLUGMAN · Loss Distributions
HOLLANDER and WOLFE · Nonparametric Statistical Methods
HOSMER and LEMESHOW · Applied Logistic Regression
HØYLAND and RAUSAND · System Reliability Theory: Models and Statistical Methods
HUBERTY · Applied Discriminant Analysis
JACKSON · A User's Guide to Principle Components
JOHN · Statistical Methods in Engineering and Quality Assurance
JOHNSON · Multivariate Statistical Simulation
JOHNSON and KOTZ · Distributions in Statistics
Continuous Multivariate Distributions
JOHNSON, KOTZ, and BALAKRISHNAN · Continuous Univariate Distributions, Volume 1, *Second Edition*
JOHNSON, KOTZ, and BALAKRISHNAN · Continuous Univariate Distributions, Volume 2, *Second Edition*
JOHNSON, KOTZ, and BALAKRISHNAN · Discrete Multivariate Distributions
JOHNSON, KOTZ, and KEMP · Univariate Discrete Distributions, *Second Edition*
JUREČKOVÁ and SEN · Robust Statistical Procedures: Aymptotics and Interrelations
KADANE · Bayesian Methods and Ethics in a Clinical Trial Design
KADANE AND SCHUM · A Probabilistic Analysis of the Sacco and Vanzetti Evidence
KALBFLEISCH and PRENTICE · The Statistical Analysis of Failure Time Data
KELLY · Reversability and Stochastic Networks
KHURI, MATHEW, and SINHA · Statistical Tests for Mixed Linear Models
KLUGMAN, PANJER, and WILLMOT · Loss Models: From Data to Decisions
KLUGMAN, PANJER, and WILLMOT · Solutions Manual to Accompany Loss Models: From Data to Decisions
KOVALENKO, KUZNETZOV, and PEGG · Mathematical Theory of Reliability of Time-Dependent Systems with Practical Applications
LAD · Operational Subjective Statistical Methods: A Mathematical, Philosophical, and Historical Introduction
LANGE, RYAN, BILLARD, BRILLINGER, CONQUEST, and GREENHOUSE · Case Studies in Biometry
LAWLESS · Statistical Models and Methods for Lifetime Data
LEE · Statistical Methods for Survival Data Analysis, *Second Edition*
LEPAGE and BILLARD · Exploring the Limits of Bootstrap
LINHART and ZUCCHINI · Model Selection
LITTLE and RUBIN · Statistical Analysis with Missing Data
MAGNUS and NEUDECKER · Matrix Differential Calculus with Applications in Statistics and Econometrics
MALLER and ZHOU · Survival Analysis with Long Term Survivors
MANN, SCHAFER, and SINGPURWALLA · Methods for Statistical Analysis of Reliability and Life Data

*Now available in a lower priced paperback edition in the Wiley Classics Library.

Applied Probability and Statistics (Continued)

McLACHLAN and KRISHNAN · The EM Algorithm and Extensions
McLACHLAN · Discriminant Analysis and Statistical Pattern Recognition
McNEIL · Epidemiological Research Methods
MILLER · Survival Analysis
MONTGOMERY and PECK · Introduction to Linear Regression Analysis, *Second Edition*
MYERS and MONTGOMERY · Response Surface Methodology: Process and Product in Optimization Using Designed Experiments
NELSON · Accelerated Testing, Statistical Models, Test Plans, and Data Analyses
NELSON · Applied Life Data Analysis
OCHI · Applied Probability and Stochastic Processes in Engineering and Physical Sciences
OKABE, BOOTS, and SUGIHARA · Spatial Tesselations: Concepts and Applications of Voronoi Diagrams
PANKRATZ · Forecasting with Dynamic Regression Models
PANKRATZ · Forecasting with Univariate Box-Jenkins Models: Concepts and Cases
PIANTADOSI · Clinical Trials: A Methodologic Perspective
PORT · Theoretical Probability for Applications
PUTERMAN · Markov Decision Processes: Discrete Stochastic Dynamic Programming
RACHEV · Probability Metrics and the Stability of Stochastic Models
RÉNYI · A Diary on Information Theory
RIPLEY · Spatial Statistics
RIPLEY · Stochastic Simulation
ROUSSEEUW and LEROY · Robust Regression and Outlier Detection
RUBIN · Multiple Imputation for Nonresponse in Surveys
RUBINSTEIN · Simulation and the Monte Carlo Method
RUBINSTEIN, MELAMED, and SHAPIRO · Modern Simulation and Modeling
RYAN · Statistical Methods for Quality Improvement
SCHUSS · Theory and Applications of Stochastic Differential Equations
SCOTT · Multivariate Density Estimation: Theory, Practice, and Visualization
*SEARLE · Linear Models
SEARLE · Linear Models for Unbalanced Data
SEARLE, CASELLA, and McCULLOCH · Variance Components
STOYAN, KENDALL, and MECKE · Stochastic Geometry and Its Applications, *Second Edition*
STOYAN and STOYAN · Fractals, Random Shapes and Point Fields: Methods of Geometrical Statistics
THOMPSON · Empirical Model Building
THOMPSON · Sampling
TIJMS · Stochastic Modeling and Analysis: A Computational Approach
TIJMS · Stochastic Models: An Algorithmic Approach
TITTERINGTON, SMITH, and MAKOV · Statistical Analysis of Finite Mixture Distributions
UPTON and FINGLETON · Spatial Data Analysis by Example, Volume 1: Point Pattern and Quantitative Data
UPTON and FINGLETON · Spatial Data Analysis by Example, Volume II: Categorical and Directional Data
VAN RIJCKEVORSEL and DE LEEUW · Component and Correspondence Analysis
WEISBERG · Applied Linear Regression, *Second Edition*
WESTFALL and YOUNG · Resampling-Based Multiple Testing: Examples and Methods for *p*-Value Adjustment
WHITTLE · Systems in Stochastic Equilibrium
WOODING · Planning Pharmaceutical Clinical Trials: Basic Statistical Principles
WOOLSON · Statistical Methods for the Analysis of Biomedical Data
*ZELLNER · An Introduction to Bayesian Inference in Econometrics

*Now available in a lower priced paperback edition in the Wiley Classics Library.

Texts and References Section

AGRESTI · An Introduction to Categorical Data Analysis
ANDERSON · An Introduction to Multivariate Statistical Analysis, *Second Edition*
ANDERSON and LOYNES · The Teaching of Practical Statistics
ARMITAGE and COLTON · Encyclopedia of Biostatistics: Volumes 1 to 6 with Index
BARTOSZYNSKI and NIEWIADOMSKA-BUGAJ · Probability and Statistical Inference
BERRY, CHALONER, and GEWEKE · Bayesian Analysis in Statistics and Econometrics: Essays in Honor of Arnold Zellner
BHATTACHARYA and JOHNSON · Statistical Concepts and Methods
BILLINGSLEY · Probability and Measure, *Second Edition*
BOX · R. A. Fisher, the Life of a Scientist
BOX, HUNTER, and HUNTER · Statistics for Experimenters: An Introduction to Design, Data Analysis, and Model Building
BOX and LUCEÑO · Statistical Control by Monitoring and Feedback Adjustment
BROWN and HOLLANDER · Statistics: A Biomedical Introduction
CHATTERJEE and PRICE · Regression Analysis by Example, *Second Edition*
COOK and WEISBERG · An Introduction to Regression Graphics
COX · A Handbook of Introductory Statistical Methods
DILLON and GOLDSTEIN · Multivariate Analysis: Methods and Applications
DODGE and ROMIG · Sampling Inspection Tables, *Second Edition*
DRAPER and SMITH · Applied Regression Analysis, *Third Edition*
DUDEWICZ and MISHRA · Modern Mathematical Statistics
DUNN · Basic Statistics: A Primer for the Biomedical Sciences, *Second Edition*
FISHER and VAN BELLE · Biostatistics: A Methodology for the Health Sciences
FREEMAN and SMITH · Aspects of Uncertainty: A Tribute to D. V. Lindley
GROSS and HARRIS · Fundamentals of Queueing Theory, *Third Edition*
HALD · A History of Probability and Statistics and their Applications Before 1750
HALD · A History of Mathematical Statistics from 1750 to 1930
HELLER · MACSYMA for Statisticians
HOEL · Introduction to Mathematical Statistics, *Fifth Edition*
JOHNSON and BALAKRISHNAN · Advances in the Theory and Practice of Statistics: A Volume in Honor of Samuel Kotz
JOHNSON and KOTZ (editors) · Leading Personalities in Statistical Sciences: From the Seventeenth Century to the Present
JUDGE, GRIFFITHS, HILL, LÜTKEPOHL, and LEE · The Theory and Practice of Econometrics, *Second Edition*
KHURI · Advanced Calculus with Applications in Statistics
KOTZ and JOHNSON (editors) · Encyclopedia of Statistical Sciences: Volumes 1 to 9 wtih Index
KOTZ and JOHNSON (editors) · Encyclopedia of Statistical Sciences: Supplement Volume
KOTZ, REED, and BANKS (editors) · Encyclopedia of Statistical Sciences: Update Volume 1
KOTZ, REED, and BANKS (editors) · Encyclopedia of Statistical Sciences: Update Volume 2
LAMPERTI · Probability: A Survey of the Mathematical Theory, *Second Edition*
LARSON · Introduction to Probability Theory and Statistical Inference, *Third Edition*
LE · Applied Survival Analysis
MALLOWS · Design, Data, and Analysis by Some Friends of Cuthbert Daniel
MARDIA · The Art of Statistical Science: A Tribute to G. S. Watson
MASON, GUNST, and HESS · Statistical Design and Analysis of Experiments with Applications to Engineering and Science
MURRAY · X-STAT 2.0 Statistical Experimentation, Design Data Analysis, and Nonlinear Optimization

*Now available in a lower priced paperback edition in the Wiley Classics Library.

Texts amd References (Continued)

PURI, VILAPLANA, and WERTZ · New Perspectives in Theoretical and Applied Statistics
RENCHER · Methods of Multivariate Analysis
RENCHER · Multivariate Statistical Inference with Applications
ROSS · Introduction to Probability and Statistics for Engineers and Scientists
ROHATGI · An Introduction to Probability Theory and Mathematical Statistics
RYAN · Modern Regression Methods
SCHOTT · Matrix Analysis for Statistics
SEARLE · Matrix Algebra Useful for Statistics
STYAN · The Collected Papers of T. W. Anderson: 1943–1985
TIERNEY · LISP-STAT: An Object-Oriented Environment for Statistical Computing and Dynamic Graphics
WONNACOTT and WONNACOTT · Econometrics, *Second Edition*

WILEY SERIES IN PROBABILITY AND STATISTICS

ESTABLISHED BY WALTER A. SHEWHART AND SAMUEL S. WILKS

Editors
Robert M. Groves, Graham Kalton, J. N. K. Rao, Norbert Schwarz, Christopher Skinner

Survey Methodology Section

BIEMER, GROVES, LYBERG, MATHIOWETZ, and SUDMAN · Measurement Errors in Surveys
COCHRAN · Sampling Techniques, *Third Edition*
COX, BINDER, CHINNAPPA, CHRISTIANSON, COLLEDGE, and KOTT (editors) · Business Survey Methods
*DEMING · Sample Design in Business Research
DILLMAN · Mail and Telephone Surveys: The Total Design Method
GROVES · Survey Errors and Survey Costs
GROVES, BIEMER, LYBERG, MASSEY, NICHOLLS, and WAKSBERG · Telephone Survey Methodology
*HANSEN, HURWITZ, and MADOW · Sample Survey Methods and Theory, Volume I: Methods and Applications
*HANSEN, HURWITZ, and MADOW · Sample Survey Methods and Theory, Volume II: Theory
KASPRZYK, DUNCAN, KALTON, and SINGH · Panel Surveys
KISH · Statistical Design for Research
*KISH · Survey Sampling
LESSLER and KALSBEEK · Nonsampling Error in Surveys
LEVY and LEMESHOW · Sampling of Populations: Methods and Applications
LYBERG, BIEMER, COLLINS, de LEEUW, DIPPO, SCHWARZ, TREWIN (editors) · Survey Measurement and Process Quality
SKINNER, HOLT, and SMITH · Analysis of Complex Surveys

*Now available in a lower priced paperback edition in the Wiley Classics Library.